The Building Conservation Directory 2003

A guide to specialist suppliers, consultants and craftsmen in traditional building conservation, refurbishment and design

CATHEDRAL
COMMUNICATIONS LIMITED

THE BUILDING CONSERVATION DIRECTORY 2003
The Eleventh Edition of the Directory

Published Summer 2003

ISBN 1 900915 26 X

PUBLISHED BY
Cathedral Communications Limited
High Street, Tisbury,
Wiltshire, England SP3 6HA
Tel 01747 871717
Fax 01747 871718
E-mail bcd@cathcomm.demon.co.uk
Website www.buildingconservation.com

MANAGING DIRECTOR
Gordon Sorensen

EXECUTIVE EDITOR
Jonathan Taylor

MARKETING DIRECTOR
Elizabeth Coyle-Camp

ADVERTISING DEPARTMENT
Tom Keenan
Anthony Male
Nicholas Rainsford

PRODUCTION & ADMINISTRATION
Jane Martin
Lydia Porter
Edward Green

TYPESETTING & DESIGN
xendo, London

PRINTING
Optichrome, Woking

The paper used for *The Building Conservation Directory* is Satimat Silk from wood fibre originating from sustainable forests.

THE BUILDING CONSERVATION DIRECTORY

The Building Conservation Directory provides specifiers of works to historic buildings, their contents and surroundings a starting point in the search for appropriate products and services, and expert advice. Approximately 1,000 different companies and organisations are represented in this edition.

The many technical editorial articles are written by leaders in this field and cover a wide range of practical issues. The articles are not intended to be comprehensive but rather to raise awareness and stimulate dialogue amongst those involved with old buildings. Other reference information provided points the way on current legislation, continuing education and sources for further information and advice.

How to find what you need

1. The table of contents opposite together with the products and services index and the advertisers index at the back should point you to the product, service or supplier you are looking for. Product Selector tables, in yellow, listing suppliers and their products and services head up each main section.
2. Follow the index or Product Selector page reference to the appropriate section or company and start the specification process. It may be helpful to contact more than one supplier. And please remember to tell each that you found them in *The Building Conservation Directory*.
3. If you still can't find what or who you need, don't despair. You can visit our regularly updated website at **www.buildingconservation.com** or ring us on **01747 871717** and we'll try to put you in touch with a supplier who can help.

All suppliers in the Directory pay a fee to be included and although Cathedral Communications does not formally 'approve' or 'recommend' them we do screen out inappropriate suppliers and products to maintain the established integrity of the Directory. This ensures that it remains a useful and credible forum in which appropriate suppliers can promote their businesses. Directory users should seek more detailed information and advice from suppliers before undertaking any sensitive project.

We are always looking for ideas to improve the Directory so please write to let us know if you have any suggestions for improvements to its content or presentation which will help you in your work with old buildings.

And don't forget, our website at **www.buildingconservation.com** is the primary Internet gateway to the building conservation and restoration industry. Please see the inside back cover of this edition for more information, or why not click in have a look for yourself?

ACKNOWLEDGEMENTS

We are most grateful to all those who have contributed to this, the eleventh edition of *The Building Conservation Directory*, without whose help its publication would not have been possible.

In particular, we would like to thank George Ferguson, RIBA President for contributing this year's Foreword, Clare Lawrence for her help with editing, all our advertisers for their continuing support and all those who have contributed articles, illustrations and information or who have helped with production, including:

Julie Aalen, Weald & Downland Open Air Museum
Malcolm Airs, IHBC
Jim Allen, Ellis & Moore
Phillip Allen
Tom Bilson, Courtauld Institute
Carol Brown, Historic Scotland
Peter Christian, Charterbuild
Ian Constantinides, St Blaise Ltd
Michael Copeman
Jim Cornell, Railway Heritage Trust
Michael Davies, Davies Sutton Architecture
Richard Davies, COTAC
Lesley Durbin, Jackfield Tile Conservation Studio
John France and Helen Sudell
Pat Gibbons, Scottish Lime Centre Trust
Neil Grieve, University of Dundee
John Griffiths
Stafford Holmes, Rodney Melville & Partners
Lynne Humphries
Alistair Hunt, University of Bath
Rob John, Office of the Deputy Prime Minister
Robin Kent
Bob Kindred MBE, Ipswich Borough Council
Peter King, The Conservation Rooflight Company
Jonathan Lamb, Lamb's Bricks & Arches
John Letts, Historic Thatch Management Ltd
Paul Livesey, Castle Cement Ltd
Ian Lund, Kennet District Council
Neil May, Natural Building Technologies
Allyson McDermott
James Morse, Light & Design Associates
Charles Mynors
Trevor Proudfoot, Cliveden Conservation
Keith Quantrill
Richard Green Estate Agents
Robin Rolfe, COTAC
Sefton Park Palmhouse Preservation Trust
Ian Sims, STATS Limited
Wayne Stanton, PricewaterhouseCoopers
Robert Tavendale, Period Property UK
Hilary Taylor, HTLA Ltd
Robin Rolfe, COTAC
Jane Travis
Gordon Urquhart, Glasgow Conservation Trust West
Elizabeth White, Nimbus Conservation Limited
Catherine Woolfitt, Ingram Consultancy Ltd
Tim Yates, Building Research Establishment
David Yeomans

CONTENTS

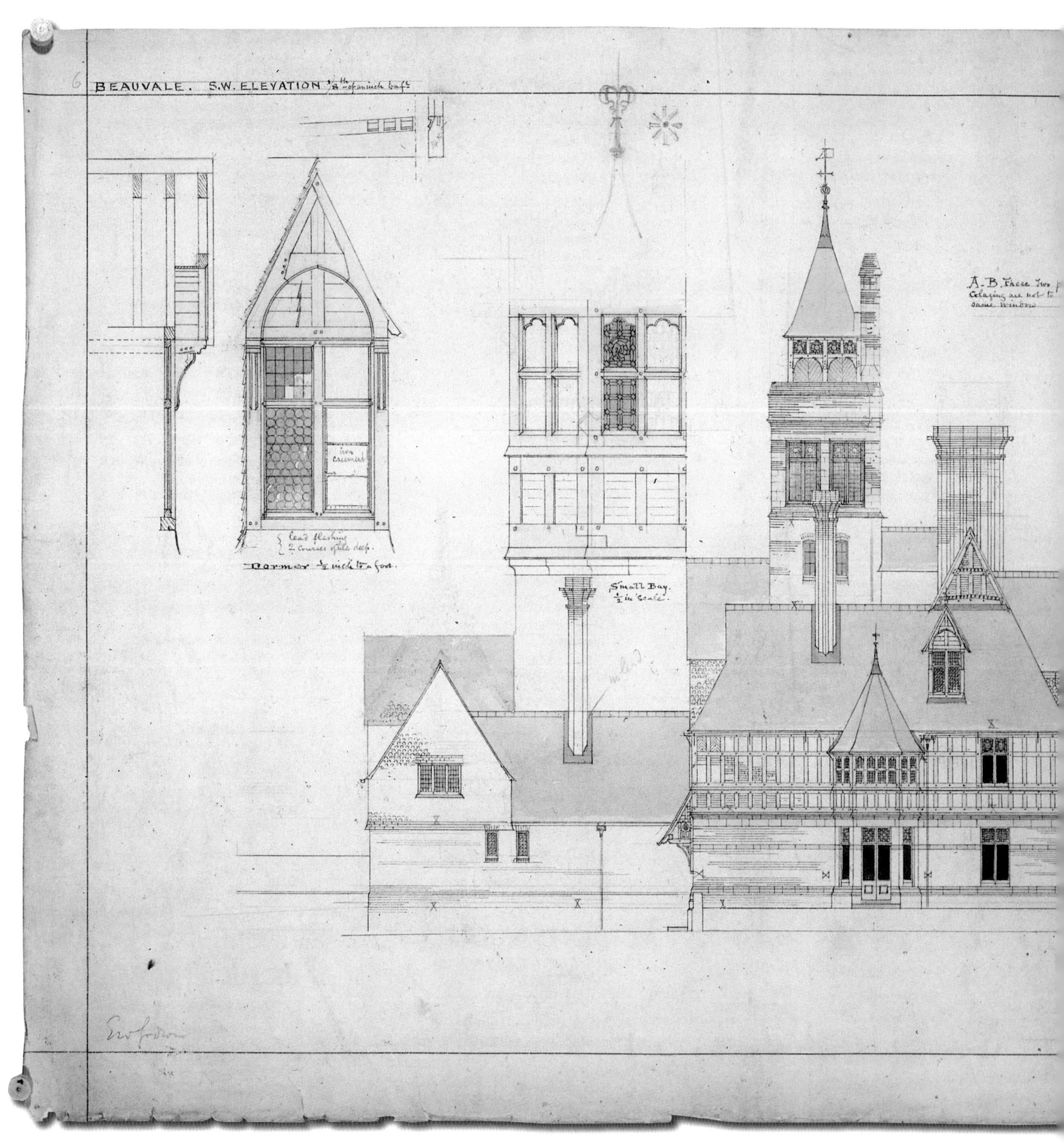

EH Godwin's design for Beauvale, Nottinghamshire; one of the treasures from the RIBA drawings collection to be displayed at the V&A

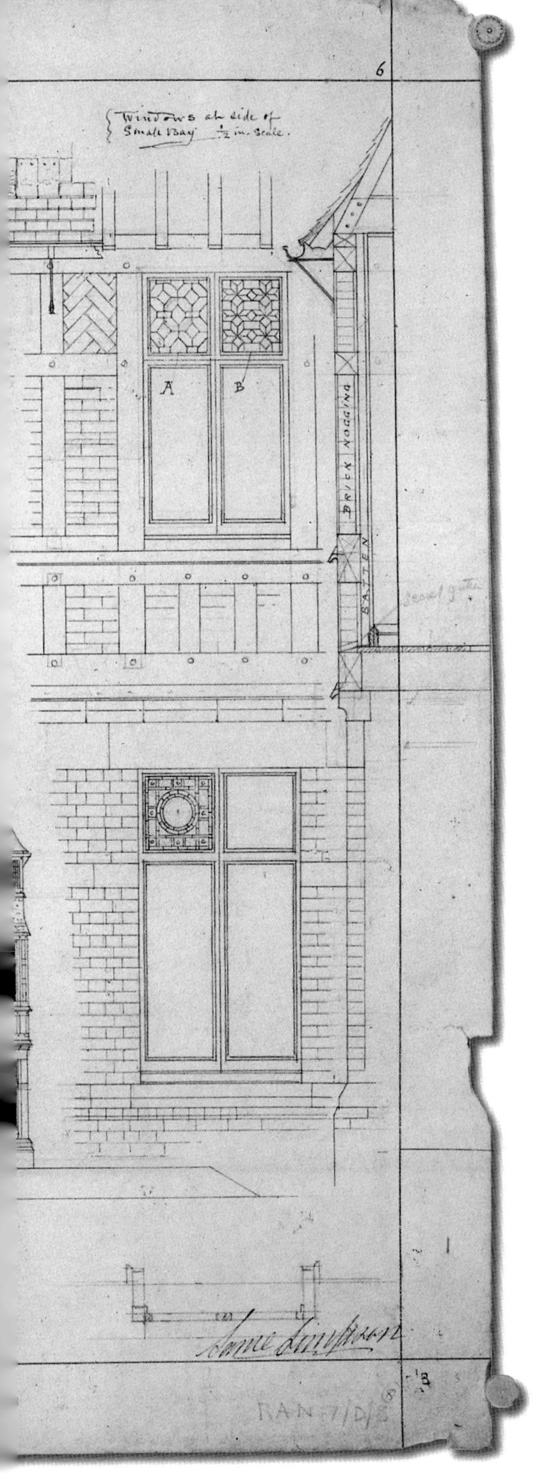

FOREWORD

EDWARD WILLIAM GODWIN, that great Victorian architect and aesthetic movement hero of mine, advised his students to "arm themselves against ugliness and sham". The best way they could do so was to learn from history and 'critical enquiry'. This is of particular relevance in the work we do with historic buildings where ignorance and lack of enquiry results in so much damage.

The *Building Conservation Directory* plays a vital role in the dissemination of knowledge through fascinating articles and hard information relevant to everyone involved in the building conservation game. It forms an important part of my practice's library, and, I am pleased to report, bears the honourable scars of constant use.

Knowledge and research are essential precursors to work on historic buildings. It is with this in mind that the RIBA has entered into an exciting partnership with the Victoria & Albert Museum, entitled *Architecture for All*, to house its great drawings collection in a new archive to be purpose built at the V&A and to create a major new architectural gallery for both permanent and temporary exhibitions. For far too long, access to this treasure house of some 600,000 drawings – the greatest collection of its type in the world – has been a mystery to many, and generally restricted to scholars because of inadequate resources, display and viewing space.

This will all change by the end of 2004 when, with major help from the Heritage Lottery Fund matched by a range of generous benefactors, small and large, we shall be opening the new joint facility for the benefit of all. While based in London, it should be seen as the launch pad for a nationwide initiative to increase the understanding and appreciation of architecture, especially amongst the young. A better informed audience will always challenge us, and our clients, to produce better and better results. I hope that everyone who cares about the future of our past will feel able to contribute, and to make use of this amazing national resource. This joint venture between the RIBA and V&A is nicely exemplified by the work of EW Godwin, whose exquisite drawings of buildings, interiors and furniture tell a remarkable story about this extraordinarily original and versatile designer. Combining the two collections will bring almost all his surviving drawings under one roof making it an invaluable resource to scholars and practitioners.

The overriding theme of my RIBA Presidency will be the 'Spirit of Architecture and of Place', coupled of course with the education and critical enquiry that is so necessary if we are to be able to understand, appreciate and take care of our heritage. I applaud everything that is done to assist us in achieving these aims and I am confident that this directory will help us to arm ourselves against 'ugliness and sham'.

George Ferguson
President, Royal Institute of British Architects
RIBA

DESIGN IN THE HISTORIC ENVIRONMENT

MICHAEL DAVIES

On this small island that is Britain it seems more difficult to build than ever. There are huge demands on our towns, cities and countryside. A growing population and a seemingly endless army of experts ready to shoot down any hint of development, protected by mountains of preservationist policy which seem to stifle the very idea of imaginative design. So it is not surprising that when development is proposed in an historic environment these difficulties are magnified many times.

It is right that we should be concerned for our historic buildings. Lessons have been learned from the 1960s and '70s where a *Brave New World* threatened to remove all traces of our past. Since then the planning control system has been developed to try to ensure that the many historic buildings and areas which have survived throughout the country are conserved, not lost or damaged, and that any new development in these areas is sensitive to its historic environment. Unfortunately, the word 'sensitive' is a very subjective term coined by conservationists; nevertheless it is an excellent watchword, serving to raise awareness of the issues at stake. But have we overreacted to the mistakes of the '60s and '70s? Have we become too conservative, losing the confidence to build well in an historic context? For some people, one of the most appealing aspects of an historic building is seeing the various phases of development, not least because they help to show how the building has evolved over time, each adding to the building in its own style. So, if we were to add to the past by building an extension, or by making minor alterations to the inside or outside of an historic building, or simply by creating a new building next door, how should we leave our mark, and how will our buildings be judged by future generations?

STYLE, SCALE AND PROPORTION

The criteria of 'style, scale and proportion' generally represents the basis upon which design in the historic environment is judged. Most design guidance will tell you that the scale and proportion of the new building should be subservient to the old. There should be respect for the historic status of the existing, and designers should adopt a sense of awareness to the historic circumstance of their surroundings. This is a sound basis upon which to design. The scale and proportion of new buildings can have a varied affect upon the neighbouring buildings. If the new building dominates the existing, the historic character might also be diminished, while a relatively indifferent design might heighten the historic qualities of the existing building. However, these prescriptions for 'good' design are no substitute for the skill of the designer. Even the most subservient design can ruin the appearance of a beautiful old building or street. If only all buildings could reflect the hand of skilful design.

QUALITY COUNTS

In recent years much has been written about the quality of architecture. Prince Charles popularised the debate in 1988 with his *Vision of Britain* – a passionate cry from the heart with which the British public could identify, if not the professional fraternity. Here the Prince searched for the answer to why all the buildings he liked were *old* and none of them *modern*. The *Vision of Britain* did go on to

CONTEXTUAL

1) PASTICHE

Richmond Riverside Development, by Erith & Terry
A very skilful approach that requires an academic understanding of the period. Every detail and choice of material is an essay in the historic language of architecture. This building could easily be mistaken for the 'real thing', but if the detail and materials are watered down it will result in a poor imitation.

2) TRADITIONAL

Televillage, Crickhowell, by Powys County Council architects
A safe option that is often encouraged by the planning authorities, as it tends to follow the local vernacular. Much of its form, detailing and materials are borrowed from the past but have evolved into a watered down version. It takes little imagination and skill to produce a solution that 'fits in'. If handled sensitively it can produce some pleasing results.

3) SUBTLE

Cathedral Library extension, Hereford, by Whitfield Partners
Probably the most universally accepted approach to design in the historic environment. It is a conservationist's approach, where a light touch is required. Note the use of historic references and traditional materials, yet it is still subtly modern. It combines a respect for its surroundings with subtle detailing that confirms its place in the present.

identify some good modern examples, but the clear message was one of tradition and the past. This led to the Prince of Wales building a new village in Dorset as an exemplar of his gospel. Walking into Poundbury is like walking into the past. However, if you compare this with another new town – Milton Keynes (planned upon a modern grid), the appeal of Poundbury becomes clearer. So it is not surprising that we hark back to the comfort of what we know and love – it is safe and enduring; it satisfies our needs. But is this the answer? After all, most villages and towns were built up over many years and evolved their own peculiar character.

We often struggle to find an identity in a style as if this is the panacea for good design, and yet some of the most outrageous designs have succeeded in the most sensitive context. The *Sagrada Familia*, Gaudi's inspired cathedral in Barcelona, is outrageously original and a triumph of modern design: the quality of its design speaks volumes and it dominates its setting. Will the same be said of Daniel Libeskind's extremely bold extension to the Victoria & Albert Museum? The success of quality is often the test of time. What we might see as a success today might not be seen in the same light in the future.

John Ruskin had the foresight to realise the endurance of good quality by proclaiming "When we build let us think we build forever" (the *Lamp of Memory*).

Edward Cullinan believes it is a lack of aesthetic sensibility by planning committees, government agencies, and pressure groups which have accepted that new buildings should reflect the old buildings surrounding them, resulting in a mild mannered architecture composed of simplified or watered down components, lifted from the past. He believes that this insults both the past and the present and enhances neither.

The Corinthian Villa by Erith & Terry – a new house in an historic style which works because of its accuracy, attention to detail and exquisite craftsmanship.

CONTEMPORARY

4) MODERN

Visitor Centre, Caerphilly Castle, by Davies Sutton Architecture Ltd

This approach displays a modern design that is clearly of its time, but still respects its historic environment. It will have a strong and clear philosophy which draws its inspiration from the past. It might assemble local traditional materials in a modern way or, use modern materials in historical forms, for example. This requires a skilful hand and a good understanding of its historical surroundings.

5) ARROGANT

Extension to the Victoria & Albert Museum, London, by Daniel Libeskind

The tension created between old and new can be quite breathtaking, but requires great skill and vision to pull it off. A bold approach that needs an enormous leap of faith by all those involved, from client to planning authorities. This may be considered a 'building of the future' and will inevitably receive mixed reviews.

A MODEL

There is therefore more than one way to design in the historic environment, and much will depend upon other influences, such as the aspirations of the building owner, cost, the aesthetic sensibilities of the planners, the skill of the designer, and so on.

The model at the bottom of the page attempts to set out five different approaches to design and gives examples of real buildings which might arguably fall into each category. The model can be used as an initial discussion tool to visualise and debate the various approaches and decide which one is suitable.

As in age and politics, design for the historic environment is polarised by two extremes: the very historic and the very modern. Then everything else fits somewhere in between on a sliding scale, and it is possible to place any building on the scale to determine its stylistic relationship with its surroundings.

The 'Pastiche' approach (1) is where a building or extension is created as an historic essay based upon academic learning. Invariably this is very difficult to pull off, and there are nearly always some concessions to modernity. In the case of our example at Richmond, modern open plan offices sit behind a very fine replica façade; its downside is that large expanses of suspended ceiling are easily visible when standing outside the building. The

VISITOR CENTRE, CAERPHILLY

AT CAERPHILLY CASTLE, the new visitor centre has provided a solution to placing a new building in the grounds of a scheduled ancient monument. The building is unashamedly modern and of its time, and yet it is sensitive to its surroundings by looking to its past for inspiration. While the castle is constructed of massive stone walls, it once also contained other structures of timber frame construction, such as small buildings, fighting platforms and lean-to roofs or pentices. Being of timber, these structures have long since rotted away, although Cadw has rebuilt for interpretive purposes several siege engines and a hourd, a timber structure which protected people on the castle walls from arrows. The dominating oak framed structure of the new building fits well within this theme, and nestles against a protective stone wall near the entrance to the castle. As the new building was never part of the original castle, it is alien, an invader, and as such its roof can be seen rising up to attack the inner gatehouse, reintroducing drama to a once dramatic environment in more dangerous times.

The extensive use of timber also satisfies another 'conservation' issue – one of sustainability and the environment. Timber from a locally managed resource has a very low embodied energy factor and can ultimately be recycled when the building has expired. By placing the building against the north wall and facing directly south it is protected from the cold northerly winds, and the large area of glass generates large amounts of free heat from the sun. The dark natural slate floor absorbs the sun's heat and radiates back into the building when the temperature drops. The underfloor heating is operated from a heat pump, which is connected to the vast moat surrounding the castle, extracting more clean, free energy from the environment.

'Traditional' approach (2) is probably the most common and is arguably that which has 'watered down components lifted from the past'. It could also represent the modern vernacular of speculative house building.

From the other end of the spectrum, the 'Arrogant' approach (5) is immensely confident and pays little regard for its historic context. For this to succeed requires the most skilful designer, and many people would always find this unacceptable. The 'Modern' approach (4) provides an unambiguous building clearly of its time drawing its inspiration from the past and respectful of its historic context. When skilfully handled this is arguably the ideal approach. The 'Subtle' approach (3) requires a light hand and a deft touch. This approach probably pays the most respect to its historic context and is often adopted where a quiet, gentle approach is appropriate, one which allows the historic environment to speak loudest.

There is no doubt that design, like art, is subjective, and trying to understand the meaning and the process of design is difficult, let alone attempting to prescribe what is good design, and what is acceptable. Sir Henry Wotton (1568–1638), translating from the writings of the first century Roman architect Vitruvius, passed down an enduring interpretation of what represents good design; "In Architecture, as in all other operative arts, the end must direct the operation. The end is to build well. Well building hath three conditions – commodity, firmness and delight".

It is therefore left to the individual to decide how to approach design in the historic environment. But beware the opinions of others. Like many other things in life, it is the diversity of opinion and personalities that enrich our lives, and so it is with architecture. The historic environment is capable of absorbing the many personalities of our new and old buildings, and there is room for all the many different approaches to design in the appropriate place. However, the one thing that must prevail is quality – quality of design and quality of materials. We should not just see it as 'building'; we are creating 'architecture'. Ironically, it was the most famous modernist of all who captures the meaning of design and beauty in architecture when he said; "You employ stone, wood and concrete, and with these materials you build houses and palaces; that is construction. Ingenuity is at work. But suddenly you touch my heart, you do me good, I am happy and I say: 'This is beautiful'. That is architecture. Art enters in." – Le Corbusier 1927.

Recommended Reading

Warren, John, Worthington, John and Taylor, Sue (Eds), *Context: New Buildings in Historic Settings*. Butterworth-Heinemann, Oxford 1998

Quality in Town and Country, Department of the Environment discussion document, July 1994

Power of Place: The Future of the Historic Environment. The Historic Environment Steering Group, English Heritage, 1994

HRH The Prince of Wales, *A Vision of Britain*. Doubleday, London, 1989

National Planning Policy Guidance:

PPG01 *General Policy and Principles*, February 1997

PPG07 *Countryside: Environmental Quality and Economic and Social Development*, February 1997

PPG15 *Planning and the Historic Environment*, September 1994

MICHAEL DAVIES BSc(Hons) BArch DipCons(AA) IHBC AABC RIBA is a chartered architect and partner in Davies Sutton Architecture Ltd (see page 23). He has 15 years experience in conservation and is a member and caseworker for the SPAB. His practice specialises in conservation, rescuing buildings from ruin, and designing modern new buildings for the historic environment. E-mail md@davies-sutton.co.uk

DYCE SYMBOL STONES being recorded by Colin Muir

Chapter 1

PROFESSIONAL SERVICES

PROFESSIONAL SERVICES BY REGION

Regions are organised geographically, from north to south. For a company offering services in a particular region, please check under both the UK and the individual region.

UK		Page
A E Thornton-Firkin & Partners	qs	47
A P R Services Ltd	ms	14
A R P Lorimer and Associates	ar cm po qs	20
Abbey Heritage Ltd	he	165
Acanthus Associated Architectural Practices	ar la po ur	19
AMBO Architects	ar id po	20
Anderson & Glenn	ar ho la po	20
Anthony Short and Partners	ar id ms su	20
Arrol & Snell Ltd	ae ar he la ms po su	20
Austin Trueman Associates	st	38
Avanti Architects Limited	ar	21
Bill Harvey Associates	st	38
Building Design Partnership	ar pc qs st su ur	21
CgMs Consulting	ae pc	34
Charles Knowles Design	ar	22
Cherished Land Limited	xu	183
Cliveden Conservation Workshop Ltd	ma	40
The Conservation Studio	pc po ur	35
Court Design and Conservation	su	44
Cube Property Services Ltd	cm he su	44
Daniells Harrison Chartered Surveyors	ae hi ms su	44
David Gibson Architects	ar	23
David Harvey Architects	ar po su	23
David Narro Associates	st	38
Demaus Building Diagnostics Ltd	tt	40
Donald Insall Associates Ltd	ar cm hi id ms pc po ur	24
Drivers Jonas	cm ms pc su	45
Ede Surveyors	cm he ms po su	45
English Heritage – National Monuments Record	hi ms ph	15
FaberMaunsell Limited	be cm fs he hs st	190
G B Geotechnics Ltd	nd st su	40
Gifford and Partners	ae bs cm fs st	39
Giles Quarme & Associates	ar hi ms pc su	25
Griff Davies Architectural Design and Conservation Ltd	ar hi su pc po	45
HGP Conservation	ar cm pc po	25
Hare & Humphreys Ltd	he	198
Harry Cursham	he nd ph	35
Hart Brothers Engineering Ltd	cm	145
Heritage Testing Ltd	ma	40
Hilary Taylor Landscape Associates Ltd	la po	15
Hirst Conservation	po	ifc
Historic Buildings Conservation Limited	po	61
The Historical Research Agency	hi	14
HOK Conservation and Cultural Heritage	ar hi id po	26
The House Historians	hi	14
Hutton+Rostron Environmental Investigations Limited	cm dn he nd su tt	40
Ingram Consultancy Ltd	ar he ma po su	35
International Fire Consultants Limited	fs hs tt	188
Julian Harrap Architects	ar hi id la su	26
KAW Design	id	200
King Sturge Heritage	cm pc su ur	45
La Playa	in	187
Latham Architects	ar la ur	27
The Leather Conservation Centre	po	200
LPOC Insurance Services	in	187
Mann Williams	hs st su	39
Martin Stancliffe Architects	ar cm su	27
McCurdy & Co	st	73
MRDA	ar he ms po	28
Network Archaeology Ltd	ae hi ma ms su	12
Niall Phillips Architects	ar po ur	28
Nicholas Jacob Architects	ar	28
Norman + Dawbarn Architects	ar be id st	28
Oxford Archaeology	ae hi ms po xu	12
Plowman Craven & Associates	ms ph su	14
PriceWaterhouseCoopers	vt	48
RBPM General Ltd	in	187
Resurgam	ar cm he ma nd po	35
RHWL	ar pc ur	29
Ridout Associates	nd su	40
Robert Kilgour & Associates	ar	29
Robin Kent Architecture and Conservation	ar hi su	29
Rope Tech International Limited	he nd ph	84
Rose of Jericho	ma	175
Ian Russell	nd st su	39
Ryder & Dutton	hi nd po	47
Scottish Lime Centre Trust	he ma	35
Sibtec Scientific, Sibert Technology Ltd	tt	40
Shambrooks	cm hs qs su	48
Simpson & Brown Architects	ar hi id ms po	30
Structural Perspectives	ae hi ms po	13
TFT Cultural Heritage	cm hi pc po su	46
Thomas Ford & Partners	ar cm hs po su	31
W R Dunn & Co	ar cm hs po su	46
Ward & Dale Smith, Chartered Building Surveyors	su	46
Weald & Downland Open Air Museum	hi	226
Wessex Archaeology, Conservation Management Section	ae hi po	13

SCOTLAND		
A O C Archaeology Group	ae dn ma ms po su	12
Elliott & Company Structural Engineers	st	38
Gibbon, Lawson, McKee Limited	ar cm fs su	24
Gray, Marshall & Associates	ar hs	25
Johnston & Wright	ar	26
Lincoln & Campbell Associates Ltd	ar	27

NORTH		
A O C Archaeology Group	ae dn ma ms po su	12
Blackett-Ord Consulting Engineers	ar hi st su	38
Brock Carmichael Architects	ar la pc po ur	21
Cambridge Dating Unit	dn	14
Capstone Consulting Engineers	st	38
CoDA Conservation	ar st	22
Elaine Rigby Architects	ar	24
Hodkinson Mallinson Ltd	su	46
Johnston & Wright	ar	26
Press & Starkey	qs	48
Robinsons Preservation Limited	nd tt	182
Spence & Dower Architects	ar	30

ISLE OF MAN		
Brock Carmichael Architects	ar la pc po ur	21
Johnston & Wright	ar	26

YORKSHIRE		
A O C Archaeology Group	ae dn ma ms po su	12
Allen Tod Architects	ar	20
Blackett-Ord Consulting Engineers	ar hi st su	38
Brock Carmichael Architects	ar la pc po ur	21
Byrom Clark Roberts	ar st su	22
Capstone Consulting Engineers	st	38
Cambridge Dating Unit	dn	14
Carden & Godfrey Architects	ar id po	22
CoDA Conservation	ar st	22
David Lewis Associates	ar su	23
Johnston & Wright	ar	26
Press & Starkey	qs	48
Richard Crooks Partnership	ar cm hs	29
Robinsons Preservation Limited	nd tt	182
Fred Tandy	st	39
Womersley's Limited	cm he ms po	175

NORTH WEST		
A O C Archaeology Group	ae dn ma ms po su	12
Anthony Blacklay & Associates	ar la	20
Anthony Short and Partners	ar ms su	20
Blackett-Ord Consulting Engineers	ar hi st su	38
Brock Carmichael Architects	ar la pc po ur	21
Byrom Clark Roberts	ar st su	22
Cambridge Dating Unit	dn	14
Capstone Consulting Engineers	st	38
CoDA Conservation	ar st	22
Edmund Kirby Architects	ar ur su	24
Field Archaeology Centre	ae	12
Graham Holland Associates	ar	24
Haigh Architects	ar	25
Hodkinson Mallinson Ltd	su	46
James Brotherhood & Associates	ar cm id ms pc	26
Johnston & Wright	ar	26
Lightwright Associates	be fs	190
Lloyd Evans Prichard	ar cm po su	27
Press & Starkey	qs	48
Robinsons Preservation Limited	nd tt	182
Fred Tandy	st	39
Wm Langshaw & Sons	cm	68

WALES		
Acanthus Clews Architects	ar cm po su	19
Brock Carmichael Architects	ar la pc po ur	21
David Harvey Architects	ar po su	23
Davies Sutton Architecture Limited	ar he po su	23
Dean & Cheason Associates	ar	24
Ellis & Moore	st	39
The Heritage Practice	ar	26
Hodkinson Mallinson Ltd	su	46
Mildred, Howells & Co	qs	47
Press & Starkey	qs	48
Robin Wolley Chartered Architect	ar	30
Stainburn Taylor Architects	ar	30

WEST MIDLANDS		
A O C Archaeology Group	ae dn ma ms po su	12
Acanthus Clews Architects	ar cm po su	19
Anthony Short and Partners	ar ms su	20
Applied Surveying and Design Group	fs hs st su	44
Baily Garner	ar cm qs su	44
Brock Carmichael Architects	ar la pc po ur	21
Cambridge Dating Unit	dn	14
Chedburn Design & Conservation	ar hi ms po su	22
David Brown and Partners	ar cm id	23
Davies Sutton Architecture Limited	ar he po su	23
Hall & Ensom, Chartered Building Surveyors	su	46
Hawkes Edwards & Cave	ar	26
The Heritage Practice	ar	26
Johnston & Wright	ar	26
King Sumners Partnership	cm hs qs	47
Lightwright Associates	be fs	190
Mildred, Howells & Co	qs	47
Peter Yiangou Associates	ar su	29
Phoenix Beard	cm hs ms su	46
Press & Starkey	qs	48
Richard Crooks Partnership	ar cm hs	29
Robert Bloxham-Jones Associates	be	190
Robinsons Preservation Limited	nd tt	182
Stainburn Taylor Architects	ar	30
Fred Tandy	st	39

EAST MIDLANDS		
A O C Archaeology Group	ae dn ma ms po su	12
Acanthus Clews Architects	ar cm po su	19
Anthony Short and Partners	ar ms su	20
Baily Garner	ar cm qs su	44
Brock Carmichael Architects	ar la pc po ur	21
Cambridge Dating Unit	dn	14
Capstone Consulting Engineers	st	38
King Sumners Partnership	cm hs qs	47
Press & Starkey	qs	48
Robinsons Preservation Limited	nd tt	182
Stainburn Taylor Architects	ar	30
Fred Tandy	st	39
The Victor Farrar Partnership	ar he ms po su	31
WCP	ar id lg su	31

PROFESSIONAL SERVICES BY REGION

EAST ANGLIA

Company	Services	Page
Acanthus Clews Architects	ar cm po su	19
Baily Garner	ar cm qs su	44
Bedford Timber Preservation Company	nd	181
Cambridge Dating Unit	dn	14
Carden & Godfrey Architects	ar id po	22
David Brown and Partners	ar cm id	23
David Pitts Chartered Architects	ar	24
Derek Rogers Associates	ar ms su	24
Ellis & Moore	st	39
Feilden + Mawson	ar hi id pc su	24
Fothergill	st	39
H-H-Heritage	nd st su	39
Katie Thornburrow Architects	ar	26
King Sumners Partnership	cm hs qs	47
Lightwright Associates	be fs	190
Mansfield Thomas & Partners	ar hi la pc qs su	28
Martin Ashley Architects	ar id	27
The Morton Partnership Ltd	ar st su	39
Peter Codling Architects	ar su	29
Press & Starkey	qs	48
Rickards Conservation	ms po su	46
Roderick Shelton	ar	30
Roger Mears Architects	ar	31
The Victor Farrar Partnership	ar he ms po su	31
WCP	ar id lg su	31
Michael Walton	ar id vt	31

HOME COUNTIES NORTH

Company	Services	Page
Acanthus Clews Architects	ar cm po su	19
Baily Garner	ar cm qs su	44
Bedford Timber Preservation Company	nd	181
Blampied and Partners Ltd	ar cm	21
The Budgen Partnership	st	38
Cambridge Dating Unit	dn	13
Carden & Godfrey Architects	ar id po	22
Chedburn Design & Conservation	ar hi ms po su	22
Christopher Rayner Architects	ar	22
David Brown and Partners	ar cm id	23
David Pitts Chartered Architects	ar	24
Derek Rogers Associates	ar ms su	24
Ellis & Moore	st	39
Feilden + Mawson	ar hi id pc su	24
Fothergill	st	39
Gilmore Hankey Kirke Ltd	ar cm id la po	25
Hall & Ensom, Chartered Building Surveyors	su	46
H-H-Heritage	nd st su	39
Katie Thornburrow Architects	ar	26
King Sumners Partnership	cm hs qs	47
Mansfield Thomas & Partners	ar hi la pc su	28
Martin Ashley Architects	ar id	27
Michael Drury Architects	ar	28
The Morton Partnership Ltd	ar st su	39
Peter Yiangou Associates	ar su	29
Press & Starkey	qs	48
Reliable Effective Research	hi	14
Rickards Conservation	ms po su	46
Roderick Shelton	ar	30
Roger Joyce Associates	ar	30
Roger Mears Architects	ar	31
Timothy J Shepherd Historic Brickwork Specialist	ma	113
The Victor Farrar Partnership	ar he ms po su	31
WCP	ar id lg su	31
Michael Walton	ar id vt	31
John Wardle	st	39

GREATER LONDON

Company	Services	Page
Acanthus Clews Architects	ar cm po su	19
Architectural Archaeology	ae	12
Babtie Murdoch Green	cm qs su	47
Baily Garner	ar cm qs su	44
Blampied and Partners Ltd	ar cm	21
The Budgen Partnership	st	38
Peregrine Bryant	ar	21
Cambridge Dating Unit	dn	14
Cameron Taylor Bedford	cm hs st su	38
Carden & Godfrey Architects	ar id po	22
Caroe & Partners	ae ar hi po	22
Chedburn Design & Conservation	ar hi ms po su	22
Christopher Rayner Architects	ar	22
David Ashton Hill Architects	ar la po	22
David Brown and Partners	ar cm id	23
Derek Plummer Architect	ar	23
James Dunnett Architects	ar	24
Ellis & Moore	st	39
Gilmore Hankey Kirke Ltd	ar cm id la po	24
Hall & Ensom, Chartered Building Surveyors	su	46
The Halpern Partnership Ltd	ar po	25
John D Clarke & Partners	ar	26
King Sumners Partnership	cm hs qs	47
Mackenzie Wheeler	ar id	28
Michael Drury Architects	ar	28
Press & Starkey	qs	48
Reliable Effective Research	hi	14
Richard Griffiths Architects	ar cm id pc po	29
Rickards Conservation	ms po su	46
Roger Joyce Associates	ar	30
Roger Mears Architects	ar	31
Timothy J Shepherd Historic Brickwork Specialist	ma	113
John Wardle	st	39
Watkinson & Cosgrave	su	46
WCP	ar id lg su	31
Michael Walton	ar id vt	31

HOME COUNTIES SOUTH

Company	Services	Page
Acanthus Clews Architects	ar cm po su	19
Adrian Cox Associates	st	38
The Architectural History Practice Limited	hi po	14
Bailey Partnership	cm so	44
Baily Garner	ar cm qs su	44
Blampied and Partners Ltd	ar cm	21
The Budgen Partnership	st	38
Cambridge Dating Unit	dn	14
Carden & Godfrey Architects	ar id po	22
Chedburn Design & Conservation	ar hi ms po su	22
Christopher Rayner Architects	ar	22
Clague	ar la su	22
Daniel Forshaw Design and Conservation Architects	ar	22
David Brown and Partners	ar cm id	23
Derek Rogers Associates	ar ms su	24
Alan Dickinson Chartered Building Surveyor	ae hi ms su	44
Ellis & Moore	st	39
Gilmore Hankey Kirke Ltd	ar cm id la po	25
John D Clarke & Partners	ar	26
King Sumners Partnership	cm hs qs	47
KingsLand Surveyors Limited	ms	14
T C R MacMillan-Scott	ar	28
Michael Drury Architects	ar	28
P W P Architects	ar cm ms pc su	28
Paul Tanner Associates	st	39
Press & Starkey	qs	48
PRM Archaeology	ae	12
Reliable Effective Research	hi	14
Rickards Conservation	ms po su	46
Roger Joyce Associates	ar	30
Roger Mears Architects	ar	31
Timothy J Shepherd Historic Brickwork Specialist	ma	113
Stuart Page Architects	ar id po	30
Anthony Swaine	ar he po	31
John Wardle	st	39

SOUTH WEST

Company	Services	Page
A S I Heritage Consultants	ae ms po xu	12
Acanthus Clews Architects	ar cm po su	19
Architecton	ar he id po	20
Baily Garner	ar cm qs su	44
Bare Leaning & Bare	hs qs	47
Barlow Schofield Partnership	ar	21
Bosence & Co	cm su	44
Cambridge Dating Unit	dn	14
Chedburn Design & Conservation	ar hi ms po su	22
Chris Romain Architecture	ar	22
Davies Sutton Architecture Limited	ar he po su	23
Ellis & Moore	st	39
Exeter Archaeology	ae	12
Gilmore Hankey Kirke Ltd	ar cm id la po	24
The Heritage Practice	ar	26
Johnston & Wright	ar	26
Jonathan Rhind Architects	ar he id	26
King Sumners Partnership	cm hs qs	47
Michael Drury Architects	ar	28
Mildred, Howells & Co	qs	47
Michael Pearce	ar pc po	34
Peter Yiangou Associates	ar su	29
Press & Starkey	qs	48
Robert Bloxham-Jones Associates	be	190
Robert Seymour Conservation	ar	29
Michael Trevallion	ar	31
Watson Bertram & Fell	ar su	31

SOUTH

Company	Services	Page
A S I Heritage Consultants	ae ms po xu	12
Acanthus Clews Architects	ar cm po su	19
The Architectural History Practice Limited	hi po	14
Bailey Partnership	cm so	44
Baily Garner	ar cm qs su	44
Blampied and Partners Ltd	ar cm	21
Bosence & Co	cm su	44
Cambridge Dating Unit	dn	14
Carden & Godfrey Architects	ar id po	22
Chedburn Design & Conservation	ar hi ms po su	22
Chris Romain Architecture	ar	22
Christopher Rayner Architects	ar	22
Clague	ar la su	22
Daniel Forshaw Design and Conservation Architects	ar	22
David Brown and Partners	ar cm id	23
Alan Dickinson Chartered Building Surveyor	ae hi ms su	44
Ellis & Moore	st	39
The Heritage Practice	ar	26
John D Clarke & Partners	ar	26
King Sumners Partnership	cm hs qs	47
KingsLand Surveyors Limited	ms	14
Julian R A Livingstone Chartered Architect	ar	27
T C R MacMillan-Scott	ar	28
Michael Drury Architects	ar	28
Mildred, Howells & Co	qs	47
P W P Architects	ar cm ms pc su	28
Paul Tanner Associates	st	39
Press & Starkey	qs	48
Radley House Partnership	ar cm po	29
Reliable Effective Research	hi	14
Rickards Conservation	ms po su	46
Roger Joyce Associates	ar	30
Timothy J Shepherd Historic Brickwork Specialist	ma	113
Stuart Page Architects	ar id po	30

KEY

ae	archaeologists	ma	materials analysis
ar	architectural services	ms	measured surveys
be	building services consulting engineers	nd	non-destructive investigations
cm	contract management	pc	planning consultants
dn	dendrochronology	ph	photographic services
fs	fire safety consultants	po	conservation plans and policy consultants
he	heritage technologists	qs	quantity surveyors
hi	historical researchers	st	structural engineers
ho	horticultural consultants	su	surveyors
hs	health and safety consultants	tt	structural timber testing
id	interior designers and consultants	ur	urban regeneration
in	insurance	vt	VAT consultants
la	landscape architects	xu	exhumation of human remains
lg	legal services		

ARCHAEOLOGISTS

▶ A O C ARCHAEOLOGY GROUP

Edgefield Road Industrial Estate, Loanhead, Midlothian EH20 9SY
Tel 0131 440 3593 Fax 0131 440 3422
E-mail admin@aocscot.co.uk Website www.aocarchaeology.com
ARCHAEOLOGISTS: AOC Archaeology Group offers a professional service in historic and listed building recording and is highly skilled in managing the building requirements as identified through NPPG18/NNP15. The specialisms available include building surveying, paint analysis, guidance on conservation of artefacts and dendrochronology. AOC Archaeology Group has more than ten years experience in the analysis of materials from standing buildings, archaeological excavations and objects in museums collections, and has offices throughout the United Kingdom.

▶ A S I HERITAGE CONSULTANTS

Furlong House, 61 East Street, Warminster, Wiltshire BA12 9BZ
Tel/Fax 01985 847791
Contact Michael Heaton BTech PgDip (BuildCons) MIFA IHBC
E-mail info@asi-heritageconsultants.co.uk
Website www.asi-heritageconsultants.co.uk
PROFESSIONAL ARCHAEOLOGICAL SUPPORT: Michael Heaton is the historic buildings specialist at ASI Heritage Consultants, a multi-disciplinary heritage practice established in 1996, that provides professional archaeological support in the planning and management of works to archaeological sites and historic buildings across the south of England.

▶ ARCHITECTURAL ARCHAEOLOGY

15 Grove Road, Ramsgate, Kent CT11 9SH
Tel 01843 585792 E-mail twoarches@aol.com
Mark Samuel PhD MIFA
ARCHAEOLOGICAL SERVICES: Architectural Archaeology provides a range of services necessary for statutory PPG15 work or faculty applications including assessment reports, accurate recording of architectural features and supervision of recording work. Architectural Archaeology offers particular expertise in the rapid analysis of stone buildings, including dating, recognition and graphic reconstruction of architectural features employing conventional and CAD methods. In-house writing, editing and research services provided. Clients include: Dean & Chapter of St Paul's Cathedral, Museum of London, Guildford Borough Council, AOC Archaeology, Syngenta.

▶ EXETER ARCHAEOLOGY

Bradninch Place, Gandy Street, Exeter EX4 3LS
Tel 01392 665521 Fax 01392 665522 E-mail exeter.arch@exeter.gov.uk
ARCHAEOLOGISTS: Exeter Archaeology can offer a wide range of archaeological services throughout the South West region. These include recording and analysis of historic buildings of all types (urban and rural), vernacular buildings, churches and monastic buildings, fortifications and industrial buildings. Recent projects include the Royal Citadel and Royal William Yard in Plymouth, St Nicholas Priory, Exeter and Tavistock Abbey, West Devon. Exeter Archaeology regularly undertakes recording projects on parish churches in Devon, Cornwall and Somerset.

▶ FIELD ARCHAEOLOGY CENTRE

The University of Manchester, Oxford Road, Manchester M13 9PL
Tel 0161 275 2314 Fax 0161 275 2315 E-mail umfac@man.ac.uk
Website www.art.man.ac.uk/fieldarchaeologycentre
Robina McNeil FSA MIFA Greater Manchester Archaeological Unit
Dr Michael Nevell MIFA The University of Manchester Archaeological Unit
ARCHAEOLOGY, BUILDINGS, CONSERVATION SPECIALISTS: The centre houses two independent units with expertise in planning archaeology, environmental assessments, field archaeology, building archaeology, analysis, research, interpretation and management of the archaeological resource and specialising in urban regeneration, building conservation, sustainable development, buildings and monuments at risk strategies and, conservation plans. Both units have access to resources and staff across the University. In addition to strategic planning, consultancy, contract work and research the units offer an advisory and information service on their cultural heritage and are committed to a strong publication programme.

▶ GIFFORD AND PARTNERS

Carlton House, Ringwood Road, Woodlands, Southampton SO40 7HT
Tel 023 8081 7500 Fax 023 8081 7600
ARCHAEOLOGISTS: *See also: profile entry in Structural Engineers section, page 39.*

▶ HIRST CONSERVATION

Laughton, Sleaford, Lincolnshire NG34 0HE
Tel 01529 497449 Fax 01529 497518
ANALYSIS OF PLASTER AND PAINT LAYERS: *See also: display entry on the inside front cover and profile entry in Building Contractors section, page 62.*

▶ NETWORK ARCHAEOLOGY LTD

Nicola Smith, 28, West Parade, Lincoln LN1 1JT
Tel 01522 532625 Fax 01522 532631
E-mail nicolas@netarch.co.uk
▶ David Bonner, 22, High Street, Buckingham MK18 1NU
Tel 01280 816174 Fax 01280 816175
E-mail davidb@netarch.co.uk
Website www.netarch.co.uk
BUILDING ARCHAEOLOGISTS: The historic buildings team is an integral part of Network Archaeology Ltd and comprises experienced and highly qualified professionals. The team provides a comprehensive range of above and below ground archaeological services nationwide, including consultation and strategic advice, historic building investigation, analysis and interpretation, hand survey and digital recording, freehand and CAD illustration, documentary research, post-survey reporting and desk-top studies. The team undertakes both urban and rural projects involving historic buildings of all types – high status houses, vernacular buildings, churches, castles, agricultural and industrial structures.

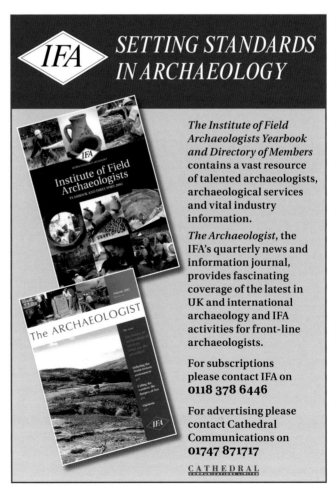

ARCHAEOLOGISTS

▶ OXFORD ARCHAEOLOGY

Janus House, Osney Mead, Oxford OX2 0ES
Tel 01865 263800 Fax 01865 793496 Website www.oau-oxford.com
Contact Julian Munby or Bob Williams
▶ OXFORD ARCHAEOLOGY NORTH
The Storey Institute, Meeting House Lane, Lancaster LA1 1TF
Tel 01524 848666 Fax 01524 848606
Contact Rachel Newman

ARCHAEOLOGISTS: Oxford Archaeology undertakes historic buildings investigations and surveys, documentary studies and archaeology. Listed building and scheduled monument consents, environmental assessments and planning. They are consultants to Historic Royal Palaces and other clients include the National Trust, English Heritage, Victoria and Albert Museum, Union Railways, local authorities, Oxford Colleges and many churches. For further information please contact Julian Munby or Bob Williams.

▶ PRM ARCHAEOLOGY

32 Herbert Road, Bexleyheath, Kent DA7 4QF
Tel/Fax 020 8306 7043 Mobile 07931 660120

SPECIALISTS IN HISTORIC BUILDINGS AND FIELD ARCHAEOLOGY: PRM Archaeology's professional and cost effective services include consultancy, project management, historic building recording - analysis, site survey, desk-top assessment, watching brief, excavation, artefact and building materials and moulded stone analysis, site conservation, CAD, historical research.

▶ WESSEX ARCHAEOLOGY, CONSERVATION MANAGEMENT SECTION

Portway House, Old Sarum Park, Salisbury, Wiltshire SP4 6EB
Tel 01722 326867 Fax 01722 337562
E-mail info@wessexarch.co.uk
Website www.wessexarch.co.uk
Contact Paul Falcini, Head of Conservation Management or John Dillon, Operations Director

ARCHAEOLOGISTS: Wessex Archaeology is a leading professional practice based in Salisbury, Wiltshire offering services throughout the UK and overseas. As one of the largest archaeological consultancies in the country, Wessex has a thriving and successful conservation management section. The company's building surveys range from simple photographic and descriptive surveys to full architectural and historical analysis. High quality digital recording and modelling techniques are frequently used for strategic management and analytical projects, as well as for building surveys. Recent projects have included medieval, post-medieval and modern buildings such as ecclesiastical structures, military installations and industrial and vernacular architecture. Recent clients include English Heritage, Defence Estates, developers, private individuals and many local authorities.

DENDROCHRONOLOGISTS

▶ CAMBRIDGE DATING UNIT

Environmental Sciences Research Centre,
Anglia University, East Road, Cambridge CB1 1PT
Tel 01223 363271 x2594
E-mail V.R.Switsur@Anglia.ac.uk orVRS1@Cam.ac.uk
Contact Roy Switsur

DENDROCHRONOLOGY AND THERMOLUMINESCENCE SPECIALISTS: Cambridge Dating Unit has had experience, since the early 1970s, in tree-ring dating and analysis of historical oak-framed buildings, wooden artefacts such as furniture and framed paintings, as well as the thermoluminescence and optically stimulated luminescence dating of pottery, burnt flints and sediments (for details of techniques see Switsur: Dating Technology, *Building Conservation Directory* 2001 pp 17–18). The Unit has university-wide access to resources and staff for studies including palaeontology, petrology and sedimentology. It has a strong policy of research and publication of dating results. Studies have covered buildings from the medieval period to the 18th century in dating listed and ecclesiastical buildings, boats and vernacular structures such as hall-houses, cottages and barns. Contracts are with English Heritage, Cambridge Archaeological Unit, Oxford Archaeological Unit, Cambridge colleges, local authorities and private individuals.

MEASURED SURVEYS

▶ A P R SERVICES LTD

Unit 6A, Chaseside Works, Chelmsford Road, Southgate, London N14 4JN
Tel 020 8447 8255 Fax 020 8882 8080
E-mail mail@aprservices.net
Website www.aprservices.net

LAND AND MEASURED BUILDING SURVEYS, 3D LASER SCANNING AND VISUALISATION: A P R Services has specialised in 3D laser scanning to complement its existing high quality surveying services. Projects include The Royal Society and Royal Institution, London, and Royal Hall, Harrogate. Surveys include cave systems, railway and road bridges.

▶ ENGLISH HERITAGE, NATIONAL MONUMENTS RECORD

Kemble Drive, Swindon, Wiltshire SN2 2GZ
Tel 01793 414802 Fax 01793 414924

MEASURED SURVEYS OF SITES AND LANDSCAPES: *See also: display entry in Archives section, page 15.*

▶ KINGSLAND SURVEYORS LIMITED

Bridge Chambers, Bridge Street, Leatherhead, Surrey KT22 8BN
Tel 01372 362059 Fax 01372 363059
Website surking.co.uk

SPECIALISTS IN ARCHITECTURAL, PHOTOGRAMMETRY SURVEY WORKS AND RECTIFIED PHOTOGRAPHY: Producing 2d and 3d digital floor plans and detailed elevations, providing a comprehensive and accurate record. Kings – architectural, land and utility surveyors, formerly known as McDowells – has over the years undertaken a wide range of survey works on major contracts, both in the United Kingdom and overseas since the company's foundation in 1963. UK clients include National Westminster Bank, NHS trust hospitals, Metropolitan Police and numerous architects.

▶ PLOWMAN CRAVEN & ASSOCIATES

141 Lower Luton Road, Harpenden, Herts AL5 5EQ
Tel 01582 765566 Fax 01582 765370

ARCHITECTURAL PHOTOGRAMMETRY, 3D LASER SCANNING, BUILDING SURVEYS AND ALL ASPECTS OF MEASURED, PHOTOGRAPHIC AND DIGITAL SURVEY: With over 35 years experience as surveyors, PCA is one of the leading firms of professional 'measured' surveyors. Clients include English Heritage, Cadw, Historic Scotland, Royal Household, Palace of Westminster, National Gallery, Church Commissioners and many others. Projects have included Windsor (before and after fire), Hampton Court and many smaller restoration tasks. Data supplied in 2D, 3D digital form.

ARCHITECTURAL PHOTOGRAPHERS

▶ HARRY CURSHAM

Parks Farm, Cambridge, Gloucester GL2 7AR
Tel 01453 890297 Fax 01453 899121
Mobile 07986 185894
E-mail harry.cursham@lineone.net

CONSULTANT REPAIRS TO OLD, VERNACULAR AND TRADITIONAL BUILDINGS: *See also: profile entry in Heritage Consultants section, page 35.*

ARCHITECTURAL HISTORIANS

▶ THE ARCHITECTURAL HISTORY PRACTICE LIMITED

Phillimore Cottage, Thorncombe Street, Nr Bramley, Surrey GU5 0LU
Tel 01483 208633 Fax 01483 208684
E-mail mail@architecturalhistory.co.uk Website www.architecturalhistory.co.uk
Directors James Anderson, Neil Burton, Clare Hartwell, Frank Kelsall

ARCHITECTURAL HISTORIANS: The Architectural History Practice is a leading architectural history consultancy with vast in-house expertise and experience. Specialist skills lie in using documentary evidence and expert analysis of the architectural history of buildings and their contexts to supplement fabric observation and archaeological investigation in order to prepare a fully rounded and properly understood building history. The firm also offers specialist advice and analysis of historic parks, gardens, and landscapes. Projects completed include conservation plans for nationally important sites; providing historical input for major development schemes; and advising commercial and private owners in cases of listed building and conservation area consent.

▶ THE HISTORICAL RESEARCH AGENCY

4 Filer Close, Peasedown St John, Bath BA2 8DQ
Tel 01761 434806 Mobile 07951 492001
E-mail kaspiers@historicalresearch.biz Website www.historicalresearch.biz

SPECIALIST IN LISTED BUILDING AND PERIOD PROPERTY RESEARCH

▶ THE HOUSE HISTORIANS

4 Springfield Cottages, Brewery Hill, Upton Cheyney, South Gloucestershire BS30 6NA
Tel 0117 932 3009 Fax 0117 932 3473
E-mail kay@thehousehistorians.co.uk Website www.thehousehistorians.co.uk
Contact Kay Ross

BUILDING HISTORIANS: The House Historians is a professional architectural history practice specialising in interpretive reports on historic buildings, including justifications for listed building consent. The practice undertakes both structural and documentary building histories, combining a comprehensive examination and analysis of the physical building, the methods and materials used in the construction and any subsequent changes with detailed documentary research of its historical development. Recent commissions include a 15th century manor house, 17th and 18th century farmhouses, 20th century lodges and updating Bath's listed buildings register for English Heritage. Clients include architects, surveyors, developers, local authorities, historical trusts, museums, television companies and private owners.

▶ McCURDY & CO

Manor Farm, Stanford Dingley, Reading, Berkshire RG7 6LS
Tel 01189 744866 Fax 01189 744375
E-mail info@mccurdyco.com Website www.mccurdyco.com

TIMBER FRAME CONSULTANTS: *See also: display entry in Timber Frame Builders section, page 73.*

▶ RELIABLE EFFECTIVE RESEARCH

17 Homestead Road, Fulham, London SW6 7DB
Fax 020 7736 3741 (Agency)

SPECIALIST IN HISTORICAL RESEARCH: Robert E Rodrigues through Reliable Effective Research (RER) provides a personal and efficient research service for those wishing to establish a better understanding of their property in its historical context. RER helps owners understand the history and form of their building using data derived from a variety of sources. Desk-top research is relied on extensively to reduce the need for disruptive site visits. Please write or fax RER to discuss how you can get to know your property better for historical reasons, or for enlightening background information for conservation project planning.

▶ WEALD AND DOWNLAND OPEN AIR MUSEUM

Singleton, Chichester, West Sussex PO18 0EU
Tel 01243 811363 Fax 01243 811475
E-mail wealddown@mistral.co.uk Website www.wealddown.co.uk

CONSERVATION TRAINING, RESEARCH AND SUPPLIES: Research library and artefact store designed for use by professionals. *See also: profile entry in Courses & Training section, page 226.*

CATHEDRAL COMMUNICATIONS LIMITED

ARCHITECTS

▶ BAILY•GARNER
146–148 Eltham Hill, London SE9 5DY
Tel 020 8294 1000 Fax 020 8294 1320
E-mail general@bailygarner.co.uk
Contact Lisa J Brooks BSc (Hons) DipBldgCons ARICS
ARCHITECTS, BUILDING SURVEYORS, QUANTITY SURVEYORS AND PROJECT MANAGERS: *See also: profile entry in Surveyors section, page 44.*

▶ BARLOW SCHOFIELD PARTNERSHIP
No 4 Wheal Agar, Tolvaddon Energy Park, Tolvaddon, Camborne, Cornwall TR14 0HX
Tel 01209 614921 Fax 01209 715639
CHARTERED DESIGN AND CONSERVATION ARCHITECTS: The practice is involved in all aspects of building conservation work, including repairs, alterations, extensions, design of new buildings in the context of historic buildings, and conversions to new uses. In addition to design services, the practice also undertakes condition surveys, feasibility studies and grant applications, and advises local authorities on conservation matters including development control. Conservation projects are led by a partner with an MA in architectural conservation, whose experience covers all grades of listed buildings together with scheduled ancient monuments. Clients include the National Trust, Cornwall Buildings Preservation Trust, MoD, local authorities and private individuals.

▶ BLAMPIED AND PARTNERS LTD
A member of the Areen Design Group of Companies
Areen House, 282 King Street, London W6 0SJ
Tel 020 8563 9175 Fax 020 8563 9176
E-mail yvette@blampied.co.uk
Website www.blampied.co.uk
CHARTERED ARCHITECTS, PROJECT MANAGERS AND PLANNING SUPERVISORS: The practice has 40 years experience of working with clients on refurbishment, alteration and conversion of listed buildings and non-listed buildings in conservation areas. Particularly experienced in hotel and residential projects and most recently involved in the sympathetic conversion of office premises back to single occupancy houses liaising closely with English Heritage and other authorities.

▶ BROCK CARMICHAEL ARCHITECTS
Federation House, Hope Street, Liverpool L1 9BS
Tel 0151 709 1087 Fax 0151 709 6418
E-mail watkins.d@brockcarmichael.co.uk
Contact David Watkins, DipArch, AADiplCons, RIBA
ARCHITECTS AND PLANNERS: Brock Carmichael Architects is an award winning practice in the conservation, refurbishment, extension and adaptation of historic and listed buildings, and the appropriate and imaginative design and integration of new buildings within sensitive and historic sites. Operating on a wide geographical basis, a comprehensive range of services is provided including building surveys and analysis, feasibility studies, conservation plans, masterplanning, full architectural and planning services and interior design. Clients include the National Trust, Stoke City council, Grosvenor, Cheshire County Council and Liverpool Anglican Cathedral.

▶ PEREGRINE BRYANT
The Courtyard, Fulham Palace, Bishop's Avenue, London SW6 6EA
Tel 020 7384 2111 Fax 020 7384 2112
E-mail peregrine@bryant.net
Website www.peregrine-bryant.co.uk
ARCHITECTURE AND BUILDING CONSERVATION: The practice specialises in the repair and restoration of listed buildings and interiors, including conversion to new uses, and the design of new architectural work. With over 30 years of experience in historic building work the practice can also provide clients with surveys, quinquennial reports, feasibility studies, preparation and application for grant aid, and historic research and analysis. Clients include the National Trust, Landmark Trust, Crown Estate and Duchy of Cornwall, as well as private owners of historic buildings.

▶ BUILDING DESIGN PARTNERSHIP
16 Brewhouse Yard, Clerkenwell, London WC1V 4LJ
Tel 020 7812 8000 Fax 020 7812 8399
E-mail t-leach@bdp.co.uk Contact Tim Leach
▶ Sunlight House, PO Box 85, Quay Street, Manchester M60 3JA
Tel 0161 834 8441 Fax 0161 832 4280
E-mail k-moth@bdp.co.uk Contact Ken Moth
Website www.bdp.co.uk
▶ Also at Sheffield, Glasgow, Belfast, Dublin
HISTORIC BUILDING CONSULTANTS, ARCHITECTS, ENGINEERS, COST CONSULTANTS: BDP is multi-specialist, its reputation based on a wide range of skills and broad knowledge of many building types. BDP understands the complexities and sensitivities involved in the conservation and the creative adaptive re-use of historic buildings, and its experience ranges from the repair of individual buildings to the regeneration of derelict areas in our historic cities. Completed projects include the award-winning Royal Opera House, the National Maritime Museum, London, and the Elizabethan Town House, Plas Mawr, Conwy, together with the Round Tower, Windsor Castle and the Museum of Science and Industry, Manchester. BDP's portfolio of current projects includes the Royal Albert Hall and the Royal College of Music, London, Liverpool Ropewalks and the Historic Dockyard, Chatham.

ARCHITECTS

▶ BYROM CLARK ROBERTS

117 Portland Street, Manchester M1 6EH
Tel 0161 236 9601 Fax 0161 236 8675
Jubilee House, West Bar Green, Sheffield S1 2BT
Tel 0114 275 7879 Fax 0114 272 8954

ARCHITECTS, SURVEYORS AND ENGINEERS: Byrom Clark Roberts' Historic Buildings Division offers comprehensive consultancy services for listed buildings and projects in conservation areas. They have extensive experience in resolving issues of 'appropriate re-use' as well as traditional maintenance and repair techniques for stone and timber framed structures. Their staff are members of the SPAB, IHBC and EASA and represent over 80 parishes. They can advise on grant aid and sources of funding, and VAT applied to listed buildings. Their work conforms with BS 7913 1988 'Principles of the conservation of historic buildings'. Please contact Alex Roberts in Sheffield or Ian Lucas in Manchester.

▶ CARDEN & GODFREY ARCHITECTS

9 Broad Court, Long Acre, London WC2B 5PY
Tel 020 7240 0444 Fax 020 7836 2244
Partners: Ian Stewart Dip Arch Dip Cons (AA) FSA RIBA
Richard Andrews MA Dip Arch RIBA
Ian Angus Dip Arch RIBA
Russell Taylor D Arch Dip Cons (AA) RIBA IHBC FRSA
E-mail mail@cardenandgodfrey.demon.co.uk

ARCHITECTS: Specialists in all aspects of historic architecture: conservation, repairs, new buildings in sensitive sites, sympathetic alterations and additions, interior design and landscape design. A sound technical knowledge with a scholarly approach to historic detail and innovative design. Clients include the National Trust, English Heritage, colleges, churches, commercial developers and private clients, on projects ranging from small to large.

▶ CAROE & PARTNERS

Penniless Porch, Market Place, Wells, Somerset BA5 2RB
Tel 01749 677561 Fax 01749 676207
E-mail wells@caroe.co.uk

ARCHITECTS: Established in 1884 the practice conserves and repairs historic buildings, designs new buildings in sensitive locations and offers a full buildings archaeology service.

▶ CHARLES KNOWLES DESIGN

80–82 Chiswick High Road, London W4 1SY
Tel 020 8742 8322 Fax 020 8742 8655

CHARTERED ARCHITECTS: Established in 1984. The practice has a reputation for high quality architectural design. From pure conservation, through refurbishment and additions to listed and historic properties, integrating contemporary design in an historical context. Philosophy: intelligent planning, good design, sound construction and the greatest attention to detail produce timeless solutions.

▶ CHEDBURN DESIGN & CONSERVATION

Bath Brewery, Toll Bridge Road, Bath BA1 7DE
Tel 01225 859999 Fax 01225 859343
E-mail chedburn@chedburn.com
Website www.chedburn.com

CONSERVATION ARCHITECTS: This small practice, with technical and CAD backup, specialises in works to historic buildings throughout the country. Their experience covers private houses to churches, public buildings to monuments, including works for the National Trust and Court Service.

▶ CHRIS ROMAIN ARCHITECTURE

45 Salisbury Road, Fordingbridge, Hants SP6 1EH
Tel 01425 650980 Fax 01425 650978
E-mail chris.romain@virgin.net
Website www.chrisromain.co.uk

ARCHITECTS: Over 25 years experience, specialising in ecclesiastical repairs and conservation and public buildings. Traditional design, drawing services (CAD), church inspections, contract administration, design and build. The geographical area covered includes Devon to Sussex and Dorset to Worcestershire.

▶ CHRISTOPHER RAYNER ARCHITECTS

Apple Cross House, 52 The Rise, Sevenoaks, Kent TN13 1RN
Tel 01732 461806 Fax 01732 461824
Principal Christopher Rayner BA MArch (California) RIBA

CHARTERED ARCHITECTS: Christopher Rayner Architects is a small architectural practice specialising in all aspects of work to churches and other historic buildings. Projects have included conservation, repairs and sympathetic alterations/extensions to Wealden hallhouses, post-medieval domestic buildings, barns, churches and buildings at risk.

▶ CLAGUE

62 Burgate, Canterbury, Kent CT1 2BH
Tel 01227 762060 Fax 01227 762149
▶ 13 North Street, Ashford, Kent TN24 8LF
Tel 01233 624354 Fax 01233 610018
▶ 19 Buckingham Gate, London SW1 6LB
Tel 020 7592 3240 Fax 020 7592 3244

ARCHITECTS AND HISTORIC BUILDINGS CONSULTANTS: Founded in 1936 and now established in three offices. Clague provides a full range of architectural and landscape design services by highly experienced conservation specialists. Services include conservation repair and alterations, alternative uses for historic buildings, new build in sensitive historic settings, surveys and measured drawings, advice on building legislation and grant applications. Clague operates an in-house quality management system. Clients range from private individuals to large corporations and the practice is appointed to many parish churches; members of the practice being approved Inspecting Architects in three Anglican Dioceses. The practice specialises in the repair and alteration of historic country houses and hotels, and military structures.

▶ CoDA CONSERVATION

No 2 Harewood Yard, Harewood, Leeds LS17 9LF
Tel 0113 288 6766 Fax 0113 288 6765

CHARTERED ARCHITECTS AND ENGINEERS: The practice offers combined architect/engineer expertise in restoring and adapting redundant old buildings for new uses, and in designing new work in historic contexts, using traditional materials and skills.

▶ DANIEL FORSHAW DESIGN AND CONSERVATION ARCHITECTS

School House, Basingstoke Road, Old Alresford, Hants SO24 9DR
Tel 01962 733017 E-mail daniel.forshaw@architects.clara.net
Website www.architects.clara.net
Contact Daniel Forshaw BA DipArch GradDipl (Cons) AA RIBA

ARCHITECTS: A conservation practice with design flair for finding strategic and detailed design solutions in historic settings. They conserve and repair across a wide spectrum of building types and ages. They are ecclesiastical architects and surveyors with appointments in Anglican and Catholic dioceses. Current work includes several reordering projects, stone repairs to a medieval tower, ancient roof frame repairs and a feasibility study for major works to a Victorian town hall.

▶ DAVID ASHTON HILL ARCHITECTS

70 Cowcross Street, London EC1M 6EJ
Tel 020 7608 1576 Fax 020 7608 1676
E-mail ashton.hill@clara.net Website www.ashton.hill.clara.net
Principal David Ashton Hill AA Dip (Cons) RIBA IHBC FRSA

ARCHITECTS: The practice specialises in the conservation and rehabilitation of historic buildings of all three listed grades, on rural and urban sites. Experience in this field over 20 years includes consultancy in conservation areas, CAPS and HERS, preparation of action and conservation plans and feasibility studies for repair and improvement work to historic buildings and churches, implementation of grant aided work for local authorities and consultation and assessment of funded work. The practice offers a full architectural service for private clients and public bodies, quinquennial surveys and management plans, design for the disabled, landscaping, garden and furniture design. They work in a multi-disciplinary team with community artists and other specialist disciplines. Projects in the past year include preparing conservation plans and method statements for Royal Clarence Victualling Yard, Gosport, Hampshire, extensions and garden structures to MH Baillie Scott House in Cambridge and alterations to the Caird Library, National Maritime Museum at Greenwich.

ARCHITECTS

▶ DAVID BROWN AND PARTNERS

51 High Street, Hampton, Middlesex TW12 2SX
Tel 020 8941 2112 Fax 020 8941 1742
E-mail info@davidbrownandpartners.co.uk

ARCHITECTS: The practice has received nine awards since 1981 for careful works to existing structures and new buildings in sensitive architectural situations. Recent award winning projects include new buildings associated with The Great Barn, Harmondsworth, Middlesex, a scheduled ancient monument within a listed site; the construction and re-ordering for Kingston University of Dorich House, Kingston, a listed three storey sculptor's studio built in 1937, and the refurbishment of the listed Castle Stables, Sunbury on Thames, to create two houses.

▶ DAVID GIBSON ARCHITECTS

35 Britannia Row, London N1 8QH
Tel 020 7226 2207 Fax 020 7226 6920
E-mail DGibArch@aol.com
Website www.DGibArch.co.uk

ARCHITECTS: David Gibson Architects is a practice committed to the art of architecture and the design of good buildings. Specialising in work to listed buildings and buildings in conservation areas, it is guided by the philosophy of sympathetic interaction between good modern interventions and existing structures. The practice has an established reputation with the national and regional heritage bodies and amenity societies and its work includes the refurbishment of listed buildings such as St Luke's Church, Battersea, Ovington Square, Vestry House Museum and Sainsbury's, Streatham. The practice is able to advise on town planning, technical, space planning and aesthetic issues for listed and other buildings.

▶ DAVID HARVEY ARCHITECTS

Abergavenny, Monmouthshire NP7 5SE
Tel 01873 857232 Fax 01873 859344
E-mail dharchs@aol.com

CHARTERED ARCHITECTS: Established in 1980, the practice has extensive experience of conservation work in South Wales and the Marches, including condition surveys, quinquennial inspections, the restoration, repair and adaptation of listed buildings, industrial archeological projects and the design of new buildings in sensitive locations. Recent work includes repairs to St John's Church, Cillybebyll and St Catwgs, Llangattock. The practice also acts as a monitor for the Heritage Lottery Fund and has undertaken town scheme re-assessments and conservation area appraisals for local authorities. A member of SPAB and IHBC, the practice offers its clients a committed, efficient and personal service.

▶ DAVID LEWIS ASSOCIATES

Delf View House, Church Street, Eyam, Derbyshire S32 5QH
Tel 01433 630030 Fax 01433 631972
E-mail partners@architects.ac
Website www.architects.ac
Contact David Lewis B Arch MA (Architectural Building Conservation) RIBA

CHARTERED ARCHITECTS AND CONSERVATION SPECIALISTS: The practice, established in 1978, has an enviable reputation for careful restoration and conservation of historic buildings, sympathetic conversions and the careful design of new buildings and extensions in historic contexts. Recent and current work includes conservation of Tudor windows to a Grade I country house in Cheshire, conversion of a Grade II* half-timbered house and associated buildings to a hotel in Gloucester, conservation and extension to a medieval building in the Lincoln Conservation Area, alteration to new uses of historic barns at a Jacobean country house in Staffordshire besides the conservation of Arts & Crafts properties in Scunthorpe and Pontefract. Expert witness services continue to be provided for planning appeals and public inquiries.

▶ DAVID PITTS CHARTERED ARCHITECTS

12a The Waits, St Ives, Cambs PE27 5BY
Tel 01480 466213 Fax 01480 493330
E-mail dpittsriba@btconnect.com
▶ PO Box 8, Oundle, Northants PE8 5JQ
Tel 01780 470170 Fax 01780 470800

CHARTERED ARCHITECTS: Established since 1976, the practice specialises in the conservation and repair of historic buildings, including agricultural, horticultural, residential and ecclesiastical fabric, their upgrading, refurbishment and alteration as necessitated by the client brief and current usage.

▶ DAVIES SUTTON ARCHITECTURE LIMITED

30 Cowbridge Road, Pontyclun, Glamorgan CF72 9EE
Tel 01443 225205 Fax 01443 238965
E-mail office@davies-sutton.co.uk
Website www.davies-sutton.co.uk

CONSERVATION ARCHITECTS: Specialising in the care and conservation of historic buildings and churches to preserve their patina and character, Davies Sutton Architecture aspires to being Wales' leading historic buildings practice, providing sensitive and practical conservation, and new buildings in a sensitive context. The practice has a commitment to quality, attention to detail, good management and continuous learning. Members and field workers for SPAB, providing a highly personal and dedicated service.

▶ DEAN & CHEASON ASSOCIATES

Old St Peter's, Peterstone, Cardiff, Wales CF3 2TR
Tel/Fax 01633 689066 or 02920 700949
E-mail rspencdean@aol.com

ARCHITECTURE AND BUILDING CONSERVATION: Award winning practice having extensive experience of sensitive conservation, refurbishment and adaptation of historic buildings and structures. Work ranges from a 13th century castle at Llantrisant to industrial archaeology and non-conformist chapels. The practice offers a comprehensive, efficient and personal service including detailed surveys, funding plans, liaison with Cadw and acts as Planning Supervisors. Members of the Institute of Historic Building Conservation, SPAB, Ancient Monuments Society and Agoriad Community Architects.

▶ DEREK PLUMMER CHARTERED ARCHITECT

28 Abbott Close, Hampton TW12 3XR
Tel 020 8979 7443 Fax 020 8941 7374

ARCHITECT: Established in 1987 in the Borough of Richmond upon Thames. An aim is to help to find viable uses for buildings of significance whilst respecting their integrity. Last year the practice received the Richmond Society Award for the restoration of Grove Gardens Chapel and a Borough Sustainable Design Award Commendation.

▶ DEREK ROGERS ASSOCIATES

Hill Farm Studios, Chilton Road, Chearsley, Buckinghamshire HP18 0DN
Tel 01844 202020 Fax 01844 202050
E-mail dra@derekrogers.co.uk

CHARTERED ARCHITECTS: The practice has extensive experience in repairs, alterations and extensions to historic buildings, be they listed or simply of local vernacular construction. New buildings, designed in an historic context, is a major area of their work as is condition survey or measured survey and recording of standing structures. Buildings and street scene improvement in conservation areas, using CAD, linked to electronic survey when appropriate, is as much a part of their work as is energy-conscious and historically sympathetic design. Apart from the historic aspects of their work, their design skills encompass all aspects of housing design, arcadia, education and conference and management training buildings, medical and commercial projects.

ARCHITECTS

▶ DONALD INSALL ASSOCIATES LTD

19 West Eaton Place, Eaton Square, London SW1X 8LT
Tel 020 7245 9888 Fax 020 7235 4370
Website www.insall-lon.co.uk
▶ with branch offices in Canterbury, Shrewsbury, Chester, Cambridge and Bath

ARCHITECTS AND PLANNING CONSULTANTS: The practice has over 40 years experience in the care and adaptation of historic buildings and towns, as well as the design of new buildings for sensitive sites. Its projects have now been recognised in more than 100 design and construction awards. London and branch offices provide a comprehensive service across a wide geographical area, including building surveys and site analysis, conservation plans, development and feasibility studies, grant aid, planning and listed building consent negotiations, fire protection and security advice, services integration in old buildings, all basic architectural services, and contemporary and historic interior design.

▶ JAMES DUNNETT MA, Dipl Arch (Cantab), RIBA

142 Barnsbury Road, London N1 0ER
Tel 020 7833 3451 Fax 020 7833 2126

ARCHITECT: Experienced in the conservation, extension, and improvement of 18th, 19th and 20th century houses, and listed social housing. Winner of Kensington and Chelsea Environment Award for Restoration and Conversion 1994; SPAB member; committee member of the Twentieth Century Society, co-chair of DOCOMOMO-UK.

▶ EDMUND KIRBY ARCHITECTS

8 King Street, Manchester M2 6AQ
Tel 0161 832 3667 Fax 0161 832 3795 E-mail davidcullearn@edmundkirby-m.co.uk
▶ 6th Floor, India Buildings, Water Street, Liverpool L2 0TZ
Tel 0151 236 4552 Fax 0151 236 4024 E-mail mikedavies@edmundkirby.co.uk

DESIGN AND CONSERVATION ARCHITECTS: An established practice with extensive experience of the conservation of existing buildings and the design of new buildings in an historical context. This conservation work relates to buildings of Grade II, II* or I status and has involved liaison with English Heritage, The Heritage Lottery Fund, Cadw, various trusts and amenity societies. The practice has received various awards and commendations for the excellence of its design work, and it operates in all parts of the United Kingdom.

▶ ELAINE RIGBY ARCHITECTS

33 Chapel Street, Appleby-in-Westmorland, Cumbria CA16 6QR
Tel/Fax 017683 52572 E-mail elaine.rigby@virgin.net
Contact Elaine Rigby, MA(Conservation Studies York) DIP ARCH IHBC RIBA

ARCHITECTS, CONSULTING ENGINEERS AND LANDSCAPE CONSULTANTS: Experts on the repair and conservation of historic buildings, structures and landscapes, particularly medieval and 17th century vernacular buildings, ruins and churches. Elaine Rigby Architects applies a strong conservation philosophy of conservative repair using traditional materials, and has considerable expertise in the use of lime and stonework consolidation.

▶ FEILDEN + MAWSON

36 Grosvenor Gardens, London SW1W 0EB
Tel 020 7730 8880 Fax 020 7730 8881 E-mail london@feildenandmawson.com
▶ 1 Ferry Road, Norwich NR1 1SU
Tel 01603 629571 Fax 01603 633569 E-mail norwich@feildenandmawson.com
▶ Horningsea Road, Fen Ditton, Cambridge CB5 8SZ
Tel 01223 294017 Fax 01223 293458 E-mail cambridge@feildenandmawson.com

ARCHITECTS, HISTORIC BUILDING CONSULTANTS, PROJECT MANAGERS: The firm has a strong reputation for innovation in conservation and is committed to finding creative ways of adapting and restoring historic buildings to ensure ongoing or new use. Leaders in the conservation field since 1957, they work on a wide variety of historic architecture ranging from major government institutions to domestic scale buildings for clients including Somerset House Trust, HM Treasury, Cabinet Office, Foreign and Commonwealth Office, Crown Estate, English Heritage, National Trust, Historic Royal Palaces, Bank of England, OGCbuying.solutions, Court Service, Royal Hospital Chelsea, Keble College Oxford, University of Cambridge, King's College London, Burford, Ballymore and Stanhope.

▶ GIBBON, LAWSON, McKEE LIMITED

41A Thistle Street Lane South West, Edinburgh EH2 1EW
Tel 0131 225 4235 Fax 0131 220 0499 E-mail dg@glmglm.co.uk

ARCHITECTS: The adaptation, repair and refurbishment of Scottish buildings is the firm's main work. Recent projects include a high class refurbishment of a Grade B Highland shooting lodge, the remodelling and extension of a steading in the Borders as a spectacular house, and the creation of a number of bathrooms in a Grade A house, including one cleverly concealed by a purpose-made mahogany wardrobe. Farm building conversions have become a major activity. Clients include estates, individuals, trusts, schools, hotels and other organisations. The firm works as an integrated team which includes building surveyors and project managers to deliver well designed and well executed projects. *See also: profile entries in Surveyors section, page 45 and Fire Protection section, page 188.*

▶ GILES QUARME & ASSOCIATES

41 Cardigan Street, London SE11 5PF
Tel 020 7582 0748 Fax 020 7587 3678
E-mail mail@quarme.com Website www.quarme.com

CONSERVATION ARCHITECTS AND PLANNING CONSULTANTS: The team is led by Giles Quarme, RIBA FRSA, recently appointed Surveyor to the Fabric of Sir Christopher Wren's Royal Naval College, Greenwich and Dr Archie Walls, ARIAS, FSA. They are architects and historians with considerable planning experience including acting as expert witness at public inquiries. The practice has worked on a wide variety of listed and historic buildings ranging from churches and hospitals to museums and offices. It prides itself on providing the same care and attention to detail in the repair of small historic buildings as for large country mansions. It aims to provide radical solutions which combine new ideas with traditional cost conscious conservation methods. They were responsible for the Princess Diana Museum at Althorp and advising Foster and Partners on the historic aspects of their millennium design of the British Museum. They have recently completed the Indian High Commissioner's residence and are currently working on Chilham Castle and Hawksmore's St Mary Woolnoth. *See also: display entry in this section, page 25.*

▶ GILMORE HANKEY KIRKE LTD

GHK House, Heckfield Place, 526 Fulham Road, London SW6 5NR
Tel 020 7471 8000 Fax 020 7736 0784 E-mail architects@ghkint.com
▶ Residence 2, Royal William Yard, Plymouth PL1 3RP
Tel 01752 262 244 Fax 01752 262 299 E-mail ghk.ply@dial.pipex.com
Website www.ghkint.com/architects

ARCHITECTS AND CONSERVATION SPECIALISTS: Established in 1973, GHK has extensive experience in the conservation, repair, refurbishment and reuse of listed and historic properties as well as the design of new buildings for sensitive and historic sites in both the UK and overseas. Current work includes the refurbishment of Grade I listed barristers chambers, the refurbishment of a Grade I listed house in Kensington and the design of a new house in Barbados. Recent awards include: RICS 2000 Conservation Award for Lulworth Castle, Dorset; Natural Stone Award 2002 for a new build country house in Gerrards Cross and restoration work at the Royal William Yard, Plymouth.

▶ GRAHAM HOLLAND ASSOCIATES

4 King Street, Knutsford, Cheshire WA16 6DL
Tel 01565 651066 Fax 01565 755265
▶ Plas Draw, Ruthin, Denbighshire LL15 1RT
Tel 01824 704709 Fax 01824 704912

ARCHITECTS AND HISTORIC BUILDING CONSULTANTS: The practice's policy is for sensitive design and conservation. Formed in 1986, now with three associates, technical and secretarial staff, principally in the care and repair of historic buildings and new buildings in historic settings. Graham Holland is a member of the Chester and Liverpool DACs and Cathedral Architect at Bangor. Clients include the Historic Chapels Trust, National Trust, Duchy of Lancaster, Church of England, Church in Wales, Methodist, URC and Unitarian churches, and private estates. Work has included: Hawarden, Gwydir and Halton Castles, Ecclesfield St Mary, Todmorden Unitarian Chapel & Lodge, Balderstone St Mary and Tutbury Priory.

ARCHITECTS

▶ GRAY, MARSHALL & ASSOCIATES

23 Stafford Street, Edinburgh EH3 7BJ
Tel 0131 225 2123 Fax 0131 225 8345

CHARTERED ARCHITECTS: Established in 1972, the practice has a proven track record in conservation work. Recent work includes General Register House, Edinburgh, the Cathedral of The Isles, Cumbrae, and Wigtown County Buildings. Working throughout Scotland, clients include building preservation trusts, churches, Edinburgh World Heritage Trust, housing associations and private owners. Gray, Marshall & Associates have considerable experience of listed buildings, including sympathetic adaptation and reuse, phased restoration projects, complex funding packages, conservation area Townscape Heritage initiative appraisals, church quinquennials and conservation plans.

▶ HGP CONSERVATION

HGP Greentree Allchurch Evans Limited, Furzehall Farm, Wickham Road, Fareham, Hampshire PO16 7JH
Tel 01329 283225 Fax 01329 237004 E-mail @hgp-architects.co.uk
Website www.hgp-architects.co.uk

ARCHITECTS AND TOWN PLANNERS: Founded in 1968, HGP Conservation's services include design and specification for repair, conversion and extension of major and smaller historic buildings, feasibility studies, conservation plans, decay diagnosis, surveys and recording, quinquennial reporting, expert witness services and new building design in historic contexts. HGPC provides innovative solutions to the often difficult problems associated with the restoration and economic use of listed buildings. It benefits from being part of HGP Architects employing 50 staff in two offices. The Fareham office is in Grade II farm buildings (The Times/RICS Conservation Award, 1982). Projects include the conversion of 1862 Grade II/II* Eastney Barracks, Portsmouth and an 1860 Grade II Fort in Devon to apartments, the adaptation of a 1786 scheduled ancient monument storehouse, Portsmouth Dockyard, into the Royal Naval Museum, an Art Deco cinema in Leicester Square, London, and masterplanning at Gunwharf Quays Portsmouth, including £2.5 million repairs to the 1814 scheduled Vulcan Building. Recent commissions include repairs to a large Berkshire country house involving many Grade I/II buildings and a Grade II* historic garden as well as works to a number of 18th and 19th century military buildings around Portsmouth Harbour.

▶ HAIGH ARCHITECTS

29 Lowther Street, Kendal, Cumbria LA9 4DH
Tel 01539 720560 Fax 01539 723570 E-mail rh@haigharchitects.co.uk
Website www.haigharchitects.co.uk
Contacts EM Bottomley B Arch ARIBA / CR Haigh BA Dip Arch RIBA

DESIGN AND CONSERVATION ARCHITECTS: The practice specialises in restoring and adapting old redundant buildings for new uses in historic contexts, using traditional materials and skills to blend the new with the old. A sensitive approach is taken to the repair, restoration and conservation of historic buildings, from country houses to traditional cottages, churches to museums. It has received several Civic Trust Awards and Commendations. Clients include private owners of historic properties, the National Trust, Churches Conservation Trust, Diocese of Carlisle, English Heritage and civic society building preservation trusts.

▶ THE HALPERN PARTNERSHIP LTD

The Royle Studios, 41 Wenlock Road, London N1 7SH
Tel 020 7251 0781 Fax 020 7251 9204
E-mail eleni.makri@halpern.uk.com

ARCHITECTURE, PLANNING, URBAN DESIGN, HERITAGE: Halpern offers comprehensive, qualified and accredited conservation expertise, which has achieved national awards. Its services include the repair and re-use of listed buildings, innovative designs in conservation areas, the appraisal and management of the historic environment, townscape regeneration, buildings at risk, grants, planning and listed building consent negotiations, appeals. The multidisciplinary practice benefits from over 50 years experience in realising its designs into successful buildings, and from a new team of experts who bring in their own significant portfolios and an up-to-date approach in dealing with the built environment. Current conservation projects include significant 20th century buildings and conservation areas.

ARCHITECTS

▶ HAWKES EDWARDS & CAVE

1 Old Town, Stratford-upon-Avon, Warwickshire CV37 6BG
Tel 01789 298877
Website www.hawkesedwards.com
ARCHITECTS: Over 45 years of experience in dealing with listed historic buildings and churches, where a sensitive, skilled and creative approach is required, both in repair and alteration. They act for, amongst others, the Churches Conservation Trust and the Ancient Monuments Society.

▶ THE HERITAGE PRACTICE

27 Grange Road, Saltford, Nr Bristol BS31 3AH
Tel/Fax 01225 400066
ARCHITECTS: A specialist practice working on historic buildings, waterways and garden structures. Commissions include: Environment Agency, British Waterways, Duchy of Cornwall and National Trust.

▶ HOK CONSERVATION AND CULTURAL HERITAGE

HOK International Limited
216 Oxford Street, London W1C 1DB
Tel 020 7637 2006 Fax 020 7636 1987
Website www.hokeurope.com
ARCHITECTS: Projects cover every aspect of tangible and intangible heritage. Historic buildings, museums, galleries, libraries, sites of natural beauty, traditional handicrafts, music and dance – all are regarded as irreplaceable components in a nation's cultural heritage to be made accessible for people from all backgrounds, ages and abilities. Projects include: 'Airspace' at Duxford, Ashton Court, British Museum Study Centre, Cabinet War Rooms, Cabinet Office, Churchill Museum, Darwin Centre Natural History Museum, Edward III Manor House, Lime Kiln Burgess Park, Forbury Gardens, Kensal Green Cemetery, King's Library British Museum, Northwood Park, Nunhead Cemetery, Palace of Westminster, sustainable tourism master plan for the Kingdom of Saudi Arabia and projects for the Crown Estate and Royal Household.

▶ INGRAM CONSULTANCY LTD

Manor Farm House, Chicklade, Hindon, Salisbury, Wiltshire SP3 5SU
Tel 01747 820170 Fax 01747 820175
E-mail enquiries@ingram-consultancy.co.uk
Website www.ingram-consultancy.co.uk
ARCHITECTS, SURVEYORS AND CONSERVATION CONSULTANTS: *See also: profile entry in Heritage Consultants section, page 35.*

▶ JAMES BROTHERHOOD & ASSOCIATES LIMITED

Golly Farm, Golly, Burton, Rossett, Wrexham LL12 0AL
Tel 01244 579000 Fax 01244 571133
E-mail jba-architects.co.uk
Website www.jba-architects.co.uk
CONSERVATION ARCHITECTS: Energetic and successful practice with the resources to undertake development projects throughout the British Isles. Specialists in the adaptation and care of historic buildings and in the integration of new buildings in conservation areas. Based on over 25 years experience, they achieve practical and high quality solutions, acceptable to clients, local authorities, English Heritage, Cadw and the Historic Monuments Commission. Full architectural services, feasibility and development studies and planning negotiations. James Brotherhood is registered as an Architect Accredited in Building Conservation and the firm has received top Europa Nostra, conservation and civic awards and its work is included in HRH The Prince of Wales', 'A Vision of Britain'. Full CAD facilities.

▶ JOHN D CLARKE & PARTNERS

2 West Terrace, Eastbourne, East Sussex BN21 4QX
Tel 01323 411506 Fax 01323 410064
E-mail admin@johndclarkeandpartners.co.uk
Website www.johndclarkeandpartners.co.uk
CHARTERED ARCHITECTS AND HISTORIC BUILDING CONSULTANTS: The practice was established in 1909 and one of the first projects was the restoration of the 13th century Lamb Inn at Eastbourne. The practice has developed a reputation for sensitive conservation and repair of historic buildings as part of a portfolio which includes work for local authorities, building societies, hotels and churches. The practice looks after Harveys Brewery in Lewes winning a Civic Trust Commendation for its Brewery Tower extension, and has won a Friend of Lewes Award for the restoration of a large Jacobean fire damaged house. The practice deals with a great variety of historic work from the very large, down to the restoration of a Victorian cast iron drinking fountain.

▶ JOHNSTON & WRIGHT

15 Castle Street, Carlisle, Cumbria CA3 8TD
Tel 01228 525161 Fax 01228 515559
E-mail jw@jwarchitects.co.uk
Website www.jwarchitects.co.uk
CHARTERED ARCHITECTS: Founded in 1885 and with over 50 awards for good design, Johnston & Wright has an extensive track record in repair, restoration and regeneration of historic buildings and their surroundings. Clients include, private owners of historic properties, National Trust, English Heritage, churches, commercial developers and local authorities.

▶ JONATHAN RHIND ARCHITECTS

The Old Rectory, Shirwell, Barnstaple, Devon EX31 4JU
Tel 01271 850416 Fax 01271 850445
E-mail jonathan@jonathan-rhind.co.uk
Website www.jonathan-rhind.co.uk
CONSERVATION ARCHITECTS: Detailed local knowledge of English Heritage/conservation/planning issues and skilled/reliable builders. High quality design and imaginative solutions for the upgrade and re-use of all historic buildings. Advice on best conservation practice supported by continued professional education. Initial guidance on project planning and budget.

▶ JULIAN HARRAP ARCHITECTS

95 Kingsland Road, London E2 8AG
Tel 020 7729 5111 Fax 020 7739 8306
DESIGN AND CONSERVATION ARCHITECTS: Julian Harrap Architects is a medium sized specialist practice offering a full range of architectural services for the repair and restoration of historic buildings, estates and landscapes and for new buildings in historic settings. The practice works with buildings of all ages from scheduled ancient monuments to 20th century icons. All projects large or small are planned within a careful management structure. Established in 1975 Julian Harrap Architects has a reputation for scholarly conservation and attention to fine detail. Clients include the National Trust, English Heritage, The Royal Academy of Arts, Sir John Soane's Museum, and many notable churches. The practice has received national and international awards.

▶ KATIE THORNBURROW ARCHITECTS

Unit 5, 25 Gwydir Street, Cambridge CB1 2LG
Tel 01223 307555 Fax 01223 307666
E-mail katie@kt-architects.com
CHARTERED ARCHITECTS: Architectural practice specialising in the conservation, restoration and extension of historic buildings. The practice has completed works in London, Kent, and throughout East Anglia. Current works include a house dating from the 14th century, two barn conversions, and a Cambridge college. The practice works closely with Cambridge Architectural Research Ltd on environmental matters in historic buildings, conservation plans and researches into conservation issues.

ARCHITECTS

▶ LATHAM ARCHITECTS

St Michael's, Queen Street, Derby DE1 3SU
Tel 01332 365777 Fax 01332 290314

DESIGN AND CONSERVATION ARCHITECTS, PLANNING CONSULTANTS, URBAN DESIGNERS, LANDSCAPE ARCHITECTS: A highly experienced firm, working with scheduled monuments, historic and ecclesiastical buildings, including structural and condition surveys, inspections, conservation and repair work and creative-reuse of redundant buildings. Winners of many conservation awards and increasingly involved in the wider planning issues of conservation and finding uses for old buildings, together with designing modern buildings in sensitive areas. Clients include The Royal Household, Duchees of Lancaster and Cornwall, MoD, English Heritage, National Trust, local authorities and many small agencies, including historic building preservation trusts.

▶ LINCOLN & CAMPBELL ASSOCIATES LTD

11 Carlton Street, Edinburgh EH4 1NE
Tel 0131 332 4888 Fax 0131 343 1519
▶ 47 Roderick Road, London NW3 2NP
Tel 020 7485 7442 Fax 020 7267 9779
E-mail office@lincoln-campbell.com

ARCHITECTURE AND CONSERVATION: Directors Oran Campbell and Richard Lincoln each have 30 years experience of historic buildings repair and conservation. Workload has included historic buildings, new institutions, houses and garden design in Scotland, England, Spain, Australia, the Indian subcontinent and the United States. New or scholarly traditional design in detail, always aiming for harmony with setting, atmosphere and landscape. Other languages: Spanish and Italian.

▶ JULIAN R A LIVINGSTONE CHARTERED ARCHITECT RIBA IHBC BAHons DipArch(Leic) GradDiplConservation(AA)

Dahlia Cottage, Vicarage Lane, Upper Swanmore, Hampshire SO32 2QT
Tel 01489 893399 Fax 01489 895021
Mobile 07720 758764
E-mail julian.livingstone@btopenworld.com

CHARTERED ARCHITECT AND HISTORIC BUILDINGS CONSULTANT: Specialising in the sensitive repair, conversion, extension and conservation of country and town houses in private ownership. Julian Livingstone concentrates on domestic buildings, producing designs with an emphasis on imaginative solutions, within the constraints of historic buildings and materials, to delight, excite and stimulate. He advises clients on the most appropriate methods of caring for their buildings and has received several awards for sympathetic and imaginative conversion of agricultural buildings to residential use. Julian Livingstone serves on the Portsmouth Diocesan Advisory Committee, is a member of SPAB and ASCHB.

▶ LLOYD EVANS PRICHARD

5, The Parsonage, Manchester M3 2HS
Tel 0161 834 6251 Fax 0161 832 1785
E-mail post@lep-architect.co.uk

CHARTERED ARCHITECTS: Lloyd Evans Prichard is a firm of general practitioners with proven skills in conservation and restoration. Backed by the broad experience of the professional staff, Lloyd Evans Prichard is able to combine expertise in historic building work and conservation plans with an extensive commercial background to tailor services offered to individual client requirements, including CDM and project management. Lloyd Evans Prichard's successful approach to conservation work, under the guidance of Director John Prichard, Surveyor to the Diocese of Manchester, has led to commissions from: English Heritage; the Holy Name Church, Manchester; Capesthorne Hall; St Michael's, Ashton-under-Lyne, John Rylands Library, Manchester, Manchester City Council at Heaton Hall and Park and Carlisle Castle.

▶ MACKENZIE WHEELER

Embankment Studios, The Embankment, Putney, London SW15 1LB
Tel 0208 785 3044 Fax 0208 785 4442 E-mail admin@mackenziewheeler.co.uk

ARCHITECTS AND DESIGNERS: The practice has established a particular reputation and expertise in the repair and restoration of historic buildings for use by the hotel, restaurant and licensed trades. Such a use often offers the last chance to save a building at risk and generate sufficient funds to repair the historic fabric in a sensitive and scholarly manner, thereby ensuring the long term future of the building and facilitating public access to study and enjoy the results. The conflicting demands of conservation, modern regulations and market forces are reconciled with intelligence and imagination in the best interests of client, user and historic building.

▶ T C R MacMILLAN-SCOTT

11 Lansdowne Road, Alton, Hants GU34 2HB
Tel/Fax 01420 549233

CHARTERED ARCHITECT

▶ MANSFIELD THOMAS & PARTNERS

81 Albany Street, Regent's Park, London NW1 4BT
Tel 020 7224 4446 Fax 020 7935 8991
▶ Little Heath Farm, Berkhamsted, Hertfordshire HP4 2RY
Tel/Fax 01442 864951

ARCHITECTS: This small practice offers a full range of conservation, architectural, town planning, surveying, quantity surveying, landscaping and historical investigation with the benefit of over 50 years experience. Committed to achieving high quality design standards for each client's complex needs – to the best solution for each commission – combining artistic expression disciplined to the brief, to 'value for money' and to the programme. Recent projects include Crown Estate quinquennial surveys, restoration, refurbishment and repairs to many Grade I listed buildings including Ockwell's Manor, Maidenhead; Upton Court, Slough; Childwicksbury, St Albans; Cumberland and Chester Terraces and York Terrace East, Regent's Park; 7 Palace Green and 15 Kensington Palace Gardens; and Trevor Hall and Bettisfield Park, North Wales etc.

▶ MICHAEL DRURY ARCHITECTS

St Ann's Gate, The Close, Salisbury, Wiltshire SP1 2EB
Tel 01722 555200 Fax 01722 555201
E-mail info@stannsgate.com

ARCHITECTS, CONSERVATION AND DESIGN: Over the past 25 years Michael Drury Architects has established a reputation for imaginative design solutions within sensitive historic environments. Such work ranges from conservation planning, management of long-term repair campaigns, access consultancy, careful adaptations of existing buildings, church re-orderings, completely new design work and pure conservation and repair. Whatever the project, the practice's approach to each one is specifically tailored to the situation. Existing clients include owners of listed private properties, parish churches and cathedrals, the National Trust and the Churches Conservation Trust.

▶ NIALL PHILLIPS ARCHITECTS

35 King Street, Bristol BS1 4DZ
Tel 0117 927 7396 Fax 0117 927 7594
E-mail info@niallphillips.freeserve.co.uk

DESIGN AND CONSERVATION ARCHITECTS: Niall Phillips Architects specialises in the conservation and imaginative reuse of historic buildings and the design of new buildings in historically sensitive areas. With considerable experience and expertise in identifying development mechanisms to maximise grant aid, enabling contribution and other resources, the practice has steered many otherwise 'unsaveable' buildings to secure and successful futures. A full range of architectural services is offered on sites across the country, along with conservation consultancy, regeneration and development planning. Projects include major art galleries and museums, conversion of historic industrial buildings and specialist facilities such as the English Heritage Ancient Monuments Laboratory. The practice has won Civic Trust and RIBA awards.

▶ NICHOLAS JACOB ARCHITECTS

89 Berners Street, Ipswich, Suffolk, IP1 3LN
Tel 01473 221150 Fax 01473 255550 E-mail nicholas.jacob@njarchitects.co.uk
Website www.njarchitects.co.uk

ARCHITECTS: The practice specialises in the repair, conservation and sensitive alterations or extension of ecclesiastical and secular historic buildings in East Anglia; imaginative design solutions for sensitive locations.

▶ NORMAN + DAWBARN ARCHITECTS

9 Kean Street, Covent Garden, London WC2B 4AY
Tel 08702 409988 Fax 020 7836 5558 E-mail mail@n-d.co.uk
Website www.n-d.co.uk

ARCHITECTS AND ENGINEERS: Established in 1934, Norman + Dawbarn undertakes architectural and interior design projects for the conservation and repair of historic and listed buildings in the UK, Europe and overseas. Commissions include renovation, extensions, building surveys and new build projects within sensitive conservation areas. Clients include The Crown Estates and The Royal Parks Authority, as well as private and commercial organisations.

▶ P W P ARCHITECTS

Newnham House, 61 South Street, Havant, Hants PO9 1BZ
Tel 023 9248 2494 Fax 023 9248 1152
E-mail design@pwp-architects.com
Contact Jeremy Sayer or John Organ

ARCHITECTS, SURVEYORS, PLANNING CONSULTANTS: Established in 1921 the practice handles the conservation, restoration and re-development of historic buildings in sites of special landscape interest, working with a broad range of conservators, landscape architects, archaeologists and environmentalists. PWP's clients use the firm repeatedly because it meticulously balances budget and time constraints against essential and urgent conservation needs. The practice has special expertise adapting and extending Grade I and Grade II buildings at risk, and negotiating viable new uses with English Heritage and the amenity societies. The practice carries out developments across the country and undertakes feasibility studies and concept designs for UK based clients in France.

ARCHITECTS

▶ PETER CODLING ARCHITECTS

7 The Old Church, St Matthews Road, Norwich, Norfolk NR1 1SP
E-mail pcodling@globalnet.co.uk
Tel 01603 660408 Fax 01603 630339

ARCHITECTS: Church repairs, re-ordering and extensions; quinquennial reports. Repair and conversion of buildings of all ages and types. Housing for individual clients and special needs groups.

▶ PETER YIANGOU ASSOCIATES

Puckham Barn, Whittington, nr Cheltenham, Glos GL54 4EX
Tel 01242 821031 Fax 01242 820193
E-mail peter@yiangou.com
Website www.yiangou.com

ARCHITECTS AND SURVEYORS: PYA has covered the Cotswolds and surrounding counties for 20 years building up in the process considerable experience with country properties, conservation areas, listed buildings and natural stone construction. Projects include new country and manor houses in stone, indoor pools, extensions and repair of listed buildings, new Oxford College buildings, small quality housing developments, museums, corporate HQ in traditional buildings; please see website for examples. Full CAD capability with ten full time staff.

▶ RADLEY HOUSE PARTNERSHIP

Radley House, St Cross Road, Winchester, Hants SO23 9HX
Tel 01962 842228 Fax 01962 842401
E-mail architects@radleyhouse.co.uk

CHARTERED ARCHITECTS AND SURVEYORS: Founded in 1933, the firm has gained extensive experience in conservation and historic buildings, including extensions and new design in an historic context. Clients include English Heritage, Historic Royal Palaces, several well-known country houses and their estates, numerous churches, and private clients.

▶ RICHARD CROOKS PARTNERSHIP

14 Calverley Lane, Horsforth, Leeds LS18 4DZ
Tel 0113 281 8080 Fax 0113 258 4070

CHARTERED ARCHITECTS: Richard Crooks Partnership provides partner led architectural, project management and planning supervisory services. Recent projects include the repair and conservation of ecclesiastical and historic buildings; quinquennial inspections; extensions and re-ordering of churches; alterations, extensions and refurbishment of domestic, retail, office, healthcare and other commercial buildings as well as new build projects. Richard Crooks is included in the Register of Architects Accredited in Building Conservation and is a member of EASA and SPAB.

▶ RICHARD GRIFFITHS ARCHITECTS

14–16 Cowcross Street, London EC1M 6DG
Tel 020 7251 6334 Fax 020 7490 2251
E-mail admin@rgarchitects.com
Website www.rgarchitects.com

ARCHITECTS: The practice combines the highest standards of conservation work with the highest quality of new building in an historic context, and has been awarded RIBA, Civic Trust and Europa Nostra awards. Richard Griffiths Architects has an outstanding reputation for the creative adaptation of historic buildings, providing new community uses and access for all, including Sutton House for the National Trust, the Southwark Cathedral Millennium project and Lambeth Palace glazed courtyard. Richard Griffiths Architects also carries out conservation and development plans to map out a sustainable future for historic buildings and to form the basis for successful Heritage Lottery funding applications.

▶ ROBERT KILGOUR & ASSOCIATES

Cranwood, Blakeshall, Kidderminster DY11 5XW
Tel 01562 851201 Fax 01562 852939
E-mail rkilgour@blakeshall.u-net.com

ARCHITECT: The practice specialises in the repair and conservation of historic buildings. Clients include the Dean and Chapter of Derby Cathedral, the Churches Conservation Trust, churches in the dioceses of Worcester and Coventry and the National Trust, working at Powis Castle and Croft Castle.

▶ ROBERT SEYMOUR CONSERVATION

The Merchants House, 10 High Street, Totnes, Devon TQ9 5RY
Tel 01803 868568 Fax 01803 834722
E-mail seymourarchitect@aol.com
Website www.robertseymourarchitects.co.uk
▶ with branch offices in London and Dartmouth

CONSERVATION ARCHITECTS AND HISTORIC BUILDING CONSULTANTS: The practice has over 25 years experience carrying out sympathetic, appropriate repairs to a wide range of historic and listed buildings. It has strong links with English Heritage and SPAB, working with private clients, local authorities, charitable trusts, almshouse associations, churches and other groups. Detailed surveys, evaluations and repair programmes undertaken, often in sensitive urban conservation areas, throughout the south of England.

▶ ROBIN KENT ARCHITECTURE & CONSERVATION

Newtown Street, Duns, Berwickshire TD11 3AS
Tel 01361 884401 Fax 01361 884402
E-mail rk@robinkent.com
Website www.robinkent.com
Contact Robin Kent BAHons DiplArch(Oxford) MACons(York) RIBA ARIAS IHBC NRAC

ARCHITECTURE AND CONSERVATION CONSULTANCY: Founded 1981 and established in Berwickshire in 1997, specialising in building design in sensitive locations, ancient monument and listed building conservation work, conservation area character appraisals, historic building evaluations, building surveys and investigations, disability access and conservation research. Conservation accredited to highest level. Registered access consultant.

ARCHITECTS

▶ ROBIN WOLLEY CHARTERED ARCHITECT

17 Well Street, Ruthin, Denbighshire, LL15 1AE
Tel 01824 703279 Fax 01824 705523
E-mail robinwolley@ruthin99.freeserve.co.uk

CHARTERED ARCHITECT AND HISTORIC BUILDINGS CONSULTANT: Extensive experience in the conservation, repair and regeneration of buildings. Quinquennial inspections, surveys, feasibility studies, HLF, Cadw and English Heritage grant assisted projects. Conservation area appraisals and local authority planning consultancy. Listed building advice. Clients include Dioceses of Chester, Liverpool and St Asaph, Historic Chapels Trust, United Reformed and Methodist Churches. Surveyor to St Asaph Cathedral.

▶ ROGER JOYCE ASSOCIATES

39 Bouverie Square, Folkestone, Kent CT20 1BA
Tel 01303 246400 Fax 01303 246455
E-mail info@rogerjoyceassociates.co.uk
Website www.rogerjoyceassociates.co.uk
Contact Roger A Joyce DipArch (Cant) DipConservation (AA) RIBA

ARCHITECT: Workload includes repair and refurbishment of historic buildings and design for new uses. The practice has won awards for conversions of redundant agricultural buildings and new build. The principal is a Diocesan approved inspecting architect. Works include Grade I listed churches, Institutional clients including further education colleges, Blue Circle, Union Rail, Kent Trust for Nature Conservation. Roger Joyce serves on the Executive Committee of Kent Building Preservation Trust, amenity groups, the Education Committee of ICOMOS UK, member of SPAB, EASA, IHBC and ACA, and also lectures at the Lille School of Architecture.

▶ ROGER MEARS ARCHITECTS

2 Compton Terrace, London N1 2UN
Tel 020 7359 8222 Fax 020 7354 5208
E-mail rma@rogermears.com
Website www.rogermears.com

CONSERVATION ARCHITECTS: Founded in 1980, the practice has built up a reputation for sensitive work to historic and domestic buildings, guided by the principles of the SPAB. Past work includes alterations, repairs and extensions to listed houses in London, Oxfordshire, Wiltshire and Dorset. Among them are Tudor House, Cheyne Walk, listed Grade II* (formerly Rossetti's house and studio) and a terrace of Grade I listed houses in Newington Green, London, dating from 1658. Current work includes refurbishment of a variety of houses in London and of a Grade I listed coach house in Richmond-upon-Thames. *See also: display entry on page 31.*

▶ RODERICK SHELTON

Stambourne Hall, Stambourne, Halstead, Essex CO9 4NR
Tel 01440 785786 Fax 01440 785427
Contact Roderick Shelton MA(Cantab) BSc(Eng) BArch(Nottm) DiplCons(AA) ACGI MCMI MaPS IHBC RegArch

ARCHITECT, HISTORIC BUILDINGS CONSULTANT AND PLANNING SUPERVISOR: An award winning independent architectural practice dedicated to the scholarly conservation and repair of our architectural heritage following SPAB principles and ecological philosophy. Listed by the Law Society as an expert witness in respect of listed buildings and building conservation.

▶ SIMPSON & BROWN ARCHITECTS

St Ninian's Manse, Quayside Street, Edinburgh EH6 6EJ
Tel 0131 555 4678 Fax 0131 553 4576
E-mail admin@simpsonandbrown.co.uk
Website www.simpsonandbrown.co.uk

ARCHITECTS, LANDSCAPE ARCHITECTS AND HISTORIC BUILDING CONSULTANTS: Established in 1977, Simpson & Brown has extensive experience in the repair, restoration, conservation and alteration of historic buildings. Recent projects include Sir Robert Rowand Anderson's magnificent 1883 Catholic Apostolic Church in Edinburgh to form new headquarters for the SCVO and the award-winning conservation of Stirling Tolbooth (with Richard Murphy Architects). New-build work combines sound ecological principles, experience of traditional materials and good quality contemporary design, such as the recently completed visitor facility at Rowardennan, Loch Lomond (with Richard Shorter Architect) and the visitor centre at Arbroath Abbey. The practice also provides research and consultancy services, advice on planning and funding applications, as well as conservation plans and feasibility studies.

▶ SPENCE & DOWER ARCHITECTS

Column Yard, Cambo, Morpeth, Northumberland NE61 4AY
Tel 01670 774448 Fax 01670 774446

CHARTERED ARCHITECTS: The practice, established in 1946, has been responsible for work to both grand and modest historic houses, castles, churches and historic industrial buildings in the north of England. Grade I Elsdon Tower major re-ordering, the consolidation of Harbottle Castle, Woodhouses Bastle and Low Cleughs Bastle all within the Northumberland National Park. Robin Dower is responsible for the inspection and maintenance of a number of churches in Newcastle Diocese and is Chairman of Durham Cathedral Fabric Advisory Committee. The practice has a framework agreement with English Heritage for work in the North East Region, and is currently working on the conservation of Belsay Hall.

▶ STAINBURN TAYLOR ARCHITECTS

Bideford House, Church Lane, Ledbury HR8 1DW
Tel 01531 634848 Fax 01531 633273
E-mail architects@stainburn-taylor.co.uk

ARCHITECTS: A practice with a wide range of work encompassing historic ecclesiastical and secular projects which range from a cathedral through to private houses. The practice has developed considerable expertise in the diagnosis and conservation of historic buildings. Clients include English Heritage, the National Trust, Eastnor Castle Estates, Downton Estate together with numerous churches in Herefordshire, Worcestershire and Shropshire plus Gloucester Cathedral.

▶ STUART PAGE ARCHITECTS

Forge House, The Green, Langton Green, Tunbridge Wells, Kent TN3 0JB
Tel 01892 862548 Fax 01892 863919
E-mail stuart.page@member.riba.org

ARCHITECTS AND INTERIOR DESIGNERS: Historic buildings require economic and appropriate uses to ensure their survival. Stuart Page Architects undertakes architectural and interior design projects for new buildings and the conservation and repair of historic and listed buildings. The practice believes the architect's role to be especially important when working in conservation areas or with historic buildings to ensure sympathetic buildings integrated with their surroundings and which satisfy the client's brief. Current projects include work for the National Trust, Historic Royal Palaces, English Heritage and private owners of historic buildings.

CATHEDRAL COMMUNICATIONS LIMITED

ARCHITECTS

▶ ANTHONY SWAINE FSA, FRIBA, FASI, IHBC

The Bastion Tower, 16 Pound Lane, Canterbury, Kent CT1 2BZ
Tel 01227 462680 Fax 01227 472743

CONSERVATION ARCHITECT AND CONSULTANT: Private practice: restoration of historic buildings, conservation areas and conservation in general. Past advisor: Thanet District Council and other authorities for historic buildings. Member of Council and Technical Panel of Ancient Monuments Society. Representative for International Council of Monuments and Sites. Patron: Venice in Peril. Author: *Faversham Conserved* and *Margate Old Town*. Past architectural consultant for Historic Churches Preservation Trust. Responsible during last war for care of Canterbury Cathedral. Past part-time teacher – History of Architecture and Construction. Past part-time listing, Ministry of Housing, local government. Past consultant Churches Conservation Trust, member of Friends of Friendless Churches, lecturing.

▶ THOMAS FORD & PARTNERS

177 Kirkdale, Sydenham, London SE26 4QH
Tel 020 8659 3250 Fax 020 8659 3146
E-mail tfp@thomasford.co.uk
Partners: Paul Sharrock, BSc, DipArch(UCL), RIBA
Daniel Golberg, MPhil(Nottm), BArch(Nottm), RIBA, ACI Arb, FRSA
Clive England, BA Hons, DipArch(Sheffield), RIBA

CHARTERED ARCHITECTS AND SURVEYORS: Established in 1926, the practice has extensive experience of historic building projects up to £14 million including churches, domestic buildings, museums, palaces, military buildings and structures. Work includes feasibility studies, quinquennial inspections, conservation, repair, extensions, remodelling and new buildings in historic settings. The practice's portfolio includes many scheduled monuments, Grade I and Grade II* buildings of national and international significance. Clients include English Heritage, National Trust, Historic Royal Palaces Agency, Royal Household, Ministry of Defence and many museums and churches.

▶ THE VICTOR FARRAR PARTNERSHIP

57 St Peters Street, Bedford MK40 2PR
Tel 01234 353012 Fax 01234 363473

CHARTERED ARCHITECTS AND SURVEYORS: Established by Victor Farrar in 1962, the firm has proven expertise in all facets of ancient building conservation, repair and sympathetic extension. Ecclesiastical work, stone and timber frame repairs are specialities.

▶ W R DUNN & CO

27 Front Street, Acomb, York YO24 3BW
Tel 01904 784421 Fax 01904 784679

CHARTERED BUILDING SURVEYORS, ARCHITECTS AND HISTORIC BUILDINGS CONSULTANTS: *See also: profile entry in Surveyors section, page 46.*

▶ WCP (The Whitworth Co-Partnership with Boniface Associates)

18 Hatter Street, Bury St Edmunds, Suffolk IP33 1NE
Tel 01284 760421 Fax 01284 704734
E-mail info@whitcp.co.uk
Website www.wcp-architects.com
Contact Matthew Stearn

CHARTERED ARCHITECTS AND SURVEYORS: Founded in 1963 and reconstituted in 1985. Services include traditional design of extensions and alterations to existing buildings, new build, surveys and defect analysis. Conservation and repair of historic buildings is a significant part of the practice's work. Each project is approached with care and sensitivity, bringing together a range of expertise appropriate for the project. Of the partners, Philip Orchard (a Lethaby Scholar) and Matthew Stearn are chartered architects, with Tony Redman and Stephen Boniface chartered surveyors (both RICS accredited for building conservation). Clients include private individuals, corporate bodies and English Heritage. The practice has CAD experience, used as appropriate, and is quality assured.

▶ MICHAEL WALTON MA Dip Arch (Cantab), RIBA

Oak Farm Barn, Woodditton Road, Kirtling, Newmarket CB8 9PG
Tel/Fax 01638 730007

ARCHITECT: Michael Walton specialises in sympathetic alterations, extensions and repairs to listed buildings and interiors, primarily in East Anglia.

▶ WATSON BERTRAM & FELL

5 Gay Street, Bath, Somerset BA1 2PH
Tel 01225 337273 Fax 01225 448537
E-mail mail@wbf-bath.co.uk

ARCHITECTS AND SURVEYORS: Watson Bertram & Fell specialises in the restoration and alteration of listed buildings or new buildings in conservation areas. Projects include Abbotsbury, Dorset – estate policy for restoration of village (European Heritage Award); Sloane Club, London – extension in conservation area (RBKC Highly Commended Conservation Award); conversion of two houses into Royal Crescent Hotel (Civic Trust Commendation); barn conversion, Tellisford, Somerset (Environment Week Design Award); construction of new cottages, Newton St Loe, Bath (BANES Building Design Award).

ARCHITECTURAL TECHNOLOGISTS

▶ HARRY CURSHAM

Parks Farm, Cambridge, Gloucester GL2 7AR
Tel 01453 890297 Fax 01453 899121
Mobile 07986 185894
E-mail harry.cursham@lineone.net

CONSULTANT REPAIRS TO OLD, VERNACULAR AND TRADITIONAL BUILDINGS: *See also: profile entry in Heritage Consultants section, page 35.*

PLANNING IN THE HISTORIC ENVIRONMENT

JONATHAN TAYLOR

Historic buildings face a constant threat of alteration, whether caused by the need to adapt a structure to suit changes in use, to accommodate more significant changes, or simply as a result of repair and maintenance requirements. Buildings constructed even 30 years ago face substantial changes in use and design to meet new comfort requirements (the addition of more bathrooms for example), technological requirements (computer cabling of offices for example), disability requirements (the introduction of ramps, for example, everywhere from public libraries to high street shops to meet the requirements of the Disability Discrimination Act by 2004) and many others. Some may be imposed by changing economic circumstances locally (the effects of an out of town shopping centre for example), nationally (such as the decline of manufacturing industries in the UK), and internationally (making historic farm buildings, for example, unsuitable for modern farming practices and machinery). Never in the history of the UK have so many buildings faced so much change, so quickly, and on such a scale.

In recent decades, the strengthening of planning controls over the alteration of listed historic buildings and the increasing willingness of planning authorities to use these controls has provided a much-needed moderating force, helping to stave off the decline of some of our finest townscapes and historic buildings. In some places that decline has even been reversed, assisted by grant schemes such as the Townscape Heritage Initiative.

Although many still see this growth of control as an infringement of liberty, restricting creative expression and the sanctity of the Englishman's castle, few people wish to see the destruction of their historic environment. In the *State of the Historic Environment Report* (see inset box) English Heritage presented a mass of statistics which clearly shows how much people from all walks of life really do care about the historic environment and that the national economy actually depends on its conservation.

THE PLANNING INFRASTUCTURE

In England and Wales the main legal requirements affecting the conservation of historic buildings are set out in the *Planning (Listed Buildings and Conservation Areas) Act 1990*. In Scotland the equivalent Act is the *Planning (Listed Buildings and Conservation Areas) (Scotland) Act 1997*.

These two acts are supplemented by various government guidance including, in England, *Planning Policy Guidance Note 15: Planning and the Historic Environment*. (PPG15 as it is commonly known, is to be combined with PPG16, which covers archaeology, and reissued in a new format by the Office of the Deputy Prime Minister, allegedly without any change in policy.) Equivalent documents for Wales are the Welsh Office Circulars 61/96 and 1/98 *Planning and the Historic Environment*, and for Scotland the equivalent is the *Memorandum of Guidance on Listed Buildings and Conservation Areas*. These documents guide local planning authorities in their decision-making, and are used to interpret planning law.

In the event of a refusal or an enforcement, an applicant can lodge an appeal with either the Planning Inspectorate (England or Wales) or the Inquiry Reporters' Unit of the Scottish Executive. The appeal is made to the relevant government minister, but they are almost invariably conducted and determined by an inspector appointed by the minister.

LISTED BUILDINGS

...no person shall execute or cause to be executed any works for the demolition or alteration of a listed building or for its alteration or extension in any manner which would affect its character as a building of special architectural or historic interest unless the works are authorised.

Section 7, Planning (Listed Buildings and Conservation Areas) Act 1990

A building included on the statutory *List of Buildings of Special Architectural or Historic Interest* is described as a 'listed' building, and special consent is required for any alteration which affects that special character which

THE STATE OF THE HISTORIC ENVIRONMENT

In 2002, English Heritage highlighted the importance of conserving the historic environment not only to our economy but also to ordinary people. In one example given it was concluded that "only five per cent of people whose homes are listed say that they would prefer to live in a new building. And this is not because only grand houses are listed: 47 per cent of people who live in listed buildings are the professionals and senior managers in social classes A and B, but 44 per cent are in classes C1, C2 and D". In Liverpool it was shown that 80 per cent of people believed that "it was important to think about preserving modern buildings for future generations", and another example given states that "95 per cent of people think that the historic environment is important because it gives them places to visit and things to do".

Businesses too prefer historic office buildings: "In the last five years, listed office buildings achieved annualised rates of total return of 15.1 per cent, 1.5 percentage points better than unlisted ones".

English Heritage's statistics also showed how important the historic environment is to our economy: "Overall, it has been estimated that 40 per cent of employment in tourism depends directly on a high quality environment, rising to between 60 and 70 per cent in rural areas".

Although many of the statistics gathered clearly related to England, a similar picture is likely to be found throughout the UK.

CATHEDRAL COMMUNICATIONS LIMITED

makes it listable. This includes the removal or alteration of any historic feature, inside or out, which is considered to form part of the fabric of the building. 'Listed building consent' is also required for alterations to any structure within its grounds or 'curtilage' which was built before 1 July 1948.

The list includes approximately 440,000 entries, but as some list entries include several buildings at the same address, the total number of listed buildings is larger – perhaps 600,000 – amounting to almost two per cent of our total housing stock. The listings are graded according to the architectural or historic importance of each building, Grade I being the most important in England and Wales and category A being the most important in Scotland. The grade or category generally reflects the age and rarity of the building, but many other factors are also taken into account, such as technological innovation, townscape value or connection with a particular historical event.

CONSERVATION AREAS

Local authorities are required under the *Planning (Listed Buildings and Conservation Areas) Act 1990* to designate as conservation areas any area 'of special architectural or historic interest' with a character or appearance which merits preservation or enhancement.

There are approximately 10,000 conservation areas in the UK, and the range is vast. Usually buildings or townscapes form the focus, but often they include the garden and landscape settings; at Wirksworth in Derbyshire the conservation area includes the valley setting of the town, extending to the skyline. In some cases conservation areas include development sites and other places which detract from the setting of an historic core, drawing attention to their importance and to promote their enhancement.

Conservation and enhancement of a conservation area works in three ways:

- the local authority can control demolition in a conservation area through the need for 'conservation area consent', and it can control other forms of development by applying ordinary planning controls in a manner which takes into account and protects the special character of the area
- special restrictions known as 'Article 4 directions' can be introduced by the local authority to protect those architectural details and features of houses on which much of the character of an area depends
- enhancement measures carried out by the local authority directly or indirectly, by other organisations such as a civic society, or through the involvement of the whole community, including residents, local businesses, local government and voluntary organisations in practical schemes of enhancement.

The development of public awareness is a vital component in the success of any conservation area, not only so that owners are aware of the restrictions which apply to their buildings (whether listed or not), but also to promote pride in their buildings and their surroundings. Conservation area designation alone is not enough.

WHAT IS CONSERVATION?'

'Conservation' is an extremely broad concept which encompasses almost any action to ensure the continuation of something or to reduce the rate at which it is being destroyed. Within the field of historic buildings and the built environment the term encompasses repair and maintenance work and other measures designed to preserve buildings in their existing condition, including their contents and their settings. However, unlike the term 'preservation', this term may also be used to encompass more proactive measures to ensure a building's survival, such as the adaptation and conversion of a redundant historic building to suit a new use, where its original use is obsolete.

'Restoration' is not the same as 'conservation', although the two terms are frequently confused. Restoration involves either putting back something that is no longer there, such as reinstating a missing finial, or altering the form or condition of an object (such as a building or artefact) to resemble its form or condition at a particular point in its history. According to the influential Venice Charter of 1964, the aim of restoration "is to preserve and reveal the aesthetic and historic value of the monument and is based on respect for the original material and authentic documents. It must stop at the point where conjecture begins".

The main principles which define good conservation are relatively straightforward:

Minimal intervention – alter as little as possible – and conserve as found

Maintain structural integrity – carry out repair and maintenance work such as cleaning out gutters when necessary, and use only appropriate materials and techniques in all repairs and alterations

Reversibility – where additions are necessary these should be made in such a way that they can be removed later if necessary, leaving the original in the same condition as it was before

Honesty – a damaged or missing element cannot be 'restored' if its original form is unknown; where there is insufficient information to be sure of the original form of the object or element of a building, any repair work or reconstruction should be honestly expressed so that it can be clearly seen to be new work.

WHAT MAKES A PLANNING APPLICATION ACCEPTABLE?

Contrary to popular belief, planning legislation is designed to manage change in the historic environment, not to prevent it. In the vast majority of cases, historic buildings can be successfully adapted if the owner's requirements are carefully analysed with an open mind; the line of least resistance is not always the most obvious design solution.

Designers, given a brief for a building in a townscape by their client, tend to consider how their building will be seen in the townscape, rather than how their building will affect the view of any other building or element in the townscape. Likewise, a shopkeeper will tend to see the success of shop design and signage in terms of how they will enhance his or her premises and attract business, rather than how they will affect the character of its setting.

Small changes, such as the removal of the glazing bars from a window, or the introduction of a piece of street furniture may have little impact on their own, but the cumulative impact of many such incidental changes can have a much more significant effect, destroying the character of the surroundings.

The fundamental issue affecting the conservation of the historic environment is character, and much of PPG15 and its equivalents in Scotland and Wales is devoted to this issue:

... the special character of a place may derive from many factors, including: the grouping of buildings; their scale and relationship with outdoor spaces; the network of routes and nodal spaces; the mix and relative importance of focus and background buildings; vistas and visual compositions; hierarchies of public and private space; materials used in buildings and other surfaces (pavements, roads, garden walls, railings, ...); architectural detailing (of windows, doors, eaves, gates, kerbs, ...); patterns of use; colours; hard and soft landscaping; street furniture; and so on.

Welsh Office Circular 61/96

Any proposal affecting a building which is listed or in a conservation area must start with an appraisal of what gives it and its surroundings their character, before going on to consider how the new facilities required can be achieved at the very least without harming that character. (The yardstick set by the legislation is actually 'preservation or enhancement'.) For large schemes and schemes which are being funded by Heritage Lottery Fund grants, this usually means preparing a substantial document called a 'conservation plan' (see *Conservation Plans* by Kate Clark at www.buildingconservation.com/articles.htm). For smaller applications the local authority will be able to advise on the information required. In addition to photographs, plans and elevations showing the proposal, applications are greatly assisted by a statement which outlines the aims of the work and its scope, a character statement which outlines what makes the building and its surroundings special, and a method statement which details how the work is to be carried out, focusing on issues which are likely to be of greatest concern to the local authority's conservation officer.

The best schemes are underpinned by a thorough understanding of what gives the place its special historic or architectural interest. Not only does this help the application to run smoothly, but it also helps to achieve the best possible scheme.

REVIEW OF HERITAGE LEGISLATION AND POLICY

The Government is developing plans to update the system of legislation, designation and management which protects our heritage in England and Wales. "The aim" according to the website of the Department for Culture, Media and Sport (DCMS), "is to improve and refocus the way in which England's historic environment receives statutory protection".

The Heritage Protection Review is looking at the inadequacies of the current system to protect urban and rural landscapes, sites, settings and groups of buildings; the merits of greater openness where buildings are being considered for listing (or 'designation'); and the 'demarcation' between listing and the 'consent regimes'. (One option put forward by the DCMS is that list descriptions 'define value', presumably to give guidance on the importance attached to individual elements. Another option is to explore the use of management agreements and 'class actions' which would remove the need for applications for certain works.)

DCMS, in partnership with English Heritage, has published an early paper setting out the issues for debate under the heritage designation review; a formal Consultation Paper is to be launched on 17 July and consultation is expected to end on 31 October 2003. A white paper will then be released in early 2004 with a view to introducing the new legislation before the end of 2004.

In addition, a number of other reviews and initiatives are being carried out by the Office of the Deputy Prime Minister which could have implications for the historic environment. These stem from the Planning Statement Sustainable Communities: Delivering through planning, published in July 2002. The amalgamation of PPGs 15 and 16 into a single Planning Policy Statement has already been agreed (although not the date for its release). Other projects include:

A review of The General Permitted Development Order (the measure which enables certain types of work to be carried out without any application for planning permission, such as some demolition work and the removal of features from unlisted homes in conservation areas). The report from the researchers conducting the review is expected later this summer.

The unification of consent regimes, which is looking at the scope for having a single consent regime instead of the current separate regimes for planning permission, listed building consent and conservation area consent. Other consent regimes such as scheduled monument consent and the building regulations are also being considered as part of this project.

The development of prescribed applications forms setting out the information that must accompany a range of applications made to local planning authorities, including planning applications, and applications for listed building and conservation area consent.

A review of the arrangements for statutory consultation over planning applications, but with no presumption that the list of consultees should be reduced.

In taking forward these various strands, it will not be easy to balance the conflicting requirements of the 'stakeholders'. Many of the voices calling for reform are those of developers who would like to see a reduction in bureaucracy and red tape, and a significant decrease in the time taken to decide consent applications. On the other hand, conservation area residents and voluntary organisations are calling for greater restrictions to close damaging loop-holes in the existing regime, such as the right of many home-owners in conservation areas to strip their buildings of their character by removing historic features, and the lack of control over demolition work outside conservation areas.

For further information on progress with these various reviews and initiatives, keep an eye on the ODPM website, www.odpm.gsi.gov.uk, and the DCMS website, www.culture.gsi.gov.uk.

Recommended Reading

Mynors, Charles; *Listed Buildings, Conservation Areas and Monuments, Third Edition*. Sweet and Maxwell, 1999

Government guidance:

England: *Planning Policy Guidance Note 15: Planning and the Historic Environment* and

Planning Policy Guidance Note 16: Archaeology and Planning

Wales: *61/96 Planning and the Historic Environment: Historic Buildings and Conservation Areas*

1/98 Planning and the Historic Environment: Directions by the Secretary of State for Wales

Scotland: *Memorandum of Guidance on Listed Buildings and Conservation Areas*

PLANNING CONSULTANTS

▶ MICHAEL PEARCE ARIBA MRTPI FRSA IHBC

The Lodge, 52 Hollows Close, Salisbury SP2 8JX
Tel/Fax 01722 334355

CONSERVATION CONSULTANT: Clients include English Heritage local planning authorities, landed estates and developers. At one time chief planner of English Heritage. He has advised on the future of a number of important historic buildings and sites, and has appeared at many public local inquiries. He works closely with other leading professional firms.

HERITAGE CONSULTANTS

▶ A O C ARCHAEOLOGY GROUP

Edgefield Road Industrial Estate, Loanhead, Midlothian EH20 9SY
Tel 0131 440 3593 Fax 0131 440 3422
E-mail admin@aocscot.co.uk
Website www.aocarchaeology.com

HERITAGE CONSULTANTS: AOC Archaeology Group offers a professional service in heritage consultancy. It has offices throughout the United Kingdom where the group's skilled personnel offer services in such areas as desk based assessments, environmental impact assessments, heritage management strategies, scheduled monument consent applications, appeals and SMC public enquiries, as well as conservation planning for long-term heritage management and monitoring projects.

▶ BYROM CLARK ROBERTS

117 Portland Street, Manchester M1 6EH
Tel 0161 236 9601 Fax 0161 236 8675
▶ Jubilee House, West Bar Green, Sheffield S1 2BT
Tel 0114 275 7879 Fax 0114 272 8954

ARCHITECTS, SURVEYORS AND ENGINEERS: *See also: profile entry in Architects section, page 22.*

HERITAGE CONSULTANTS

▶ HARRY CURSHAM

Parks Farm, Cambridge, Gloucester GL2 7AR
Tel 01453 890297 Fax 01453 899121 Mobile 07986 185894
E-mail harry.cursham@lineone.net

CONSULTANT REPAIRS TO OLD, VERNACULAR AND TRADITIONAL BUILDINGS: After 20 years of repairing and conserving historic buildings Harry Cursham offers to the trade and private clients a unique insight into the solutions and practical methods of repair. Aesthetics, lime work and structures are his subjects. He works in an advisory capacity both here and abroad. Simple practical solutions are always the best.

▶ INGRAM CONSULTANCY LTD

Manor Farm House, Chicklade, Hindon, Salisbury, Wiltshire SP3 5SU
Tel 01747 820170 Fax 01747 820175
E-mail enquiries@ingram-consultancy.co.uk
Website www.ingram-consultancy.co.uk

CONSULTANTS IN THE REPAIR AND CONSERVATION OF HISTORIC BUILDINGS AND ARCHAEOLOGICAL SITES: Ingram Consultancy is dedicated to the survey, recording, analysis and practical conservation of historic buildings, ancient monuments and their sites. Principal activities include condition surveys, production of specialist tender documents for conservation and repair work, conservation planning for historic sites, historic building assessment and building recording. Stone, terracotta and lime-based materials are particular specialities, often involving the execution of exemplar works. Ingram Consultancy designs and delivers Building Conservation Masterclasses for West Dean College. Clients include English Heritage, The Royal Household Property Section, The States of Guernsey, local authorities, and architectural and surveying practices.

▶ RESURGAM®

Netley House, Gomshall, Surrey GU5 9QA
Tel 01483 203221 Fax 01483 202911
E-mail ei@handr.co.uk Website www.handr.co.uk
Contact Christopher Marsh RIBA

ARCHITECTURAL, HISTORICAL AND TECHNICAL CONSERVATION CONSULTANTS: Resurgam, a division of H+R Environmental Investigations Limited, consists of a group of independent experts and scientists. Resurgam carries out research, conservation plans, condition surveys and analysis of traditional buildings and sites, combining extensive architectural and construction experience with innovative investigative technology. Stone, mortar, plaster and decorative finishes are specialities. Remedial specifications, schedules, bills of quantities and tender procurement are provided. Resurgam provides conservation consultancy and management for refurbishment projects. Clients include The Royal Household, English Heritage, National Trust, UNESCO, Crown Estate, architects, surveyors and property managers. *See also: Hutton+Rostron profile entry in Damp & Timber Decay section, page 182.*

▶ SCOTTISH LIME CENTRE TRUST

The Schoolhouse, Rocks Road, Charlestown, Fife KY11 3EN
Tel 01383 872722 Fax 01383 872744
E-mail info@scotlime.org
Website www.scotlime.org

TRADITIONAL BUILDINGS CONSULTANTS: Specialist advice for traditional buildings including diagnosis, investigation of problems, with recommendations and specifications for repair. In-house laboratory for lime mortar analysis, investigation and matching service. *See also: display entry in Courses & Training section, page 226.*

▶ WESSEX ARCHAEOLOGY, CONSERVATION MANAGEMENT SECTION

Portway House, Old Sarum Park, Salisbury, Wiltshire SP4 6EB
Tel 01722 326867 Fax 01722 337562
E-mail info@wessexarch.co.uk
Website www.wessexarch.co.uk
Contact Paul Falcini, Head of Conservation Management or John Dillon, Operations Director

HERITAGE CONSULTANTS: *See also: profile entry in Archaeologists section, page 13.*

PROJECT MANAGERS

▶ BYROM CLARK ROBERTS

117 Portland Street, Manchester M1 6EH
Tel 0161 236 9601 Fax 0161 236 8675
▶ Jubilee House, West Bar Green, Sheffield S1 2BT
Tel 0114 275 7879 Fax 0114 272 8954

ARCHITECTS, SURVEYORS AND ENGINEERS: *See also: profile entry in Architects section, page 22.*

▶ GIFFORD AND PARTNERS

Carlton House, Ringwood Road, Woodlands, Southampton SO40 7HT
Tel 023 8081 7500 Fax 023 8081 7600

PROJECT MANAGEMENT: *See also: profile entry in Structural Engineers section, page 39.*

▶ RICHARD CROOKS PARTNERSHIP

14 Calverley Lane, Horsforth, Leeds LS18 4DZ
Tel 0113 281 8080 Fax 0113 258 4070

CHARTERED ARCHITECTS: *See also: profile entry in Architects section, page 29.*

▶ W R DUNN & CO

27 Front Street, Acomb, York YO24 3BW
Tel 01904 784421 Fax 01904 784679

CHARTERED BUILDING SURVEYORS, ARCHITECTS AND HISTORIC BUILDINGS CONSULTANTS: *See also: profile entry in Surveyors section, page 46.*

CRACKING

JIM ALLEN

Whether dealing with the conservation of historic buildings, or more mundane commercial or institutional buildings, it is imperative that the cause of cracking in masonry structures is well understood. Perhaps one of the most significant differences between working on relatively modern structures and historic buildings is that often, in the latter case, it will not be possible to deal with the root cause. However, this does not in any way reduce the need for understanding, because a suitable repair strategy can only be determined once the cause of the problem has been identified.

Cracks arise because the real world is not static: whether at a macro or a micro level, materials respond to changes in their environment by trying to move. There are several generic causes for movement (a more specific listing would run to several pages):

Ground movement (beneath foundations) Examples include shrinkage in clay sub-soils (often tree related), loss of fine particles from granular material (caused by drain failure or ground water movement for example), land slip and even the activities of burrowing animals.

Foundation failure Consolidation of rubble foundations, decay of soft clay brick, and attack of concrete by aggressive chemicals all fall into this category.

Decay of superstructure Most structures are composite, being made up of different materials each with its own life expectancy under any given environmental exposure. When one material fails prior to another, movement and cracking arise. A classic example would be the decay of timber wall plates in masonry, or the corrosion of iron cramps in stone walls.

Moisture movement Materials either expand or contract as they gain or lose moisture. In some cases (the expansion of new clay bricks for instance) the movement is irreversible but, if anticipated originally, the construction may have been designed to accommodate it. In others, movement caused by variations in moisture content are reversible and because of their cyclical nature, often harmless, but they can lead to progressive movement, often out of plane. Seasonal changes in timber are a good example, and of course as timber is a natural material, movement will not be uniform.

Thermal movement As temperature increases or decreases, materials either expand or contract. If such movements are prevented, very great stresses can occur. This form of movement is also cyclical and can lead to the progressive deterioration of structures.

Inherent defects Structures, particularly historic structures, often lack (by today's standards) sufficient lateral restraint of vertical elements, resulting in lateral movement. Often progressive alteration of structures can concentrate load in areas ill-suited to carry it. Movements associated with such defects are often progressive and unless arrested can eventually end in structural failure.

A cracked gauged brick arch (photograph reproduced by kind permission of Charterbuild)

Inappropriate specification Modern repair techniques are often too strong or too rigid for the repair of structures built with pliable lime-based mortars. Pointing with Portland cement-rich mortar is a good example, as strong mixes do not behave elastically and, if overstressed, the new mortar can fracture or, more usually, the surrounding masonry fractures.

Deflection under load Suspended structures such as floors tend to deform under load, and even vertical elements subject to load will compress by a small amount. Any infill (which is by definition non-structural) must be detailed to accommodate such movements or cracking will occur. A classic defect of 1960s buildings is delamination of brick slips caused by the deflection of reinforced concrete supporting structures.

While not exhaustive, this list serves to illustrate the complex interactions that can occur, and the difficulties that may arise in gaining an understanding of cause.

REASONS FOR CRACKS

Cracking is an inevitable response to the inability of a structure to accommodate the movement to which it is subjected. There are two issues to be considered when assessing the reasons for cracking: the first is the nature and significance of the cause of movement (as described above); the second is the ability of the structure to accommodate movement. The latter will depend on the nature of the material or, in the case of composite materials like masonry, the nature of the combination of materials used in the structure.

For new buildings, designers can try to eliminate many of the causes of cracking and design tolerance for those factors that remain either by introducing joints, or by choosing movement-tolerant materials. The latter is the traditional approach to design, however materials such as fat lime (lime putty) and hydraulic lime can now be specified to give structures the ability to absorb considerable movement.

Existing buildings are what they are, and when considering performance an assessment must be made as to the likely cause of movement and the degree of the movement-tolerance inherent in any given structure. The cracks may be predominantly associated with an external cause (as in the case of subsidence) or with the material itself (an over-specified cement render for example).

STRATEGY FOR ASSESSMENT

Where historic buildings are concerned it is essential to employ someone with the experience and expertise to recognise the patterns of movement taking place, and who will be able to plan a programme of investigation to verify their assessments.

Research is essential and can save hours of expensive time, and minimise the need for the disruption that investigations can cause.

A specialist structural engineer will be able to form an opinion on the basis of a visual inspection, but even when good archival information is available (from a previous structural survey, for example), some site investigation will still be required to tease out the various interdependencies, and to establish the most important factors at work.

This can be an iterative process as investigations can raise more questions than they answer.

CATHEDRAL COMMUNICATIONS LIMITED

STRATEGY FOR REPAIR

It is important to establish a brief before embarking on any programme of repair. On the assumption that both the cause of movement and the degree of movement-tolerance have been established, a range of options will be available. These may encompass major intervention on the one hand (and a good prospect of avoiding major future problems) to limited cosmetic action in the knowledge that further maintenance work will be required on a regular basis. The most important consideration is whether movement is likely to be progressive. If the structural condition is going to deteriorate, then for a structure to be preserved, intervention becomes essential.

Having agreed an approach to underlying problems, the result of the intervention must be assessed. For instance, new foundations can be expected to settle following underpinning, as the ground beneath consolidates. This is particularly an issue where clay forms the sub-strata. Premature or inappropriate repair of crack damage can be an expensive mistake.

In effect, the likelihood and the nature of future movements need to be assessed for the new condition created by intervention. Detailing will be dependent on the kind of movement likely to occur in the future. In most cases structures will still be subject to moisture content changes and variations in temperature, as short and long term changes in environment occur.

REPAIR DETAILS

Cracking creates an inherent weakness in a structure. This is why a superficial approach to repair, limited to filling and decorative work, will almost always fail. Even the smallest of movements will lead to further cracking.

In engineering terms, this is explained by the size of cracks in relation to the rest of the structure. Cracks are usually narrow, and if movement is concentrated at these positions of weakness, the strain will be high. The stress created by movement is a function of strain, and will vary depending on a parameter called the Young's modulus which, in effect, describes the stiffness of a material.

For this reason, materials with high values of Young's modulus, such as a rich Portland cement (OPC) mix, should not be used to repair a crack in a masonry wall, because the stresses that arise will cause the repair to fail. Materials with a low modulus, or with the ability to deform in a plastic manner, will offer better prospects for success. A high calcium hydraulic lime, for example, would be much better than OPC in an area of high strain.

Increasing the width of a repair will also reduce strain and, therefore, stress. In practical terms this can be achieved by cutting out cracks and substituting a formed joint, but this is major work and may not be acceptable visually or in terms of cost. Rebonding masonry using hydraulic or fat limes is usually preferable, as this will give the necessary plasticity, providing a mechanism to accommodate strain.

One alternative is to insert corrosion-resistant metal reinforcement (usually stainless steel) into bed joints to redistribute strain and therefore stress over a wider area, thus reducing the risk of failure. Care must be taken to vary the width of reinforcement across the crack as abruptly terminating a bar in a wall or mesh in a render in parallel lines on either side of the crack can also cause stress concentrations leading to further cracks.

Such techniques require the basic form of construction to be robust and in relatively good condition. This is not always the case, and sometimes it is necessary to carry out remedial work to stabilise a structure before it can be repaired. To assess this need, the basic form of construction must be understood, and the degree of distortion estimated. Generating cross sections to show the deflection and deformation can be helpful, allowing the extent of voids and other defects to be estimated and taken into account.

Where voids are suspected, the use of instruments such as boroscopes can be invaluable to see inside structures to verify the extent of the problems. There are other more sophisticated methods of non-destructive investigation that can be employed, including radar and thermography, but these can be expensive and complex to interpret. These methods are discussed further in *Non-Destructive Investigations*, a *Building Conservation Directory* article by Robert Demaus now available on the website www.buildingconservation.com.

The response to voids is usually either reconstruction (often unavoidable) or grouting to fill them with cementitious material. Grouting requires the voids to be accessed by drilling holes in the structure through which the grout can be poured. There is no single 'best way' to approach grouting, but some methods are better than others. In particular:

- use trials to find the best grout mix
- use rounded fine aggregates rather than sharp sands
- make several holes to access voids
- start below and work up; grout appearing in the next hole is a sign of success
- pre-wetting of the wall will prevent moisture-loss from the grout and reduce clogging of the grout points with dust
- grout can exert high levels of hydrostatic pressure, so limit the pour height and the pressure head used
- where walls are in poor condition it may be preferable to slowly feed the grout into the wall using clay applied to the wall and shaped into cups
- grout can take time to penetrate walls
- use a grout mix that is compatible with the fabric of the wall
- consider the future performance of the wall when specifying the work.

These comments are from an engineering perspective. There will be other factors influencing the specification for repair, and it may be appropriate to depart from the 'ideal' engineering solution.

The successful repair of cracks in masonry structures requires a fundamental understanding of the reasons for the cracking. Intervention must respect the structure, and it is important to assess the likelihood of progressive movement. An appropriate repair will accommodate future movements in the wall, but may require stabilisation works if it is to be successful.

***12 London Road, Chippenham**: (top) the front elevation showing new and old stone; (middle) detail over a door showing a lintel cut to shape to accommodate settlement; and (bottom lateral movement in the front elevation due to the decay of embedded timbers both to front and side elevation (note the buttressing).*

A wall showing settlement where it was built over an old well

JIM ALLEN PhD BEng MICE MIOSH is a consulting engineer and a partner in Ellis and Moore. His research experience in cementitious building materials has been applied to both repair and refurbishment projects, particularly listed buildings, and he is collaborating with Bristol University and major industrial partners in the development of hydraulic lime for main stream construction applications.

STRUCTURAL ENGINEERS

▶ ADRIAN COX ASSOCIATES

The Studio, 3 Bayham Road, Sevenoaks, Kent TN13 3XA
Tel 01732 462640 Fax 01732 740893
E-mail engs@adriancox.co.uk
Website www.adriancox.co.uk

CONSULTING CIVIL AND STRUCTURAL ENGINEERS: A small team of experienced conservation engineers, giving priority to good communication by avoiding jargon. Committed to achieving the best for buildings by blending the latest computer analysis methods with traditional repair techniques and materials. Projects have included numerous historic buildings including houses, museums, churches and schools.

▶ AUSTIN TRUEMAN ASSOCIATES

8 Spicer Street, St Albans, Herts AL3 4PQ
Tel 01727 858752 Fax 01727 852376
E-mail engineers@austintrueman.co.uk
Website www.austintrueman.co.uk

CIVIL AND STRUCTURAL ENGINEERS: Established in 1973 Austin Trueman Associates specialises in the renovation and repair of existing buildings especially those with an architectural and or historic interest. The firm operates a philosophy of sympathetic engineering through a unique blend of technical flair and creative problem solving. Projects include Brocket House, Knebworth House, Hatfield House, Somerset House, The Royal Naval College and The Palace of Westminster. Offices in London and St Albans and associated offices in Jersey, Paris and Prague. Member firm of the Association of Consulting Engineers and British Consultants Bureau.

▶ BILL HARVEY ASSOCIATES

85 Pennsylvania Road, Exeter EX4 6DW
Tel 01392 499934 Fax 0870 458 0295
E-mail bill@obvis.com

STRUCTURAL ENGINEER: Experience ranges from domestic scale to large bridges. Assessment and conservation of masonry and timber structures. Specialist analysis of complex masonry and other structures; usually able to prove the structure is stable. Clients include Historic Scotland, English Heritage, the Landmark Trust, numerous local authorities.

▶ BLACKETT-ORD CONSULTING ENGINEERS

33 Chapel Street, Appleby-in-Westmorland, Cumbria CA16 6QR
Tel/Fax 017683 52572
Contact Charles Blackett-Ord, CEng FICE FConsE and Elaine Rigby, DipArch, MA (Con Studies, York) RIBA
E-mail mail@blackett-ordconsulting.co.uk
Website www.blackett-ordconsulting.co.uk

CIVIL AND STRUCTURAL ENGINEERS AND ARCHITECTS: The practice works throughout the country on repair and conservation of historic buildings and ancient monuments, including listed railway viaducts, 18th century grand houses, churches, medieval buildings and ruins. The practice's projects have won an RICS Conservation Award and Ian Allan Railway Heritage and Civic Trust Awards. The use of traditional materials both in repairs and new work is encouraged and the practice has considerable expertise, in particular, in the use of lime mortars and renders.

▶ THE BUDGEN PARTNERSHIP

56 Lisson Street, London NW1 5DF
Tel 0207 224 8887 Fax 0207 224 8883
E-mail budgenp@aol.com
Website www.budgenpartnership.com

STRUCTURAL ENGINEERING CONSULTANTS: The Budgen Partnership was founded in 1960 acting for clients in all sectors. They have earned a significant reputation for works to historic Grade I and Grade II listed buildings. Involvement with many fine buildings over the years has engendered a close working relationship with English Heritage. Important works include the Foreign Office, Royal Albert Hall, Parliament Hill Mansions, St Pancras Chambers, Handel Museum and many smaller projects. They have received awards and commendations in recognition of design excellence in both new building and conservation work.

▶ BYROM CLARK ROBERTS

117 Portland Street, Manchester M1 6EH
Tel 0161 236 9601 Fax 0161 236 8675
▶ Jubilee House, West Bar Green, Sheffield S1 2BT
Tel 0114 275 7879 Fax 0114 272 8954

ARCHITECTS, SURVEYORS AND ENGINEERS: *See also: profile entry in Architects section, page 22.*

▶ CAMERON TAYLOR BEDFORD

Lorne Close, London NW8 7JJ
Tel 020 7262 7744 Fax 020 7724 0917
E-mail c.richardson@camerontaylor.co.uk

CONSERVATION ENGINEERS AND STRUCTURAL ARCHAEOLOGISTS: Conservation-accredited specialists in the survey, repair and development of buildings, large and small. Fire damage assessment. Temporary works and scaffolding designs. Expert advice for planning inquiries and litigation. Clients include the National Trust, the Church Commissioners, and the Crown Estate. CTB's dedicated conservation team is known to English Heritage, and led by Clive Richardson, Engineer to Westminster Abbey and Visiting Lecturer in building conservation at the Architectural Association. Offices also at Birmingham, Bury St Edmunds, Chelmsford, Norwich and Great Yarmouth.

▶ CAPSTONE CONSULTING ENGINEERS

38 Main Street, Menston, Ilkley, West Yorkshire LS29 6LL
Tel 01943 876300
E-mail conservation@capstone-uk.com
Website www.capstone-uk.com
Contact John Ruddy BEng(Hons) MA(Conservation Studies York) CEng MICE MIStructE

CONSERVATION ENGINEERS: Capstone offers a flexible and professional service on a broad range of projects, large and small. Innovation and creativity play key roles in providing good quality conservation-led solutions. Recent work encompasses repair, adaptation and new-build with ecclesiastical, domestic, civic, rural and historic park buildings and structures.

▶ DAVID NARRO ASSOCIATES

36 Argyle Place, Edinburgh EH9 1JT
Tel 0131 229 5553 Fax 0131 229 5090
E-mail mail@davidnarro.co.uk
Website www.davidnarro.co.uk

CONSULTING STRUCTURAL AND CIVIL ENGINEERS: David Narro Associates was established with the aim of providing a high quality service from committed and experienced staff. A desire to extend the experience of the practice has led to expertise in fields such as conservation of ancient monuments and the repair and restoration of listed buildings. The skills of its engineers are shown to best advantage when designing innovative solutions to engineering problems. The challenges posed by old buildings requiring sensitive non-invasive strengthening or alteration are met with a flexible and inventive approach. The initial contact can be the most fruitful and this is always entrusted to senior personnel.

▶ ELLIOTT & COMPANY, STRUCTURAL ENGINEERS

51 Niddry Street, Edinburgh EH1 1LG
Tel 0131 558 9797 Fax 0131 558 9696
E-mail structures@ecoeng.co.uk

STRUCTURAL ENGINEERS: Award winning small specialised practice offering bespoke service on the conservation and alterations of historic structures and ancient monuments throughout Scotland. Extensive knowledge of traditional building methods ensures a considered assessment of the existing structure and highlights any original or imposed defects to be addressed with minimal intervention. Close liaison with other members of the design team ensures that implications of structural alterations and repairs are considered in their wider context. Projects include Stanley Mills (Historic Scotland), Newhailes House (National Trust for Scotland), Lady Victoria Colliery (Scottish Mining Museum Trust) and Dowanhill Church (Four Acres Charitable Trust).

STRUCTURAL ENGINEERS

▶ ELLIS & MOORE

9th Floor, Hill House, Highgate Hill, London N19 5NA
Tel 020 7281 4821 Fax 020 7263 6613

CIVIL AND STRUCTURAL ENGINEERS: The practice is involved in the repair and refurbishment of listed buildings in London and the West Country. Services include structural investigations, organising testing and preparing reports identifying defects with recommendations on remedial works. Lime mortars and renders are used in repair work where possible.

▶ FOTHERGILL

62 Hill Street, Richmond, Surrey TW9 1TW
Tel 020 8948 4165 Fax 020 8948 5105
Website www.fothergills.demon.co.uk

▶ GIFFORD AND PARTNERS

Carlton House, Ringwood Road, Woodlands, Southampton SO40 7HT
Tel 023 8081 7500 Fax 023 8081 7600

CONSERVATION ENGINEERS AND ARCHAEOLOGISTS: Award winning specialist expertise in structural, mechanical and electrical services engineering and archaeology for historic buildings. This unique blend of multi-disciplinary engineering and archaeological services allows Gifford and Partners to provide a fully integrated service to their clients. Numerous commissions on buildings of national importance have been secured with clients in the Public and Private sectors including English Heritage, National Trust, Historic Royal Palaces Agency, the Royal Household, Ministry of Defence, church PCCs, along with a number of key developers and private owners. Specialisms include: post fire repair, fire safety services, building fabric conservation, environmental control, electronic security and surveillance, energy conservation, public health engineering, measured surveys, interpretative studies for conservation-based research and analysis, and conservation and development planning.

▶ H-H-HERITAGE

PO Box 276, Bedford MK41 9DZ
Tel 01234 344233 Fax 01234 365274

STRUCTURAL ENGINEERS AND SURVEYORS: The practice carries out work throughout England and Wales and is concerned with the sympathetic repair and conservation of historic buildings, particularly vernacular structures. Minimal intervention is the philosophy employed wherever possible with close liaison at all times with other team members. The services offered include condition and defect surveys, non-destructive testing, detailed repair schemes, liaison with local authority conservation officers and English Heritage, contract supervision, insurance claim administration, expert witness reports. H-H-Heritage is dedicated to solutions using traditional methods and materials.

▶ MANN WILLIAMS

4 Palace Yard Mews, Bath BA1 2NH
Tel 01225 464419 Fax 01225 448651
▶ 98 Cardiff Road, Llandaff, Cardiff CF5 2DT
Tel 02920 554193 Fax 02920 554196
Website www.mannwilliams.co.uk

CONSULTING STRUCTURAL AND CIVIL ENGINEERS: Working nationally on historic structures and ancient monuments since 1986, Mann Williams has built an impressive portfolio of projects for clients such as the National Trust and Landmark Trust. Projects range from cathedrals at Winchester, Exeter and St David's, plus Avebury Stones and Corfe Castle, through to the Bath Spa Roman complex and significant timber framed structures. With its new office in Cardiff, Mann Williams is better able to serve its clients in South and West Wales. Please visit the firm's website at www.mannwilliams.co.uk.

▶ THE MORTON PARTNERSHIP LTD

The Old Cavalier, 89 Dunbridge Street, Bethnal Green, London E2 6JJ
Tel 020 7729 4459 Fax 020 7729 4458
E-mail london@themortonpartnership.co.uk
▶ Arcadia House, 19 Market Place, Halesworth, Suffolk IP19 8BB
Tel 01986 875651 Fax 01986 875085
E-mail halesworth@themortonpartnership.co.uk
Website www.themortonpartnership.co.uk

STRUCTURAL AND CIVIL ENGINEERS: Brian Morton founded this practice in 1966. It is now almost completely involved in minimum repair solutions to preserve historic buildings. Current work includes: work to Canterbury Cathedral, Pell Wall Hall by Sir John Soane, the new tower at Bury St Edmonds Cathedral, work to many parish churches, barns and domestic buildings, work for the Crown Estate in Regents Park, and National Trust properties. The practice carries out a considerable amount of work for local authorities, and is well known to all the national amenity groups. Work to small buildings is an important part of its work. Services include preliminary advice, structural surveys and presentation of the most cost effective solution to the proper repair of historic buildings and structures.

▶ PAUL TANNER ASSOCIATES

14 St Clement Street, Winchester, Hampshire SO23 9HH
Tel 01962 859800 Fax 01962 856452
E-mail paul@ptanner.co.uk
Website www.ptanner.co.uk
Contact Paul Tanner BSc, CEng, FIStructE, MICE

STRUCTURAL AND CIVIL ENGINEERS: A small practice offering a personal service. Experience in sensitive repairs to a variety of historical buildings and in a range of materials including traditional timber frame, brickwork, stonework and ironwork. Also reinforced concrete and steelwork repairs in modern buildings. Corporate, institutional and private clients.

▶ IAN RUSSELL CEng MICE MIStructE

Shulbrede Priory, Lynchmere, Haslemere, Surrey GU27 3NQ
Tel 01428 653049 Fax 01428 645068

CIVIL AND STRUCTURAL ENGINEERING CONSULTANCY WITH BUILDING CONSERVATION: Specialising since 1983 in repairs and alterations to historic buildings and structures where using sympathetic materials and methods is important. Recent projects: listed offices, London EC1; Michelham Priory and Lewes Castle, Sussex; Council House, Chichester. Current work: Marlipins Museum, Shoreham, repairs/extension; church repairs, barn stabilisation, Hampshire.

▶ FRED TANDY

9 Cannock Drive, Heaton Mersey, Stockport, Cheshire SK4 3JB
Tel 0161 432 4416 Fax 0161 442 1283

▶ JOHN WARDLE BSc CEng MIStructE

5 Lotus Road, Biggin Hill, Kent TN16 3JL
Tel 01959 540696
E-mail jwardlese@aol.com

CONSULTANT STRUCTURAL ENGINEER: Experienced in conservation of older buildings/structures (listed or not). Timber framed, stone or brick. Castles, churches, houses, barns and canal structures. Projects include: Howbury moated site (scheduled ancient monument), East London; Caledonian Canal, Scotland; and The Jealous Wall, Southern Ireland. Reconstruction of a roof structure at Dower House, Bayham Abbey, Kent.

NON-DESTRUCTIVE INVESTIGATIONS

GBG
Care & Judgement
G B Geotechnics Ltd
Downing Park Swaffham Bulbeck
Cambridge CB5 0NB ENGLAND
Tel: (+44) 01223 812464
Fax: (+44) 01223 812462
e-mail works@gbg.co.uk
Structural Assessments by GBG

▶ DEMAUS BUILDING DIAGNOSTICS LTD

Stagbatch Farm, Leominster, Herefordshire HR6 9DA
Tel 01568 615662 E-mail info@demaus.co.uk
Website www.demaus.co.uk
INFRARED THERMOGRAPHY, ULTRASOUND, MICRODRILLING AND ENDOSCOPY: Demaus Building Diagnostics Ltd specialises in the non-destructive investigation and assessment of historic buildings using a wide range of advanced techniques. Thermographic imaging, using the latest most sensitive equipment provides a very wide range of information quickly and economically. Services also include the location and accurate measurement of decay in concealed structural timber, the assessment of fire damage, the identification of concealed structural alterations and failures, and advice on repair and conservation. Clients range from English Heritage, The Royal Household, the National Trust and Department of National Heritage to private individuals.

▶ HARRY CURSHAM

Parks Farm, Cambridge, Gloucester GL2 7AR
Tel 01453 890297 Fax 01453 899121 Mobile 07986 185894
E-mail harry.cursham@lineone.net
CONSULTANT REPAIRS TO OLD, VERNACULAR AND TRADITIONAL BUILDINGS: *See also: profile entry in Heritage Consultants section, page 35.*

▶ HUTTON+ROSTRON ENVIRONMENTAL INVESTIGATIONS LIMITED

Netley House, Gomshall, Surrey GU5 9QA
Tel 01483 203221 Fax 01483 202911 E-mail ei@handr.co.uk
Website www.handr.co.uk
Contact Tim Hutton MA MSc MRCVS
CONSULTANTS ON BUILDING FAILURES AND ENVIRONMENTS: *See also: profile entry in Damp & Timber Decay section, page 182, and Resurgam profile entry in Heritage Consultants section, page 35.*

▶ RIDOUT ASSOCIATES

147a Worcester Road, Hagley, Stourbridge, West Midlands DY9 0NW
Tel 01562 885135 Fax 01562 885312
E-mail ridout-associates@lineone.net
Website www.ridoutassociates.co.uk
NON-DESTRUCTIVE TIMBER SURVEYS: *See also: profile entry in Damp & Timber Decay section, page 182.*

MATERIALS ANALYSIS

▶ CLIVEDEN CONSERVATION WORKSHOP LTD

Head Office – The Tennis Courts, Cliveden Estate,
Taplow, Maidenhead, Berkshire SL6 0JA
Tel 01628 604721 Fax 01628 660379
SCULPTURE, STONE AND WALL PAINTINGS CONSERVATION, MORTAR ANALYSIS: *See also: profile entry in Stone section, page 102.*

▶ HERITAGE TESTING LTD

Unit 43, The Old Brickworks, Station Road, Plumpton Green,
Lewes, East Sussex BN7 3DF
Tel 01273 891785 Fax 01273 891443
E-mail consultant@heritagetesting.co.uk / heritagetesting@connectingbusiness.com
Website www.heritagetesting.co.uk
SPECIALISED CONSULTANCY AND LABORATORY TESTING SERVICE: In support of construction, conservation and preservation, restoration and remediation. The unique nature of each historic building or structure investigated and the wide range and type of (often local) materials and craftsmanship employed requires an 'individually matched' approach to each customer's requirements. The company's portfolio includes Westminster Palace (stone identification), St Mary Le Bow church basement crypt (deterioration of internal stonework and mortar surfaces due to soluble salt crystallisation), The Royal Pavilion, Brighton (paint and composite materials analysis), King's Cross Tunnel (historic mortar analysis), and many other listed structures, including historic churches, castles, forts, fountains, ruin consolidations and international archaeological research projects.

▶ HIRST CONSERVATION

Laughton, Sleaford, Lincolnshire NG34 0HE
Tel 01529 497449 Fax 01529 497518
ANALYSIS OF PLASTER AND PAINT LAYERS: *See also: display entry on the inside front cover and profile entry in Building Contractors section, page 62.*

DECAY DETECTING EQUIPMENT

▶ SIBTEC SCIENTIFIC, SIBERT TECHNOLOGY LTD

2a Merrow Business Centre, Merrow Lane, Guildford, Surrey GU4 7WA
Tel 01483 440724 Fax 01483 440727
E-mail ndt@sibtec.com
Website www.sibtec.com
MANUFACTURERS AND SUPPLIERS OF DECAY DETECTION EQUIPMENT

FINANCIAL AID FOR HISTORIC BUILDINGS

NEIL GRIEVE

Sefton Park Palm House, Merseyside, rescued and restored by a voluntary preservation trust with the aid of a variety of grants including £2.4 million from the Heritage Lottery Fund. (Photograph courtesy of Sefton Park Palm House Preservation Trust)

After "Do I need consent?" the next most common question asked by owners of historic buildings has to be "Will there be any grant available?". Yet the subject of finance remains complicated and neglected: there is never enough money and never any automatic entitlement to a grant and, while there are a number of good publications which list sources, nothing written so far adequately describes the application process. The situation is so complex that it frequently begs for the involvement of a person who David Pearce calls the "conservation entrepreneur – skilled at assembling money and employing persuasive arguments" (see *Recommended Reading* below). This article examines the background to the lack of resources, looks at some of the potential sources and examines a number of the ground rules before attempting to draw everything together by exploring a specific case study.

BACKGROUND

The public purse is the main source of funding so the question needs to be asked, is this right? On balance it would seem that it is. Work on historic buildings is frequently more expensive and we have strong legal control which requires that listed buildings are properly looked after, so on a simple carrot and stick basis some financial incentive seems fair. In addition there are benefits to the construction industry and to tourism; the green argument in terms of embodied energy and the use of sustainable materials is quite well developed; and it is clear that grants lever out money from other sources; furthermore, there is seldom any real criticism of things conservation-led.

Nevertheless, government funding remains limited, and it is increasingly difficult to find money for simple repair and maintenance, or for more expansive re-use projects. Probably underpinning the problem is the notion that historic buildings can, and should, look after themselves. This is partly because most people will view proper repair and maintenance as so sensible that they assume the means must be there to carry out the work required, and partly because historic buildings still smack of privilege and are frequently seen as belonging to people with adequate resources. Both arguments are nonsensical. The government does not give the conservation of the built environment any real priority; its expenditure on the heritage is miniscule and highly focused compared say with the health service and it is unlikely that the profile can ever be raised enough to change the situation significantly. Also, successive governments have seen conflict between conservation and development so may be loath to offer greater resources. There is even a feeling shared by some conservationists that grant aid is not necessarily a good thing as it encourages dependency; more resources could lead to unwelcome change, and many of the best buildings have survived because no one has had money to spend on them. The establishment of the Heritage Lottery Fund has helped but it is no pot of gold and has clearly demonstrated that there is no such a thing as a single source of grant (and there never will be).

SOURCES AND ELIGIBILITY

While potentially there is any number of different grants, they all come from five basic sources:

- international/European organisations
- central government via either *public* or *public sector* funding
- local government
- charities
- private sector.

International organisations concerned with conservation worldwide principally include UNESCO, ICOMOS and ICCROM. Each plays a vital role in areas such as building international solidarity to protect heritage, establishing international standards or, particularly in the case of ICCROM, by providing technical advice. None can offer grants. Likewise within Europe, the Council of Europe and Europa Nostra between them co-ordinate a range of educational and participatory programmes and make awards, but do not offer grants. The one significant source is the European Union itself. The EU has a number of programmes which can provide financial support for building conservation including education, training and economic regeneration. The most potentially beneficial EU programme, the European Regional Development Fund (ERDF), involves three tiers of government working together to agree priorities which can result in sources of finance for historic buildings.

Central government departmental statutory bodies such as the ministries of defence, agriculture, education, and social security have the potential to put money into the built heritage, as do non-departmental statutory bodies such as the Countryside Commission, the Museums Council and Scottish Natural Heritage. Schemes such as job creation, housing provision, economic regeneration and the regeneration of former coal mining areas, farm diversification schemes and urban programmes can all provide sources of grants for historic buildings. In addition, the government has established task forces, enterprise companies and other bodies, all with remits that can include projects involving historic buildings. Indeed, it has been estimated that there are over 200 possible central government sources of finance, some of which, admittedly, have rather tenuous heritage links. However, by far the most important and obvious sources are English Heritage, Historic Scotland, Cadw (Wales), and the Environment and Heritage Service (Northern Ireland). This is all *public sector funding*.

Public funding on the other hand, is funding provided by government to be administered by bodies outside government. Of these probably the best known are the National Heritage Memorial Fund and the Architectural Heritage Fund, the latter acting as a parent body supporting building preservation trusts. Somewhere in between the two types of funding is the National Lottery created in 1994 to provide extra funds for good causes one of which is Heritage. The Heritage Lottery Fund, the body set up to administer these grants through a number of special programmes, is now the main lifeline for many conservation projects.

In addition to all of this, while the government has so far resisted the notion of direct tax breaks, it has made concessions (controversially) on Value Added Tax and initiated the Landfill Tax Credit Scheme which can benefit historic buildings.

Local government can make finance available to historic buildings under the Listed Buildings and Conservation Areas Acts, and can also make finance available, albeit much less directly, under powers contained in such legislation as The Housing Act. Occasionally, central and local government will work together to fund initiatives such as the Edinburgh World Heritage Trust which has grant making powers.

The charitable sector is supported by the public through bequests and donations and by the government, and it pays out in excess of £1 billion per year. Expenditure tends to be shrouded in some secrecy, possibly because there is no wish to have applicants – successful or otherwise – drawing comparisons with similar organisations which may have fared better. There are several hundred charitable trusts, at least 18 of which have assets of over £10 million. Some 40 well-known trusts and many more small local trusts are concerned with the heritage.

The private sector includes banks, pension funds and companies, as well as private individuals who may have funds which can be made available for the conservation of the historic environment under certain circumstances.

All of these bodies have their uses and each offers finance in different ways. While grant is best, soft loans are frequently made; costs may also be shared in projects to the benefit of two or more parties and some projects have been moved forward on the basis of equity participation where shares are issued to provide capital.

Overall, a project involving a historic building can be eligible for assistance from several different sources and can attract several different types of assistance depending on, for example:

- what the project is, and whether it involves restoration, preservation, conversion or other types of eligible work
- who owns the building or artefact – most non-government grant sources are only available to charitable or not-for-profit organisations
- the status of the building or artefact, and how important the project is to the heritage
- what the project is designed to do – for example, to create housing or a museum
- what the benefits are in terms of access, education, employment or similar, depending on the aims of the grant-making organisation
- who will run the project – for example, a private individual, a government body, or a trust – and on what basis the professionals involved are selected.

All grant-making bodies have their own guidelines but it is possible to identify a number of standard restrictions and conditions.

- Applications should be made well in advance of work starting. It is unlikely that grant will ever be paid retrospectively and conditions will probably be imposed relating to when work can commence and a date for completion. All consents will have to be in place before work starts.
- Grant will usually be paid on satisfactory completion of the work, but on larger projects stage payments are usually made but there will be retentions, probably of ten per cent. In the event of a building being sold or performance indicators not being met, there may be some claw back.
- If more than one public body is able to give a grant there will be a combined maximum percentage that cannot be exceeded.
- Awards will only be made if the project is viable and, increasingly, projects are assessed on some form of prioritising – how much the building is at risk for example – and means testing can be applied.
- Conditions will be imposed, the most common of which include the need for public access and adequate insurance.

Every project is different but it is possible to illustrate some of what has been said using a current project of Tayside Building Preservation Trust. The Trust is a company limited by guarantee with charitable status which exists to restore and return to productive use buildings of architectural or historic interest which are at risk in Tayside.

Recommended Reading

Funds for Historic Buildings in England and Wales: A Directory of Sources, The Architectural Heritage Fund, London 1998

Pearce, David, *Conservation Today*, Routledge, London 1989

Pearce, Michael, *Conservation A Credit Account*, Save Britain's Heritage, London 1988

Sources of Financial Help for Scotland's Historic Buildings, Scottish Civic Trust 2003

Weir, Hilary, *How to Rescue a Ruin*, Architectural Heritage Fund, London 1997

The main building fronting the High Street, a tenement of c1600. The small red door below the Timpson sign provides access to Gray's Close.

GARDYNE'S LAND

A CASE STUDY

Gardyne's Land is a generic term for three buildings grouped around a courtyard in the centre of Dundee. Two of the buildings face onto the high street – one is a tenement of *c*1640, the other is a Victorian retail outlet of *c*1870 – while to the rear is a merchant's house of *c*1560, whose first owner was John Gardyne, after whom the complex is named. The buildings are listed Category A and are prominent within the city centre which is an 'outstanding' conservation area. The earliest building is one of only a few recognisably ancient urban buildings left in Scotland.

The merchant's house is virtually a unique construction. Its walls are thin, only ten inches in places and, being five storeys high, its structural integrity depends to a very large extent on the use of timber. The older of the high street buildings has a unique timber frame; this building might represent a late crossover to stone from the timber construction common in late medieval Scotland, which possibly survived longer in Dundee because of the shipbuilding tradition and the availability of good timber imported from the Baltic.

Investigation revealed that little remains of Gardyne's pre-18th century interiors, although it does have good authentic work of the early 1720s, and it is highly atmospheric throughout.

An illustration of the front elevation of John Gardyne's House in 1895, printed in A C Lamb's Dundee: its quaint and historic buildings. *This elevation is still complete, and part of it is shown in the next picture.*

When the Trust first became interested in the buildings in 1995, the upper floors had lain empty for approximately 25 years. The ground floor was used as a shop by a company which leased the whole building from the Prudential Assurance Company. After the Trust completed a major feasibility study in 1996, there followed a lengthy period of negotiation with the owners and various authorities before the Trust finally acquired the buildings for £1 in 1999. At the same time an endowment of £270,000 was secured from the tenant to enable urgent works to be undertaken, as the company had failed to maintain the property despite having a full repairing lease. Phase one works were successfully completed by mid 2001.

The major funding bodies such as Historic Scotland and the Heritage Lottery Fund indicated that they could not contribute to the restoration of the property without an appropriate, sustainable use being found for the building, preferably one which embraces issues of community involvement, economic regeneration, education and access to the heritage. Accordingly, it was decided to convert the buildings to a youth hostel which Dundee lacks, and an agreement is in place with the Scottish Youth Hostels Association which will buy the buildings from the Trust once the work is completed. The requirements of the SYHA for staff accommodation, storage and suchlike will ensure that all the spaces within the building, some of which are rather quirky in their layout, will be used.

The total cost of phase one works including build costs and fees was £291,852. In addition to the endowment, against this expenditure, grants totalling £173,000 were obtained from the sources listed below (see table). This sum is now providing development funding for phase two, the total project costs for which are estimated at £3.5 million (see table).

An oblique view of the same elevation today (on the left) from Gray's Close. Here the architecture presents an amazing, powerfully atmospheric contrast with the formality of its High Street frontage.

GRANTS FOR PHASE ONE

Historic Scotland	£62,433
European Union (ERDF)	66,602
Dundee City Council (façade enhancement grant)	36,000
Landfill Tax Credit Scheme	8,200
Total	**£173,235**

PROJECT FUNDING FOR PHASE TWO

Historic Scotland	£500,000
Scottish Enterprise Tayside	240,000
Dundee City Council	15,000
European Union (ERDF)	763,864
Charitable trusts/commercial sponsorship	175,000
Sale of property to Scottish Youth Hostels Association	350,000
BPT volunteers input	84,000
Council in kind contributions	10,500
Heritage Lottery Fund	1,361,093
Total	**£3,499,457**

Some of the total project income is still assumed but it is based on experience from past projects, preliminary discussions with the various funding bodies, and the expectation that some of the bodies having contributed to phase one costs might reasonably be expected to contribute to phase two. Historic Scotland set a ceiling on phase two of £500,000. A loan from the Architectural Heritage Fund of £350,000, which can be paid back once the property is sold to the SYHA, will be used to pump-prime the construction period of phase two. The selling price has been determined by the district valuer. The Trust has set itself the target of raising £175,000 from other trusts and businesses, a figure which represents some five per cent of the total cost. Following some 200 requests for assistance, contributions have been received from 11 charitable trusts and four local businesses.

The ERDF grant has been calculated on the basis of 25 per cent of the complete costs less a number of selected items which are not eligible for their consideration including the selling price. All prices exclude VAT. The scheme includes a manager's flat, the cost of which can be zero-rated, beyond which the Trust has 'opted to tax'. This means that it will claim back VAT on construction costs and pass the VAT liability on to the SYHA which will have to pay VAT on the purchase price of £350,000. This represents a considerable saving for the Trust and a considerable loss to Customs & Excise.

The Trust will be the project manager for the second phase of the project. The Trust does not exist to make a profit, but needs to meet various running costs and, generally, hopes to realise some capital from its work which it can then roll forward into future projects. The cost of volunteer input is accepted by the Heritage Lottery Fund and has been calculated on the basis of three years of the chief executive's time, at 80 days per year, which takes the project up to the start on site of phase two works: thereafter the project manager's fee will be claimed. An 18-month contract period is due to commence January 2004.

NEIL GRIEVE is a former local authority conservation officer who now lectures in Dundee University's School of Town and Regional Planning where he is year tutor to the school's long running postgraduate course on European Urban Conservation. The course enjoys a strong relationship with Tayside Building Preservation Trust of which he is chief executive. He is a member of the IHBC and the RTPI, and a Fellow of the Society of Antiquaries of Scotland.

A glimpse of the building's timber frame. This very rare survivor has, so far, proved impossible to date.

A detail of the panelled room. Once restored this will be used as an exhibition room, open to the public, and accommodating everything learned about the building from the project. Public access is a condition of most grants.

SURVEYORS

▶ APPLIED SURVEYING AND DESIGN GROUP

17 Barbourne Road, Worcestershire WR1 1RS
Tel 01905 619458 Fax 01905 731201
E-mail info@asdg.co.uk
Website www.asdg.co.uk

BUILDING DESIGNERS, ENGINEERS AND CHARTERED SURVEYORS: Established 1988. The practice provides professional services to commercial, industrial, agricultural and residential clients, including pre-acquisition surveys and advice, preparation of feasibility studies, measured and building surveys, scheme drawings and specifications for renovation, refurbishment and repair. The practice also undertakes structural analysis, design and engineering, statutory consent applications and grant sourcing, site management and certification. Particular experience with timber framed structures and 20th century conservation in and around Herefordshire, Worcestershire and the Cotswolds. Additional services include party walls, dilapidations and condition surveys, solutions to building defects, hazard and risk assessments and planning supervision.

▶ BAILEY PARTNERSHIP

Amhurst House, 22 London Road, Riverhead, Sevenoaks, Kent TN13 2BW
Tel 01732 455522 Fax 01732 460607
E-mail r.sutch@baileypartnership.co.uk
▶ Plymouth office Tel 01752 229259 Fax 01752 224280

HISTORIC BUILDINGS CONSULTANCY: This specialist consultancy is soundly based upon many years of experience in dealing sympathetically with listed buildings and scheduled ancient monuments. A comprehensive range of professional services is available covering aspects of surveying, repair and conservation, adaptation and management. The practice has experience in dealing with a wide range of building types including: residential, commercial, industrial, ecclesiastical, prisons, palaces, military, farm buildings, museums, standing ruins and archaeological sites. The consultancy is headed by Richard Sutch BSc Dip Bldg Cons MRICS.

▶ BAILY•GARNER

146–148 Eltham Hill, London SE9 5DY
Tel 020 8294 1000 Fax 020 8294 1320
E-mail general@bailygarner.co.uk
Contact Lisa J Brooks BSc (Hons) DipBldgCons ARICS

BUILDING SURVEYORS, ARCHITECTS, QUANTITY SURVEYORS AND PROJECT MANAGERS: Baily•Garner was established in 1976 and has a commendable track record in the care and conservation of historic buildings. As a multi-disciplinary practice they are able to offer clients a comprehensive range of services from the preparation of feasibility reports, surveys, detailed design and specification, contract administration, through to party wall matters, listed building consent and planning applications. Their multi-disciplinary approach enables them to maintain a much closer control of projects both in terms of adherence to programme and cost whilst providing clients with a 'one stop' service.

▶ BOSENCE & CO

Oxenham Farm, Sigford, Newton Abbot, Devon TQ12 6LF
Tel/Fax 01626 821609
Contact Oliver Bosence MA DipBldgCons(RICS)

CONSERVATION CONSULTANT AND PROJECT MANAGER: Extensive experience with traditional buildings in the South West, especially masonry repair, slatework and lime renders; vernacular to polite; drainage to decoration; practical guidance on specifications, programming, project and site management; defects analysis and reports; lectures and demonstrations of repair techniques and conservation craft skills.

▶ BYROM CLARK ROBERTS

117 Portland Street, Manchester M1 6EH
Tel 0161 236 9601 Fax 0161 236 8675
▶ Jubilee House, West Bar Green, Sheffield S1 2BT
Tel 0114 275 7879 Fax 0114 272 8954

ARCHITECTS, SURVEYORS AND ENGINEERS: *See also: profile entry in Architects section, page 22.*

▶ COURT DESIGN AND CONSERVATION

The Coach House, Lightcliffe, Private Road, Staplegrove, Taunton, Somerset TA2 6AJ
Tel 01823 272155 Fax 01823 272197
E-mail brdcourt@globalnet.co.uk
Website www.courtdesign.co.uk

HISTORIC BUILDING CONSERVATION CONSULTANCY: A conservation-based philosophy is adopted with a sympathetic approach to the care and repair of historic buildings, aided by the use of traditional working methods and materials. Sensitive design for alterations and refurbishment, including new build where appropriate is also undertaken. A full range of professional services is provided including pre-purchase surveys, repair schedules, preparation of measured surveys, construction drawings, specification and monitoring of building works, obtaining listed building, planning and building control approvals. The principal, Andrew Hayes, is a chartered building surveyor, holds the RICS postgraduate diploma in building conservation and is a member of SPAB.

▶ CUBE PROPERTY SERVICES LTD

Unit 1, Greenwood Court, Taylor Business Park, Warrington WA3 6DD
Tel 01925 767713 Fax 01925 768819
Website www.cubepsl.co.uk

HISTORIC BUILDING CONSULTANTS: Cube is a multi-disciplinary surveying practice, which offers advice on all aspects of historic and listed buildings including maintenance, refurbishment, extension and adaptation. Working nationally, services include: feasibility studies; condition surveys; quinquennial surveys; cost consultancy; planning consultancy; detailed building design; contract administration; project management and planning supervision. The historic building consultancy is led by Stuart Prescott BSc DipBldgCons IHBC MRICS, who has experience working on world heritage sites, scheduled ancient monuments and Grade I listed buildings.

▶ DANIELLS HARRISON CHARTERED SURVEYORS

Post Office House, Stockton, Warminster, Wiltshire BA12 0SE
Tel 01985 850067 Fax 01985 850082
▶ Also at Poole, Fareham and Newport, Isle of Wight
E-mail buildings@dhcs.co.uk
Website www.dhcs.co.uk

CHARTERED BUILDING SURVEYORS AND BUILDING HISTORIANS: Multi-discipline surveying consultancy working across England and Wales for commercial, private and public sector clients. They specialise in historic building conservation and restoration, together with interpretative surveys, both above and below ground archaeological investigations, measured and record surveys, analysis and background historical research. New complementary designs for all types of work from internal refurbishments to new extensions or conversions. Recent projects include fire and structural damage restoration, historic building conversions, archaeological recording and building dismantling, pre design stage integrated standing structure analysis, town centre and group building surveys, historic environmental impact studies, project management and historic property management.

▶ ALAN DICKINSON, MRICS

1 The Grove, Rye, East Sussex TN31 7ND
Tel 01797 225139 Fax 01797 227956
Website www.alandickinson.com

CHARTERED BUILDING SURVEYOR, HISTORIC BUILDINGS CONSULTANT: Sussex and Kent based sole practitioner specialising in surveys, restoration and alterations to period buildings backed by an understanding of historical development. Measured surveys, archival studies and archaeological analysis are also offered for planning and research purposes.

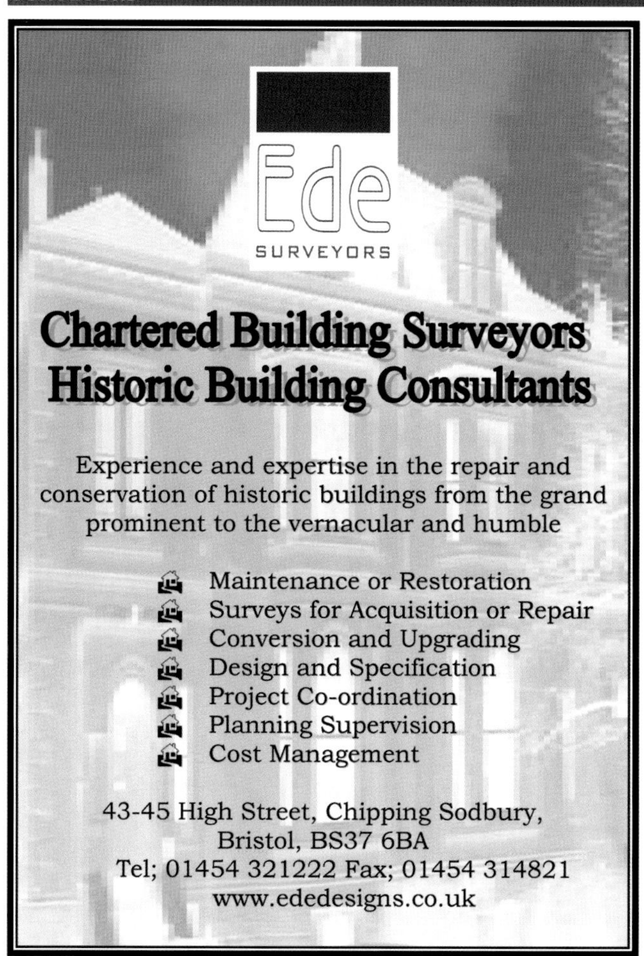

Ede
SURVEYORS
Chartered Building Surveyors
Historic Building Consultants
Experience and expertise in the repair and conservation of historic buildings from the grand prominent to the vernacular and humble
Maintenance or Restoration
Surveys for Acquisition or Repair
Conversion and Upgrading
Design and Specification
Project Co-ordination
Planning Supervision
Cost Management
43-45 High Street, Chipping Sodbury,
Bristol, BS37 6BA
Tel; 01454 321222 Fax; 01454 314821
www.ededesigns.co.uk

GRIFF DAVIES
ARCHITECTURAL DESIGN
AND CONSERVATION LTD

Llyshendy New Quay, Ceredigion SA45 9PS
Tel 01545 560261 Fax 01545 561348
Email enquiries@griffdavies.co.uk
http://www.griffdavies.co.uk

A PERSONAL SERVICE SUPPORTED BY A MULTI DISCIPLINARY PROFESSIONAL TEAM

CONSERVATION OF HISTORIC AND LISTED BUILDINGS, CHANGE OF USE, EXTENSIONS, PLANNING CONSULTATIONS AND APPLICATIONS FOR GRANT ASSISTANCE

DESIGN OF NEW BUILDINGS IN ARCHITECTURALLY SENSITIVE LOCATIONS

HISTORICAL RESEARCH AND AUTHENTICITY CARRIED OUT

Griffith Morgan Davies Dip.S.CEM, Dip Bldg Cons (RICS) MRICS MRTPI, IHBC
Chartered Building Surveyor
Chartered Town Planner
Architectural Technologist & Historic Building Conservator

MEMBER OF THE RICS BUILDING CONSERVATION FORUM

BUILDING CONSERVATION WALES
CADWRAETH ADEILAD CYMRU

▶ DRIVERS JONAS

6 Grosvenor Street, London W1K 4DJ
Tel 020 7896 8339 Fax 020 7896 7902

CHARTERED SURVEYORS: Drivers Jonas is a multi-disciplined firm covering all aspects of property consultancy. Their strong Building Unit includes specialists in conservation, led by Peter Bickerstaff BSc DipBldgCons FRICS (RICS Accredited). Major projects include the refurbishment of North Mymms Park, work and surveys at Hampton Court Palace and the re-presentation of John Wesley's House as a museum. They specialise in planned maintenance programming for large complexes of historic buildings such as Winchester and Atlantic Colleges, Westminster Cathedral; and in fire insurance assessments of historic buildings including Leeds Castle, Heveningham Hall and Tonbridge School. Other services include: dilapidations, party walls and rights of light, planning consultancy and property management.

▶ GIBBON, LAWSON, McKEE LIMITED

41A Thistle Street Lane South West
Edinburgh EH2 1EW
Tel 0131 225 4235 Fax 0131 220 0499
E-mail dg@glmglm.co.uk

BUILDING SURVEYORS: Every year the firm undertakes condition surveys of a significant proportion of the major properties coming onto the market in Scotland and is asked, particularly, to investigate and provide costed recommendations where the condition of the property is of special significance or difficult to assess. The firm also undertakes building conservation and refurbishment projects. *See also: profile entries in Architects section, page 24 and Fire Protection section, page 188.*

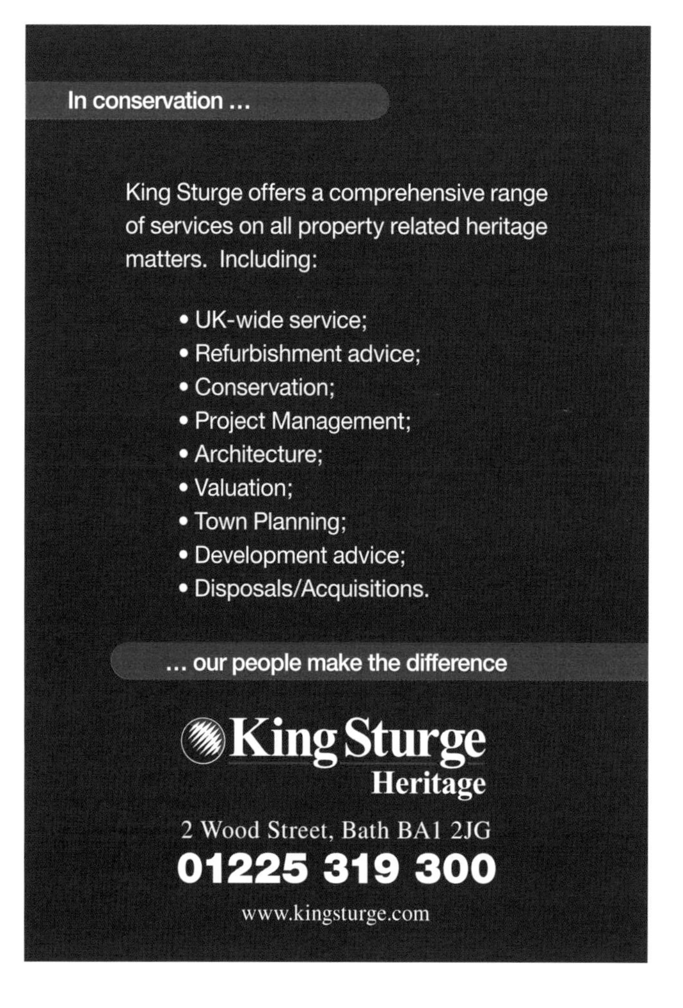

SURVEYORS

▶ HALL & ENSOM, CHARTERED BUILDING SURVEYORS

3 Meadow Court, High Street, Witney, Oxon OX28 6ER
Tel 01993 774995 Fax 01993 779004
E-mail mail@hallandensom.co.uk
Website www.hallandensom.co.uk

CHARTERED BUILDING SURVEYORS, ARCHITECTURAL DESIGN SERVICES: Listed buildings are a permanent feature of the practice workload and the directors, Martin Hall and David Ensom, have over 40 years experience in building surveying, much of it in relation to historic buildings. Hall & Ensom's services include surveys, maintenance planning, repairs, alterations and major refurbishment projects. The practice has worked on all grades of protected buildings, from simple repairs to £1 million refurbishment projects, and can provide a service tailored to your needs including quantity surveying and services engineering. Hall & Ensom is a small practice providing a personal service throughout the Cotswolds and Thames Valley.

▶ HODKINSON MALLINSON LTD

1 Derby Place, Hoole, Chester CH2 3NP
Tel 01244 329505 Fax 01244 312403
E-mail surveyors@hodmal.co.uk

CHARTERED BUILDING SURVEYORS: The practice has expertise in the conservation and restoration of historic buildings including designs for contemporary structures and for appropriate forms of repair with minimal intervention. All aspects are dealt with, including the preparation of detailed condition surveys, architectural drawings and project administration.

▶ PHOENIX BEARD

The Exchange, 19 Newhall Street, Birmingham B3 3PJ
Tel 0121 200 4500 Fax 0121 200 4599
E-mail richard.budd@phoenix-beard.co.uk
Website www.phoenix-beard.co.uk

BUILDING SURVEYING, BUILDING CONSERVATION, PROJECT MANAGEMENT AND PLANNING SUPERVISORS: Phoenix Beard is a nation-wide building consultancy with building conservation experience. The firm can survey, provide cost-effective technical advice, liase with other professionals and provide project management services and supervise repair, refurbishment and alteration work to historic buildings with in-house architectural support. Phoenix Beard is an approved quinquennial surveyor to the Diocese of Birmingham. Advice on grants and alternative methods of funding is available.

▶ RICKARDS CONSERVATION

105 St John's Hill, Sevenoaks, Kent TN13 3PE
Tel 01732 741677 Fax 01732 740149

HISTORIC BUILDINGS CONSERVATION CONSULTANCY: Rickards Conservation is an independent professional building conservation practice providing a comprehensive range of services to care for historic and listed buildings, including churches and ruins. Specialists in structural and condition surveys, defects analysis and repair. Stephen Rickards GradDiplCons(AA) FRICS IHBC ARPS is a Chartered Building Surveyor, Accredited by the RICS for conservation work, and holds the Architectural Association postgraduate Diploma in Building Conservation. He was responsible for conservation work praised by the Civic Trust in Chatham Historic Dockyard. Clients include the SPAB, National Trust and PCCs. Conservation Award winner in 1999.

▶ TFT CULTURAL HERITAGE

211 Piccadilly, London W1J 9HF
Tel 020 7917 9590 Fax 020 7917 9591
E-mail sbond@tftheritage.com
Website www.tftheritage.com

HERITAGE CONSULTANCY: TFT Cultural Heritage is a specialist division of Tuffin Ferraby & Taylor, working throughout the UK. Its services include: condition surveys, sensitive repair, strategic management, conservation planning, and development advice and master planning in historic areas. Its canvas is broad and diverse: city streets, towns and villages; World Heritage Sites and conservation areas; individual buildings, archaeological sites and landed estates; parkland and landscapes of outstanding interest. Such spaces and places are our legacy from the past; yet, they must work to survive. The practice promotes solutions that will respect their value and significance, whilst enriching their contribution to society.

▶ WCP (The Whitworth Co-Partnership with Boniface Associates)

Unit 9, Hastingwood Business Centre, Hastingwood Road, Hastingwood, Essex CM17 9GC
Tel 01279 421500 Fax 01279 421509
E-mail boniface@ukgateway.net
Website www.wcp-surveyors.com
Contact Stephen Boniface

HISTORIC BUILDINGS CONSULTANCY: *See also: profile entry in Architects section, page 31.*

▶ W R DUNN & CO

27 Front Street, Acomb, York YO24 3BW
Tel 01904 784421 Fax 01904 784679

CHARTERED BUILDING SURVEYORS, ARCHITECTS AND HISTORIC BUILDING CONSULTANTS: Established in 1986, W R Dunn & Co undertake commissions throughout the UK on behalf of English Heritage, the Home Office, the MoD, the Church of England Church Commissioners, the Archbishop of York, Harewood House Trust Ltd, several local authorities and councils and many private and commercial clients. The practice principal is an RICS Accredited Surveyor in Building Conservation and is included on the GHBAU Consultants Register. Past schemes include re-roofing a Grade I listed stately home, repairs/alterations to a Grade I listed stable block and repairs to castles, lodges, halls, monuments, walled gardens, dwellings and other structures. The practice also provides project management services plus health and safety advice.

▶ WARD & DALE SMITH, CHARTERED BUILDING SURVEYORS

The Walker Hall, Market Square, Evesham, Worcs WR11 4RW
Tel 01386 446623 Fax 01386 48215
E-mail wds@ricsonline.org

HISTORIC BUILDING CONSULTANCY INCLUDING CONDITION SURVEYS: *See also: profile entry in Damp & Timber Decay section, page 182.*

▶ WATKINSON & COSGRAVE

Fleet House, 62 Highgate Road, London NW5 1PA
Tel 020 7485 6016 Fax 020 7284 4058
E-mail info@watcos.co.uk

CHARTERED BUILDING SURVEYORS AND STRUCTURAL ENGINEERS: Watkinson & Cosgrave has 40 years experience of repairing and restoring buildings, including initial appraisals, structural surveys, feasibility studies, design, specification, supervision of works and contract management.

▶ WATSON BERTRAM & FELL

5 Gay Street, Bath, Somerset BA1 2PH
Tel 01225 337273 Fax 01225 448537
E-mail mail@wbf-bath.co.uk

ARCHITECTS AND SURVEYORS: *See also: profile entry in Architects section, page 31.*

STUCCO REPAIRS AT THE CHURCH OF ST JAMES THE LESS, ST PETER'S PORT, GUERNSEY
by Nimbus Conservation Limited and Ingram Consultancy Limited
Photograph by kind permission of the States of Guernsey Board of Administration

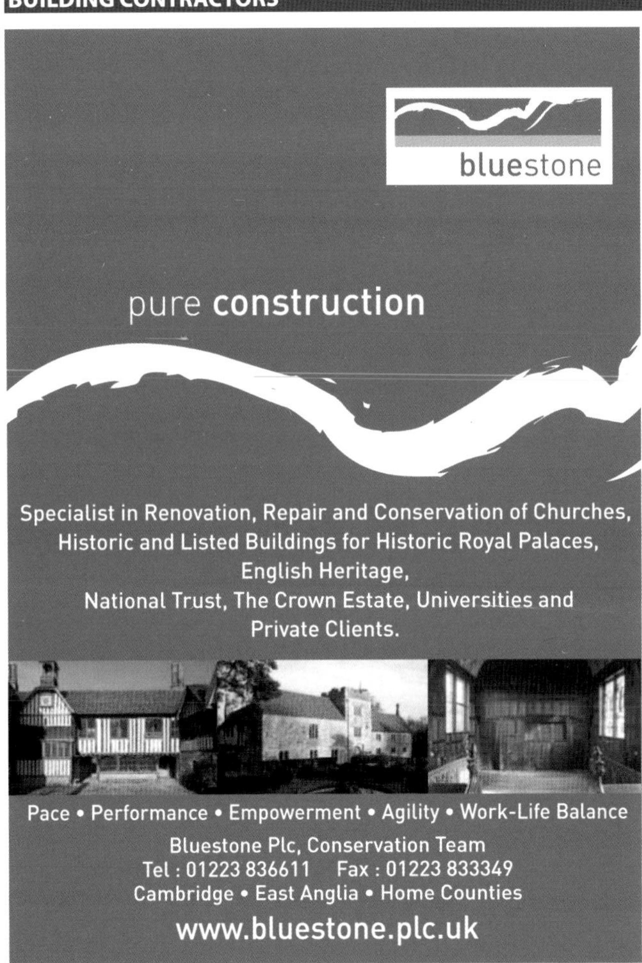

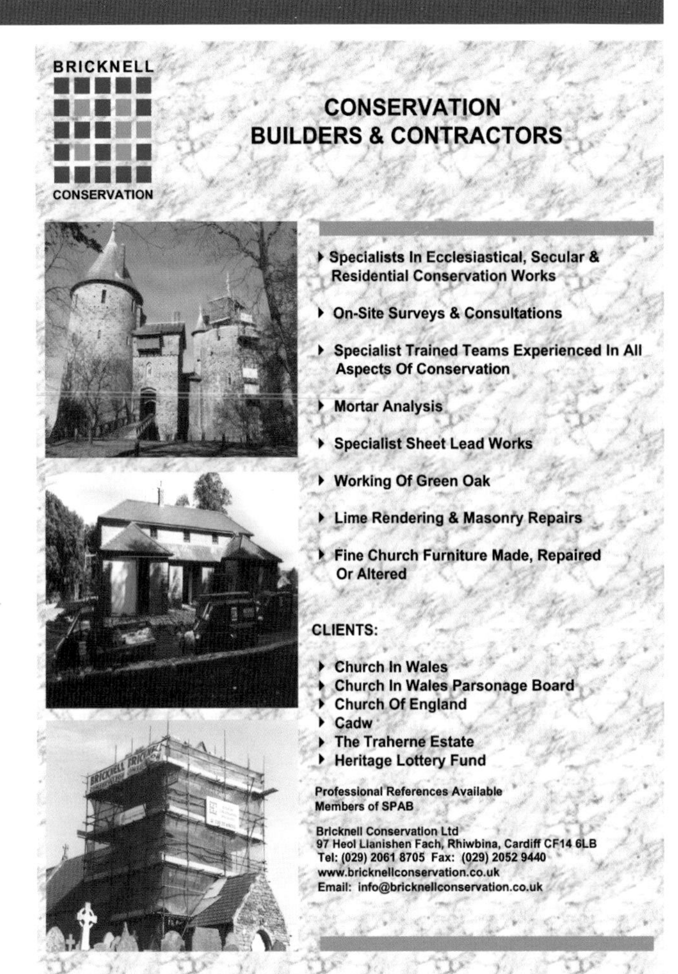

Chester

CHESTER MASONRY GROUP

SPECIALISTS IN THE RESTORATION AND CONSERVATION OF HISTORIC BUILDINGS, CASTLES AND CHURCHES NATIONWIDE

Acting as Principal Contractor backed up by our own in-house skills including:
STONEWORK
ROOFING
LEADWORK
JOINERY

Hawarden Industrial Park
Manor Lane
Hawarden
Nr Chester
CH5 3PZ

Tel 01244 536132 Fax 01244 520217
E-mail admin@chestermasonry.co.uk
Website www.chestermasonry.co.uk

Corbel

Conservation, Restoration, Repair, Alteration

Your opportunity to access specialist conservators who have cared for the country's finest cathedrals, monuments and historic buildings.

Clear and knowledgeable advice from people you'll find easy to work with. You'll get the results you want, in the best way for your property.

Complete range of services: From initial survey, report and then management of your project, we provide all the skilled craftsmen you'll need.

Lime Plaster, Renders.
Stone-work & cleaning.
Joinery; Roof-work; Lead.
Timber & Damp treatments.
Rope access; reports.
Interior décor.
Gilding & Iron-work.
Iron detection.
Gardens.

Corbel Conservation Ltd
176 Greenway Road, Taunton,
Somerset TA2 6LH
Tel/Fax: 01823 332766
Mobile: 07957 645577
Email: corbel.info@virgin.net

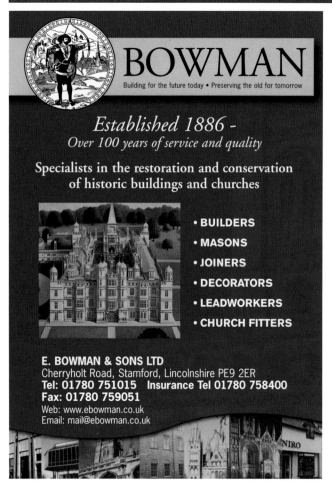

▶ BETWEEN TIME LTD

Bachelors Hall, High Street, Stanstead Abbotts, Nr Ware, Herts SG12 8AB
Tel 01920 877822 Fax 01920 877933 Contact John Lloyd

BUILDING CONSERVATION CONTRACTORS: Specialising in the traditional repair of listed buildings and period properties throughout Hertfordshire, North London and West Essex. The teams of skilled craftsmen work with lime renders, mortars, plasters and limewashes, renewing oak timber framing, architectural joinery, brickwork, leadwork and old tile roofing. Great care is taken to match original materials, and use traditional methods where appropriate. Total project management up to £500,000.

▶ BLUESTONE PLC

Conservation Team, Babraham Road, Sawston, Cambridge, Cambs CB2 4LJ
Tel 01223 836611 Fax 01223 833349
E-mail cambridge@bluestone.plc.uk
Website www.bluestone.plc.uk

CONSTRUCTION SPECIALIST – CONSERVATION, RESTORATION, RENOVATION: Bluestone has a reputation for sympathetic work in repair, reinstatement, conservation and refurbishment of historic and listed buildings and for new-build in classical styles, in an area encompassing Cambridge, East Anglia and the Home Counties within the contract value range of £10,000 to £10 million. Clients include the National Trust, English Heritage, Crown Estate Commissioners, Historic Royal Palaces, The Cadogan Estate, church PCCs and dioceses, local authorities, county councils, universities, developers and private clients. Directly employed labour skills and the building services company give Bluestone the flexibility and expertise to adapt to the particular demands of each project.
See also: display entry in this section, page 59.

▶ BOSHERS (CHOLSEY) LTD

6 Reading Road, Cholsey, Wallingford, Oxon OX10 9HN
Tel 01491 651242 Fax 01491 651800 Contact CWL Bosher

BUILDING CONTRACTORS: This family company has been working on historic buildings in Berkshire, Oxfordshire and neighbouring counties for over 175 years and employs many of its own tradesmen and apprentices. Boshers actively seeks to be involved in work of a high quality and of a wide variety, especially to listed and other buildings. Various works have won awards. They have their own extensive joinery and timber milling facilities and are also able to provide specialist plasterwork, decoration and leadwork to the highest standards.

▶ BURLEIGH STONE CLEANING & RESTORATION CO LTD

The Old Stables, 56 Balliol Road, Bootle, Merseyside L20 7EJ
Tel 0151 922 3366 Fax 0151 922 3377
E-mail info@burleighstone.co.uk
Website www.burleighstone.co.uk

BUILDING CONTRACTORS: A comprehensive high quality service in historical and listed buildings, archaeological contracting and advice.
See also: display entry in Masonry Cleaning section, page 166.

▶ BURROWS DAVIES LIMITED

The Stoneyard, West End, Strensall, York YO32 5WH
Tel 01904 491849 Fax 01904 491910

MASONRY AND RESTORATION SPECIALISTS: *See also: profile entry in Stone section, page 102.*

▶ BUSBY'S BUILDERS

Buzwood, Basingstoke Road, Old Alresford, Hants SO24 9DL
Tel/Fax 01962 732076

SPECIALIST BUILDING CONTRACTORS: A small family firm with experienced craftsman used to working on older properties and listed buildings including churches. Recent contracts include the complete renovation and refurbishment of a thatched timber barn at Armsworth; extension and renovations to an 1867 rendered lodge; re-instatement of three main period rooms at Tichborne House following dry rot eradication; a new cottage in the style of, and together with, the re-building of barns and garages at Cheriton. Busby's works within a 25 mile radius of Old Alresford.

▶ C J ELLMORE & CO LTD

Henshaw Works, Henshaw Lane, Yeadon, Leeds LS19 7RZ
Tel 0113 250 2881 Fax 0113 239 1227
E-mail mail@ellmore.co.uk
Website www.ellmore.co.uk

BUILDING CONTRACTORS: Ellmore Construction is a family run business, established in 1972. They have a well equipped joiners shop capable of handling large and specialist joinery contracts supported by a fully skilled site workforce, including stone masons and carpenters. The company undertakes refurbishment and restoration of historic buildings including churches, cathedrals, town halls, monuments, retail and financial premises in addition to new-build and maintenance work throughout the North of England. The company is interested in contracts up to the value of £2 million. For more information please contact Steve Ellmore.

▶ C R CRANE & SON LTD

Manor Farm, Main Road, Nether Broughton, Leics LE14 3HB
Tel 01664 823366 Fax 01664 823534
Website www.crcrane.co.uk

SPECIALIST BUILDING AND JOINERY CONTRACTORS: Chartered Builders established 1910 by the current Managing Director's grandfather. Winner of conservation awards, specialising in traditional repair works using SPAB and English Heritage methods to churches, cathedrals, barns, follies – they undertake contracts up to £2 million. Their apprentice trained craftsmen are experienced in timber framing, leadwork, masonry, ironworks, brickwork, lime mortars and plasters, backed up by CIOB qualified staff. Consultancy service for historical research, structural surveys, planning, feasibility studies and cost/value analysis. Their in-house joinery works produces traditional windows, doors, panelling, staircases and ecclesiastical joinery.

BUILDING CONTRACTORS

▶ CARREK LIMITED

Mason's Yard, Wells Cathedral, Wells, Somerset BA5 2PA
Tel 01749 689000 Fax 01749 689089
Website www.carrek.co.uk

HISTORIC BUILDING REPAIR COMPANY: Carrek provides a full range of skills and expertise for the conservation and repair of historic buildings, monuments and sculpture. Specialist trades include; stone and plaster conservation, sensitive cleaning including, JOS and DOFF systems, lime rendering, plain and decorative plastering, stonemasonry and carving, carpentry and traditional joinery and leadwork. Carrek will assist with the preparation of reports and specifications and act as consultants for the analysis of historic mortars, plasters and paints. Clients include Churches Conservation Trust, National Trust, English Heritage and numerous private individuals. Recent or current contracts include: St Pauls, Portland Square, Bristol; St Catherine's Chapel, Abbotsbury; Pendennis Castle, Falmouth; Marlborough Mound; Godolphin House, Cornwall.

▶ CHALK DOWN LIME LTD

102 Fairlight Road, Hastings, East Sussex TN35 5EL
Tel 01424 443301 Fax 01580 830096 Mobile 0771 873 8708
E-mail chalkdownlime@supanet.com

MAINTENANCE AND REPAIR OF HISTORIC BUILDINGS: *See also: profile entry in the Mortars & Renders section, page 172.*

▶ CORBEL CONSERVATION LTD

176 Greenway Road, Taunton, Somerset TA2 6LH
Tel/Fax 01823 332766 Mobile 07957 645577
E-mail corbel.info@virgin.net

BUILDING CONTRACTORS: *See also display entry in this section, page 59.*

▶ D S SHERIDAN BUILDING SERVICES

'Jode' Lodge Road, Messing, Colchester, Essex CO5 9TU
Tel 01621 816792 Fax 01621 816093 Mobile 07801 505418
E-mail dssbuild@aol.com

BUILDING CONTRACTORS: A small family firm established 25 years. Experienced in conservation and renovation of older properties and listed buildings, and restoring timber framed houses and roofs, building sympathetic extensions and loft conversions, tiling and slating. Recent projects include the restoration of listed Oveshot Mill at Colne Engaine. The company also undertakes smaller works to houses and cottages in local villages within a 25-mile radius.

▶ DAVID BALL RESTORATION (LONDON) LIMITED

104A Consort Road, London SE15 2PR
Tel 020 7277 7775 Fax 020 7635 0556 E-mail mail@dbr.uk.com

SPECIALIST BUILDING CONTRACTORS: *See also: display entry and profile entry in Stone section, page 103.*

▶ DUNNE AND CO LTD

Ashbrooke, Chalkhouse Green, near Reading, Berks RG4 9AN
Tel 0118 972 2364 Fax 0118 972 1120
E-mail dunneandco@aol.com

BUILDING AND RESTORATION: A long established company employing a large team of specialist crafts people including masons, joiners, roofers and decorators specialising in the restoration and renovation of period and listed buildings using traditional materials. Mainly working in Berkshire and Oxfordshire, Dunne and Co has carried out projects in France and The Channel Islands. Excellent references available.

▶ GALLET CONSTRUCTION LTD

Unit B2, Boston Industrial Centre, Norfolk Street, Boston, Lincs PE21 9HG
Tel 01205 358807 Fax 01205 361275
E-mail mail@galletconstruction.co.uk

CONSERVATION AND GENERAL BUILDING WORKS: The three directors have over 70 years practical building industry experience between them, managing and working hands-on on many interesting and challenging conservation and restoration projects, exterior and interior. Gallet Construction works mainly with listed, vernacular and ecclesiastical buildings, including sympathetic alterations. All employees hold CITB certification at the appropriate levels.

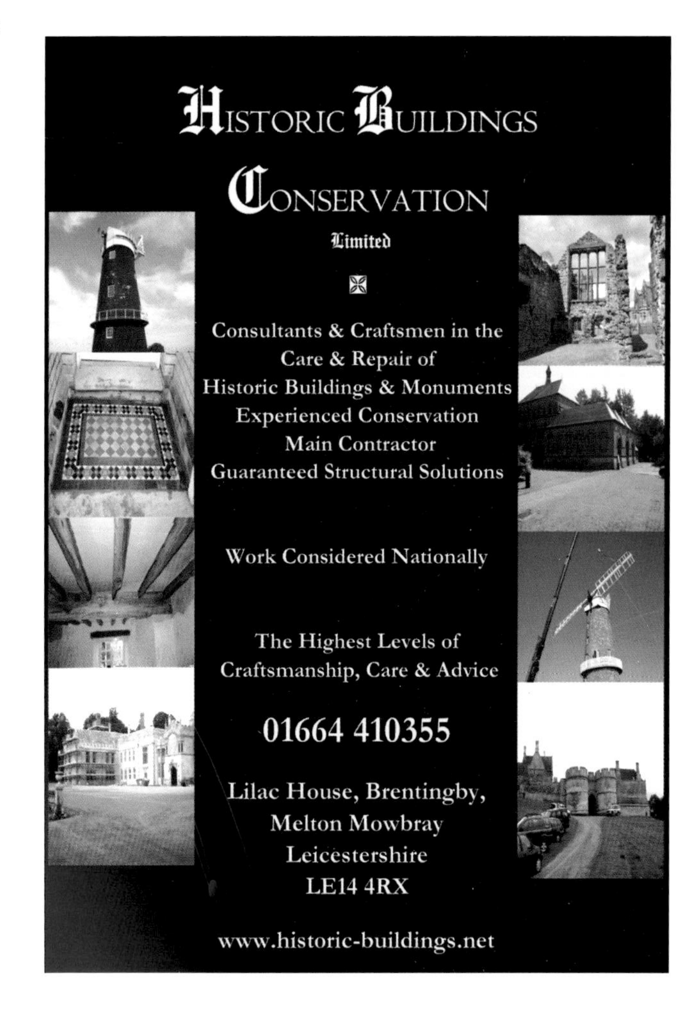

Historic Buildings
Conservation
Limited
Consultants & Craftsmen in the
Care & Repair of
Historic Buildings & Monuments
Experienced Conservation
Main Contractor
Guaranteed Structural Solutions
Work Considered Nationally
The Highest Levels of
Craftsmanship, Care & Advice
01664 410355
Lilac House, Brentingby,
Melton Mowbray
Leicestershire
LE14 4RX
www.historic-buildings.net

Specialists in all renovation, maintenance and repair work of listed buildings including:
• New oak frames
• Green oak timber frame repairs
• Lime rendering
• Barn conversions
• Lathes
• Purpose made doors/sash windows
• Brick repairs
• Lime repointing
building for conservation
jgmatthews Ltd
Home Farm, Hatfield Park, Hatfield, Herts AL9 5NH
Tel: 01707 262351
Fax: 01707 264813
www.building-conservation.co.uk
in association with
Hatfield House Oak

▶ HALL CONSTRUCTION

Unit 6, Tannery Yard, Witney Street, Burford, Oxfordshire OX18 4DQ
Tel 01993 822110 Fax 01993 823880
Website www.hallconstruction.co.uk

CHARTERED BUILDING COMPANY: *See also: display entry in this section, page 61.*

▶ HASLEMERE BUILDERS LIMITED

Cylinders Lane, Fisher Street, Northchapel, West Sussex GU28 9EL
Tel 01428 707282 Fax 01428 708040

BUILDING CONTRACTORS: A small family run business carrying out a variety of work from small repairs to major projects. Relying on traditional methods and high standards of craftsmanship they are able to provide a comprehensive service covering all aspects and trades. Working mainly in West Sussex, Surrey and North East Hampshire they have been involved in the extension, repair, restoration and refurbishment of period, listed and historic buildings. Haslemere Builders has been a member of the CIOB Chartered Building Company Scheme since 1992 and was the main contractor on an 18th century Mill restoration project which was awarded a commendation in the Civic Trust Awards.

▶ HIRST CONSERVATION

Laughton, Sleaford, Lincolnshire NG34 0HE
Tel 01529 497449 Fax 01529 497518
E-mail hirst@hirst-conservation.com
Website www.hirst-conservation.com

SPECIALIST BUILDING AND ART CONSERVATORS: Consultancy and conservation work to painted and applied decoration on plaster, stone, canvas, wood and metal substrates. Restoration and recreation of historic decorative schemes. Also specialist building works including joinery, sculpture, marble, stonework, stone cleaning, stucco, pargetting, wall and floor plasters. Surveys, specifications and analysis services available. Hirst Conservation's policy is to provide a conservation service that is second to none. The company takes great pride in ensuring that it remains at the forefront of contemporary conservation ethics and thinking. The highly professional and dedicated team represents many different conservation skills and disciplines, and through its combined knowledge and experience is constantly striving to enhance current and develop future conservation practices. *See also: display entry on the inside front cover.*

▶ HOLLOWAY WHITE ALLOM

43 South Audley Street, Grosvenor Square, London W1K 2PU
Tel 020 7499 3962 Fax 020 7629 1571
E-mail derek.a.ednie@hwaltd.co.uk

HIGH QUALITY BUILDING CONTRACTORS FOR FIRST CLASS REFURBISHMENT: Benefit from Holloway White Allom's more than 100 directly employed skilled craftsmen and women, and ten apprentices, all trained in traditional trades and crafts for top quality refurbishment and restoration, fine decorative finishes and meticulous attention to detail. Their high reputation for excellent work on listed buildings, historic monuments, country estates and prestigious London homes, means that exacting standards will be met. Based in Mayfair since 1902, the company's century of experience and delighted client base enables the client to choose confidently.

▶ J & W KIRBY

37, Slack Lane, Crofton, Wakefield, W Yorks WF4 1HH
Tel/Fax 01924 862713

TIMBER FRAME CONSERVATION: *See also: profile entry in Timber Frame Builders section, page 72.*

▶ J G MATTHEWS LIMITED

Home Farm, Hatfield Park, Hatfield, Hertfordshire AL9 5NH
Tel 01707 262351 Fax 01707 264813
Website info@buildingconservation.co.uk

BUILDING CONTRACTORS: J G Matthews, managed by a chartered surveyor, has over 25 years experience with old buildings. The qualified and experienced workforce carries out repairs and maintenance to listed buildings and timber frames and artfully crafts barn conversions. The company works with lime render and skilfully carries out brick repairs using lime mortars. J G Matthews also specialises in purpose made joinery including doors, windows and sash repairs. Members of Hertfordshire Building Preservation Trust and Society for the Protection of Ancient Buildings. Clients include Gascoyne Cecil Estates (Hatfield House), Diocese of St Albans, Hertfordshire BPT. *See also: display entry on this page.*

▶ JACKSURE CONSTRUCTION

5 Trowscoed Avenue, Leckhampton, Cheltenham, Gloucestershire GL53 7BP
Tel/Fax 01242 221742 Mobile 07932 746579

TRADITIONAL BUILDING CONTRACTORS: Jacksure Construction is an owner-operated business based in Cheltenham which offers high quality building contracting services using traditional methods and materials where possible. Refurbishment of historic buildings is a speciality, with an emphasis on working with lime mortars and lime renders. Sympathetic new-build work is also undertaken and especially new-build stone masonry. Jacksure is experienced in nearly all types of traditional stonewalling including dry stone walling, and also repointing existing walls using matching traditional materials. Roof construction and tiling also undertaken.

▶ KEN BIGGS CONTRACTORS LIMITED

High Street, High Littleton, Bristol BS39 6HP
Tel 01761 470743 Fax 01761 471428
BUILDING CONTRACTORS: *See also: display entry on this page.*

▶ LONGLEY A division of Kier Regional Ltd

East Park, Crawley, West Sussex RH10 6AP
Tel 01273 561212 Fax 01273 564333
BUILDING RESTORATION AND CONSERVATION IN THE CENTRAL HOME COUNTIES: Longley has over 130 years experience working to maintain the nation's heritage buildings. Appropriate skills and technical expertise have been used on projects such as the Hampton Court Palace restoration following the tragic fire, the re-roofing of the National Trust's Abbey at Romsey, Newhaven Fort restoration in conjunction with English Heritage, and the rebuilding of Tonbridge School Chapel after the fire which destroyed the listed structure. Contracts range from £100,000 to £5 million and can benefit from Wallis joinery and stonework as Longley is also part of the Wallis division within the Kier Group. For further information contact Graham Todd, General Manager. *See also: Wallis display entry in this section, page 54.*

▶ MAGENTA BUILDING CONSERVATION LTD

Milton Mills, Milton Abbas, Blandford, Dorset DT11 0BQ
Tel 01258 880016 / 01305 871708 Fax 01258 881544
BUILDING CONTRACTORS AND STONEMASONS: Magenta Building Conservation Limited specialises in the repair of historic and ecclesiastical buildings using traditional building methods and materials. Since the formation of the company in 1995, Magenta has worked on many historic buildings and churches in the South of England. The company works directly for English Heritage and the National Trust and regularly works for a number of church architects. Magenta acts as a main contractor or specialist sub-contractor in the following trades: stonemasonry, brickwork, lime plastering, conservation joinery and heavy oak carpentry, leadwork, millwrights and engineers. IRATA rope access technicians. Magenta also carries out one-off new builds related to traditional construction.

▶ MAYSAND LIMITED

109–111 Windsor Road, Oldham, Lancs OL8 1RH
Tel 0161 628 8888 Fax 0161 627 0996
E-mail sales@maysand.co.uk
MASONRY AND TIMBER CONSERVATION CONTRACTORS: *See also: display entry in Stone section, page 106.*

▶ McCURDY & CO

Manor Farm, Stanford Dingley, Reading, Berkshire RG7 6LS
Tel 01189 744866 Fax 01189 744375
E-mail info@mccurdyco.com
Website www.mccurdyco.com
TIMBER FRAME BUILDERS: *See also: display entry in Timber Frame Builders section, page 73.*

▶ MELCOMBE REGIS CONSTRUCTION LTD

The Old Forge, Wyke Square, Weymouth, Dorset DT4 9XP
Tel/Fax 01305 773239
BUILDING CONTRACTOR AND HIGH QUALITY REFURBISHMENT, REPAIR AND RESTORATION: A competitive and dynamic traditional building contractor with a full range of restoration, repair, refurbishment and new construction services. An experienced and informed management with responsible, responsive construction teams. Melcombe Regis Construction delivers a high level of care and performance recognisable by results. Recent projects include: Whitcombe Manor, refurbishment of 18th century church, restoration of 19th century Open Air Singing Theatre at Larmer Tree Gardens (Rushmore Estates), variety of works to Grade I and Grade II listed buildings, new construction of traditional stonework houses. Contact Melcombe Regis for an instant service from drawings to completion including planning consents, if required. *See also: display entry this section, page 64.*

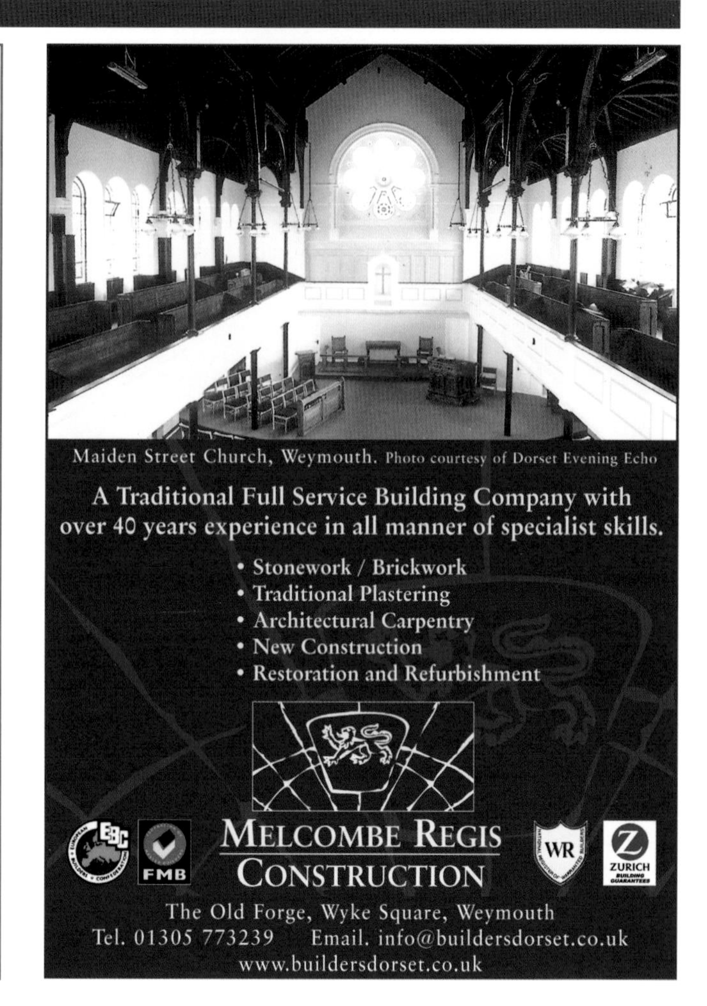

▶ R J SMITH & CO

Manor Court, Bloswood Lane, Whitchurch, Hants RG28 7BN
Tel 01256 892276 Fax 01256 893993
E-mail email@rjsmith.co.uk
Website www.rjsmith.co.uk
SPECIALISTS IN THE CONSERVATION OF HISTORIC BUILDINGS

▶ RICHARD COLES BUILDING & PROJECT MANAGEMENT

The Briary, Plough Lane, Ewhurst, Surrey GU6 7SG
Tel 01483 548856 Fax 01483 548458 Mobile 07801 259949
E-mail richard@rcoles.co.uk
CHARTERED BUILDING COMPANY: Richard Coles BSc MCIOB studied building technology and management at university before embarking on a career in construction. This training led to the formation of his own company in the mid 1980s and corporate membership of the Chartered Institute of Building. The company specialises in conservation and traditional construction projects ranging from £100,000 to £1 million. Its team of loyal and dedicated tradesmen relies on traditional methods and high standards of craftsmanship to provide a comprehensive service. Recent projects include the reconstruction of an antique barn to provide a clubhouse and restaurant facilities for Hurtwood Park Polo Club, and works to Pinkhurst Farm in Surrey, the county where most of the company's work is carried out.

▶ ST BLAISE LTD

Westhill Barn, Evershot, Dorchester, Dorset DT2 0LD
Tel 01935 83662 Fax 01935 83017
E-mail info@stblaise.co.uk

▶ SANDY & CO (CONTRACTORS) LIMITED

Grey Friars Place, Stafford ST16 2SD
Tel 01785 258164 Fax 01785 256526
E-mail info@sandy.co.uk
Website www.sandy.co.uk
HISTORIC BUILDING CONTRACTORS: Established in 1903, Sandy & Co is a well known firm of high quality building contractors which specialises in work to historic buildings and churches. Conservation, restoration and repair services are provided throughout the Midlands and across the United Kingdom for large and interesting projects. Sandy & Co works with top architects, local government and private individuals. Recent projects have included 17/19 High Street, Kinver for which they won the Carpenters' Award, and works to Windsor Castle and the Tower of London. Please ring Peter Godwin to discuss your project requirements.

▶ SIMMONDS OF WROTHAM

The Square, Wrotham, Sevenoaks, Kent TN15 7AH
Tel 01732 883079 Fax 01732 884055
E-mail robert@whsimmonds.co.uk
RESTORATION BUILDING CONTRACTORS AND DEVELOPERS: Simmonds employs a long standing team of craftsmen, experienced in matching traditional methods of construction with current practice. Barn and oast conversions are a speciality. Recent projects include work to a Grade I listed church and several other churches, the complete renovation of a listed 16th century farmhouse, with its listed outbuildings of a thatched barn, oast house and stables. Operating in London, Kent and surrounding areas.

▶ SPLITLATH LTD

Forest Lane, Craswall, Herefordshire HR2 0PL
Tel 01981 510611 Fax 01981 510342
E-mail forestlawn@compuserve.com
OAK FRAME REPAIRS: *See also: display entry in Building Contractors section, page 67.*

BUILDING CONTRACTORS

▶ T J EVERS LTD

New Road, Tiptree, Colchester, Essex CO5 0HQ
Tel 01621 815787 Fax 01621 818085
E-mail office@tjevers.co.uk

BUILDING RESTORATION AND CONSERVATION, SPECIALIST JOINERY MANUFACTURERS: Established in 1918, a traditional contractor offering a quality service in the conservation and restoration of historic and ancient buildings. The high calibre of their craft skills, traditional and modern construction techniques, coupled with quality management skills, they believe enables them to offer a service that is second to none. T J Evers' in-house specialist joinery division produces period and bespoke joinery to the highest standard of craftsmanship for many clients, including English Heritage. Small and major works departments enable them to carry out contracts both large and small.

▶ TIMOTHY WILLIAMS (BUILDERS) LIMITED

St Nicholas Barn, Church Road, Lower Oddington, Moreton in Marsh, Gloucestershire GL56 0XE
Tel 01451 832554 Fax 01451 832257
E-mail twbuild@aol.com
Contact Tim Williams BSc (Hons) ACIOB
▶ 27 Palace Gate, Kensington, London W8 5LS
Tel 020 7590 7588 Contact Neil Garrett ACIOB

SPECIALIST BUILDING AND JOINERY CONTRACTORS: Fourth generation family business specialising in high quality building renovation, refurbishment and the installation of purpose made joinery to domestic period and listed property in the Cotswolds and Central London. Adept at managing all trades in the construction process from lime plasterwork to stone slate roofs. Client list and project portfolio available on request.

▶ TRADITIONAL BUILDINGS LIMITED

Willow House, Lower Road, Woodchurch, Kent TN26 3SQ
Tel 01227 451795 Fax 01227 478797
Mobile 07786 434 008/9

BUILDING CONSERVATION CONTRACTORS: Traditional Buildings seeks to maintain the building craft standards of the past and is staffed by experienced craftsmen dedicated to traditional skills. The company repairs and rebuilds historic brickwork, masonry, timber frames and roofs. Other activities include traditional plastering, renders, stuccos, slating, tiling and lead work. Recent work has included repairs to Grade II* and II listed London churches and repairs to barns for the National Trust and private owners. New work includes the construction of serpentine flint walls, the rebuilding of a brick barn destroyed in gales and a timber frame barn conversion.

▶ TREASURE & SON LTD

Temeside, Ludlow, Shropshire SY8 1JW
Tel 01584 872161 Fax 01584 874876

CONSERVATION AND RESTORATION CONTRACTORS: Established in 1747, Treasure & Son is a family firm specialising in all aspects of building, from work on scheduled monuments to the restoration of Georgian and half-timbered houses. The company employs 75 time-served craftsmen and has worked recently at: Warwick Castle; Church of St Bartholomew, Richards Castle; Ludlow Castle; Knowle Lime Kilns; and many private houses in the West Midlands and border country areas. In 1997, as main contractor, Treasure & Son Ltd won the Building of the Year Award for its work at the Mappa Mundi and Chained Library in Hereford. The company has its own in-house joinery shop which received The Carpenters Award in 1997 for the high quality joinery at the Mappa Mundi building.

▶ STONEGUARD

St Martins House, The Runway, Ruislip, Middlesex HA4 6SG
Tel 0870 241 6366 Fax 020 8839 9988
▶ Bath, Tel 01225 754025
▶ Birmingham, Tel 01543 307298
▶ Manchester, Tel 0161 773 3398
▶ Stirling, Tel 01786 831144
E-mail sales@stoneguard.co.uk
Website www.stoneguard.co.uk

BUILDING RESTORATION AND CONSERVATION: Stoneguard is one of the country's leading refurbishment contractors, operating nationwide from offices in London, Manchester, Bath, Birmingham and Stirling with over 120 operatives and skilled craftsmen. Stoneguard provides a complete masonry contracting service covering all aspects of work from a simple face-lift involving masonry repair and sensitive cleaning, to a complete structural refurbishment.

▶ WALLIS

47 Homesdale Road, Bromley, Kent BR2 9TN
Tel 020 8464 3377 Fax 020 8464 5847

BUILDING CONTRACTORS: Over many years Wallis, with its associated companies Wallis Joinery and Broadmead Cast Stone and now incorporating Longley, has earned an exceptional reputation for the quality and excellence of its restoration and refurbishment work. In recent years Wallis has carried out work in the Foreign and Commonwealth Office (Old Public Offices Whitehall), Marlborough House, Windsor Castle, The National Portrait Gallery, The Natural History Museum, The British Museum, London Oratory, The Guildhall, Somerset House, Ightham Mote, Canada House, Eltham Palace and Hampton Court Palace. In 1991 the Old Public Offices, Whitehall, won the coveted Europa Nostra Award "For the magnificent and meticulous restoration to the original design of one of the finest examples of Victorian architecture in the United Kingdom". As part of the Kier Group, Wallis combines the advantages of a substantial and secure organisation with the skills and resources capable of undertaking the most demanding projects large or small in its chosen field. Wallis undertakes contracts throughout London and the South East of England and has offices in London and Maidstone. *See also: display entry on page 54.*

▶ WATES GROUP

Wates House, Station Approach, Leatherhead, Surrey KT22 7SW
Tel 01372 861000 Fax 01372 861375
Website www.wates.co.uk

RESTORATION AND REFURBISHMENT OF LISTED AND HISTORIC BUILDINGS: *See also: display entry in this section, page 68.*

▶ WILLIAM SAPCOTE & SONS LTD

87 Camden Street, Birmingham, West Midlands B1 3DE
Tel 0121 233 1200 Fax 0121 236 2731
E-mail keith.learoyd@sapcote.co.uk

BUILDING CONTRACTORS, RESTORATION SPECIALISTS: Established in 1853, a family owned business throughout its history, William Sapcote & Sons Ltd has an excellent reputation in Birmingham for high quality building projects. Repair, reinstatement and refurbishment of historic and ecclesiastical buildings has been a specialist arm of the company since the 1950s. Considerable experience at office and site level, together with directly employed craftsmen and apprentices brings repeat business from all aspects of the restoration field, local councils and authorities, English Heritage, National Trust and local churches. Operating in the Midlands the firm specialises in stonework and timber framing, and is an approved JOS and DOFF cleaner and Cintec installer. *See also: display entry in this section, page 67.*

TIMBER FRAME BUILDERS

▶ SELECTED OAK

47 Darwin Drive, Tonbridge, Kent TN10 4SA
Tel 01732 355061 Fax 01732 771610
Website www.selectedoak.co.uk

TIMBER FRAME BUILDERS: Selected Oak has been established for over a decade constructing traditional oak structures. These vary from the trend in oak garages and outbuildings to individually designed structures. Their main strength is to construct with precision large sections of oak, strategically placing them to form trusses, roof structures and many other aspects of the oak building industry. Their reputation is built upon attention to detail and a commitment to delivering excellence throughout a project.

▶ SPLITLATH LTD

Forest Lane, Craswall, Herefordshire HR2 0PL
Tel 01981 510611 Fax 01981 510342
E-mail forestlawn@compuserve.com

OAK FRAME REPAIRS: *See also: display entry in Building Contractors section, page 67.*

▶ TRADITIONAL CARPENTRY & JOINERY

9 The Red House, Main Road, Yockleton, Shrewsbury, Shropshire SY5 9PH
Tel/Fax 01743 821641 Mobile 07971 284555
E-mail traditional_joinery@hotmail.com
Contact Robert, Stephen or Pat Steele

CARPENTRY AND JOINERY: A small family-run business with personal involvement by the owners in all aspects of work undertaken and in quality control. Traditional Carpentry & Joinery specialises in sympathetic restoration and maintenance of listed, historical and half-timbered buildings. Carpentry and purpose-made joinery is expertly crafted in hardwood and softwood.

▶ WEALD AND DOWNLAND OPEN AIR MUSEUM

Singleton, Chichester, West Sussex PO18 0EU
Tel 01243 811363 Fax 01243 811475
E-mail wealddown@mistral.co.uk
Website www.wealddown.co.uk

CONSERVATION SUPPLIES AND SERVICES: The Museum offers various advice and supplies for use in timber frame building and traditional building methods. *See also: profile entry in Courses & Training section, page 226.*

MILLWRIGHTS

▶ DOROTHEA RESTORATIONS LTD

INCORPORATING ERNEST HOLE (ENGINEERS) LTD
Northern office & works: New Road, Whaley Bridge, High Peak, Derbyshire SK23 7JG
Tel 01663 733544 Fax 01663 734521
E-mail north@dorothearestorations.com
Website www.dorothearestorations.com
▶ Southern office & works: Riverside Business Park, St Anne's Road, St Anne's Park, Bristol BS4 4ED
Tel 0117 971 5337 Fax 0117 977 1677
E-mail south@dorothearestorations.com
Website www.dorothearestorations.com

RESTORATION OF ARCHITECTURAL METALWORK, TRADITIONAL MACHINERY AND MILLS: *See also: display entry in Architectural Metalwork section, page 144.*

Brick making by hand at Lambs Bricks and Arches
Photograph by Jonathan Taylor

Chapter 3
STRUCTURE & FABRIC

ROOFING CONTRACTORS				
A C Wallbridge & Co Ltd	sj			84
A V Brown			me	86
Adam Cooper, Master Thatcher		th		81
Aire Valley Roofing	rc sr			82
Andrew Rees		th		81
Anelays - William Anelay Ltd	rc		me	86
Anglia Lead Limited			me	87
Award	rc		me	86
B & S Fowler Master Thatcher		th		81
B J N Roofing Ltd	rc			82
Bardsley & Brown		th		81
Barry Milne Thatcher		th		81
Between Time Ltd			me	58
Bluestone plc	rc			60
Bosence & Co	rc			44
Boshers (Cholsey) Ltd	rc			60
Bradford Roofing Contractors Ltd	rc sr			82
Bryan Hodges	rc sr			82
Busby's Builders	rc			60
C E L Ltd	rc			87
Carrek Limited	rc		me	61
Church Conservation Limited	ax rc sj		me	84
Coyle Timber Products	rc			83
D S Sheridan Building Services	rc			61
E Bowman & Sons Ltd			me	60
E G Swingler & Sons	rc			82
Geoff Thams Master Thatcher		th		81
H & W Sellors Ltd	rc			104
Hall Construction	rc			61
Haslemere Builders Limited	rc			62
Holloway White Allom	rc			62
J G Matthews Ltd	rc			62
John Williams & Co	rc		me	83
Karl Terry Roofing Contractors Ltd	rc		me	83
Linford-Bridgeman Limited	rc			64
M J Read		th		81
Master Thatchers of Oxford		th		81
Mowlem Rattee & Kett	rc			106
Norman Underwood			me	87
Northwest Lead			me	88
Osiris Lead Limited			me	88
R W Armstrong & Sons Limited	rc			66
Rope Tech International Limited	ax sj			84
Russell & Buckingham		th		81
Salmon (Plumbing) Limited	rc		me	88
Simmonds of Wrotham	rc			64
Timothy Williams (Builders) Ltd	sr			65
Wallis	rc			54
Weald & Downland Open Air Museum		th		81
Welsh Heritage Thatching		th		81
Wessex Thatchers	rc	th		81
William Langshaw & Sons	rc			68

KEY

ax aerial access
me metal sheet roofing
rc roofing contractors
sj steeplejacks
sr stone slate roofing
th thatch

ROOFING PRODUCTS			
Aldershaw Handmade Clay Tiles Ltd	rt		85
Alumasc Architectural Rainwater Systems		rd	91
Babylon Tile Works	bt rt		85
Best Demolition	as		155
The Bulmer Brick and Tile Company	rt		118
Bursledon Brickworks Conservation Centre	bt		172
C E L Ltd	rf	ld rd	87
Carpenter Oak & Woodland Ltd	bt os		72
The Cast Iron Company		rd rl	143
Castaway Cast Products and Woodware		rd	142
Cathedral Works Organisation (Chichester) Limited	bt		102
Chalk Down Lime Ltd	bt		172
The Cleft Wood Co	bt		85
Clement Windows Group Ltd		rl	89
The Conservation Rooflight Company		rl	89
The Cornish Lime Company Ltd	bt		174
Coyle Timber Products	bt os		86
Daniel Platt Limited	rf rt		85
E T Clay Products Limited	pt rs rt		118
Eternit Building Materials	rt		86
The Fine Iron Company		wv	145
G & N Marshman	bt		86
Hargreaves Foundry Ltd		rd	91
Harrison Thompson & Co Ltd		rd	92
Heritage Structural Ventilation Ltd	ve		89
Dreadnought Tiles	rt		85
J & J W Longbottom Ltd		rd	92
John Boddy Timber Ltd	os		123
Keymer Tiles Limited	bt rt		85
Marsh Brothers Engineering Services Ltd		rl	146
Mike Wye & Associates	bt		174
Minchinhampton Architectural Salvage	as		155
Northwest Lead		rd	88
Osiris Lead Limited		ld	88
Ransfords Reclaimed Building Supplies	as		155
Saint-Gobain Pipelines Plc		rd	92
Sandtoft Roof Tiles	rf rs rt		86
Smith of Derby Clockmakers		wv	147
Smithbrook Building Products Ltd	rt		86
Solopark Plc	as		156
The Standard Patent Glazing Co Limited		rl	89
Terca	rf rt		119
The Traditional Lime Co	bt		176
Tudor Roof Tile Co Ltd	rt		86
Ty-Mawr Lime Ltd	bt		176
Vale Garden Houses Ltd		rl	136
Walcot Reclamation Ltd	as		157
Weald & Downland Open Air Museum	bt		226
West Meon Pottery		pt	86
Whippletree Hardwoods	bt os		124
William Blyth	rt		86

KEY

bt battens, laths and tile pegs
os oak shingles
rf roof features
rs roofing slates
rt clay roof tiles
sr stone roofing slates
as reclaimed roofing materials
ld lead sheet
pt chimney pots
rd roof drainage
rl roof lights and lantern lights
ve roof vents
wv weathervanes

THATCHERS

❶ ADAM COOPER, MASTER THATCHER
The Grieves Cottage, Pert Farm, Northwaterbridge, Laurencekirk AB30 1QP
Tel/Fax 01674 840 538
New thatch, rethatch and repairs. Free estimates and advice.

❷ ANDREW REES
20 Rowntree Way, Saffron Walden, Essex CB11 4DG
Tel 01799 513743 Mobile 07712 840596
A highly experienced thatcher, using long straw and water reed. Local council approved. Please ring for a free estimate locally.

❸ B & S FOWLER MASTER THATCHERS
54 Westland Road, Faringdon, Oxon SN7 7EY
Tel 01367 242185 Mobile 07770 593681
www.roofthatching.co.uk
Complete rethatches, ridges and patching, insurance work undertaken, free estimates. Established for 25 years. Member of the Oxfordshire, Bucks and Berks Master Thatchers Association.

❹ BARDSLEY & BROWN
1 Marlston Cottages, Marlston, Berkshire RG18 9UN
Tel 01635 201546/01635 255149
High quality craftsmanship from a well established business. All thatching work quoted for. Best thatched house (OBBMTA) 2000 winners

❺ BARRY MILNE THATCHER
15 Fleetwood Crescent, Banks, Southport, Lancs PR9 8HF
Tel 01704 231510
Specialist in combed wheat reed and water reed. Established 25 years. Secretary and Member of North of England and Scotland Master Thatchers Association.

❻ DAVID KENWARD THATCHING
10 Domneva Road, Minster, Thanet, Kent CT12 4DR
Tel 01843 821970 Mobile 07989 688871
Long straw thatcher.

❼ GEOFF THAME MASTER THATCHER
Garden Cottage, Fairacre, Axminster EX13 5BH
Tel 01297 33500
A consortium of master thatchers covering the Devon/Dorset/Somerset area. All are current members of their county master thatchers associations.

❽ M J READ
3 Ridge Cottages, Chilmark, Salisbury, Wiltshire SP3 5BS
Tel 01722 716631
Third generation thatcher in combed wheat reed, long straw and Norfolk reed. Timber repairs undertaken. All work guaranteed. Guild member.

❾ MASTER THATCHERS OF OXFORD
17 The Green, Steventon, Abingdon, Oxon OX13 6RR
Tel/Fax 01235 832286 Mobile 07966 229418
E-mail bodthedog@aol.com
Specialists in both straw and water reed. Only quality nitrate tested materials used. Fire retardant treatments. Warranty underwritten by the NFU Mutual. Trained and experienced staff. Conservation advice.

❿ RUSSELL & BUCKINGHAM
37 Hanborough Close, Eynsham, Witney, Oxford OX29 4NR
Tel 01865 883818 Mobile 0860 919257
Superior work carried out by traditionally trained craftsmen with over 30 years experience. Only quality selected materials used.

⓫ WEALD AND DOWNLAND OPEN AIR MUSEUM
Singleton, Chichester, West Sussex PO18 0EU
Tel 01243 811363 Fax 01243 811475
E-mail wealddown@mistral.co.uk
www.wealddown.co.uk
Specially grown traditional varieties of longstraw and wheat reed supplied, also hazel spars produced on site by our sparmaker in residence.

⓬ WELSH HERITAGE THATCHING
8 Brynbach Road, Brynbamman, Ammanford, Carmarthenshire SA18 1BH
Tel/Fax 01269 824585 Mobile 07703 379917
E-mail craftsman@thatching-uk.com
www.thatching-uk.com
Thatch renovations and repairs, thatched new build, museum thatch maintenance, authentic thatch for film sets, thatched garden buildings. Free advice and estimates.

⓭ WESSEX THATCHERS
2 Trotts Lane, Eling, Totton, Hampshire SO40 4UE
Tel 02380 667637 Mobile 0790 1597814
E-mail thatcher@dial.pipex.com
Complete thatching service. Council approved. Water reed. Wheat reed. Longstraw. Proven quality and reliability. Superior work by experienced craftsmen.

▶ AIRE VALLEY ROOFING

20 Round Hill Avenue, Cottingley, Bingley BD16 1PQ
Tel 01274 568878/01274 560840 Fax 01274 560840

SPECIALIST IN YORKSHIRE STONE SLATE: Aire Valley Roofing has been specialising in all aspects of traditional Yorkshire Stone random slating for over 20 years, mainly to buildings in Yorkshire and surrounding districts. All works are sympathetically carried out using the company's stocks of the best reclaimed materials by an experienced team of slaters using traditional methods. Included along with Aire Valley's stone slate works are random Westmoreland, Burlington and Welsh slate. Contracts undertaken predominantly in the £5,000 to £100,000 price range. Aire Valley Roofing is a member of the Federation of Master Builders and is a Warranted Builder. All enquiries welcomed.

▶ B J N ROOFING (CONTRACTORS) LTD

Gladstone House, Gladstone Road, Horsham, West Sussex RH12 2NN
Tel 01403 255155 Fax 01403 211794

ROOFING CONTRACTORS: The company has been established for 30 years and is a long standing member of the National Federation of Roofing Contractors. The business specialises in the re-roofing of listed, period and church buildings. Traditional Horsham stone, slating, tiling, shingling and associated leadworks. Areas covered include Sussex and parts of Surrey and Kent. Contract values range from £5,000 to £150,000. Local authority approved.

▶ BRYAN HODGES

1 Castle View, Clyro, Hereford HR3 5SZ
Tel 01497 820733
Mobile 07977 805229
Website www.stonetileroofing.com

SPECIALIST IN STONE TILE ROOFING AND DRY STONE WALLING

▶ COYLE TIMBER PRODUCTS
Bassett Farm, Claverton, Bath BA2 7BJ
Tel 01225 427409 Fax 01225 789979
E-mail info@coyletimber.co.uk
Website www.coyletimber.co.uk
Contact Joe Coyle
SPECIALIST SHINGLE ROOFING CONTRACTORS: *See also: profile entry in General Building Materials section, page 154.*

▶ E G SWINGLER & SONS
27/29 Weedon Road, Northampton NN5 5BE
Tel 01604 755055 Fax 01604 587506
SPECIALIST ROOFING: E G Swingler & Sons, established in 1964, is a family business which for the past 20 years has worked almost exclusively on historic and listed buildings. Within a 50 mile range of Northampton, they offer the full range of traditional building services with special emphases on traditional slate, tile and lead roofing commissions, in the £10,000 to £145,000 range. There are currently nine in-house craftsmen who work under the close supervision of Robert Swingler. The craftsmen's work is supported by a ready supply of the best reclaimed and new materials. Recently the company has completed the re-roofing of listed buildings in Norfolk, Leicestershire and Warwickshire. *See also: display entry in this section, page 82.*

STEEPLEJACKS & LIGHTNING PROTECTION

▶ A C WALLBRIDGE & CO LTD

Windsor Road, Salisbury, Wiltshire SP2 7DX
Tel 01722 322750 Fax 01722 328593

STEEPLEJACKS, LIGHTNING CONDUCTOR ENGINEERS: Wallbridge has been providing comprehensive and professional lightning conductor installation and maintenance services for the past 25 years. Its work is carried out by the company's own staff of trade certified engineers. All of the work carried out by Wallbridge complies with the current Health and Safety Regulations and to the BS6651 standard. Wallbridge is a full member of the National Federation of Lightning Conductor Engineers and is party to and works within the industry's agreed methods of safety and quality control. This ensures workmanship to the correct BS standards.

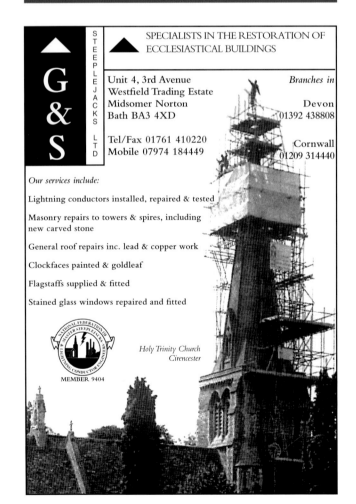

▶ ALDERSHAW HANDMADE CLAY TILES LTD

Pokehold Wood, Kent Street, Sedlescombe, East Sussex TN33 0SD
Tel 01424 756777 Fax 01424 756888 E-mail tiles@aldershaw.co.uk
Website www.aldershaw.co.uk

HANDMADE CLAY TILES: Truly handmade clay tiles and fittings in the widest range of colours are manufactured from Aldershaw's own Wadhurst clays in the heart of the Sussex countryside. Aldershaw's roofs have a patina normally only associated with a bygone age. The company specialises in restoration, in Kent peg tiles, mathematical tiles, fireplace briquettes and Gault clay tiles for the Cambridge and Ely areas. Special sizes and shapes are no problem. Sussex Terracotta produces handmade floor tiles in a vast range of colours shapes and sizes. The tiles are made to look mature from the day they are laid. Recent work includes 1,000 yards in the Crypt of St Paul's Cathedral.

▶ BABYLON TILE WORKS

Babylon Lane, Hawkenbury, nr Staplehurst, Tonbridge, Kent TN12 0EG
Tel 01622 843018 Fax 01622 843398
Website www.babylontileworks.co.uk

KENT PEG TILES AND FITTINGS: Babylon Tile Works produces its unique Kent peg tiles using traditional moulding methods with modern quality control to ensure the highest tile quality and performance. The tiles are made from Babylon's own Kent clay which is available on site. Two natural colours are available – Terracotta and Antique – giving you the warm rich colours of the Weald. Please contact Melvin Gash for further information and friendly specification advice.

▶ THE CLEFT WOOD COMPANY

1 Littlecote Cottages, Littlecote, Dunton, Near Winslow, Bucks MK18 3LN
Tel 01525 240434 Fax 01525 240434
E-mail enquiries@cleftwood.com
Website www.cleftwood.com

RIVEN OAK LATHS, ROOFING BATTENS, TILE PEGS, HEWN BEAMS AND HAZEL WATTLE HURDLES. *See also: profile entry in Plasterwork section, page 213.*

CLAY TILES & ROOF FEATURES

▶ COYLE TIMBER PRODUCTS

Bassett Farm, Claverton, Bath BA2 7BJ
Tel 01225 427409 Fax 01225 789979
E-mail info@coyletimber.co.uk
Website www.coyletimber.co.uk
Contact Joe Coyle

RIVEN OAK, CHESTNUT, SCOTS PINE LATHS, RIVEN BATTENS AND STAINLESS STEEL NAILS: *See also: profile entry in General Building Materials section, page 154.*

▶ ETERNIT BUILDING MATERIALS

Whaddon Road, Meldreth, Royston, Herts SG8 5RL
Tel 01763 264600 Fax 01763 262338
E-mail marketing@eternit.co.uk
Website www.eternit.co.uk

CLAY ROOF TILES AND FITTINGS: Eternit Building Materials is one of the UK's leading manufacturers of clay plain tiles and fittings. The collection includes single and double-cambered clay plain tiles, handcrafted clay plain tiles, large format interlocking pantiles, ventilation systems and a full range of fittings, including decorative ridges and finials. Eternit's product portfolio is supported by a comprehensive selection of literature, a 'free sample' service and a national sales network. Due to the extensive range of colours and products available, complete design flexibility is offered.

▶ G & N MARSHMAN

1 Nell Ball, Plaistow, Billingshurst, West Sussex RH14 0QB
Tel 01798 342427

MANUFACTURERS OF RIVEN OAK, CHESTNUT PLASTERERS' LATHS AND TILE AND STONE SLATE BATTENS: *See also: profile entry in Plasterwork section, page 216.*

▶ SANDTOFT ROOF TILES

Sandtoft, Doncaster, South Yorkshire DN8 5SY
Tel 01427 871200 Fax 01427 871222
Contact Nigel Pittman
E-mail heritage@sandtoft.co.uk

ROOF TILES: Sandtoft Roof Tiles has a history of excellence as rich and respected as the tiles it produces. Founded in 1904, not only does Sandtoft boast the largest product portfolio, but it is one of the largest independent tile producers in the UK, employing nearly 400 people at five manufacturing plants. Sandtoft's Heritage Service has provided replacement tiles for the finest historic roofs in England, including the Victoria and Albert Museum, Cambridge University, London Zoo, Eton College and the Verulamium Museum in St Albans. Combining traditional techniques with modern technology, its dedicated craftsmen recreate even the most difficult of historic tiles, from Roman under-and-overs to bespoke handmade hips, valleys and ridges.

▶ SMITHBROOK BUILDINGS PRODUCTS LIMITED

Pollingfold Farm Barn, Ellen's Green, Rudgwick, Horsham RH12 3AS
Tel 01403 824170 Fax 01403 824171

GLAZED CLAY PANTILES, GLAZED BRICKS, ENCAUSTIC TILES: These are some of the speciality clay products offered by Smithbrook Building Products. Of particular interest to conservation officers and architects are the glazed pantiles to match colours used predominantly on 1930s buildings, and non-interlocking black glazed pantiles used extensively in East Anglia. Unobtrusive modern ventilation systems are also available. Unders and overs ex stock.

▶ TUDOR ROOF TILE CO LTD

Denge Marsh Road, Lydd, Kent TN29 9JH
Tel 01797 320202 Fax 01797 320700
E-mail info@tudorrooftiles.co.uk
Website www.tudorrooftiles.co.uk

HAND MADE CLAY TILES: Tudor Roof Tile Company, situated at Lydd, are manufacturers of the highest quality genuine hand made clay tiles. Production comprises the full range of plain and peg tiles and fittings plus ornamentals. Tudor's craftsmen work to the very highest standard. They are a BS EN ISO 9002 registered company and their tiles are strength tested to comply with BS EN 538:1994 and tested to comply with BS EN 539-2:1998 for resistance to frost. Tudor tiles have been used for the widest range of buildings from refurbishment of listed buildings to new build, from historic houses to supermarket complexes. Tudor produce a unique invisible 'under tile venting system' which is completely invisible externally preserving the charm of a hand made clay tile roof. Tudor's employees provide their customers with the highest grade of service.

▶ WEST MEON POTTERY

Church Lane, West Meon, Petersfield, Hampshire GU32 1JW
Tel 01730 829434

ARCHITECTURAL CERAMICS: *See also: profile entry in Architectural Terracotta section, page 110.*

▶ WILLIAM BLYTH

Hoe Hill, Pasture Lane North, Barton on Humber, North Lincolnshire DN18 5RB
Tel 01652 632175 Fax 01652 660966
Contact Mr R Coulam

HAND-MADE CLAY ROOF TILES AND FITTINGS: William Blyth has provided 150 years of service to the building trade, during which it has been owned and managed by the same family. William Blyth takes pride in producing a range of quality, hand-made single lap tiles, plain tiles, and fittings, using traditional materials and expertise. Age old methods of drying and firing the natural clay give a tile which has excellent weathering properties and a long life expectancy – a tile which will enhance any building old or new.

METAL SHEET ROOFING

▶ A V BROWN

82, Poolbrook Road, Malvern, Worcestershire WR14 3JD
Tel/Fax 01684 562969 Mobile 0860 625447

LEAD ROOFING CONTRACTOR: Established in 1976, this family firm has undertaken lead and copper roofing for over 20 years. Their workforce of craftsmen has worked on churches, castles and all types of houses throughout their area. Clients include local authorities, architects, builders, water authorities and owners of listed buildings. All types of work are undertaken including cast milled lead and copper sheet roofing.

▶ ANELAYS

William Anelay Limited, Murton Way, Osbaldwick, York YO19 5UW
Tel 01904 412624 Fax 01904 413535
E-mail office@williamanelay.co.uk
Website www.williamanelay.co.uk

BUILDING AND RESTORATION CONTRACTORS: *See also: profile entry in Building Contractors section, page 58.*

▶ AWARD SPECIALISED LEADWORK & PLUMBING

91 Queen Edith's Way, Cambridge CB1 8PL
Tel/Fax 01223 527036
Mobile 07974 389235/07715 172347
E-mail awardlead@hotmail.com

METAL SHEET ROOFING SPECIALISTS: Having worked on most of Cambridge's colleges and English Heritage sites, Award has a wealth of experience in all forms of leadwork, from traditional (including ecclesiastical restoration) to modern design application. Approved by the Guild of Master Craftsmen, Award always provides a competitive quote coupled with skilled workmanship. Please contact the company to discuss your project.

METAL SHEET ROOFING

Anglia Lead Ltd
specialists in the casting and laying of lead sheet roofing for over 125 years
Sheet lead, cast in the traditional way by Anglia Lead, crowns buildings throughout the United Kingdom – from Castles to Cathedrals and from humble Parish Churches to the Mother of Parliaments.
Telephone
01603 626856
Fax: 01603 619171
E-mail: anglialead@paston.co.uk

CEL
Winners of The Copper Development John Smith Craftmanship Award 2001
Architectural Metal Roofing
Specialist Contractors for the installation of Lead, Copper, Stainless Steel & Zinc Roofing for the restoration and refurbishment of Heritage and Ecclesiastical Projects.
Canopies · Domes · Spires
Flat or Sloping Roofs
Horizontal or Vertical Applications
A tried and tested contractor with the ability to design and install their materials using traditional skills with the sensitivity required for such projects.
Manufacturers & Suppliers of Traditional Sand Cast Lead Sheet, Cast Lead, Rainwater Goods, Decorative Mouldings & Features.
C.E.L. LTD
Progress House, 256 Station Road, Whittlesey, Peterborough PE7 2HA
Telephone 01733 206633 Fax 01733 206644
METAL ROOFING
MRCA
CONTRACTORS ASSOCIATION

OVER 175 YEARS OF CRAFTSMANSHIP AND RESTORATION

Norman & Underwood are one of the largest and oldest independent family companies engaged in restoration work on heritage buildings in the country. We have supplied and installed traditional, sand cast lead to Salisbury Cathedral, Westminster Abbey, Hampton Court Palace, Chatsworth House and many of the major cathedrals and heritage buildings in the UK. We also work in Copper, Zinc, Aluminium and Stainless Steel on both restoration and new build contracts. The same levels of craftsmanship are also applied to Stained Glass and Glazed Screening projects in chapels, churches and cathedrals throughout the country.

ESTABLISHED 1825

11 - 27 FREESCHOOL LANE, LEICESTER LE1 4FX TEL 0116 251 5000
FAX 0116 253 2669 EMAIL INFO@NANDU.CO.UK WEB WWW.NANDU.CO.UK

STRUCTURE & FABRIC
3.1 Roofing

▶ C E L LTD

Progress House, 256 Station Road, Whittlesey, Peterborough PE7 2HA
Tel 01733 206633 Fax 01733 206644

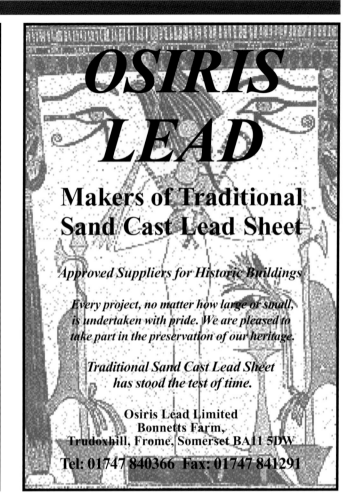

TRADITIONAL METAL ROOFING: Manufacturers of traditional sand cast lead sheet and flashings also reproduction lead rainwater goods, decorative mouldings, plaques and much more. Skilled craftsmen from the company's contracting division use authentic metal construction materials including lead, copper, zinc and stainless steel for restoration or replacement of roofs and other areas for all types of ecclesiastical and heritage buildings. *See also: display entry in Metal Sheet Roofing section, page 87.*

▶ E BOWMAN AND SONS LTD

Cherryholt Road, Stamford, Lincolnshire PE9 2ER
Tel 01780 751015 Fax 01780 759051
E-mail mail@ebowman.co.uk
Website www.ebowman.co.uk

SPECIALIST BUILDING CONTRACTORS: *See also: display entry in Building Contractors section, page 60.*

ROOF LIGHTS

improve Your Quality of Light

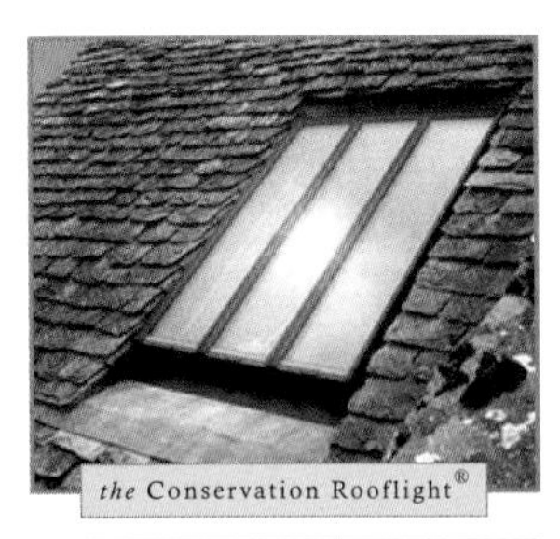
the Conservation Rooflight®

Pyramid Rooflights

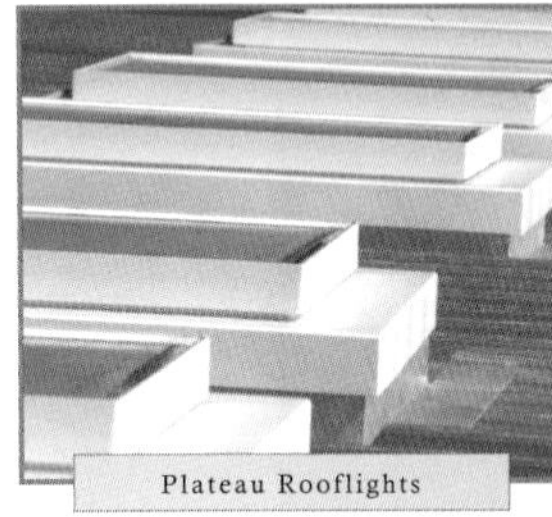
Plateau Rooflights

Special Rooflights

Substantially exceed the performance requirements of BS6375:Part1:1989 - weathertightness. Patented thermally decoupled lining, ***Thermoliner***® maximises thermal performance. 'K' glass as standard gives an average installed U-Value of 1.7W/m^2°K.

t: 01993 830 613
f: 01993 831 066
e: info@metalwindow.co.uk
w: www.conservationrooflight.com

the Conservation Rooflight Company

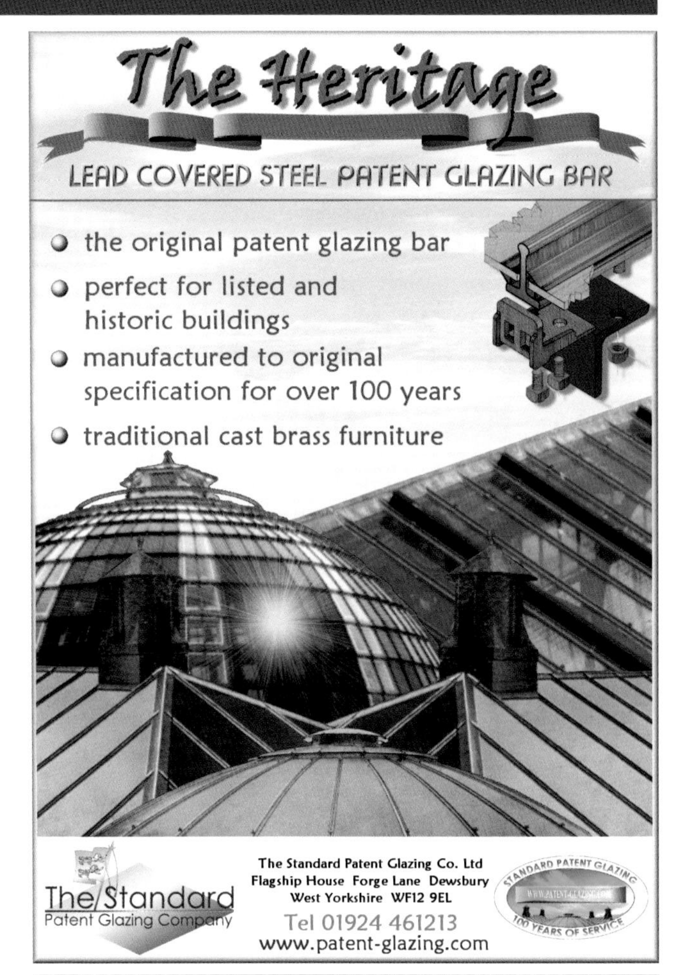

The Heritage
LEAD COVERED STEEL PATENT GLAZING BAR
the original patent glazing bar
perfect for listed and historic buildings
manufactured to original specification for over 100 years
traditional cast brass furniture
The Standard Patent Glazing Company
The Standard Patent Glazing Co. Ltd
Flagship House Forge Lane Dewsbury
West Yorkshire WF12 9EL
Tel 01924 461213
www.patent-glazing.com
STANDARD PATENT GLAZING
100 YEARS OF SERVICE

▶ **CLEMENT WINDOWS GROUP LTD**
Clement House, Haslemere, Surrey GU27 1HR
Tel 01428 643393
Website www.clementwg.co.uk
ROOF LIGHTS, METAL WINDOWS AND LANTERN LIGHTS:
See also: display entry in Metal Windows section, page 135.

For conservators of existing roof lights, see also Architectural Metalwork, pages 141–147

ROOF VENTS

BUILDINGS MUST BREATHE TO SURVIVE
lapVent®
Roof Space Ventilation
FOR ALL TYPES OF ROOF OLD AND NEW
RESTORATION - NEW BUILD - REPAIR
'Completely unexposed' and non-load bearing
Prevents structural alteration/material damage
The roof's inbuilt defenses remain un-breached
Unique safety and storm protection features
BRE tested – 20,000mm² filtered free-vent area
Universally adaptable – slate, stone, clay etc
Made from best high impact 'recycled' material
Cost effective input with a 'lifetime' guarantee
Retrofit inside at anytime or during construction
Multi task Mech. ext. version for direct ductwork

Heritage Structural Ventilation Ltd
33 Queens Road – Bournemouth – Dorset BH2 6BN
Fax/Phone: 01202 769 255 – Email: info@lapvent.co.uk
Website: www.lapvent.co.uk

ROOF LIGHTING

PETER KING

The appearance of roof lighting is a relatively recent occurrence in our architectural history. Yet flip the pages of any architectural journal or view any city or urban area from above today and it will quickly become apparent that lighting from above is now an important aspect of our architectural legacy. It is difficult to find any large new public or commercial building without a prominent roof lighting feature nowadays, and domestic roofscapes are often peppered with rooflights. This was not always the case. Roof lighting has gone from non-existence to ubiquity in the span of about 300 years.

In Britain and throughout most of Europe skylights did not appear until the mid 1700s when the developments in the process of glass manufacture made available relatively cheap glass sheets of adequate size. For rooflights, large panes which could span from top to bottom without a break were preferable, as small panes connected together by horizontal glazing bars (or lead cames) could invite leaks.

The earliest and simplest form of rooflight was a sheet of glass inserted into a tiled roof in lieu of a roof tile. This had the advantage of being relatively cheap, but it could not be cleaned easily, it was difficult to repair if broken and could not be opened to provide ventilation. Consequently this form of roof lighting was used primarily in secondary or agricultural buildings.

The earliest openable rooflights used in domestic architecture appear to have been timber framed with a lead-clad timber kerb and an opening casement (also lead-clad) which overhung the kerb on all sides. The lower edge of the glass would be left free of framework in order to allow rainwater to run off the rooflight quickly. Rooflights of this type can be still be found on some Georgian and Regency residential buildings, usually tucked away around a side or rear elevation in a position where a dormer window would be architecturally undesirable or physically problematic.

Rooflights did not become common in ordinary houses until the late 19th century, partly because excise duties, which were imposed on glass by weight in the mid 18th century, favoured small, thin panes of glass. Dormer and gable windows and lantern lights (in which the glazing is vertical) therefore provided the principal means of lighting attic rooms and staircases respectively, throughout the 18th century, and they remained so throughout the 19th century, long after rooflights had become common. Following the removal of excise duties in 1845, which coincided with a number of technological improvements in the manufacture of glass, there was an immediate and dramatic change in the appearance of windows as glazing bars were removed, and a less noticeable increase in rooflights, which remained functional necessities, often tucked out of sight.

Mass-produced rooflights of cast iron became available in the mid 19th century and their use in domestic architecture has continued up to modern times. Large-scale overhead glazing was achieved by means of the patent glazing which was similarly devised,

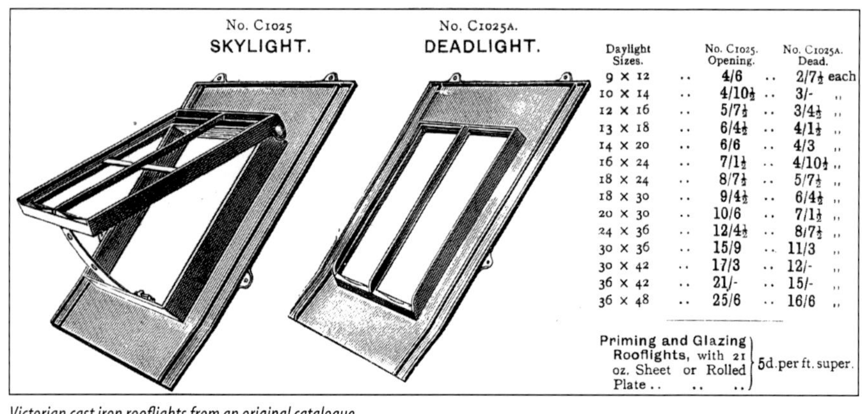

No. C1025
SKYLIGHT.

No. C1025A.
DEADLIGHT.

Daylight Sizes.	No. C1025. Opening.	No. C1025A. Dead.
9 × 12	4/6	2/7½ each
10 × 14	4/10½	3/- ,,
12 × 16	5/7½	3/4½ ,,
13 × 18	6/4½	4/1½ ,,
14 × 20	6/6	4/3 ,,
16 × 24	7/1½	4/10½ ,,
18 × 24	8/7½	5/7½ ,,
18 × 30	9/4½	6/4½ ,,
20 × 30	10/6	7/1½ ,,
24 × 36	12/4½	8/7½ ,,
30 × 36	15/9	11/3 ,,
30 × 42	17/3	12/- ,,
36 × 42	21/-	15/- ,,
36 × 48	25/6	16/6 ,,

Priming and Glazing Rooflights, with 21 oz. Sheet or Rolled Plate .. .. .. } 5d. per ft. super.

Victorian cast iron rooflights from an original catalogue

The stairway of an Edwardian terraced house lit by a rooflight

developed and confidently used by the Victorians in all manner of structures from railway stations to museums.

By the late 19th century, glass had become relatively inexpensive and was available in good quality large sheets. The Victorians self-confidently built on a scale previously unheard of and with a collective panache and flair that has seldom been matched since. Building types that had never existed previously had evolved during this period, such as the great railway stations, which utilised vast areas of overhead glazing. Other deep-plan public building types of this time, such as the Leeds Corn Exchange and University of Oxford Museum, would not have been possible without the use of overhead glazing, and the most famous Victorian building, the Crystal Palace of 1851, employed glazing on an unprecedented scale.

In domestic buildings, original cast iron rooflights are most commonly found over the stairwells in late Victorian and Edwardian terraced houses. To reduce the draught caused by air circulating across the cold surface, these rooflights often had secondary glazing fixed at ceiling level.

THE FUTURE OF TOP-LIGHTING

Today, rooflights are more common and more popular than ever before, particularly in attic conversions and in refurbished historic buildings, where they are seen as a practical and unobtrusive means of letting light into roof spaces. Their popularity is due in part to the improvements made in the thermal and weathering performance of metal and timber rooflights which have brought the small-unit rooflight up to date, and they are also much less expensive to introduce than dormer windows (roughly 25%–30% of the cost), and size-for-size they admit more light. However it must be said that rooflights are still seldom seen as other than as a necessity in domestic architecture, and they take an architecturally subservient role to other elements. Rooflights have been and probably always will be seen in domestic architecture as a device to admit light and ventilation to a space as inconspicuously as possible.

Although top-lighting has become widespread in both the domestic and public realms, lighting from above is rarely used by designers for its own particular qualities. In domestic work it is generally used for utility purposes only, while in public buildings it is invariably 'mixed' with side-light. As anyone who has visited a top-lit gallery will know, natural top-lighting, when separated from side-lighting, generates in our country (with its almost permanent cloud 'filter') an ethereal, almost luminous effect. It may be that our eyes are used to seeing objects in side-light, and top-light – like theatrical low light – throws our senses slightly out of kilter.

The value of top lighting was widely recognised by 18th century architects like Robert Adam, who used lanterns over grand staircases and halls of grand houses and stately homes to great effect, establishing a tradition continued in buildings as diverse the Dulwich Picture Gallery and the Johnson Wax Building. Hopefully, apart from offering us economy and planning convenience, top-lighting through rooflights will, in the right hands, continue to bring us the unique sensation of light from above.

PETER KING RIBA is an architect and principal of the Carden-King Partnership which operates from Kelmscott village in Oxfordshire. He is also chairman of the Conservation Rooflight Company and designer of the Conservation Rooflight™ range of rooflights.

THE ART OF HERITAGE RAINWATER SYSTEMS

WITH 50 YEARS EXPERIENCE IN THE MANUFACTURE AND SUPPLY OF ARCHITECTURAL RAINWATER SYSTEMS, ALUMASC'S EXPERTISE IS SECOND TO NONE. THE HERITAGE RAINWATER RANGE IS A COMPLETE SELECTION OF THE MOST POPULAR PROFILES, FITTINGS AND ACCESSORIES, AVAILABLE IN CAST ALUMINIUM AND CAST IRON FOR BOTH FAST TRACK AND CRAFT BASED DESIGNS. THROUGH A COMBINATION OF CONVENTIONAL STYLING AND MODERN PRODUCTION TECHNIQUES, ALUMASC HAVE ENGINEERED A DISTINGUISHED, CLASSICAL RANGE TO MEET THE DEMANDS OF ANY CONTEMPORARY PROJECT. **FREEPHONE 0808 1002008**.

ALUMASC – BRINGING MORE TO THE ART OF BUILDING

ARCHITECTURAL RAINWATER SYSTEMS

The Cast Iron Rainwater Systems for all your needs

Hargreaves Foundry can provide a range of standard **Cast Iron Rainwater** products or additionally non-stock items can be made to order to meet the needs of any project.

The standards for **Rainwater** have remained remarkably constant for many years which means that replacing or integrating new with old is made easy. The physical properties of cast iron are sustained throughout its lifetime maintaining its effectiveness, under normal conditions and coupled with proper maintenance, we would expect these products to last for over 100 years.

Benefits

- Immense strength, durability and long life
- Low maintenance, fire resistance and 100% recyclable
- Cost effective
- The only material genuinely suitable for conservation or restoration work
- Available from builders' merchants nationwide

HARGREAVES FOUNDRY DRAINAGE LTD
GENERAL IRONFOUNDERS

Hargreaves Foundry Drainage Limited, Water Lane, South Parade, Halifax, West Yorkshire HX3 9HG
TEL: ·44 (0) 1422 330607 FAX: ·44 (0) 1422 320349
EMAIL: info@hargreavesfoundry.co.uk

▶ J & J W LONGBOTTOM LIMITED

(inc Sloan & Davidson Ltd)
Bridge Foundry, Holmfirth, Huddersfield HD9 7AW
Tel 01484 682141 Fax 01484 681513

CAST IRON RAINWATER AND SOIL: A traditional foundry, producing a comprehensive range, comprising pipes, gutters, and all the necessary fittings for rainwater, soil (BS 416) and smoke. An extremely extensive pattern range of moulded gutters (half round, ogee, box, valley) (including curves) and of ornamental rainwater heads is produced. Also air bricks and gratings. Substantial stocks of all products are continually maintained and prompt delivery throughout the UK can be effected. Their comprehensive catalogue is available on request.

▶ SAINT-GOBAIN PIPELINES

Sinclair Works, PO Box 3, Ketley, Telford, Shropshire TF1 5AD
Tel 01952 262500 Fax 01952 262555
Literature Hotline 0800 028 2134
Website www.saint-gobain-pipelines.co.uk (soil and drain)

CAST IRON RAINWATER SYSTEMS: The classical range of traditional cast iron rainwater and gutter systems has been manufactured at Sinclair Works for over 100 years. The range includes a choice of standard half round, beaded and deep half round, OG, Notts OG, moulded no 46, and box gutter systems, with downpipes available in round and rectangular profiles. Most of the profiles are offered in a series of diameters, and all offer a full range of fittings. A new gutter jointing kit has recently been introduced to the half round gutter profile, speeding up installation on site. Classical 'Plus' is now available in a black factory applied finished coating for immediate installation.
See also: display entry above.

MASONRY SERVICES

A D Calvert Architectural Stone Supplies Ltd	sm ss		100
A F Jones (Stonemasons)	sm ss		100
Abbey Heritage Ltd	sm ss	tc	101
Abbey Masonry and Restoration Limited	sm ss		100
Graciela Ainsworth	ss		95
Anelays - William Anelay Ltd	sm	br	58
Architectural Stone Conservation	sm		101
Back to Earth	ea sm		58
Balmoral Stone Ltd	sm ss		101
Beckwith Tuckpointing & Restoration		br	177
Between Time Ltd		br	58
Bluestone plc	sm		59
Boden & Ward	sm ss		101
Boshers (Cholsey) Ltd	sm	br	60
Bricknell Conservation Limited	sm		59
Bryan Hodges Stone Tile Roofing	dw		82
Burleigh Stone Cleaning & Restoration Co Ltd	sm	br tc	166
Burnaby Stone Care Ltd	sm		166
Burrows Davies Limited	sm		102
Busby's Builders	sm	br	60
C Ginn Building Restoration	sm		102
C J Ellmore & Co Ltd	sm	br	60
C R Crane & Son Ltd	sm	br	60
Carrek Limited	sm	br	61
Carthy Conservation Ltd	sm ss	tc	102
Cathedral Works Organisation (Chichester) Limited	sm ss		102
Chalk Down Lime Ltd	sm		172
Chester Masonry Group	sm		59
Church Conservation Limited	sm		84
Cliveden Conservation Workshop Ltd	sm ss		102
Coe Stone Ltd	sm		103
Corbel Conservation Ltd	sm		59
David Ball Restoration (London) Limited	sm	br tc	103
E Bowman and Sons Ltd	sm		103
Fairhaven of Anglesey Abbey	sm		103
Filkins Stone Co	sm		104
Grall Plaster & Stone Specialists	sm		214
Hall Construction	dw sm	br	61
Haslemere Builders Limited	sm	br	62
Hirst Conservation	ea ss		ifc
Historic Buildings Conservation Limited	dw ea sm		61
Holden Conservation Ltd	sm ss	tc	95
Holloway White Allom	sm	br	62
Invicta Stone Ltd	sm		104
J G Matthews Ltd	sm	br	62
J Joslin (Contractors) Ltd	sm		105
James Lugg Stoneworks	sm		104
Ken Biggs Contractors Ltd	sm		63
Linford-Bridgeman Limited	sm	br	64
Gerard C J Lynch		br	113
Magenta Building Conservation Ltd	sm		63
Mather & Ellis Ltd	sm		105
Mathias Builders		br	113
Maysand Limited	sm	br tc	106
Millway Builders	sm		65
Minerva Stone Conservation	ea sm		106
Mowlem Rattee & Kett	sm ss	br	106
Mott Graves Projects Ltd	sm		106
Nicholas West Brickwork	br		113
Nicolas Boyes Stone Conservation	ss		106
Nimbus Conservation Limited	sm ss	br tc	107
Paye Stonework	sm ss	tc	107
Plowden & Smith Ltd	ss		203
Priest Restoration	sm	br tc	107
Rope Tech International Limited	ss		84
Rupert Harris Conservation	ss		203
Richard Noviss	sm ss		106
Russcott Conservation and Masonry	sm		107
Sarabian Limited	sm	br	168
Timothy J Shepherd Historic Brickwork Specialist		br	113
Simmonds of Wrotham	sm	br	64
Spencer & Richman	ss		189
Splitlath Ltd	sm		67
StoneCo Limited	sm		108
Stoneguard	sm	tc	108
Stonewest Ltd	sm ss	br tc	108
Taylor Pearce Restoration Services Limited	ss		94
Timothy Williams (Builders) Ltd	sm		65
Wallis	sm	br	66
Weldon Stone Enterprises Ltd	sm		109
Wells Cathedral Stonemasons Ltd	sm ss		109
Wells Masonry Services Ltd	sm		108
Westland London	ss		189
William Sapcote & Sons Ltd	sm		67
William Taylor Stonemasons	sm		109
Wm Langshaw & Sons	sm		68
Womersley's Limited	sm	br	176

KEY

br brick work
dw drystone walling
ea cob and earth construction
sm stone masons
ss stone carvers and sculptors
tc terracotta

MASONRY PRODUCTS

A D Calvert Architectural Stone Supplies Ltd		sq			100
Abbey Masonry & Restoration Ltd		sq			100
Aldershaw Handmade Clay Tiles Ltd	mt				85
The Bath Stone Group		sq			101
Best Demolition	as				155
Boden & Ward Stonemasons Ltd		sq			101
Brickfind (UK)	bk				118
Broadmead Cast Stone			cs		110
The Bulmer Brick and Tile Company	bk mt				118
Bursledon Brickworks Conservation Centre				to	172
Cathedral Works Organisation (Chichester) Limited		sq			102
Chalk Down Lime Ltd		sq			172
Chilstone			cs		110
The Cornish Lime Company Ltd				to	174
The Downs Stone Company Ltd		sq			104
Drummond's Architectural Antiques Limited	as				155
E Bowman and Sons Ltd		sq			103
E T Clay Products Limited	bk		cs		118
Filkins Stone Co		sq			104
H G Clarke & Son		sq			104
Haddonstone Limited			cs		110
Hammill Brick Ltd	bk				118
Hanson Brick	bk				118
Kent Balusters			cs		110
Ketley Brick Company	bk				118
Lambs Bricks and Arches	bk				120
Lambs Terracotta & Faience	at				111
LASSCO	as				155
The Lime Centre				to	174
Manchester Brick & Precast Limited	bk				113
Mather & Ellis Ltd		sq			105
Minchinhampton Architectural Salvage	as				155
Minerva Stone Conservation		sq			106
Mike Wye Associates				ea	174
Mongers Architectural Salvage					156
Priors Reclamation					156
Ransfords Reclaimed Building Supplies	as				156
Retrouvius Architectural Salvage	as				157
Shaws of Darwen	at				112
SALVO	as				155
Smithbrook Building Products Ltd	bk				86
Solopark Plc	as				156
Sussex Brick Ltd	bk mt				119
Terca	bk				119
Thorverton Stone Company Ltd			cs		110
Twyford Lime Products				ea	176
Walcot Reclamation Ltd	as				157
Weldon Stone Enterprises Ltd		sq			109
Wells Cathedral Stonemasons Ltd		sq			109
West Meon Pottery	at bk				113
Westland London					189
Womersley's Limited				to	176
York Handmade Brick Company Limited	bk				119

KEY

as architectural salvage
at terracotta
bk bricks
cs cast stone
ea cob and earth
mt mathematical tiles
sq stone
to masonry tools

STRUCTURE & FABRIC
3.2 Masonry

STATUARY

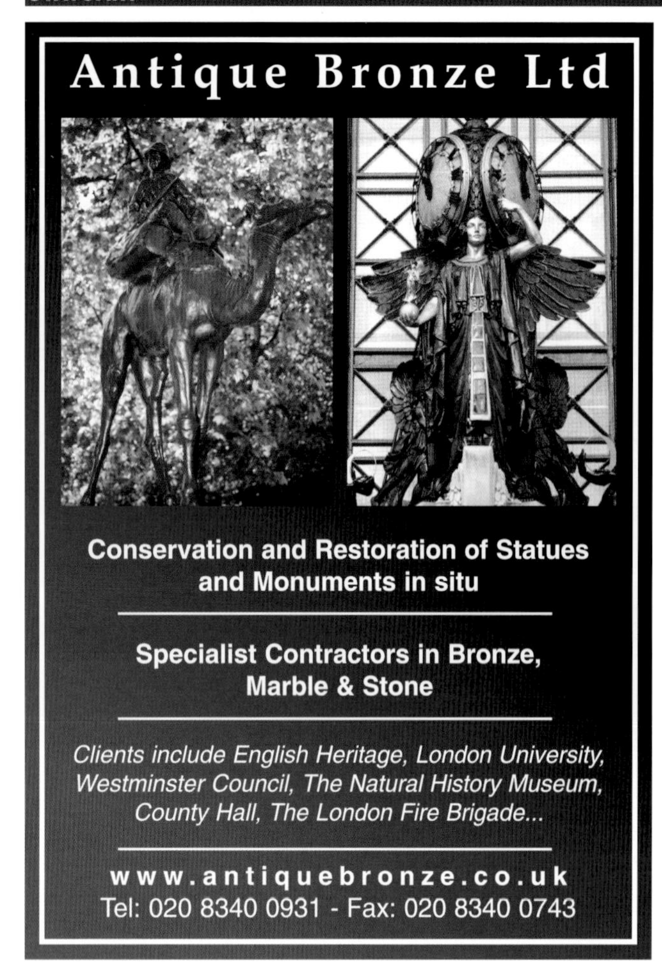

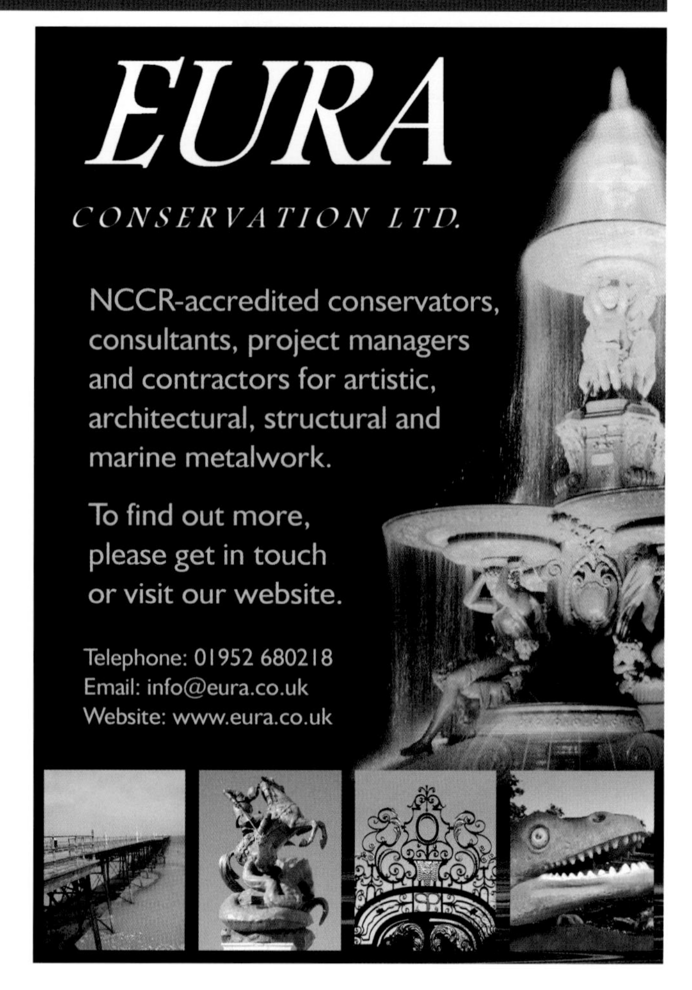

▶ CHILSTONE
Victoria Park, Fordcombe Road, Langton Green, Tunbridge Wells, Kent TN3 0RE
Tel 01892 740866 Fax 01892 740249
STONEMASONRY: Handmade reconstructed stone statuary. *See also: profile entry in Cast Stone section, page 110.*

▶ EURA CONSERVATION LTD
Unit H10, Halesfield 19, Telford, Shropshire TF7 4QT
Tel 01952 680218 Fax 01952 585044
E-mail mail@eura.co.uk
ACCREDITED CONSERVATORS AND CONTRACTORS FOR ARTISTIC, ARCHITECTURAL AND MARINE METALWORK: *See also: display entry in Statuary section, this page and profile entry in Architectural Metalwork section, page 142.*

▶ HADDONSTONE LIMITED
The Forge House, East Haddon, Northampton NN6 8DB
Tel 01604 770711 Fax 01604 770027
E-mail info@haddonstone.co.uk
Website www.haddonstone.co.uk
STANDARD RANGE INCLUDES BUSTS AND FULL FIGURES IN CLASSICAL AND TRADITIONAL STYLES: *See also: display entry in Cast Stone section, page 110.*

▶ HIRST CONSERVATION
Laughton, Sleaford, Lincolnshire NG34 0HE
Tel 01529 497449 Fax 01529 497518
MONUMENTS AND SCULPTURE CONSERVATION: *See also: display entry on the inside front cover and profile entry in Building Contractors section, page 62.*

▶ NIMBUS CONSERVATION LIMITED
Eastgate, Christchurch Street East, Frome, Somerset BA11 1QD
Tel 01373 474646 Fax 01373 474648
E-mail enquiries@nimbusconservationlimited.com
Website www.nimbusconservation.com
STONE CONSERVATION AND MASONRY: *See also: display entry and Profile entry in Stone section, pages 107 and 106, and profile entry in Masonry Cleaning section, page 166.*

▶ PLOWDEN & SMITH
190 St Ann's Hill, London SW18 2RT
Tel 020 8874 4005 Fax 020 8874 7248
E-mail richard.rogers@plowden-smith.com
Website www.plowden-smith.com
CONSERVATION AND RESTORATION: *See also: display entry in Fine Art Conservation section, page 203.*

▶ TAYLOR PEARCE RESTORATION SERVICES LIMITED
Fishers Court, Besson Street, London SE14 5AF
Tel 020 7252 9800 Fax 020 7277 8169
STATUARY AND ARCHITECTURAL ORNAMENTS CONSERVATORS: Specialists in the restoration, conservation and installation of stone statuary and ornaments, architectural statuary, architectural ceramics, mosaic work and church monuments. Taylor Pearce has a reputation for professional conservation practice executed to the highest standards. Projects have been carried out for the V & A, The Royal Academy, English Heritage, National Trust, Imperial War Museum, Kew Gardens, Westminster Abbey (Chapter House annunciation group), Westminster Cathedral and The Royal Collection. Because the company's projects are executed to museum standards within commercial parameters it also numbers top architectural and surveying practices amongst its regular clients.

SCULPTURE

▶ GRACIELA AINSWORTH
Units 4 & 10 Bonnington Mill Business Centre, 72 Newhaven Road, Leith, Edinburgh EH6 5QG
Tel 0131 555 1294 Fax 0131 467 7080
E-mail graciela@graciela-ainsworth.com
SCULPTURE CONSERVATION AND RESTORATION, SCULPTURE CARVING COMMISSIONS: Dedicated to the conservation and restoration of sculpture, ornaments, monuments, memorials, museum artefacts, architectural ornament and historic building fabric. Work is conducted in specially designed workshops or on-site in museums, galleries, churches, and historic buildings and gardens. Scale of projects undertaken ranges from fine pieces from the Oriental Gallery in Durham, marble busts from the Wallace Monument and Playfair Library, through to the statuary of Drummond Castle Gardens and Glamis Castle; and up to the 32 foot William Wallace statue in Drybourgh, St Giles Cathedral and the decorative façade of Jenners in Edinburgh. Regular clients include The National Galleries of Scotland, The University of Edinburgh, Glasgow Museums, Perth and Kinross Council, and notable private clients.

▶ CLIVEDEN CONSERVATION WORKSHOP LTD
Head Office – The Tennis Courts, Cliveden Estate, Taplow, Maidenhead, Berkshire SL6 0JA
Tel 01628 604721 Fax 01628 660379
SCULPTURE, STONE AND WALL PAINTINGS CONSERVATION: *See also: profile entry in Stone section, page 102.*

▶ CLIVEDEN CONSERVATION WORKSHOP LTD
Home Farm, Ammerdown Estate, Kilmersdon, Bath, Somerset BA3 5SN
Tel 01761 420300 Fax 01761 420400
E-mail info@clivedenconservation.co.uk
SCULPTURE, STONE, PLASTER, MOSAIC, WALL PAINTING CONSERVATION, STONE CARVING: *See also: profile entry in Stone section, page 102.*

▶ NICOLAS BOYES STONE CONSERVATION
46 Balcarres Street, Edinburgh EH10 5JQ
Tel 0131 446 0277 Fax 0131 446 0283
Website www.nb-sc.co.uk
STONE CONSERVATION SERVICES: *See also: profile entry in Stone section, page 106.*

▶ RUPERT HARRIS CONSERVATION
Studio 5, No 1 Fawe Street, London E14 6PD
Tel 020 7515 2020 Fax 020 7987 7994
E-mail enquiries@rupertharris.com Website www.rupertharris.com
ANTIQUE AND FURNITURE RESTORATION: *See also: display entry in Fine Art Conservation section, page 203.*

BRONZE STATUARY

▶ RUPERT HARRIS CONSERVATION
Studio 5, No 1 Fawe Street, London E14 6PD
Tel 020 7515 2020 Fax 020 7987 7994
E-mail enquiries@rupertharris.com
Website www.rupertharris.com
ANTIQUE AND FURNITURE RESTORATION: *See also: display entry in Fine Art Conservation section, page 203.*

MARBLE & GRANITE

▶ NOSTALGIA
Holland's Mill, Shaw Heath, Stockport, Cheshire SK3 8BH
Tel 0161 477 7706 Fax 0161 477 2267
Website www.nostalgia-uk.com
RECLAIMED ANTIQUE FIREPLACES: Nostalgia supplies and restores fine marble chimneypieces. *See also: profile entry in Fireplaces section, page 189.*

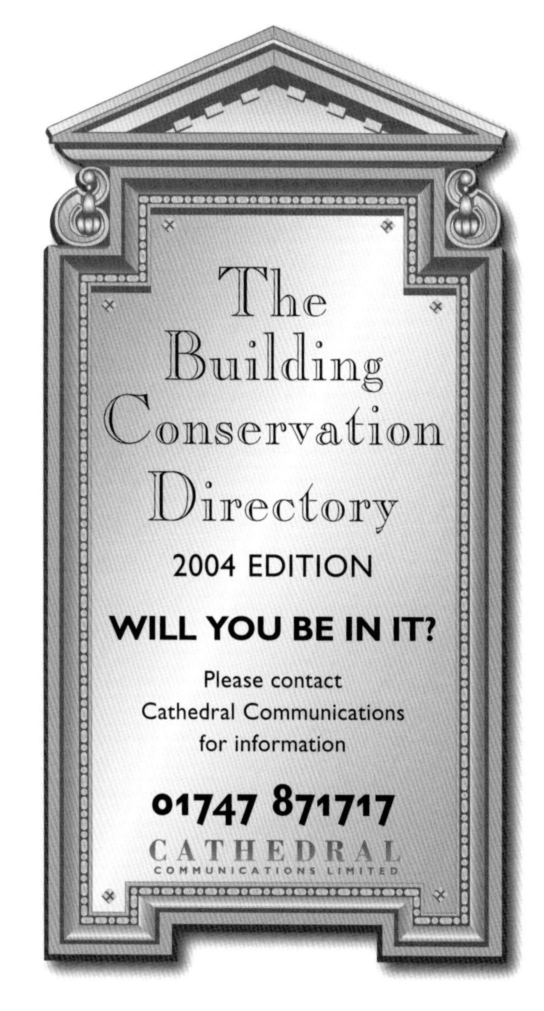

CONSERVING RAILWAY HERITAGE

JIM CORNELL

Britain's built heritage is as varied as any in the world. Its architects, designers and engineers drew upon centuries of experience to construct a huge variety of buildings and structures which provide an invaluable insight into much of the economic and social history of our land. All manner of materials have been used, including natural and local resources, those emerging from the industrialisation and, latterly, those produced by 20th century high-tech processes. These advancements together with developments in construction methods and design standards have led to an array of buildings, and to a lesser extent structures, which increasingly are becoming recognised as being of enormous value and interest to the nation.

RAILWAY HERITAGE AND ARCHITECTURE

Railway architecture, although all less than 200 years old, displays its own vast variety of features and components. The emergence of the steam railway in the early part of the 19th century had a great impact on civilisation and on industrialisation worldwide. Within less than 25 years, Richard Trevithick's experiments at Coalbrookdale led to the opening of the Stockton & Darlington Railway in 1825. Without doubt, the 19th century was the age of the railway.

Between 1825 and 1875 the network of railways spread throughout the country as rival investors built competing lines. Although not always benefiting the business bottom line, and often duplicating both routes and facilities, the result united the nation, benefiting passengers and delivering freight to even remote parts of the country.

Some 9,000 stations, 1,000 tunnels and 60,000 bridges were constructed in addition to an enormous range of ancillary buildings such as warehouses, rolling stock housing and maintenance sheds, signal boxes, water towers, offices and hotels.

The first railway stations often reflected the local architecture and would not have looked out of place in the high street of any small country town. Local materials and styles were used, with speed and economy of construction being a major design consideration, and size was more often than not relatively modest. There were, however, a few notable exceptions including Neo-classical, Gothic or even Moorish-style buildings, perhaps intended as much to reassure the first nervous passengers as to impress prospective investors.

Leading architects, engineers and contractors were attracted by the fame and prestige, not to mention fees and salaries, which were associated with this new era of opportunity. The benefit to our heritage was a rich legacy of well-designed buildings and structures which has stood the test of time – 175 years in some instances.

Bristol Temple Meads Station: the main entrance in 1999

The second half of the 19th century and the early years of the 20th saw tremendous growth in rail traffic. During this time the railways' power and prestige were expressed in the many ambitious buildings and structures that were needed to handle the ever-increasing business. At the start of this period – the age of the Crystal Palace and the Great Exhibition of 1851 in Hyde Park – significant examples of pioneering structures were completed. Their building involved the use of what was then the latest technology and an unlimited supply of cheap labour; men prepared to live on the job in far from ideal conditions, who still produced generally high standards of workmanship. Excellent examples of such achievements include Dobson's Newcastle Central Station, and Stephenson's High Level and Royal Border Bridges.

While hardly anything physical remains of the Great Exhibition, much of the railway infrastructure which brought millions of visitors to the Crystal Palace in 1851 still survives in daily use. Bristol Temple Meads,

Settle Station in 2000, illustrating the distinctive features of the Midland Railway's design, with ridge tiles, barge boards, window and door frames appearing very much the same as when they were constructed in the mid 1870s (all photographs by Railway Heritage Trust unless otherwise stated)

Great Malvern Station: the capital of one of the platform canopy columns in 1988

Paddington, St Pancras and Huddersfield railway stations compare favourably with our great country houses, cathedrals and public buildings. The great architectural practices of the day were involved; the hotel at St Pancras, for example, was the work of Sir George Gilbert Scott, famous also for the Albert Memorial in Hyde Park; while the Charing Cross Hotel was designed by Edward M Barry, the younger son of Sir Charles Barry who built the Houses of Parliament.

Civil engineers also played their part, with Brunel's Royal Albert Bridge crossing the Tamar at Saltash and Robert Stephenson's Britannia tubular structure over the Menai Strait in North Wales.

ARCHITECTURAL VARIETY

It may not be readily apparent to today's railway traveller that the lines were built by many different railway companies rather than by one entity. However, countless distinguishing features still survive which reflect the architectural house-style of a particular company and where the ravages of time have not taken their toll.

The Midland Railway's Settle to Carlisle route illustrates this particularly well. This line was driven through some of the most demanding terrain in Britain. Doubts about the line's future in the 1970s and '80s, coupled with budgetary pressures, resulted in virtually all of the station buildings remaining *in situ* although lacking attention to maintenance. Some properties were even sold to private owners. Despite the lack of maintenance (since reversed), virtually all the distinctive features of the Midland Railway's design remains, with ridge tiles, barge boards, window and door frames appearing very much the same as when they were constructed in the mid 1870s. The progressive restoration of all the buildings on the route is now bringing pleasure to those who travel by train and significant benefits to the local economy.

The design of the Midland Railway was distinctive, but it was also robust; a statement of intent and durability. Contrast that with the ornate style of Elmslie's Great Malvern Station which is unique in minor station architecture. The wrought iron ornamental roof ridge and cast iron brackets give way to the most elaborate and richly decorated column capitals to be found anywhere on the rail network.

Of the largest stations, Bristol Temple Meads forms one of the most important assemblages of historic railway buildings anywhere in the world and is listed Grade I reflecting its architectural and historical importance. Brunel's 1840 terminus gave way to Digby Wyatt's 1874 building with its magnificent clock tower, delicate canopies and terracotta chimneys, and the complex also includes Brunel's remaining passenger building, engine sheds and offices. Bounded by the City of Bristol's new Temple Quays development to the north east, well over 160 years of architectural styles are present in a relatively small geographical area.

Not all listed railway buildings date from the 19th century. Surbiton, probably the oldest suburb in Europe, owed its existence to the railway and grew around Kingston Station which opened in 1838. This station was quickly succeeded by Kingston Junction in 1845 (renamed Surbiton and Kingston in 1863) and the Southern Railway, in turn, replaced this with the present station in 1937. J Robb Scott's attempt at the 'International' style has received mixed reviews over the years including one description as 'flashy'. Pevsner, however, regarded it as "one of the first in England to acknowledge the existence of a modern style". Restoration works in recent years have confirmed the success of Surbiton's elegant design features.

The builders of our railways did not confine imaginative design to stations and other buildings. There are numerous examples of extravagant designs which were applied to tunnel portals and bridges. Most people find the west portal of Brunel's Box Tunnel near Bath amazing, but perhaps Clayton Tunnel's north portal, on the London to Brighton line near Hassocks in West Sussex, surpasses even Brunel's efforts. Dating from 1841, the portal comprises a massive two-centred arch surmounted by a headwall flanked by two hexagonal towers. Outside each tower a flanking wall extends to a small turret on each side, combining into one of the most elaborate tunnel entrances in Britain. To complete the extravagance, nestling between the towers stands a dwelling house, originally provided for the railway's policeman.

Not too far away from Clayton Tunnel is John Urpeth Rastrick's Ouse Valley Viaduct at Balcombe. Considered by many as one of the most elegant railway viaducts, it may owe some of its classical elegance to the distinguished architect David Mocotta. In addition to its slender brick piers it has stone balustrades and no less than eight stone pavilions. The original limestone came from Caen in Normandy but this source was exhausted by the time Railtrack PLC commissioned a four-year programme of repairs in 1996–7. Comparable stone was brought from Auxerre in France enabling this Grade II* listed viaduct to be fully restored to its 1841 splendour.

Through the post-war era right up until the early 1980s, the nation's concern for its built heritage was far from a priority. Railway heritage was regarded in the same manner and the situation probably made worse as the industry lurched from one financial crisis to another.

The demolition of the Great Doric Arch at Euston in 1961 was an act that caused outrage and may well be regarded as a watershed after which the nation's Victorian heritage, and its railway heritage in particular, assumed greater importance. It wasn't until 1981 however, that the then Secretary of State for the Environment decided that a record of the rich heritage of historic railway buildings should be published, and the first volume

Surbiton Station (J Robb Scott, 1937) in 1999 © Milepost 92½

Clayton Tunnel with its elaborate portal in 1997 © Milepost 92½

was duly completed by the Public Services Agency. The British Railways Board, conscious of the interest and variety of British Rail's heritage, responded positively by establishing the Railway Heritage Trust in April 1985. The Board's view that a respected and maintained heritage complements an efficient and modern railway service was a welcome vision indeed.

DEALING WITH CONSERVATION

A major challenge for the railway industry is finding solutions for historic buildings which have become totally or partly surplus to operational requirements. The problem applies to buildings rather than bridges; there are only 44 listed bridges within the non-operational estate.

Buildings, however, range from signal boxes, crossing keeper's cottages and rolling stock maintenance depots to small, medium and large stations where, following the introduction of modern technology, productivity and re-organisation initiatives, their operational usefulness has reduced significantly or altogether. For example, the introduction of new signalling technology such as Solid State Interlocking or Radio Electronic Token Block can often result in one new signalling centre removing the requirement for up to 80 traditional signal boxes. The de-staffing of rural branch line stations and rationalisation of rolling stock maintenance depots have created unused space or totally empty buildings. Often the siting of such buildings, with their close proximity to the railway, imposes restrictions on reuse for safety reasons.

Following years of decline, demand to use the railway is increasing. 21st century customers usually welcome the historic environment provided by the variety of architecture on display to them as they use station buildings, but they rightly have 21st century expectations such as lifts, escalators, customer information systems, modern ticketing facilities, lighting and signage.

These modern requirements must be carefully balanced against the calls for minimum intervention and for alterations to be reversible. Clearly there is a need to respect the character and historic interest of our built heritage. The use by the industry of architects with conservation experience is essential if win-win solutions are to be achieved.

The Railway Heritage Trust's remit can be effectively split into two parts. First, the focus is the conservation and enhancement of buildings and structures which are listed for their special architectural or historical interest or scheduled as ancient monuments. Second, the Trust acts as a catalyst assisting outside parties and the buildings' owners on the conservation of non-operational property and securing alternative uses, including the possible transfer of responsibility to local trusts or other interested organisations.

How the railway industry, supported by the Railway Heritage Trust, attempts to balance conservation and business requirements is best demonstrated by examples.

Probably the most dramatic demonstration of reversing previous unsympathetic action is the removal of the 1954 signalling centre/relay room which had been driven through Bell's south barrel roof at Newcastle. Following major engineering works, the steel framed intrusion was removed in 1998/99 and the splendid 1893 roof has now been fully restored to its original condition.

Manchester Piccadilly Station is another example. By the late 1990s, a succession of uncoordinated initiatives, many severely restricted financially, had produced a hotchpotch of station facilities which neither met customer needs nor respected the buildings' architectural and historic importance. Then an intensive three-year programme of work was commenced which led to a complete transformation of the concourse, the restoration of the train shed walls and

Newcastle Station: Bell's south barrel roof in 1997, shortly before the removal of the intrusive 1954 signalling centre/relay room

The former Doric Arch at Euston Station, London which was demolished in 1961, shown in a watercolour of 1838

A former water tank (1861) at Haltwhistle Station following restoration in 1999

former Goods Agents Offices and the dramatic 'opening' of the 1883 undercroft. All of this work now splendidly complements the earlier recladding of the train-shed roof in modern 'planar' glazing. Damaged capitals and filigree brackets of the cast iron train shed columns have been repaired and even a travelator has been introduced. Modern intervention has blended excellently with the 1883 construction in a massive and highly successful investment.

On a much smaller scale, and as a complete contrast, is the restoration of the 1861 water tank at Haltwhistle, including the provision of office accommodation within the red sandstone arched base as part of the Haltwhistle Regeneration Strategy. For almost two decades this structure had declined and been little used other than as a store for platelayer tools and materials. Restoration and reuse has delivered conservation through regeneration. Even the decorative castings on the water tank have been restored.

Reuse on a much grander scale occurred at Liverpool where the former London & North Western Railway's North Western Hotel of 1871 was re-opened by John Moores University as student accommodation in October 1996. The hotel had closed in 1933 and for the next few decades the building was used as railway offices. After it was finally abandoned in the 1970s, its condition declined rapidly. This magnificent structure has now been fully restored, apart from a relatively small part of the ground floor, with significant works being carried out to the stairway and entrance hall. The University has taken out a 150 year lease on the building, thus conserving its future.

Most of the stations on the West Highland Line between Helensburgh and Fort William became unstaffed halts in the 1980s. Several of the stations which were built in the early 1890s are in a 'Swiss chalet' style with matching pavilion-like signal boxes and are listed Grade B. Bridge of Orchy was one such example. Only minimal use was being made of one small room until in 1998 an imaginative proposal was developed to convert the station into a 15-bed bunkhouse for use by walkers, fishermen, canoeists, skiers and tourists. The excellent conversion, including the restoration of dado panelling, window frames, doors and cladding 'shingles', was achieved through a five way partnership. The accommodation, now known as the 'West Highland Way Sleeper', sits on an island platform which still receives regular train services on this route.

No building is too small to be brought back into use once it has become non-operational. A simple conversion has allowed a small business to take over the signal box at Torquay Station and this is just one example where determination by all concerned can ensure conservation in situ.

It would be wrong not to give an example of the reuse of a non-operational railway viaduct to complete this short review of conservation and restoration works. Lambley Viaduct in Northumberland was designed by Sir George Barclay-Bruce and when it was opened in 1852 it was the only major feature on the Haltwhistle to Alston branch line. The branch closed in 1976 and over the next 16 years defects started to grow and the parapets were vandalised. British Rail Property Board assembled a group of partners to fund significant devegetation and repair works. The partners included the North Pennines Heritage Trust which, on completion of the works, took over the viaduct and opened it as a footpath and cycle way which links to the South Tyne Trail.

Historic railway buildings and structures are important examples of our nation's built heritage. Significant efforts have been made over the past two decades to conserve that heritage whilst at the same time responding to modern business requirements.

JIM CORNELL CEng FREng FICE FCIT FCMI is a career railwayman with almost 44 years experience in the industry. Having trained as a civil engineer, he held a series of senior roles within British Rail including General Manager, ScotRail, Director of Civil Engineering, Managing Director, Regional Railways and Group Managing Director, BRIS. Upon leaving BR he became Executive Director, Railway Heritage Trust in 1996. He is also a Non-Executive Director, Network Rail.

Railway Heritage Trust, PO Box 686,
Melton House, 65/67 Clarendon Road,
Watford, Hertfordshire WD17 1XZ
Tel 01923 240250 Fax 01923 207079
E-mail rht@networkrail.co.uk

▶ A D CALVERT ARCHITECTURAL STONE SUPPLIES LTD

Smithy Lane, Grove Square, Leyburn, North Yorkshire DL8 5DZ
Tel 01969 622515 Fax 01969 624345
E-mail stone@calverts.co.uk
Website www.calverts.co.uk

STONE MASONRY AND ARCHITECTURAL STONE SUPPLIER: From the heart of the Yorkshire Dales A D Calvert supplies stone sawn six sides, profiled, lathed, carved and hand-finished walling, using eight types of sandstone and two limestone. This includes the company's own local Witton Fell medium grained buff coloured sandstone which is highly recommended for new and renovation work. For further information contact Andrew Calvert who will be pleased to discuss your requirements.

▶ A F JONES (STONEMASONS)

33 Bedford Road, Reading, Berkshire RG1 7EX
Tel 0118 957 3537 Fax 0118 957 4334
E-mail af.jones@ukonline.co.uk
Website www.afjones.co.uk

MASTER STONEMASONS: Established in 1858 by Arthur F Jones and continued today by G A and A G Jones, this firm has accumulated experience and expertise gained by five generations of stone masons. There are currently 20 highly skilled craftsmen, many of whom have spent their working lives with A F Jones, restoring, carving, cleaning and conserving stone facades. A F Jones offers a complete service from consultancy and specification to production and site fixing. Their style of management is non-confrontational and above all, fair and honest. A F Jones has worked for The Churches Conservation Trust, English Heritage, the National Trust and Historic Royal Palaces. *See also: display entry on this page.*

STONE

▶ ABBEY HERITAGE LTD

Midstfields, Frome Road, Writhlington, Radstock, Near Bath, Somerset BA3 5UD
Tel 01761 420145 Fax 01761 437103
E-mail stone@abbeyh.co.uk

MASONRY, RESTORATION, FACADE CLEANING AND LASER CLEANING: Specialists in facade cleaning, anti-graffiti treatments, natural stone replacement and restoration, new stone projects, terracotta, faience, granite, marble and all aspects of brickwork, Abbey Heritage works closely with leading architects, surveyors, local authorities and consultancies throughout the UK on historic and prominent national buildings. Dedication to detail and a high level of managerial input has earned Abbey Heritage a reputation for integrity and reliability, generating consistent repeat business. As principal contractor Abbey provides associated leadwork, roofing, decoration and joinery services. Leaders in the field of laser cleaning for buildings, artefacts and artworks. They offer full national and international consultancy services.

▶ ABBEY MASONRY AND RESTORATION LIMITED

Heol Parc Mawr, Cross Hands Business Park, Cross Hands, Llanelli, Carmarthenshire SA14 6RE
Tel 01269 845 084 Fax 01269 831 774
E-mail info@abbeymasonry.com Website www.abbeymasonry.com

CRAFTSMEN IN STONE MASONRY AND SPECIALISTS IN CARVING CONSERVATION: As recognised specialists in the conservation and restoration of ecclesiastical and historical buildings, Abbey Masonry have restored over 160 churches and cathedrals in England and Wales. Purpose built masonry works accommodate state of the art technology enhanced by skilled, dedicated craftsmen which combine to offer an exceptional service. They are also approved installers of Cintec anchors and registered operatives of the Jos and Doff stone cleaning systems. Contact Abbey Masonry today to discuss your requirements. *See also: display entry on page 100.*

▶ ANELAYS

William Anelay Limited, Murton Way, Osbaldwick, York YO19 5UW
Tel 01904 412624 Fax 01904 413535
E-mail office@williamanelay.co.uk Website www.williamanelay.co.uk

BUILDING AND RESTORATION CONTRACTORS: *See also: profile entry in Building Contractors section, page 58.*

▶ ARCHITECTURAL STONE CONSERVATION

Greystones, Litcham Road, Mileham, Norfolk PE32 2PS
Tel/Fax 01328 700092
E-mail ascgrp@btopenworld.com Website www.ascgrp.btinternet.co.uk

STONE SPECIALISTS: For the repair of churches, historic buildings and monuments. ASC works both as main contractor and specialist sub-contractor. Clients include: English Heritage, National Trust, The Churches Conservation Trust, local authorities, parochial church councils and private clients. Recent projects can be seen on ASC's website.

▶ THE BATH STONE GROUP

Stoke Hill Mine, Midford Lane, Limpley Stoke, Nr Bath BA2 7SP
Tel 01225 723792 Fax 01225 722129
E-mail bsg@bath-stone.co.uk Website www.bath-stone.co.uk
Contact Elaine Marson

STOKE GROUND BATH STONE: The Bath Stone Group, producing award winning Stoke Ground Bath Stone, offers an integrated service from design through to installation on site. The company's highly skilled masons and carvers, can fabricate and finish stone to exacting specifications. Stoke Ground is regularly specified by English Heritage and the National Trust, and used in restoration and new-build projects including: Buckingham Palace and Windsor Castle; Oxbridge colleges: Merton, Manchester, Brazenose and Pembroke; Stations: Liverpool Street, London and Temple Meads, Bristol; Development: Wessex Water, Bath and Britannic Assurance HQ, Birmingham; Conservation: Temple of Concorde, Stowe and Waddesdon Manor, Berks, British Waterways Board, canals and aqueducts etc. Bath Stone Group offers cost effective natural stone building solutions.

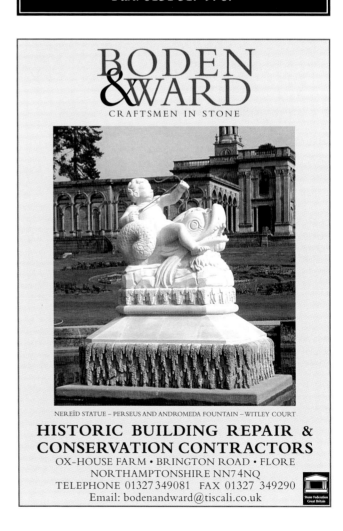

STONE

▶ BODEN & WARD STONEMASONS LTD

Ox-House F101arm, Brington Road, Flore, Northants NN7 4NQ
Tel 01327 349081 Fax 01327 349290
E-mail bodenandward@tiscali.co.uk
Website www.bodenandward.co.uk

CRAFTSMEN IN STONE: Masons and stone carvers, highly experienced in all aspects of stone repair, cleaning and restoration. Recent projects include the restoration of the Perseus and Andromeda Fountain at Witley Court for English Heritage, along with the production of two new Nereid statues each carved out of eight tonnes of Portland stone. Boden & Ward has also carried out masonry repairs at Woburn Abbey and major stonework restoration projects at North Marston and Drayton Parslow churches. The company is just embarking on the complete renewal of a rose window on St Magnus Cathedral, Kirkwall in the Orkney Islands, and has recently won another Stone Federation Great Britain award for its work on Ringshall Lodge, Little Gaddesden. Boden & Ward is an experienced principal contractor and specialist sub-contractor supplying and fixing worked and sawn stone. A stone consultancy service is also offered. *See also: display entry on page 101.*

▶ BURLEIGH STONE CLEANING & RESTORATION CO LTD

The Old Stables, 56 Balliol Road, Bootle, Merseyside L20 7EJ
Tel 0151 922 3366 Fax 0151 922 3377
E-mail info@burleighstone.co.uk
Website www.burleighstone.co.uk

STONE SPECIALISTS: A comprehensive high quality service in stone masonry, cleaning, restoration, carving and replacement. *See also: display entry in Masonry Cleaning section, page 166.*

▶ BURROWS DAVIES LIMITED

The Stoneyard, West End, Strensall, York YO32 5WH
Tel 01904 491849 Fax 01904 491910

MASONRY AND RESTORATION SPECIALISTS: Burrows Davies Limited carries out high quality masonry, conservation and restoration works on historic properties and listed buildings including churches, historic houses and monuments throughout the Midlands, North of England and Scotland. The company works primarily as principal contractor and occasionally as specialist sub-contractor and over the past few years has successfully completed various prestigious projects ranging in value from a few thousand pounds to £1 million.

▶ C GINN BUILDING RESTORATION

89 Lunedale Road, Fleet Estate, Dartford, Kent DA2 6LW
Tel 01322 290505 Fax 01322 284839
E-mail c.ginnbuildingrestoration@btinternet.com
Website www.stonecleaning-restoration.com

RESTORATION AND CLEANING SPECIALISTS: C Ginn Building Restoration has a proven track record for the repair and cleaning of exterior fabrics such as stone, stucco, faience, terracotta and brickwork. They are also experienced at supplying and fixing natural and precast stone details, and are always willing to undertake surveys together with costs for all aspects of work within their field. Whilst undertaking large projects the company is always available for the smaller projects and offers the same level of service throughout. If your project requires assistance in 'taking the weight' please contact C Ginn Building Restoration for further discussions and information required. *See also: display entry in Masonry Cleaning section, page 167.*

▶ CARREK LTD

Mason's Yard, Wells Cathedral, Wells, Somerset BA5 2PA
Tel 01749 689000 Fax 01749 689089
Website www.carrek.co.uk

HISTORIC BUILDING REPAIR COMPANY: *See also: profile entry in Building Contractors section, page 61.*

▶ CARTHY CONSERVATION LTD

18 Alexandria Road, London W13 0NR
Tel/Fax 020 8840 3294 E-mail deborahcarthy@btclick.com

CONSERVATION OF ARCHITECTURAL DETAIL AND SCULPTURE: Carthy Conservation offers an established team which carries out high quality conservation and consultancy projects on stone, terracotta, plaster, mosaic and wood. Gilding, polychrome and monochrome surfaces are also specialities. The company has established strong links with analytical laboratories and scientists both in this country and abroad who have established reputations in conservation. Work is undertaken across the United Kingdom for private clients, architects and building consultants, construction main contractors, cathedrals, churches and government agencies. This conservation practice is included on the Conservation Register maintained by the United Kingdom Institute for Conservation.

▶ CATHEDRAL WORKS ORGANISATION (CHICHESTER) LTD

Terminus Road, Chichester, West Sussex PO19 8TX
Tel 01243 784225 Fax 01243 813700
E-mail c.w.o@osborne.co.uk Website www.cwo.uk.com

SPECIALIST STONEMASONRY, RESTORATION, CLEANING AND SUPPLY: CWO has a £6 million annual turnover and employs 90 staff and skilled personnel. A holder of the Royal Warrant to HM the Queen, they have the skills, experience and resources to deal with main contracting or specialist sub-contracting in new build, restoration and conservation projects throughout the South of England, including Central and Greater London. CWO offers a complete service from specifications and drawings, to production and completion on site. In 2002/03 major new investments in machinery and production have taken place, giving advanced production facilities. Recent contracts include Queen's Gallery, Buckingham Palace; Memorial Gates, Constitution Hill; Paternoster Square; Mansion House; Temple Bar, all Central London; Christ Hospital, Horsham and St Mary's Church, Shoreham. CWO Special Works Department carries out contracts up to £100,000.

▶ CLIVEDEN CONSERVATION WORKSHOP LTD

Head Office – The Tennis Courts, Cliveden Estate, Taplow, Maidenhead, Berkshire SL6 0JA
Tel 01628 604721 Fax 01628 660379

SCULPTURE, STONE, PLASTER AND WALL PAINTINGS CONSERVATION: Established in 1982 as the National Trust Statuary Workshop, independent since 1991. Retained as the National Trust centre for the conservation of statuary, stone and plasterwork. Also serving English Heritage and the Royal Palaces. Projects extensively in country houses and churches. Reinstated decorative plaster ceilings at Edinburgh Castle and Uppark. Repair and replication of chimneypieces and monument repairs. Repair and replacement of statuary at Stowe Landscape Gardens and Osborne House. Archaeological conservation at Aphrodisias, Turkey. Conservation of wall and ceiling paintings in the chapels of Keble and Worcester Colleges (Oxford) and Royal Holloway (London University). Consultancy service, specifications, security work and materials analysis.

▶ CLIVEDEN CONSERVATION WORKSHOP LTD

Home Farm, Ammerdown Estate, Kilmersdon, Bath, Somerset BA3 5SN
Tel 01761 420300 Fax 01761 420400
E-mail info@clivedenconservation.co.uk

SCULPTURE, STONE, PLASTER, MOSAIC, WALL PAINTING CONSERVATION, STONE CARVING: The Bath workshop of Cliveden Conservation has all the skills and experience of the Maidenhead workshop – the conservation of statuary, masonry, ceramics, plasterwork, wall paintings and mosaics is undertaken as well as moving, fixing, and securing of statuary. Regular clients include the National Trust, local authorities, English Heritage, ecclesiastical and private clients. Geographically the work is on site in the South-West, the Midlands, the North of England, Wales, Scotland, Northern Ireland and Eire with occasional projects in Europe and the USA. They act either as main contractor or specialist sub-contractor. The company's stone carving section is based at its Bath workshop. *See also: head office profile entry above.*

STONE

▶ COE STONE LTD

4 Castle Street, Frome, Somerset BA11 3BN
Tel/Fax 01373 462210 Mobile 07771 887574
E-mail enquiries@olivercoe.com Website www.olivercoe.com

STONE MASONS: Coe Stone is a dedicated small company of craftsmen which specialises in the repair of stonework to historic buildings combining many years experience in the field. Coe Stone also offers exceptional carving skills undertaking commission work ranging from the reproduction of 15th century alabaster relief to design and execution of a classical garden temple. Based in Somerset, the company works principally in the south of England and on occasions as far afield as the USA. Coe Stone is pleased to offer consultation and advice on request.

▶ DAVID BALL RESTORATION (LONDON) LIMITED

104A Consort Road, London SE15 2PR
Tel 020 7277 7775 Fax 020 7635 0556
E-mail mail@dbr.uk.com

CLEANING AND REPAIR SPECIALISTS: David Ball Restoration Limited specialises in the cleaning and repair of stone, brick, stucco and terracotta facades. The company has the management and craft skills to undertake major contracts, but also acts as a specialist sub-contractor in their core business activity of cleaning and repairing commercial and public buildings. The company's comprehensive knowledge of the repair of historic buildings enables them to offer advice on traditional forms of construction as well as the latest conservation techniques. Ancillary craft operations, such as leadwork, ironwork, woodwork, roofing, painting and decoration, may also be undertaken as part of a managed contract. *See also: display entry on this page.*

▶ E BOWMAN AND SONS LTD

Cherryholt Road, Stamford, Lincolnshire PE9 2ER
Tel 01780 751015 Fax 01780 759051
E-mail mail@ebowman.co.uk Website www.ebowman.co.uk

SPECIALIST STONEMASONS AND STONE CONSULTANCY SERVICES: Masonry works undertaken to include supplying numerous limestones – Clipsham, Ancaster, Ketton, Bath and Portland along with some sandstone. Services provided vary from initial advice to clients, detailed template work, through to supplying both sawn and carved masonry along with all installation works and associated fixings. All works carried out by a dedicated team of experienced masons. Properties and projects carried out in the past range from large country houses such as Ashridge College – with detailed clunch masonry repairs and replacement to ecclesiastical projects as seen at Warkton Church with new limestone tracery windows. *See also: display entry in Building Contractors section, page 60.*

▶ F T B RESTORATION BVBA

Vaartstraat, 128 E3, 2520 Ranst, Belgium
Tel 0032 3475 1957 Fax 0032 3475 1962

MASONRY CLEANING: *See also: Arte Mundit display entry in Masonry Cleaning section, page 167.*

▶ FAIRHAVEN OF ANGLESEY ABBEY

Northfield Farm Stoneyard, Lode Road, Bottisham, Cambridgeshire CB5 9DN
Tel 01223 812555 Fax 01223 812554
Website www.fairhavengroup.co.uk

CONSERVATION AND RESTORATION OF STONE IN BUILDINGS AND OTHER ARTEFACTS: There are four departments: the conservation department has a fully equipped laboratory for materials analysis and off-site conservation work. Site conservation includes JOS cleaning, solvent and re-agent gel poulticing, shelter coating and limewashing, environmental monitoring and surveys and graffiti removal. The masonry restoration and contracting department covers churches, colleges and listed buildings and monuments within a 100 mile radius of Cambridge. The stonemasonry department has fully equipped workshops and directly employed apprentice trained craftsmen. The stone carving and sculpture department covers ornamental and figurative carving, modelling, design and letter cutting. Visit the website at www.fairhavengroup.co.uk for further details.

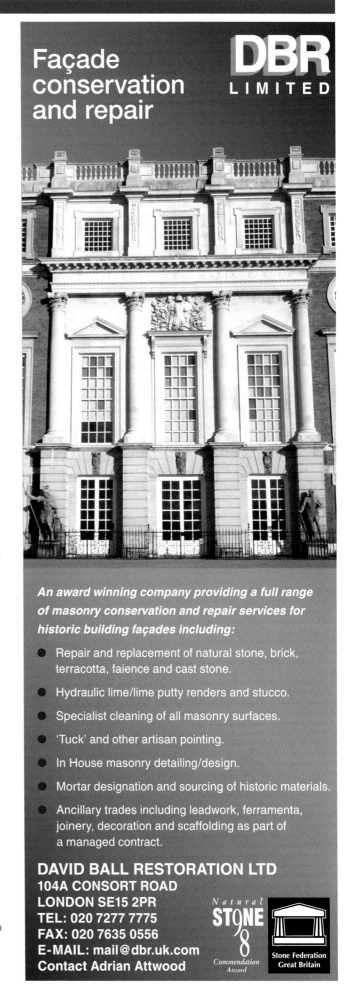

▶ FILKINS STONE COMPANY

Filkins, Nr Lechlade, Gloucestershire GL7 3JJ
Tel 01367 860660 Fax 01367 860661 E-mail stone@naturalbest.co.uk

BANKER MASONRY: Filkins Stone Company can work and supply any architectural stonework for restoration or new-work projects; ashlar through to complex mouldings, plus a large range of stone flooring. The company can also help with on-site measurement, advice on stone selection, and practical matters. Working drawings can be prepared in-house. Excellent references are available. Visits to Filkins Stone's workshops and splendid gallery are warmly welcomed.

▶ GRALL PLASTER & STONE SPECIALISTS

Forest Corner, Purlieu Lane, Godshill, Fordingbridge, Hampshire SP6 2LW
Tel 01425 655430 Mobile 07973 832 846

STONE: *See also: display entry in Plasterwork section, page 214.*

▶ H & W SELLORS LTD

Milford Works, Bakewell, Derby DE45 1DX
Tel 01629 812058 01629 815138

ROOFING, RESTORATION AND BUILDING CONTRACTORS: A family firm which was established in 1850 as traditional stone slating contractors has expanded to undertake work in building conservation and conversion. The company has been involved in a programme of maintenance work at Chatsworth House, Derbyshire including restoration of the Cascade and Cascade House. Work is undertaken on both residential and commercial properties, and a recent conversion of a timber mill to offices within the Peak District National Park has been nominated for an award.

▶ H G CLARKE & SON

Weston Underwood, Olney, Bucks MK46 5JS
Tel 01234 711358 Fax 01234 712047 Mobile 07714 408151

TOTTERNHOE LIMESTONE SUPPLIERS: Original, historic Totternhoe clunch stone. Keen prices for random sawn slabbing, suitable for extensions and new house facings in chalk stone areas. Contact anytime.

▶ INVICTA STONE LTD

Hayward Queens Road, Lydd, Kent TN29 9DF
Tel 01797 322321 Fax 01797 322321
E-mail invictastone@aol.com

STONE MASONRY

▶ J JOSLIN (CONTRACTORS) LTD

Southrah Quarry, Lower Road, Long Hanborough, Witney, Oxfordshire OX29 8LR
Tel 01993 882153 Fax 01993 882960 E-mail masons@joslins.demon.co.uk
Contact Managing Director, Neville McLean

CONSERVATION, RESTORATION, AND SPECIALIST MASONRY CLEANING: Joslins of Oxfordshire is one of Britain's leading stonemasons, with over 100 years experience. Recently awarded the Royal Warrant Joslins has an enviable reputation for quality of workmanship and reliability. Current projects include the restoration of Chichester Cathedral and Waddesdon Manor. Whether a project involves restoration, conservation or new build Joslins' skilled craftsmen can provide a complete service of the highest standards including hand carving stone to specification. Joslins successfully combines the traditional skills of hand-crafted stonemasonry with the latest technology, and is also an approved user of the Jos, Doff and Torc methods of specialist stone cleaning. *See also: display entry on page 105.*

▶ JAMES LUGG STONEMASONRY LTD

Unit 5A, Kiln Industries, Fittleworth Lane, Wisborough Green, West Sussex RH14 0ES
Tel 01403 700199 Fax 01403 700199 Mobile 07770 672880 E-mail jima1@tiscali.co.uk
Website www.lugg.uk.com

STONE: Lugg Stonemasonry provides a complete service in all areas of stonemasonry. From the design of new and contemporary buildings, to the restoration and conservation of stone. For further information, contact 01403 700199 or 07770 672880, or visit the website www.lugg.uk.com.

▶ MAYSAND LIMITED

109–111 Windsor Road, Oldham, Lancs OL8 1RH
Tel 0161 628 8888 Fax 0161 627 0996 E-mail sales@maysand.co.uk

MASONRY REPAIR, CLEANING, RESTORATION AND REPLACEMENT: *See also: display entry in this section, page 106.*

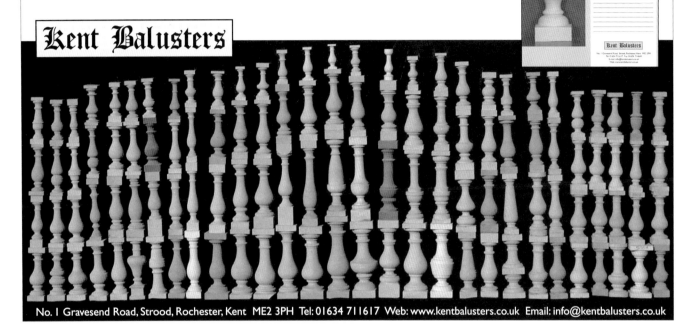

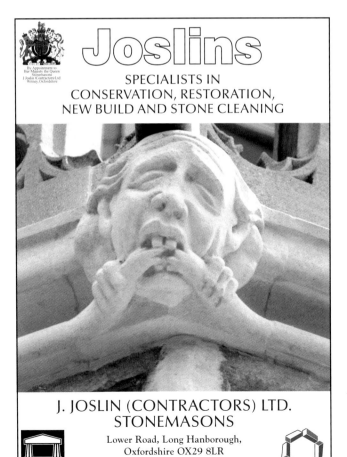

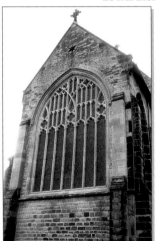

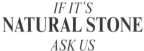

STONE

▶ MINERVA STONE CONSERVATION

Oak House, Turleigh Hill, Bradford-On-Avon, Wiltshire BA15 2HE
Tel 01225 862386 Fax 01225 862386
STONE MASONS

▶ MOWLEM RATTEE & KETT

Digital Park, Station Road, Longstanton, Cambridge CB4 5FB
Tel 01954 262600 Fax 01954 262626
SPECIALIST RESTORATION AND MASONRY CONTRACTORS: Building on 160 years of experience, Mowlem Rattee & Kett has been responsible for building, conserving and repairing some of the country's finest historic buildings. Management skills have developed alongside the traditional craft based activities including dedicated conservation staff to provide a sensitive and commercial approach to restoration, construction and refurbishment contracts. Operating as a specialist division of the Mowlem Group, few companies can demonstrate the same depth of project management experience and comprehensive track record. Key projects include work on the Houses of Parliament, Westminster Abbey, Ely Cathedral and the £9 million restoration of the Albert Memorial.

▶ NICOLAS BOYES STONE CONSERVATION

46 Balcarres Street, Edinburgh EH10 5JQ
Tel 0131 446 0277 Fax 0131 446 0283 Website www.nb-sc.co.uk
STONE CONSERVATION SERVICES: Throughout Scotland and Northern England Nicolas Boyes Stone Conservation specialises in the conservation of carved and decorative stonework – sculpture, statuary, architectural decoration and historic building fabric in traditional materials – and decorative plasterwork. Replication of subjects can be undertaken. The Phoenix Conservation Laser system is often employed to remove soiling, and environmental monitoring and preventive conservation services can be provided as part of the full service. The company conducts important conservation work from recording through to report with sensitivity and imagination, on time, within budget and with good conservation rigour.

▶ NIMBUS CONSERVATION LIMITED

Eastgate, Christchurch Street East, Frome, Somerset BA11 1QD
Tel 01373 474646 Fax 01373 474648 E-mail enquiries@nimbusconservationlimited.com
Website www.nimbusconservation.com
STONE CONSERVATION AND MASONRY: NIMBUS has been established for over 20 years as specialists in stone working on many historic secular and religious buildings and important monuments. Experienced teams conserve and clean a wide variety of materials including marble, limestone, sandstone, brick, terracotta, render, plaster and decorative and polychrome surfaces. The company's masons provide sensitive replacement carving and new commissions either on site or in Nimbus' dedicated workshop. Support is provided for the London and the West Country teams from the company's purpose built head office. Recent contracts include St Paul's and Gloucester Cathedrals, Cobham Hall, the Tower of London, Hampton Court, Oxford's Martyrs Memorial and churches, monuments and sculpture throughout Britain. *See also: display entry in this section, page 107, and profile entry in Masonry Cleaning section, page 166.*

▶ RICHARD NOVISS

63 North Street, Middle Barton, Chipping Norton, Oxon OX7 7BH
Tel 01869 347062 Mobile 07714 140574 Website www.thestonesculptor.co.uk
SCULPTOR, STONEMASON AND RESTORER

▶ PAYE STONEWORK

Stationmaster's House, Mottingham Station Approach, London SE9 4EL
Tel 020 8857 9111 Fax 020 8857 9222 E-mail kw@payestone.co.uk
Website www.payestone.co.uk
MASONRY CONSTRUCTION AND REPAIR SPECIALISTS: With its own labour resources skilled in traditional techniques, PAYE is able to undertake cleaning, adaptation, and conservation repairs of stone, brickwork, stucco, terracotta and faience to the highest standards. They frequently work on historic monuments, national public buildings, Royal Palaces, churches and commercial projects both as a principal contractor and specialist subcontractor. Recent contracts include Tower of London, Windsor Castle, Marlborough House, H M Treasury, Royal Albert Hall, Osborne House Isle of Wight. PAYE Stonework is pleased to undertake surveys, provide cost budgets and consultancy advice. Contact Karrie White or George Burrage for further information and a company brochure. *See also: display entry in this section, page 107.*

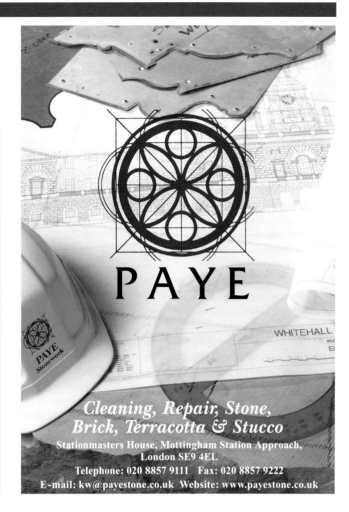

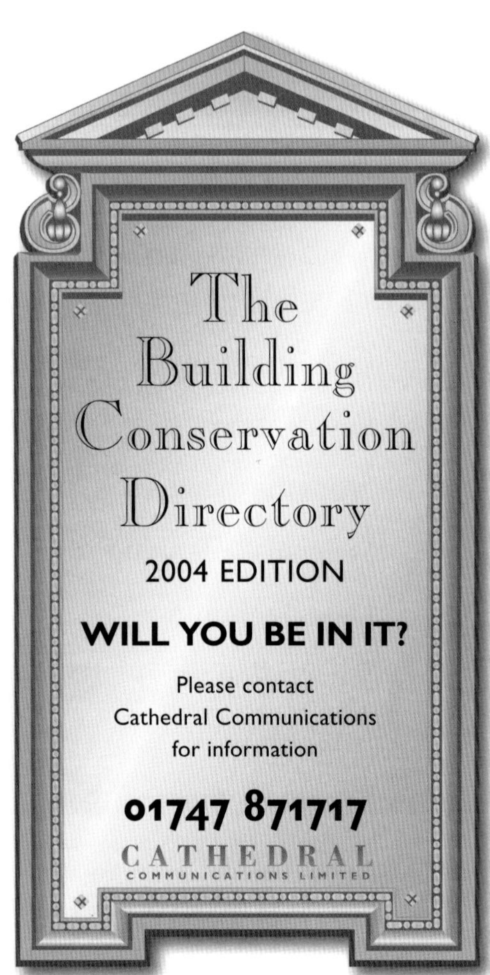

▶ PRIEST RESTORATION

96 Moyser Road, London SW16 6SH
Tel 020 8677 5660 Fax 020 8677 2550
E-mail enquiries@priestrestoration.co.uk
Website www.priestrestoration.co.uk

STONE SPECIALISTS: Priest Restoration is able to offer a comprehensive service providing a fully detailed survey identifying the cause of defects and most appropriate methods of cleaning, restoration and repair. They have recently restored prestigious buildings such as: The Mansion House, The Lyceum Theatre, Aldwych Theatre, Hatchlands Park Surrey, Aquascutum Regent Street, Langham Hotel Regent Street, Eton College, and The Royal Brompton Hospital. They have also been responsible for the following new-build projects: Louis Vuitton and Tommy Hilfiger stores in New Bond Street, stone paving and flooring at Columbus Courtyard and HSBC Tower Canary Wharf, and the Sackler Library stonework at Oxford University. Priest's management team has considerable experience within the stone and restoration industry offering expertise in cleaning and restoring; stonework, brickwork, terracotta, faience, stucco, marble and granite.

▶ RUSSCOTT CONSERVATION AND MASONRY

The Barn, Wood Lane, Eastcote, Warwickshire B92 0JL
Tel 01675 446123 Fax 0121 706 3886

SPECIALISTS IN THE REPAIR AND CONSERVATION OF HISTORIC BUILDINGS AND MONUMENTS: *See also: display entry in this section, page 107.*

▶ STONEGUARD

St Martins House, The Runway, Ruislip, Middlesex HA4 6SG
Tel 0870 241 6366 Fax 020 8839 9988
Website www.stoneguard.co.uk
▶ with offices also in Bath, Birmingham, Manchester and Stirling

MASONRY CLEANING AND PROTECTION: *See also: profile entry in Building Contractors section, page 66.*

▶ STONEWEST LTD

Lamberts Place, St James's Road, Croydon CR9 2HX
Tel 020 8684 6646 Fax 020 8684 9323
E-mail stonewest@cwcom.net
Website www.stonewest.co.uk

BUILDING CONTRACTORS AND STONE MASONS: *See also: display entry in Building Contractors section, page 67.*

▶ WELLS CATHEDRAL STONEMASONS LTD

Brunel Stoneworks, Station Road, Cheddar, Somerset BS27 3AH
Tel 01934 743544 Fax 01934 744536
E-mail wcs@stone-mason.co.uk
Website www.stone-mason.co.uk

STONEMASONS: *See also: display entry in this section, page 109.*

▶ WELLS MASONRY SERVICES LTD

Ilsom Farm, Cirencester Road, Tetbury, Glos GL8 8RX
Tel 01666 504251 Fax 01666 502285

STONE RESTORATION SERVICES: A privately owned specialist masonry company, Wells Masonry Services provides specialist stone cleaning, restoration and carving services. Its team of capable masons handles a diverse and prestigious range of structural and decorative natural stonework particularly in the repair and conservation of historic buildings. Projects range from extensive restoration at Balliol College Oxford and stonework for the Queen Mother's Commemorative Gates in Hyde Park to recent commissions at the Four Pillars Hotel, Tortworth Hotel near Bristol and Tewkesbury Abbey. Wells Masonry Services excels where quality workmanship is paramount. Wessex Stone Fireplaces, a subsidiary company, is a market leading producer of natural stone fire surrounds.

STONE

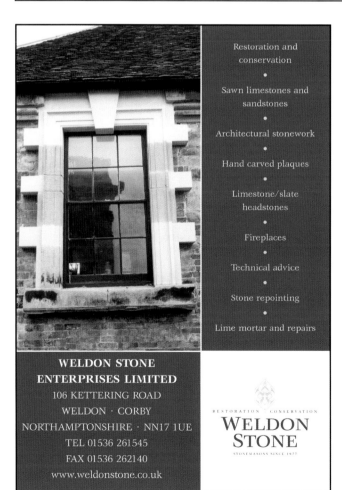

CAST STONE

▶ BROADMEAD CAST STONE

Broadmead Works, Hart Street, Maidstone, Kent ME16 8RE
Tel 01622 690960 Fax 01622 765484

RECONSTRUCTED STONEWORK: Broadmead Cast Stone has since 1920 manufactured reconstructed stonework and is an acknowledged leader in the market, undertaking both refurbishment and new build projects. The ability of Broadmead to recreate high quality stone units by the use of natural aggregates with extremely accurate surfaces and details is the basis of its market strength. By integrating its specialist knowledge with that of the designer's objectives, Broadmead is able to provide a satisfactory solution to all aesthetic and construction requirements. Broadmead has an unparalleled reputation for the quality of its work, and has attained accreditation as a Quality Assured Company to BS 5750 Part 2. *See also: Wallis display entry in Building Contractors section, page 54.*

▶ CHILSTONE

Victoria Park, Fordcombe Road, Langton Green, Tunbridge Wells, Kent TN3 0RE
Tel 01892 740866 Fax 01892 740249

RECONSTRUCTED STONE PRODUCTS: Handmade architectural stonework for classical buildings and gardens. Range includes columns, coping, balustrade, cornice, temples, statues, urns, porticos, birdbaths, sundials, fountains, gate piers and finials. Used in numerous historic gardens - Warwick and Hever Castles, Kew, Regents Park and Buckingham Palace. Also 'specials' to your design. Fully illustrated catalogue available.

▶ KENT BALUSTERS

1, Gravesend Road, Strood, Rochester, Kent ME2 3PH
Tel 01634 711617 Fax 01634 714644
E-mail info@kentbalusters.co.uk
Website www.kentbalusters.co.uk

RECONSTRUCTED STONE: Replacement baluster specialists with a range of 200 different profiles. Also standard items such as columns, caps, coping and balustrade. *See also: display entry in this section, page 105.*

▶ THORVERTON STONE COMPANY

Seychelles Farm, Upton Pyne, Exeter, Devon EX5 5HY
Tel 01392 851822 Fax 01392 851833 E-mail caststone@thorvertonstone.co.uk
Website www.thorvertonstone.co.uk

RECONSTRUCTED STONE: Manufacturers of standard and bespoke architectural masonry in reconstructed cast stone, for use in all types of building and landscapes: typical products include mullion window surrounds, cills, heads, columns, arches, porticos, steps, balustrading, coping, pier caps, ball finials, fireplaces, retaining wall sections and water channels. Thorverton's reconstructed stone is manufactured from natural aggregates and crushed stone from many regions of the country. Each piece is carefully finished by skilled craftsmen. Specialists in customer designed specials.

ARCHITECTURAL TERRACOTTA

▶ BURLEIGH STONE CLEANING & RESTORATION CO LTD

The Old Stables, 56 Balliol Road, Bootle, Merseyside L20 7EJ
Tel 0151 922 3366 Fax 0151 922 3377 E-mail info@burleighstone.co.uk
Website www.burleighstone.co.uk

ARCHITECTURAL TERRACOTTA: A high quality comprehensive service on terracotta: cleaning, repair, restoration and replacement. *See also: display entry in Masonry Cleaning section, page 166.*

▶ LAMBS TERRACOTTA & FAIENCE

Nyewood Court, Brookers Road, Billingshurst, W Sussex RH14 9RZ
Tel 01403 785141 ext 223 Fax 01403 784663 E-mail sales@lambsterracotta.com
Website www.lambsterracotta.com

ARCHITECTURAL TERRACOTTA: Lamb's Terracotta & Faience is manufactured to match your originals by taking moulds, drawings and photographs of the original pieces. Research is carried out to ensure that all aspects of the original are replicated and colour samples submitted for approval. The ethos is such that Lamb's is prepared to investigate matching any fired medium for individual pieces or multi repeat items. The company's team of sculptural technicians will replicate any style of carved or moulded ornamentation. *See also: display entry in this section, page 111, and entries in Brick section, page 114 and 120.*

▶ MAYSAND LIMITED

109–111 Windsor Road, Oldham, Lancs OL8 1RH
Tel 0161 628 8888 Fax 0161 627 0996 E-mail sales@maysand.co.uk

MASONRY REPAIR, CLEANING, RESTORATION AND REPLACEMENT: *See also: display entry in Stone section, page 106.*

▶ SHAWS OF DARWEN

Waterside, Darwen, Lancashire BB3 3NX
Tel 01254 775111 Fax 01254 873462

ARCHITECTURAL TERRACOTTA, FAIENCE AND GLAZED BRICKS: Manufacturers for over a century, Shaws offers traditional craftsmanship combined with the latest production technology for architectural restoration. For advice, surveys or quotations, please contact Jon Wilson. A division of Shires Limited. *See also: display entry in this section, page 112.*

▶ STONEGUARD

St Martins House, The Runway, Ruislip, Middlesex HA4 6SG
Tel 0870 241 6366 Fax 020 8839 9988
Website www.stoneguard.co.uk
▶ with offices also in Bath, Birmingham, Manchester and Stirling

BUILDING RESTORATION AND CONSERVATION: *See also: profile entry in Building Contractors section, page 66.*

▶ WEST MEON POTTERY

Church Lane, West Meon, Petersfield, Hampshire GU32 1JW
Tel 01730 829434

ARCHITECTURAL CERAMICS: This country workshop has the resources to reproduce practically any piece of architectural ceramic for conservation projects. Hand thrown chimney pots up to five feet tall are a speciality. Also, special bricks, terracotta blocks and mouldings, finials and ridge tiles, floor tiles and large garden pots. West Meon can mix different clays and aggregates to match the finish of an original example. Recent work includes pieces for The British Museum, Uppark House, Kensington Palace, Watts Memorial Chapel, Ishiya project, Japan; Kirby Hall; Kirklees Terrace, Glasgow; Shetland Amenity Trust, Ham House and other National Trust properties.

BRICK SERVICES

▶ GERARD C J LYNCH, LCG, Cert Ed, MA (Dist)

10 Blackthorn Grove, Woburn Sands, Milton Keynes MK17 8PZ
Tel/Fax 01908 584163
E-mail glynch@cwcom.net
Website www.brickmaster.co.uk

HISTORIC BRICKWORK CONSULTANT: Gerard Lynch is an internationally acknowledged historic brickwork consultant and master bricklayer, offering a comprehensive consultancy and on occasions, a specialist bricklaying service. He lectures on historic brickwork and its conservation, and runs master classes embracing his extensive practical skills, covering historic, modern, gauged and decorative brickwork and traditional pointing methods. Mr Lynch is frequently relied on by heritage bodies, architectural practices and lay people to provide reports on brickwork status, methods of sympathetic repair and conservation, and training for craftsmen. Author of *Gauged Brickwork a Technical Handbook*, and *Brickwork: History, Technology and Practice volumes* 1 & 2 (Donhead).

▶ MATHIAS BUILDERS

5 Elmside, Kensworth, Dunstable, Beds LU6 3RR
Tel 01582 873418 Fax 01582 517837
Mobile 07710 326625

HISTORIC BRICKWORK SPECIALISTS: Mathias is a well-established father and son partnership specialising in the repair and restoration of historic brickwork. Work includes rubbed and gauged arches, flintwork and tuck pointing using traditional methods and materials, and the use of lime mortars. Carrying out a wide spectrum of projects ranging from small privately owned properties to country estates. Examples of brickwork can be found around the Hertfordshire, Bedfordshire and Buckinghamshire area from restoring an 18th century garden wall to rebuilding Tudor chimney stacks within the Chequers Estate. Works can also be found on National Trust properties and other listed buildings.

▶ NICHOLAS WEST BRICKWORK SPECIALIST

Barnsdale Bush Road, Cuxton, Rochester, Kent ME2 1HB
Tel 01634 722943 Fax 01634 722943
Mobile 07909 564941
E-mail Nicholas@Nicholaswest.fsnet.co.uk
Website www.brickworkspecialist.co.uk

HISTORIC BRICKWORK SPECIALIST: Nicholas West Brickwork offers a specialised bricklaying service for the conservation and restoration of historical buildings including replacement brickwork and re-pointing. Nicholas West undertakes new build properties in classical styles, extensions to properties using original quality materials to blend to the existing buildings and contract work. Work is completed in the South East.

▶ TIMOTHY J SHEPHERD HISTORIC BRICKWORK SPECIALIST

101 Moffat Road, Thornton Heath, Surrey CR7 8PZ
Tel/Fax 020 8653 2438
E-mail tim@histbric.demon.co.uk

▶ STONEWEST LTD

Lamberts Place, St James's Road, Croydon CR9 2HX
Tel 020 8684 6646 Fax 020 8684 9323
E-mail stonewest@cwcom.net
Website www.stonewest.co.uk

BUILDING CONTRACTORS AND STONE MASONS*: See also: display entry in Building Contractors section, page 67.*

See also Pointing in Chapter 4, pages 176–177

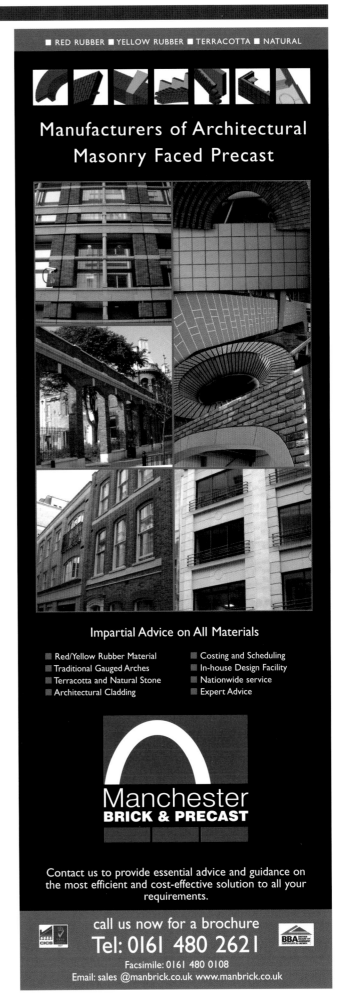

BRICK ARCHES

WINDOW HEAD DETAILS IN EXTERIOR BRICK WALLS

JONATHAN TAYLOR

In most modern brick buildings, brickwork is supported across the openings in exterior walls by either reinforced concrete, reinforced brickwork or a steel lintel discreetly hidden by a skin of brickwork. Often the head of the opening is not expressed in any way, and even in traditionally-inspired architecture, brick arches are often facings applied across a cunningly designed lintel.

This modern approach to the design of brickwork is radically different from traditional construction methods in which the lintel or arch was almost invariably expressed. Even in the simplest, most functional building types, including humble terraces and warehouses, the lintels and arches over the doors and window openings were often elaborated, often subtly, but nevertheless providing an important element of detail in the façade. The addition of a substantial cill below the window and an obvious arch, lintel or cornice above, emphasised its height and elegance, reducing the visible gap between the window above. This vertical emphasis is a vital component in the rhythm of many 18th and 19th century terraces, establishing strong vertical arrangements across the horizontal lines of roofs and windows, string courses and street or garden features, contributing to the character of the street scene as well as to the character of the individual window or doorway. The importance of such details cannot be overstated: even the simplest segmental arch detail can have a significant impact on the visible height of the window, of far greater importance than the sum of its humble components might suggest.

▲ *Polychrome brick arches at the Granary Building, Bristol (1869 by Ponton and Gough)*

LINTELS

The simplest way to support the brickwork above an opening is to use a lintel – a horizontal structural member. Stone and wood were commonly used for this purpose in the past. Stone, being relatively weak in tension, could only be used over narrow spans. However, this problem could be overcome by dividing wider windows into a number of 'lights' separated by stone posts or 'mullions'. This form of construction is typically associated with medieval architecture and the earliest windows to be glazed.

Rough timber lintels are often seen in stone cottages, largely due to the current

▶ *A rare example of an early window without either a lintel or an arch; the brickwork appears to be supported by the heavy timber mullioned window frame alone*

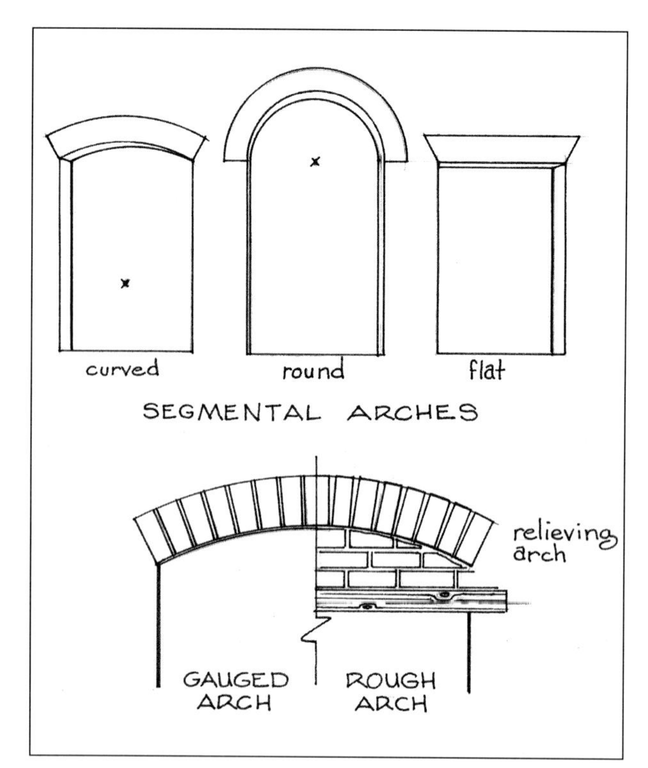

A broken segmental arch, illustrating the strength of this type of construction: for the arch to fail, the bricks themselves had to crack (photograph courtesy of Charterbuild)

fashion for stripping these buildings of their render. However, they are less commonly seen in brick buildings. After the Fire of London, exterior structural timberwork was seen as a fire risk, and even the sash windows were pushed back in their window openings so that their sash boxes (the hollow boxes on either side which contained the weights) could be tucked neatly behind the outer skin of brickwork.

The emergence of mass-produced wrought iron beams following the industrial revolution enabled much larger openings to be created in brick walls. Typical applications include industrial buildings, stables and shopfronts, particularly of the Victorian period, when the beams were most commonly disguised by a painted timber fascia. In the architecture of shopfronts, this beam or 'bressumer' became an integral part of the complex designs which evolved in the late 18th and 19th centuries incorporating cornices, pilasters and stall risers. Blinds, shutters and window grilles were often incorporated into the space below the bressumer, above the window itself.

In the late 19th and early 20th century, terracotta was often used to provide decorative details such as window and door surrounds in brick buildings. Small, solid blocks of terracotta are essentially the same as moulded bricks. However, large blocks of terracotta which appear much like large blocks of carved stone in size and shape, were usually made from hollow pots moulded on the visible face only, and filled with concrete. In this form the terracotta often acted as infill and cladding around an iron frame.

Flat soffits could also be created by using a relieving arch infilled with brickwork below and with a small timber just above the window frame. This technique was commonly used where the brick façade was rendered or covered with stucco which enabled elaborate window surrounds to be made with cornices and moulded jambs (see *Exterior Stucco* page 178).

ARCHES

Although generally considered to be a Roman development, the earliest known use of arches to span openings was around 4000BC in Babylon. However, it was the Romans who really exploited the potential of the device as a structural element, and in the UK arches have been commonly used as a means of spanning windows and doorways in every period of architecture since then, with fine examples of Romanesque and Norman carved doorway arches. The simplest form, the semicircular arch, gave way to more elaborate forms of arch including the pointed arch usually associated with gothic architecture and the later gothic revivals, and ultimately to the flat and segmental arches which emerged in the late 17th century.

An arch is composed of small elements such as bricks bonded in mortar. It works by transferring the load sideways and downwards, from the key stone at the centre to the jambs of the walls below. As the structure only works when completed, brick arches are usually constructed on a timber frame or former known as a 'centring' which is removed after the mortar has set sufficiently.

There are essentially two forms of arch construction: 'segmental' and 'rough'. In the case of a segmental arch, the bricks or stone blocks are specially made so that they are tapered to the radius (or radii in the case of an arch with more than one centre), with parallel joints between. Being wider at the top than at the bottom, these 'voussoirs', as the units are called, cannot fall out. However, in the case of a rough arch, it is the mortar which is tapered, not the bricks, and the arch relies on the cohesion of the bricks and mortar for its strength. This form of construction was generally considered inferior to the specially tapered voussoirs, both structurally and aesthetically, and rough arches are usually hidden from sight behind render. (Beware when carrying out repair and conservation work. One of the symptoms of the dearth of skilled bricklayers today is that non-specialist builders often use this form of construction when repairing or replacing much finer work.)

The voussoirs of segmental brick arches could be made by cutting ordinary bricks on site with a bolster and chisel, finishing with a bricklayer's axe or 'scutch'. More accurate results could be achieved by using 'specials' manufactured to the shape required. Brick manufacturers produce these either by cutting the brick in its green state before firing, or by using a specially shaped mould. However, accuracy was limited as some distortion during drying and firing was almost inevitable.

The late 17th and early 18th centuries were a high point in the evolution of brickwork. Manufacturing techniques improved, with blended clay, better moulding and more even firing which led to greater consistency in shape and size. One of the most interesting developments where arches are concerned was the introduction of gauged brickwork, a technique which was popularised by Sir Christopher Wren in particular. This form of brickwork was made using fine textured, soft clay bricks commonly called 'rubbers'. For the tapered voussoir of an arch, a rubber was first shaped by rubbing on a flat plane of millstone grit until it accurately fitted a specially shaped wooden box, open at the top and on one or more sides. This enabled the exposed face to be cut to a complex profile if needed, incorporating the mouldings of the arch, using a bow saw, and any irregularities were then smoothed out. The whole arch was laid out on the floor on site to the required size as the components were made, to make sure that the all the components fitted together correctly, with evenly spaced joints.

Gauged brickwork enabled the use of extremely fine joints, so that the joints did not disrupt the appearance of the fine mouldings.

A late 19th century terracotta arch in Tisbury, Wiltshire

A typical flat segmental arch with red rubbers and lime putty joints, early 19th century

Rubbed brick arches and reliefs adorning a terraced house in Collingham Gardens, London (c1880)

The technique was widely used in fine architecture of the Georgian and Victorian periods, and some of the finest examples can be seen in the elaborate brickwork of the Tudor Revival in the late 19th century (see illustration).

Plain gauged voussoirs were often bedded in pure lime putty, producing neat white radiating lines which emphasised the geometry of the arch. This effect was widely imitated using much cheaper 'tuckpointed' brickwork. For this technique, rough and cut brick arches were flush-pointed with a mortar of the same colour as the brick so that the two merged together to be almost indistinguishable. Then, while the mortar was still fresh, a thin line was struck neatly across the centre of the joint and filled with pure lime putty. Roughly shaped and distorted bricks thus acquired the neat edges of gauged brickwork. A by-product of the improvements in brick production at this time was the widespread use of segmental and flat arches.

A replacement rubbed brick and some of the tools used to make it, including the open faced box and a carburundum block used for rubbing (photograph courtesy of Nimbus Conservation Ltd)

THREATS AND CHALLENGES

Arguably the greatest threats to the fine details which give buildings their character are complacency and ignorance. Repairs, when they are carried out, are often made using inappropriate techniques which actually cause further deterioration. Brick façades are sometimes painted over, obscuring the subtle variations in colour and texture which are so important to their character, and locking in moisture. Hard cement mortars are commonly used in place of the traditional, much softer lime mortars. Both alterations can cause the brickwork to deteriorate. Other typical causes of damage include the cleaning methods commonly used, rust damage caused by metal fixings, and frost damage following saturation by, for example, overflowing gutters.

Where window lintels and arches have failed, repairing the faults without reducing the loads can be counterproductive. By reinforcing the brickwork above the opening it is possible to convert the brickwork itself into a lintel, reducing the stress in the arch or lintel below. This can be achieved quite simply by introducing narrow stainless steel bars into the mortar joints above to span the width of the opening and repointing so that they are invisible externally. Some specialist companies have also developed the technique of introducing threaded stainless steel rods through the bricks themselves. These rods are then attached to plates on either side of the opening with nuts which can be tightened to post-tension them. The brickwork thus acts in compression, while the steel carries the tensile forces.

Repointing should always be carried out using a mortar which matches the composition of the original. In older buildings this will usually mean using a lime mortar, often with a local sand or aggregate. These mortars are relatively porous and expand and contract to a degree which cannot be matched by modern cement mortars. Repointing with a hard cement mortar can introduce stresses acrss the face of the brickwork, causing the edges of the bricks to crack and crumble. They also lock moisture in, which can lead to frost damage and salt crystallisation, causing the brickwork to crumble.

As in so many aspects of historic buildings, it is important that the consultants and contractors used are all experienced in dealing with historic and traditional materials. It is not enough for a person to be a member of a professional body or trade association, if these vital features are to be understood and properly cared for.

BRICK SUPPLIERS

▶ BRICKFIND (UK)

Bentley House Farm, Copyholt Lane, Lower Bentley, Bromsgrove, Worcestershire B60 3BE
Tel 01527 540099 Fax 01527 544500 Mobile 07860 230013

SPECIALIST BRICK MATCHING – NATIONWIDE: Brickfind (UK) prides itself on personal service. With access to all UK and European manufacturers as well as others further afield, Brickfind also sources reclaimed materials. Additionally, Brickfind offers restoration solutions such as remedial tinting, re-facing and cleaning of brickwork. Special shapes, including non-standard, can be fast-tracked.

▶ THE BULMER BRICK & TILE CO LTD

Bulmer, Nr Sudbury, Suffolk CO10 7EF
Tel 01787 269232 Fax 01787 269040

CLAY BRICKS AND TILES: Situated on one of the finest seams of clay in Britain, with bricks, tiles and pots having been made at Bulmer since the Bronze Age. With this long history, the traditions and knowledge passed down to today express themselves in the specialist work of restoration. Many of the company's 19th century moulds are being used to provide replacement bricks for the original. New patterns are being added to the 3,000-plus patterns they already hold. The company produces a range of quality rubbing bricks from fine washed clays and has a cutting service for special shapes and gauged arch work.

▶ E T CLAY PRODUCTS LIMITED

7–9 Fowler Road, Hainault, Ilford, Essex IG6 3UT
Tel 0208 501 2100 Fax 0208 500 9990

BRICKS, TILES, STONE, PAVERS: Supplying South, Mid England and Wales, E T Clay Products specialise in finding the right products to suit their clients' needs, whether working to a budget or satisfying an architectural specification, for listed or heritage projects. Being independent they can call upon vast resources which are both unbiased and supportive. E T Clay Products also have their own exclusive range of Imperial sized bricks which have been produced to blend with traditional buildings, some of which are held in stock at both Croydon and Hainault depots.

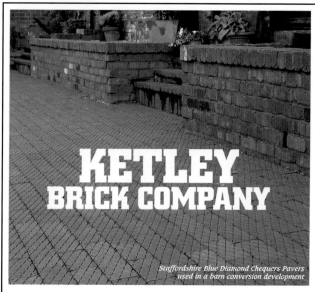

▶ HAMMILL BRICK LTD

Woodnesborough, Nr Sandwich, Kent CT13 0EJ
Tel 01304 617613 Fax 01304 611036

SPECIALIST STOCK BRICK MANUFACTURERS: The brickworks was established at the Hamlet of Hammill, near Sandwich in 1926. Production has concentrated on quality stock facing bricks, which closely match those used for centuries in Southern and Eastern England. The range consists of the Light Red Stock, Medium Multi Stock and the Selected Light Multi Stock, all produced with a fine hand-made texture. The brickworks will also blend different ranges to suit a customer's preference. The company stocks an excellent range of British Standard specials and has a CAD service to assist clients in the design of special shapes and detail work. The bricks have been used on many environmentally sensitive schemes and are ideally suited to restoration, matching of old work and new projects in conservation areas.

▶ HANSON BRICK

Marketing Department, Stewartby, Bedford MK43 9LZ
Tel 08705 258258 Fax 01234 762040
Website www.hansonbrick.com

FACING BRICKS: Hanson Brick is one of the UK's leading manufacturers of quality facing bricks. The product portfolios of London, Butterley and Desimpel include over 200 different colours and textures from handmades and stocks through to wirecut rustics. Hanson Brick also manufactures a range of clay pavers in complementary colours and surface textures. A range of 73mm bricks in tones selected to aid the renovation and extensions of old buildings is also available. Hanson Bricks range of brick arch details offers the desired benefits of reduced cost, high quality and aesthetic appeal. Special shapes are available along with a technical advisory service, with CAD facilities, to assist in the creation of brickwork features.

BRICK SUPPLIERS

BRICK SUPPLIERS

▶ KETLEY BRICK LTD

Dreadnought Works, Dreadnought Road, Pensnett, Brierley Hill, Staffordshire DY5 4TH
Tel 01384 77405 Fax 01384 74553
Website www.ketley-bricks.co.uk

BRICKS AND BRICKS PAVERS: *See also: display entry on page 119.*

▶ LAMBS BRICKS & ARCHES

Nyewood Court, Brookers Road, Billingshurst, W Sussex RH14 9RZ
Tel 01403 785141 ext 223 Fax 01403 784663
E-mail sales@lambsbricks.com
Website www.lambsbricks.com

MANUFACTURERS OF HANDMADE BRICKS AND RED RUBBER BLOCKS: Lamb's continues to manufacture traditional handmade facing bricks from a variety of clays replicating a vast number of bricks found in English villages and towns. These bricks, manufactured in the traditional way to replicate textures and sizes, give the craftsman every chance to repair and extend with seamless continuity. Lamb's own fully washed brick earth is used to produce rubbing blocks for traditional replacement or new work arches. The TLB trademark presented to Lambs by Thomas Lawrence of Bracknell Ltd is proudly displayed on the best grade blocks continuing a tradition from the last century. 21st century CAD is used to design arches and specials from site surveys, photographs and architectural drawings. *See also: display entry in this section, page 114, and entries in Architectural Terracotta section, page 110 and 111.*

▶ SUSSEX BRICK LTD

Fourteen Acre Lane, Three Oaks, Hastings, East Sussex TN35 4BN
Tel 01424 814344 Fax 01424 814707
Website www.sussexbrick.co.uk

HANDMADE BRICKS, GAUGED BRICK ARCHES AND MATHEMATICAL TILES: *See also: display entry on page 119.*

▶ WIENERBERGER LTD

Adamson House, Pomona Strand, Manchester M16 0BA
Tel 0161 873 7701
Website www.terca.co.uk

SPECIALIST BRICK MANUFACTURER: *See also: display entry on page 119.*

▶ YORK HANDMADE BRICK COMPANY LIMITED

Forest Lane, Alne, York YO61 1TU
Tel 01347 838881 Fax 01347 838885
Website www.yorkhandmade.co.uk

SPECIALIST BRICK MANUFACTURER: *See also: display entry on page 119.*

3.3 METAL, GLASS & WOOD

METALWORK

Albion Manufacturing Ltd	ci	wr	bs	141
Antique Bronze Ltd			bs	94
Architectural Metalwork Conservation		wr		141
B Levy & Co (Patterns) Ltd		wr		141
Barr & Grosvenor Ltd	ci			141
Bassett and Findley Ltd			bs	141
Best Demolition	as			155
Britannia Architectural Metalwork Ltd	ci	wr		143
C E L Ltd			pb	87
Cambrian Castings (Wales) Limited	ci	wr		142
Carrek Limited		wr		142
The Cast Iron Company	ci	wr		143
Castaway Cast Products & Woodware	ci			142
Casting Repairs Limited	ci	wr		142
Chris Topp & Company Wrought Ironworks	ci	wr		144
Crowncast Ltd	ci			142
Don Barker Ltd		wr		144
Dorothea Restorations Ltd	ci	wr		144
Eura Conservation Ltd	ci	wr	bs pb	142
The Fine Iron Company	ci	wr	bs	145
George James & Sons, Blacksmiths		wr		145
Glasgow Steel Nail Co Ltd		wr		184
Haddoncraft Forge		wr		145
Hargreaves Foundry Ltd	ci			91
Harrison Thompson & Co Ltd	ci			92
Hart Brothers Engineering Ltd	ci			145
Hodgsons Forge		wr		146
J & J W Longbottom Ltd	ci			92
J H Porter & Son Ltd	ci	wr		146
LASSCO	ci	wr		155
Marsh Brothers Engineering Services Ltd	ci	wr		146
Mather & Smith Ltd / M J Allen Group	ci			146
Mid Beds Locksmiths		wr	bs	140
Minchinhampton Architectural Salvage	as			155
Minter Reclamation	as			156
Norman & Underwood Ltd			pb	87
Northwest Lead			pb	88
Peter S Neale Blacksmiths	ci	wr		146
Plowden & Smith Ltd	ci	wr	bs gs pb	147
Ransfords Reclaimed Building Supplies	as			155
The Real Wrought Iron Company	ci	wr		147
Retrouvius Architectural Salvage	as			157
Rupert Harris Conservation	ci	wr	bs gs pb	147
Saint-Gobain Pipelines Plc		wr		92
Salmon (Plumbing) Limited			pb	88
Shepley Engineers Limited	ci	wr		147
Smith of Derby Clockmakers	ci	wr		147
Solopark plc	as			156
Structural Perspectives	ci	wr		13
Victorian Classic Style	ci			152
Walcot Reclamation Ltd	ci	wr		157

KEY

- **bs** brass
- **ci** cast iron
- **gs** gold and silver
- **pb** decorative leadwork
- **wr** wrought ironwork

GLASS

Paul Bradbury	sg	137
Bursledon Brickworks Conservation Centre	wg	172
C J L Designs	gp ll sg	137
The Cathedral Studios	sg	138
Clement Windows Group Ltd	wg	135
The Cotswold Casement Company	ll sg wg	135
The Fine Iron Company	gp	145
Goddard & Gibbs Studios Ltd	ll sg	137
Gowar Grilles Ltd	gp	141
Holdsworth Windows Ltd	ll wg	136
Illumin Glass Studio	ll sg wg	138
Jonathan Leckie Associates	gp ll sg	138
The London Crown Glass Company Ltd	wg	139
Nick Bayliss Architectural Glass	ll sg	137
Norgrove Studios	ll sg wg	138
Norman & Underwood Ltd	ll sg	87
Plowden & Smith Ltd	sg	203
Riverside Studio	gp ll sg	138
The Stained Glass Specialist	gp sg	138
Tatra Glass Co	sg wg	139
UK and European Joinery & Glassworks	sg	138

KEY

- **gp** window grilles
- **ll** leaded lights
- **sg** stained glass
- **wg** window glass

3.3 METAL, GLASS & WOOD

WINDOWS AND DOORS

A M Experimental			lk	141
Albion Manufacturing Ltd			gp	141
Antique Buildings Limited	as td			157
Bassett and Findley Ltd		mw		129
Best Demolition	as td			155
Bramah			lk wf	140
Bursledon Brickworks Conservation Centre			wg	138
C J L Designs		mw	gp	137
C R Crane & Son Ltd	td tw			60
The Cast Iron Company		mw rl		143
Castaway Cast Products and Woodware		mw		142
Casting Repairs		mw		143
Chilstone			cs	110
The Classic Window Company Ltd	tw			129
Clayton Munroe Limited			wf	139
Clement Windows Group Ltd		mw rl	wf wg	135
The Conservation Rooflight Company		mw rl		89
The Cotswold Casement Company		mw	wg	135
Crittall Windows		mw		135
Don Forbes Sash Fittings			wf	140
Dorothea Restorations Ltd		mw		144
The Fine Iron Company		mw	gp wf	145
Gowar Grilles Ltd			gp	141
Haddonstone Limited			cs	110
Hearns (Specialised) Joinery Ltd	td tw			127
Holdsworth Windows Ltd		mw	wf wg	136
Illumin Glass Studio			wg	138
J G Matthews Ltd	td tw			62
J Scott (Thrapston) Limited	td tw			132
Jonathan Leckie Associates			gp	138
Joseph Giles, Croydon			wf	139
Karters Joinery	tw			129
L Daniels & G Eldridge	td tw			126
L S Longden	td			131
Linley			wg	137
The London Crown Glass Company Ltd			wg	139
MacKinnon & Bailey			wf	140
Marsh Brothers Engineering Services Ltd		mw rl		146
Marvin Architectural	tw			131
Mather & Smith Ltd		mw		146
Mid Beds Locksmiths			lk wf	140
Minchinhampton Architectural Salvage	as td		wf	155
Minter Reclamation	as td			156
Mumford & Wood			wf	130
Nick Bayliss Architectural Glass	tw	mw		137
Norgrove Studios		mw	wg	138
Original Oak	td			211
Oxford Sash Window Company Limited	tw			130
Paul Christensen Windows and Doors	tw			130
Peter S Neale Blacksmiths		mw		146
Quality Lock Co			lk	140
R W Armstrong & Sons Limited	tw			66
Ransfords Reclaimed Building Supplies	td			155
The Real Wrought Iron Company		mw		147
Refurb-a-Sash	tw		wf	132
Restoration Windows	tw			130
Riverside Studio		mw	gp	138
The Round Wood Timber Co Ltd	td tw			128
Rundum Meir	td			132
The Sash Restoration Company	tw			133
Sashcraft of Bath	tw			133
Scotts of Thrapston	td			132
Selectglaze Ltd	tw			134
Solopark Plc	as td			156
The Stained Glass Specialist		mw	gp	138
The Standard Patent Glazing Co Limited		rl		89
Steel Window Service and Supplies Ltd		mw		136
Tankerdale Ltd	td tw			128
Tatra Glass Co			wg	139
Ternex Ltd	tw			124
Thorverton Stone Company Ltd			cs	110
Timber Tech Ltd	td tw			134
UK and European Joinery & Glassworks	tw			138
Vale Garden Houses Ltd		mw rl		136
Ventrolla Ltd	tw			133
Walden Joinery	td tw			128
The Wooden Window Company	tw			134

KEY

- as architectural salvage
- cs cast (reconstructed) stone
- gp window grilles
- lk locksmiths
- mw metal windows
- rl roof lights
- td timber doors
- tw timber windows
- wf door and window furniture
- wg window glass

FINE JOINERY

Agrell Architectural Carving			wc	125
Anelays - William Anelay Ltd	jo			58
Anthony Beech Furniture Conservation		cb		205
Anthony Hicks Limited	jo			72
Arthur Brett	jo	cb	wc	126
Between Time Ltd	jo			58
Bluestone plc	jo			59
Boshers (Cholsey) Ltd	jo			60
Bricknell Conservation Limited		cb		59
Busby's Builders	jo			60
C J Ellmore & Co Ltd	jo			60
C R Crane & Son Ltd	jo			60
Carrek Limited	jo			61
Carvers & Gilders Limited			wc	124
Church Conservation Limited	jo			84
Clive Beardall		cb		205
Michael Court		cb		126
D S Sheridan Building Services	jo			61
E Bowman & Sons Ltd	jo			60
Early Oak Specialists	jo			126
Giltwood Restorations Ltd	jo	wc		124
H J Hatfield and Sons Ltd		cb		205
Hall Construction	jo			61
Haslemere Builders Limited	jo			62
Hearns (Specialised) Joinery Ltd	jo	cb		126
Historic Buildings Conservation Limited	jo			61
Holloway White Allom	jo			62
J & W Kirby	jo			62
J G Matthews Ltd	jo			62
Jameson Joinery	jo	cb		73
Ken Biggs Contractors Ltd	jo			63
L Daniels & G Eldridge	jo			126
Linford-Bridgeman Limited	jo			64
Luard Conservation	jo		wc	198
Mark Bridges Carver & Gilder Ltd			wc	125
Mott Graves Projects Limited	jo	cb		128
N E J Stevenson	jo	cb		206
Plowden & Smith	jo	cb		128
R W Armstrong & Sons Limited	jo			66
Romark Specialist Cabinetmakers Limited		cb	wc	205
The Round Wood Timber Co Ltd	jo			128
Sandy & Co (Contractors) Ltd	jo			64
Simmonds of Wrotham	jo			64
T J Evers Ltd	jo			65
Tankerdale Ltd	jo	cb	wc	128
Ternex Ltd	jo			128
Tim Peek Woodcarving Services	jo	cb	wc	125
Timothy Williams (Builders) Ltd	jo			65
Tiptree Joinery Services	jo			128
Traditional Carpentry & Joinery	jo			74
Treasure & Son Ltd	jo			65
Vale Garden Houses Ltd	jo			152
Walden Joinery	jo			128
Wallis	jo	cb		54
Wm Langshaw & Sons	jo			68

KEY

- cb cabinet makers
- jo joinery
- wc wood carvers and turners

TIMBER SUPPLIERS

A H Peck		tp	211
Agrell Architectural Carving		tm	125
Altham Hardwood Centre Ltd	ti		72
Antique Bronze Ltd (Historic Flooring Division)		tp	211
Antique Buildings Limited	as		157
Bursleden Brickwork Conservation Centre		bt	172
Capital Crispin Veneer		ve	124
Carpenter Oak & Woodland Ltd	ti	bt	122
The Cleft Wood Co	ti	bt	213
Clive Beardall		ve	205
Country Oak (Sussex) Ltd	ti		122
Coyle Timber Products		bt	72
Drummond's Architectural Antiques Limited	as	tp	155
English Woodlands Timber Ltd	ti	tp	122
G & N Marshman		bt	216
Hatfield House Oak	ti		122
Jameson Joinery	ti	tp	123
John Boddy Timber Ltd	ti		123
L Daniels & G Eldridge		tm	126
Minchinhampton Architectural Salvage	as		155
Northwood Forestry Ltd		tp	123
Original Oak	ti	tp	211
Priory Hardwoods Limited	ti	tp	211
Ransfords Reclaimed Building Supplies	as		155
Rockingham Oak	ti		74
The Round Wood Timber Co	ti	tp	123
Solopark Plc	as		156
Tankerdale Ltd		tp ve	128
Ternex Ltd	ti	tp	124
Timber Supplies	as		156
Vastern Timber Co Ltd	ti		124
Victorian Woodworks	ti		213
Walden Joinery		tp	128
Weald & Downland Open Air Museum		bt	124
Weldon Flooring Limited	ti	tp	212
Whippletree Hardwoods	ti	bt tp	124

KEY

- as architectural salvage
- bt laths, battens & tile pegs
- ti timber suppliers
- tm timber mouldings
- tp timber and parquet flooring
- ve veneers

STAIRS & BALUSTRADES

Abbey Masonry and Restoration Limited	sm	100
Albion Manufacturing Ltd	ci	141
The Bath Stone Group	sm	101
Barr & Grosvenor Ltd	ci	141
Britannia Architectural Metalwork Ltd	ci	143
Broadmead Cast Stone	cs	110
C R Crane & Son Ltd	jo	60
Cambrian Castings (Wales) Limited	ci	142
The Cast Iron Company	ci	143
Casting Repairs	ci	143
Chilstone	cs	110
Early Oak Specialists Ltd	jo	126
The Fine Iron Company	ci	145
Haddonstone Limited	cs	110
Hearns (Specialised) Joinery Ltd	jo	126
J H Porter & Son Ltd	ci	146
Jameson Joinery	jo	73
L Daniels & G Eldridge	jo	126
LASSCO	as	155
Marsh Brothers Engineering Services Ltd	ci	146
Mather & Smith Ltd / M J Allen Group	ci	146
Minchinhampton Architectural Salvage	as	155
Minter Reclamation	as	156
Mott Graves Projects Limited	jo	106
Ransfords Reclaimed Building Supplies	as	155
Retrouvius Architectural Salvage	as	156
The Round Wood Timber Co Ltd	jo	128
Shepley Engineers Limited	ci	147
Smith of Derby Clockmakers	ci	147
Solopark	as	156
Tankerdale Ltd	jo	128
Ternex Ltd	jo	124
Thorverton Stone Company Ltd	cs	110
Tiptree Joinery Services	jo	128
Walcot Reclamation Ltd	ci	157
Walden Joinery	jo	128
Wallis	jo	128

KEY

- as architectural salvage
- cs cast (reconstructed) stone
- ci cast iron
- jo timber (fine joinery)

TIMBER SUPPLIERS

▶ ALTHAM HARDWOOD CENTRE LTD

Altham Corn Mill, Burnley Road, Altham, Accrington, Lancs BB5 5UP
Tel 01282 771618 Fax 01282 777932
Website www.oak-beams.co.uk
HAND CUT OAK BEAMS: *See also: profile entry in Timber Frame Builders section, page 72.*

▶ ANTIQUE BUILDINGS LIMITED

Hunterswood Farm, Dunsfold, Surrey GU8 4NP
Tel 01483 200477 Fax 01483 200752
OAK BEAMS AND BARN FRAMES: *See also: display entry in Reconstructed Buildings section, page 157.*

▶ BURSLEDON BRICKWORKS CONSERVATION CENTRE

Bursledon Brickworks, Coal Park Lane, Swanwick, Southampton SO31 7GW
Tel/Fax 01489 576248
E-mail bursledon@ndirect.co.uk
RIVEN CHESTNUT LATHS AND OAK PEGS: *See also: profile entry in Mortars & Renders section, page 172.*

▶ CARPENTER OAK & WOODLAND CO LTD

Hall Farm, Thickwood Lane, Colerne, Chippenham, Wiltshire SN14 8BE
Tel 01225 743089 Fax 01225 744100
E-mail info@cowco.biz
SPECIALIST TIMBER SUPPLIES: Carpenter Oak & Woodland is a one-stop shop for your special timber needs: oak shingles, for historic church spires, traditionally cleft only from English oak: riven batten and oak tile pegs for historic roofs and riven laths for traditional plasterwork. Timber frames can be surveyed, repaired, and timber supplied: sawn, planed, or hewn timber for crucks, braces and all frame components, including hand-made oak framing pegs. Specialist fixings also available: rosehead nails and stainless steel products. Carpenter Oak & Woodland can be relied on for the finest quality materials, a rapid response and continuity of supply. *See also: display entry in Timber Frame Builders section, page 72.*

▶ THE CLEFT WOOD COMPANY

1 Littlecote Cottages, Littlecote, Dunton, Near Winslow, Bucks MK18 3LN
Tel 01525 240434 Fax 01525 240434
E-mail enquiries@cleftwood.com
Website www.cleftwood.com
RIVEN OAK LATHS, ROOFING BATTENS, TILE PEGS, HEWN BEAMS AND HAZEL WATTLE HURDLES. *See also: profile entry in Plasterwork section, page 213.*

▶ COUNTRY OAK (SUSSEX) LTD

Little Washbrook Farm, Brighton Road, Hurstpierpoint, West Sussex BN6 9EF
Tel 01273 833869 Fax 01273 831742
E-mail oakean@aol.com
ANTIQUE OAK BEAMS AND EARLY FLOORBOARDS: *See also: profile entry in Timber Flooring section, page 211.*

▶ ENGLISH WOODLANDS TIMBER LIMITED

Cocking Sawmills, Cocking, Near Midhurst, West Sussex GU29 0HS
Tel 01730 816941 Fax 01730 816874
E-mail sales@ewtimber.co.uk
Website www.ewtimber.co.uk
TIMBER SUPPLIERS: English Woodlands Timber Limited supplies oak beams and air or kiln dried English hardwoods. Established in 1986, it holds an extensive range of hardwood stocks with all European species represented from ¾" to 4" thickness. A cut to size and planing service is offered alongside normal through and through plank. Oak beams are produced to order, supplied in all lengths and dimensions, including bends. English Woodlands Timber also supplies oak flooring up to 200mm wide, end matched and ready to lay. The company recently supplied oak laths used in construction of the innovative gridshell roof for the Weald and Downland Open Air Museum.

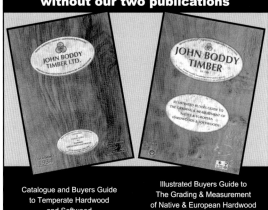

▶ JAMESON JOINERY

Hook Farm, West Chiltington Lane, Billingshurst, West Sussex RH14 9DP
Tel 01403 782868 Fax 01403 786766

TIMBER AND JOINERY FOR OLD BUILDINGS: From their well equipped sawmill and joinery works in Sussex, Jameson Joinery produces traditionally jointed oak framed buildings, together with oak and elm flooring, oak beams, doors, windows, furniture, staircases, plasterer's laths and cabinetry. All joinery work is made to order for bespoke projects – beams up to 24″ wide can be tenoned and jointed in the traditional way and hand adze-finished. Traditional window and door fittings and nails are hand-forged on-site as are traditional square headed nuts. Deliveries nation-wide. *See also: display entry in this section, page 122.*

▶ ORIGINAL OAK

Ashlands, Burwash, East Sussex TN19 7HS
Tel 01435 882228 Mobile 07771 890835

BUILDING AND RESTORATION OAK TIMBER AND FLOORING SUPPLIERS: *See also: profile entry in Timber Flooring section, page 211.*

▶ RANSFORDS CONSERVATION AND RECLAIMED BUILDING SUPPLIES

Drayton Way, Drayton Fields, Daventry, Northants NN11 5XW
Tel 01327 705310 Fax 01327 706831
Website www.ransfords.com

SUPPLIERS OF RECLAIMED OAK AND PINE FLOORBOARDS AND OLD OAK BEAMS: *See also: display entry and profile entry in Architectural Salvage section, page 155.*

TIMBER SUPPLIERS

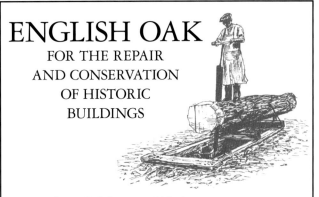

▶ TERNEX LTD
The Sawmill, 27 Ayot Green, Welwyn, Herts AL6 9BA
Tel 01707 324606 Fax 01707 334371
SAWMILL AND JOINERY COMPANY: Ternex is a long established sawmill and joinery company which produces English hardwoods, mainly oak, for beams and construction, flooring, furniture and joinery. Oak, strength-graded to BS5756:1997 is available. Ternex also produces bespoke joinery, machined timber, flooring and mouldings in English and imported hardwoods and softwoods.

▶ VICTORIAN WOODWORKS
54 River Road, Creekmouth, Barking, Essex IG11 0DW
Tel 020 8534 1000 Fax 020 8534 2000 E-mail sales@victorianwoodworks.co.uk
Website www.victorianwoodworks.co.uk
SPECIALIST HARDWOOD FLOORING MANUFACTURERS AND SUPPLIERS: *See also: profile entry in Timber Flooring section, page 213.*

▶ WEALD AND DOWNLAND OPEN AIR MUSEUM
Singleton, Chichester, West Sussex PO18 0EU
Tel 01243 811363 Fax 01243 811475 E-mail wealddown@mistral.co.uk
Website www.wealddown.co.uk
CONSERVATION SUPPLIES AND SERVICES: The Museum offers various supplies for use in timber frame building and the care of ancient buildings. *See also: profile entry in Courses & Training section, page 226.*

VENEERS

▶ CAPITAL CRISPIN VENEER
12 Bow Industrial Park, Carpenters Road, Stratford, London E15 2DZ
Tel 020 8525 0300 Fax 020 8525 0070
SUPPLIERS OF OVER 100 SPECIES OF WOOD VENEERS: Capital Crispin Veneer holds vast stocks of rare and precious varieties including many burrs and decorative veneers from all over the world. Alongside these, native and other temperate species are stocked, many in extra thicknesses for antique or constructional purposes. Whether your requirements are for an architectural project or a single piece of furniture, Capital Crispin Veneer welcomes all enquiries. Stock list and illustrated list of inlays sent on request.

WOOD CARVERS

▶ AGRELL ARCHITECTURAL CARVING
Lansdell House, 52 Lansdell Road, Mitcham, Surrey CR4 2JE
Tel 020 8646 9734 Fax 020 8646 9608 E-mail kate@agrellcarving.co.uk
Website www.agrellcarving.com
ARCHITECTURAL SCULPTURE: Agrell and Thorpe is a large capacity workshop, providing high quality architectural hand carving for religious, residential and commercial projects. Master carver, Ian Agrell, founded the company in 1978. After working on cathedrals, palaces and residences in the UK the business moved to San Francisco where it has been established for many years. There the company earned an excellent reputation for designing and undertaking a range of complex projects and styles to create the ideas of architects, designers and contractors. With a capacity of over 40,000 hours of hand carving a year, Agrell and Thorpe has recently been able to expand its sales operations across the UK and Europe, providing quality, volume and delivery regardless of project size. *See also: display entry in this section, page 125.*

▶ CARVERS & GILDERS
9 Charterhouse Works, Eltringham Street, London SW18 1TD
Tel 020 8870 7047 Fax 020 8874 0470 E-mail bcd@carversandgilders.com
Website www.carversandgilders.com
RESTORATION AND CONSERVATION: Restoration, conservation and replacement of fine decorative woodcarving and giltwood. Designers and makers of carved and gilt furniture, mirror frames, ornamental carving including chimneypieces, decorative panels and overdoors, and other decorative pieces in both period and contemporary styles. Commissions undertaken. Consultancy service offered. Members of The Master Carvers Association, Guild of Master Craftsmen. Clients include The Royal Collection, Historic Royal Palaces, English Heritage, National Trust, Harewood House Trust, Spencer House, Woburn Abbey and a wide range of British and international designers and private collectors.

WOOD CARVERS

▶ GILTWOOD RESTORATIONS LTD

Lea House, Willey Lane, Caterham, Surrey CR3 6AR
Tel/Fax 01883 349111
E-mail rococo1750@freenet.co.uk

WOODCARVING AND GILDING RESTORATION: Specialists in high quality restoration and replacement of architectural wood carvings. Giltwood Restoration's work can involve many aspects from replacement of coats of arms to carved skirtings, dado rails, cornices, balustrades, overdoors and mantelpieces. Major works have been in the Royal Palaces, House of Lords, House of Commons, Brighton Pavilion, National Trust houses and city livery companies. The company is also known in England and America for the restoration of fine carved giltwood furniture. Established in 1968, Giltwood is a small competent business capable of gathering a large skilled workforce for large commissions.

▶ TIM PEEK WOODCARVING

The Woodcarving Studio, Highfield Avenue, Booker, High Wycombe, Buckinghamshire HP12 4ET
Tel/Fax 01494 439629
E-mail timpeekwoodcarving@hotmail.com
Website www.timpeekwoodcarving.co.uk

SPECIALIST WOOD CARVING – FURNITURE, ARCHITECTURAL AND ECCLESIASTICAL: Following a family tradition Tim Peek Woodcarving was established in 1980. An accomplished freelance woodcarver, furniture maker and antique restorer, Tim Peek undertakes all aspects of wood carving, each exacted to the very highest of standards. Projects can be designed, made and finished to meet individual requirements, which Tim is always happy to discuss and provide free quotations. The studio is located close to Junction 4 of the M40. Please telephone to make arrangements to visit the studio or view a selection of his work on the website.

FINE JOINERY

ARTHUR BRETT

We offer a specialised service creating and installing classic hand carved doors, panelling, interior fitments and bespoke furniture of the highest quality. Over five generations of cabinet making experience combined with meticulous attention to detail and access to exotic woods and rare materials have made Arthur Brett a watchword for excellence.

LONDON SHOWROOM
TEL: 020 7730 7304 • FAX 020 7730 7105
EMAIL: pimlico@arthur-brett.com

www.arthur-brett.com

BIG SOLUTIONS

FROM A SMALL COMPANY

HISTORIC WOODWORK RESTORATION
COVERING LONDON AND SURROUNDING COUNTIES

"The service provided was thorough and organised, and we have been very satisfied with the quality of their workmanship"
MR RICHARD GRUBB, PROJECT ARCHITECT RHWL NEW LONDON CENTRE, COUNTY HALL

Restoring historic buildings and incorporating new elements requires a sensitive approach, careful research and innovative solutions to conserve their integrity. Each project – large or small – is personally supervised by the directors through to completion.

THE COMPLETE SOLUTION
FROM THE PROJECT SPECIALISTS

MOTT GRAVES PROJECTS LTD
MAKING HISTORY

SAMPLEOAK LANE
CHILWORTH
GUILDFORD GU4 8QW

CONTACT: JAMES MOTT
TEL: 01483 453326
www.mottgraves.co.uk

▶ ANELAYS
William Anelay Limited, Murton Way, Osbaldwick, York YO19 5UW
Tel 01904 412624 Fax 01904 413535 E-mail office@williamanelay.co.uk
Website www.williamanelay.co.uk
BUILDING AND RESTORATION CONTRACTORS: *See also: profile entry in Building Contractors section, page 58.*

▶ C R CRANE & SON LTD
Manor Farm, Main Road, Nether Broughton, Leics LE14 3HB
Tel 01664 823366 Fax 01664 823534 Website www.crcrane.co.uk
SPECIALIST BUILDING AND JOINERY CONTRACTORS: *See also: profile entry in Building Contractors section, page 60.*

▶ CARREK LTD
Mason's Yard, Wells Cathedral, Wells, Somerset BA5 2PA
Tel 01749 689000 Fax 01749 689089 Website www.carrek.co.uk
HISTORIC BUILDING REPAIR COMPANY: *See also: profile entry in Building Contractors section, page 61.*

▶ E BOWMAN AND SONS LTD
Cherryholt Road, Stamford, Lincolnshire PE9 2ER
Tel 01780 751015 Fax 01780 759051
E-mail mail@ebowman.co.uk Website www.ebowman.co.uk
SPECIALIST BUILDING CONTRACTORS: *See also: display entry in Building Contractors section, page 60.*

▶ EARLY OAK SPECIALISTS
142 Seven Sisters Road, Willingdon, Eastbourne, East Sussex BN22 0PA
Tel 01323 506972 E-mail enquiries@earlyoakspecialists.co.uk
Website www.earlyoakspecialists.co.uk

▶ HEARNS (SPECIALISED) JOINERY LIMITED
101 Waldegrave Road, Teddington TW11 8LW
Tel 020 8977 1644/020 8977 0032 Fax 020 8977 1655
E-mail hearns.joinery@btconnect.com
Website www.hearnsjoinery.co.uk
TRADITIONAL SKILLS AND MODERN TECHNOLOGY COMBINE FOR HIGH QUALITY JOINERY: Hearns joinery and woodworking facility in Teddington is purpose-built to combine the highest quality of craftsmanship to the very latest production techniques. Modern computerised turning and milling machines ensure that traditional skills and modern technology are married to provide the finest joinery products of their type. Products from Hearns are equally suited to private and commercial customers and designs range from the ornate to classic simplicity. Craftsmen made doors add beauty to any home and embellish the image of a company, as do superb quality windows of all sizes and complexity. Staircases range from easy to fit production models made from softwoods to individual hardwood designs that enhance any home. Hardwood shop fronts that offer both style and durability are also supplied as are fitted kitchens and other in-built furniture. Traditional skills and expertise are combined with modern technology to provide top quality products at highly competitive prices. *See also: display entry in this section, page 127.*

▶ L DANIELS & G ELDRIDGE
Windy Heights, High Cross, Froxfield, Petersfield, Hants GU32 1EH
Tel 01730 827472 Workshop 01730 828440
ARCHITECTURAL JOINERS AND CARPENTERS: L Daniels & G Eldridge is a family run business of three working partners, who have been in the wood working trade all their working lives. They make architectural joinery and curved work which they can fit if required, including purpose-made kitchens, stairs, doors, windows and mouldings. Their kitchens are constructed with morticed and tenoned front frames of solid timber, grooved into an 18mm melamine-faced carcase made of water-resistant board, with glued joints and doors hung on traditional brass butts. Stocks of air-dried English hardwoods are held and repair work is also undertaken.

▶ MICHAEL COURT
Old Barnes Stables, 3 Betsoms Farm, Pilgrims Way, Westerham, Kent TN16 2DS
Tel/Fax 01959 561158
PERIOD INTERIORS (ARCHITECTURAL WALL PANELLING)

FINE JOINERY

▶ MOTT GRAVES PROJECTS LIMITED

Sampleoak Lane, Chilworth, Guildford, Surrey GU4 8QW
Tel 01483 453326 Fax 01483 453325
E-mail james@mottgraves.fsbusiness.co.uk
Website www.mottgraves.co.uk

SPECIALIST JOINERY, CABINET WORK AND RESTORATION: Mott Graves provides a comprehensive woodworking service for private and professional clients, such as architects, project developers, building contractors and museums. The directors personally organise and manage each commission, carefully combining proven restoration methods with a sensitive approach to new work. They refurbished over 100 oak panelled rooms within County Hall, Westminster and most recently have restored and oak conservation area within Ingress Abbey, Dartford. Also, MGP regularly designs and manufactures high quality bespoke furniture and cabinets for fine interiors. Please contact James Mott for an information pack. *See also: display entries on page 126, in Stone section, page 106 and in Building Contractors section, page 65.*

▶ N E J STEVENSON

Church Lawford Business Centre, Limestone Hall Lane, Church Lawford, Coventry CV23 9HD
Tel 024 765 44662 Fax 024 765 45345

DESIGNERS AND MAKERS OF DISTINCTIVE COMMISSIONED FURNITURE: *See also: profile entry in Cabinet Makers section, page 206.*

▶ PLOWDEN & SMITH

190 St Ann's Hill, London SW18 2RT
Tel 020 8874 4005 Fax 020 8874 7248
E-mail richard.rogers@plowden-smith.com
Website www.plowden-smith.com

CONSERVATION AND RESTORATION*: See also: display entry in Fine Art Conservation section, page 203.*

▶ THE ROUND WOOD TIMBER CO LTD

Newick Lane, Mayfield, East Sussex TN20 6RG
Tel 01435 867072 Fax 01435 864708 E-mail info@roundwoodtimber.com

FINE JOINERY: The Round Wood Timber Company specialises in architectural joinery and oak frame construction. The company supplies fresh sawn, air and kiln dried oak, a wide range of hardwood floors and doors and a handmade range of traditional ironmongery. *See also: display entries in Timber Suppliers section, page 123, and Timber Flooring section, page 211.*

▶ TANKERDALE LIMITED

Johnson's Barns, Waterworks Road, Sheet, Petersfield, Hampshire GU32 2BY
Tel 01730 233792 Fax 01730 233922
E-mail mail@tankerdale.co.uk Website www.tankerdale.co.uk

CONSERVATION AND RESTORATION OF HISTORIC JOINERY AND FURNITURE: Established in 1977, Tankerdale offers high quality joinery and furniture conservation and restoration to private clients, institutions, museums and historic houses throughout the UK. For details of recent projects, see the website at www.tankerdale.co.uk. *See also: display entry on this page.*

▶ TERNEX LTD

The Sawmill, 27 Ayot Green, Welwyn, Herts AL6 9BA
Tel 01707 324606 Fax 01707 334371

BESPOKE JOINERY IN ENGLISH AND IMPORTED HARDWOODS AND SOFTWOODS: *See also: profile entry in Timber Suppliers section, page 124.*

▶ TIPTREE JOINERY SERVICES

New Road, Tiptree, Colchester, Essex CO5 0HQ
Tel 01621 819220 Fax 01621 815499 E-mail office@tjevers.co.uk

SPECIALIST AND BESPOKE JOINERY MANUFACTURERS, RESTORATION AND CONSERVATION: Specialists in the manufacture of high quality traditional and architectural joinery. They offer a complete service from site survey, design to manufacture and installation. They have many years experience as joinery manufacturers. The high calibre of their craft skills and traditional techniques coupled with quality management skills, they believe enables them to offer a service that is second to none. They offer the practical experience and expertise required in providing a personal and professional service. For further information please contact Ian Garrod.

▶ WALDEN JOINERY

Empstead Works, Henley-on-Thames, Oxfordshire RG9 1UF
Tel 01491 572555 Fax 01491 410849
Website www.bwt.org.uk/walden

FINE JOINERY: Walden Joinery provides a high quality joinery service, combining traditional skills with modern techniques and plant. All joinery is purpose-made to clients' exact requirements, much is manufactured from detailed drawings, often to replace existing period pieces such as mouldings, windows, doors and panelling. Their ability to recreate period styles identically is of paramount importance when working in local conservation areas. Walden's recognises the importance of environmental considerations, and takes care to source its timber only from reputable suppliers, whose environmental policies are vetted.

▶ WALLIS JOINERY

Broadmead Works, Hart Street, Maidstone, Kent ME16 8RE
Tel 01622 690960 Fax 01622 693553

SPECIALIST JOINERY: Wallis Joinery has since its formation in 1860 become one of the United Kingdom's leading specialist joinery companies offering an unparalleled standard of craftsmanship and expertise. This high standard of joinery is exemplified by being twice awarded the prestigious Carpenters Award and receiving accreditation as a Quality Assured Company to BS 5750 Part 2. Wallis Joinery is able to provide all components, from standard softwood joinery to complex restoration projects using hardwood, including purpose designed fixtures and fittings in veneers and laminates. While methods and materials constantly change in timber technology the constant factor with Wallis Joinery is the skill of its craftsmen to produce quality joinery. *See also: Wallis display entry in Building Contractors section, page 54.*

▶ **BASSETT & FINDLEY LTD**
Talbot Road North, Wellingborough, Northamptonshire NN8 1QS
Tel 01933 224898 Fax 01933 227731
E-mail info@bassettandfindley.ltd.uk
Website www.bassettandfindley.co.uk
METAL WINDOWS: *See also: display entries in Metal Windows section, page 141, and Architectural Metalwork section, page 142.*

▶ **C R CRANE & SON LTD**
Manor Farm, Main Road, Nether Broughton, Leics LE14 3HB
Tel 01664 823366 Fax 01664 823534
Website www.crcrane.co.uk
SPECIALIST BUILDING AND JOINERY CONTRACTORS: *See also: profile entry in Building Contractors section, page 60.*

▶ **CHILSTONE**
Victoria Park, Fordcombe Road, Langton Green, Tunbridge Wells, Kent TN3 0RE
Tel 01892 740866 Fax 01892 740249
WINDOWS AND DOORS: Standard and special window surrounds, sills, heads and key stones. Also porticos and door surrounds. *See also: profile entry in Cast Stone section, page 110.*

▶ **HADDONSTONE LIMITED**
The Forge House, East Haddon, Northampton NN6 8DB
Tel 01604 770711 Fax 01604 770027
E-mail info@haddonstone.co.uk
Website www.haddonstone.co.uk
STANDARD AND CUSTOM-MADE WINDOW SURROUNDS, CILLS, HEADS AND KEYSTONES. ALSO PORTICOS AND DOOR SURROUNDS: *See also: display entry in Cast Stone section, page 110.*

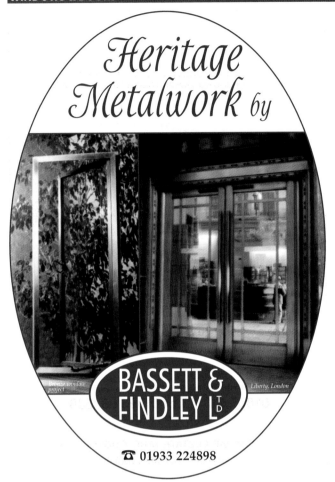

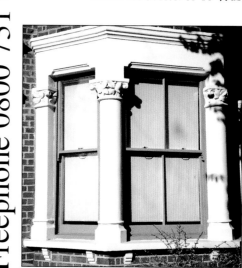

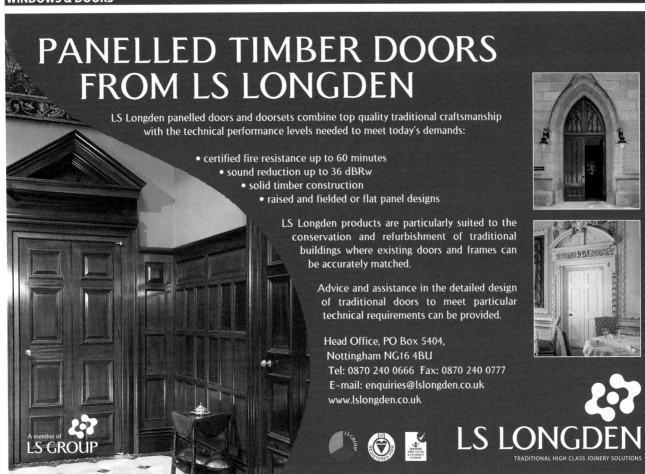

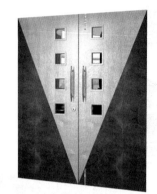

▶ MARVIN ARCHITECTURAL

Canal House, Catherine Wheel Road, Brentford, Middx TW8 8BD
Tel 020 8569 8222 Fax 020 8560 6374
E-mail sales@marvinuk.com Website www.marvin-architectural.com

SLIDING SASH WINDOWS: Marvin has been engineering and manufacturing made-to-order sliding sash windows for over 60 years. Marvin has over 11,000 windows and doors in its range and extensive custom design and size capabilities. Traditional craftsmanship and in-house technical support ensure solid wood windows of superior quality and performance. *See also: display entry in this section, page 131.*

▶ PAUL CHRISTENSEN WINDOWS & DOORS

123/125 Gloucester Place, London W1U 6JZ
Tel 020 7486 0631 Fax 01279 835108 Website www.pcvinduer.dk

WINDOWS AND DOORS: Period window and door manufacturers employing traditional joinery methods. Only the highest quality quarter-sawn timber is used for sliding sashes, casements and bespoke joinery. Designs incorporating metal windows and stone surrounds can be accommodated as conservation projects often demand. Unique systems to improve insulation without loss of period charm are available. *See also: display entry in this section, page 130.*

▶ REFURB-A-SASH

Queens Joinery, Queens Road, Twickenham, Middlesex TW1 4LJ
Tel 020 8892 1166 Fax 020 8892 0055 E-mail refurbasash@btclick.com

TRADITIONAL SASH WINDOW SPECIALISTS: London based, award-winning company offering specialist advice and services for traditional timber sash windows. Services include operational overhauls and draught proofing of existing windows, full or part replacement in single or double glazing and reinstatement of originals. Noise reduction with acoustic glass and guidance for listed building owners. All products produced by craftsmen faithfully replicating original features using traditional joinery methods. All timber is preservative treated, hand prepared and spray paint finished. A FENSA registered company. Corporate members of the Guild of Master Craftsmen and the Glass and Glazing Federation complying with their code of ethical practice.

The Sash Restoration Co.

www.sash-restoration.co.uk

Tel: 01432 359562

Traditional custom made sliding box sash windows
Single or Double Glazed • Document 'L' compliant
Hand made glass (laminated or double glazed units)
Fitting & Painting service • Specialists in Oak windows

Fax: 01432 269749 email: sales@sash-restoration.co.uk
Pigeon House Farm, Lower Breinton, Hereford, HR4 7PG

Repair, Replacement and Restoration

of traditional timber box sash, casement windows and doors

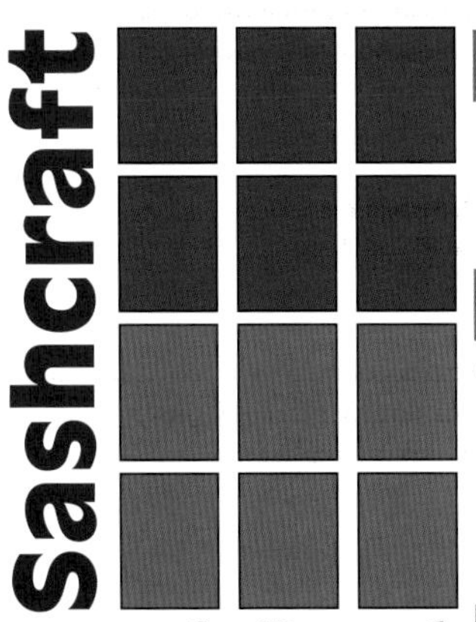

of Bath

SASHCARE

Dedicated repair system or complete restoration

SASHCURE

The ultimate draughtproofing and upgrade for sliding sash windows (can also be applied to casement window and doors)

SASHGLAZE

Bespoke timber windows and doors for period and listed properties. Sympathetic double glazing or traditional single glazing. Low maintenance factory finishing service.

Tel 01225 424434
Unit 1 Mill Farm
Kelston
Bath BA1 9AQ
info@sashcraftofbath.co.uk
www.sashcraftofbath.co.uk

Sash Window Problems?

Ventrolla offer a nation wide service to renovate and upgrade original windows. Fitted by experienced craftsmen the Ventrolla System will:

- Reduce external noise by up to 10db
- Virtually eliminate draughts and rattles
- Ensure sashes open and close with ease
- Maintain the architectural integrity of these period windows
- Offer easy cleaning and painting with the Sash Removal System option

Recognised for its use in Conservation areas and Listed Properties, windows installed with the Ventrolla System benefit from modern standards of comfort and energy efficiency.

Ventrolla®

SASH WINDOW RENOVATION SPECIALISTS
11 Hornbeam Square South, Harrogate HG2 8NB
www.ventrolla.co.uk

FREEPHONE 0800 378278

WINDOWS & DOORS

▶ SELECTAGLAZE LTD

Campfield Road, St Albans, Herts AL1 5HT
Tel 01727 837271 Fax 01727 844053
Website www.selectaglaze.co.uk

SECONDARY GLAZING AND SOUND PROOFING WINDOWS: Secondary glazing, not to be confused with double glazing, involves the installation of an additional internal glazed window. These inner windows offer a number of distinct advantages, the main one being soundproofing. Selectaglaze has over 35 years experience in the design, manufacture and installation of secondary windows and is a leading secondary glazing specialist, with particular experience in listed and period properties. The Selectaglaze range provides the highest levels of sound and thermal insulation, while increasing security and maintaining the character of the property. Please ring 01727 837271 for a copy of Selectaglaze's 'Elegant Solutions' brochure or 'Specifier's Product Guide', or visit the company's website.

▶ STEEL WINDOW SERVICE AND SUPPLIES LTD

30 Oxford Road, Finsbury Park, London N4 3EY
Tel 020 7272 2294/6391 Fax 020 7281 2309
E-mail post@steelwindows.co.uk
Website www.steelwindows.co.uk

SUPPLY AND FIX, SERVICING AND REPAIR OF STEEL WINDOWS: *See also: display entry in Metal Windows section, page 136.*

▶ THORVERTON STONE COMPANY

Seychelles Farm, Upton Pyne, Exeter, Devon EX5 5HY
Tel 01392 851822 Fax 01392 851833
E-mail caststone@thorvertonstone.co.uk
Website www.thorvertonstone.co.uk

RECONSTRUCTED STONE MULLION WINDOW SURROUNDS AND COMPONENTS: *See also: profile entry in Cast Stone section, page 110.*

▶ TIMBER TECH LTD

Unit 1, Le Bel Horizon, La Route Des Cotils, Grouville, Jersey JE3 9AP
Tel 01534 852837 Fax 01534 852847
E-mail timbertech@jerseymail.co.uk

SPECIALIST TIMBER DOUBLE GLAZED VERTICAL SLIDING SASH WINDOW MANUFACTURERS: Michael O'Connor established Timber Tech in 1996 to produce high quality timber windows for historic buildings. Products replicate the fine and unique detail of original sash windows but with the benefit of double glazing and weather sealing. Windows are manufactured using the latest innovations and technology. All windows are delivered fully factory painted or stained complete with all necessary ironmongery. Competitive pricing guarantees good value for money.

▶ THE WOODEN WINDOW COMPANY

Unit 8, The Tannery Industrial Estate, The Midlands, Holt, Trowbridge BA14 6BB
Tel 01225 783040 Fax 01225 783668

WINDOWS & DOORS: Manufacturers of quality wooden sliding sash and casement windows. Single or double glazed. Conservation conscious, Document L, traditional looking replicas.

METAL WINDOWS

▶ CRITTAL WINDOWS LIMITED

Springwood Drive, Braintree, Essex CM7 2YN
Tel 01376 324106 Fax 01376 349662
E-mail hq@crittall-windows.co.uk
Website www.crittall-windows.co.uk

STEEL WINDOWS AND DOORS: Design, manufacture and installation of replica refurbishment steel windows and doors for all types of buildings. The modern steel windows include weather-stripping, corrosion protection, polyester powder finish and can incorporate double glazed units that meet Part L of the new Building Regulations as well as leaded lights. *See also: display entry in this section, page 135.*

DECORATIVE & STAINED GLASS

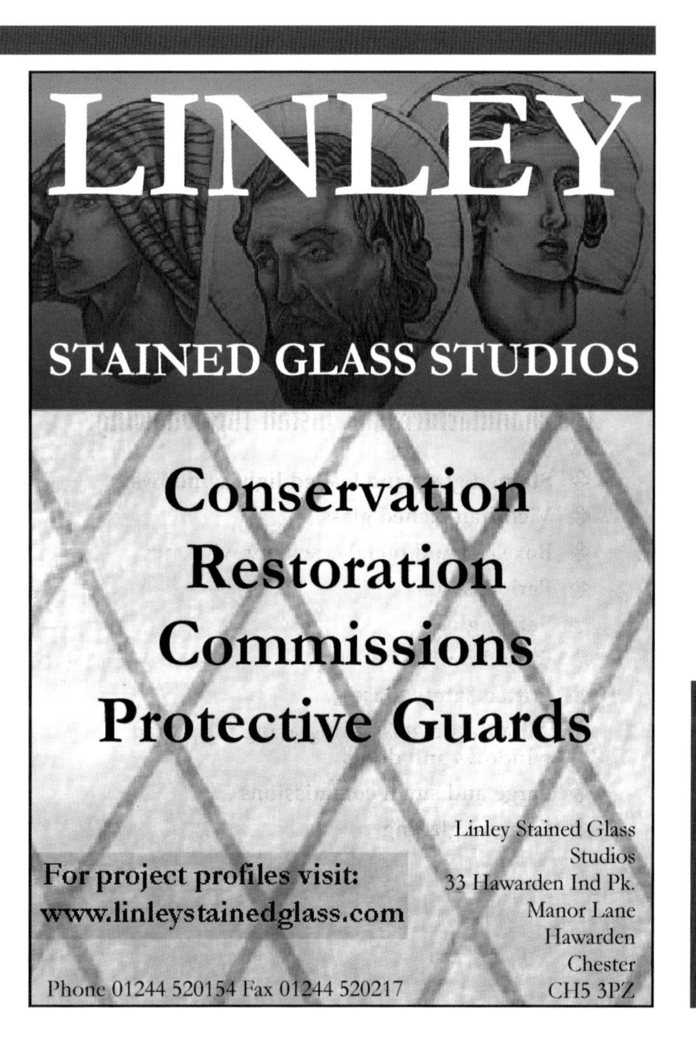

▶ PAUL BRADBURY STAINED GLASS

21 Arosa Court, 419 Wilmslow Road, Withington, Manchester M20 4LZ
Tel 07932 610852

DECORATIVE AND STAINED GLASS: Architectural glass, stained and decorative glass. Specialising in the design and manufacture of all types of stained glass and leaded lights. Repair, restoration and contemporary design. Decorative work including acid etching, sand blasting and screen printing. Available for maintenance, cleaning and installation. Free estimate and advice.

▶ C J L DESIGNS

Pantyffynnon, Felindre, Llandysul SA44 5XS
Tel 01559 371670 Mobile 07968 175069
E-mail cjldesigns@macunlimited.net

STAINED GLASS AND LEADED LIGHTS: Christopher and Bronwen Worrall design, manufacture and restore windows in all settings. They offer professional service and advice for restoring existing glass, frames and furniture and a bespoke design service for new commissions in public and domestic buildings. Specialisations include moulded quarries for Powell & Co panels at St Thomas, Colnbrook and St Nicholas, Islip; regimental colours, water ingress, vandalism and protective grilles. Also, maintenance and refurbishment for leaded glass and frames including Crittal type and hoppers. Projects include churches within the Oxford, Guildford and Winchester dioceses, private houses and listed buildings, Pembrokeshire National Park and Cadw. Other schemes include Falklands Islands Chapel, Pangbourne; the Theatre Royal, Windsor; Littlecote House, Hungerford; St Mary's Convent, Mill Hill, London.

DECORATIVE & STAINED GLASS

▶ THE CATHEDRAL STUDIOS

Dean and Chapter of Canterbury
8a The Precincts, Canterbury, Kent CT1 2EG
Tel 01227 865265 Fax 01227 865222
E-mail cathedralstudios@canterbury-cathedral.org

SPECIALISTS IN STAINED GLASS CONSERVATION AND RESTORATION. The Cathedral Studios look back on 31 years of successful conservation of the medieval, late medieval and Victorian heritage in Canterbury Cathedral and in many other churches and buildings throughout the country. The team of six highly trained and specialised conservators is headed by Dr Sebastian Strobl FMGP ACR, UKIC accredited conservator and an internationally renowned stained glass expert, who will be pleased to advise on the conservation and protection of your glass.

▶ ILLUMIN GLASS STUDIO

82 Bond Street, Macclesfield, Cheshire SK11 6QS
Tel 01625 613600

STAINED GLASS WINDOWS AND LIGHTING: Illumin manufacture and repair stained glass windows and lighting; and supply antique and other sheet glass. They also provide to order sandblasting, brilliant cutting, etching, etc. Established in 1979, Illumin cover all aspects from design to installation. They cover a 30 mile radius of Macclesfield and will travel further for an interesting and viable project. Past work has included: The Millennium Window at Great Brington Church, Northampton; Grade I listed Chethams Library, Manchester; renovating windows of a 16th century farmhouse; repairs to local churches, and; repairs and new windows for private houses of various ages.

▶ JONATHAN LECKIE ASSOCIATES

26 Pembroke Avenue, Hersham, Walton-on-Thames, Surrey KT12 4NT
Tel/Fax 01932 252567

STAINED GLASS DESIGN, MANUFACTURE, RESTORATION AND CONSERVATION: Jonathan Leckie Associates is a UKIC accredited studio, experienced in all aspects of ecclesiastical and secular stained/decorative glass. They can restore existing glass to its original condition and also produce new designs to client specification. Every stage of a new commission can be discussed from conception to final realisation. Each individual project is executed to a very high standard with an emphasis on attention to detail. The studio works closely with conservation architects, DACs and other conservation bodies such as English Heritage and the National Trust. Architects, churches, listed historic buildings and houses are amongst their main clientele.

▶ NORGROVE STUDIOS

Bentley, Redditch, Worcs B97 5UH
Tel 01527 541545 Fax 01527 403692
Website www.norgrovestudios.co.uk

DECORATIVE AND STAINED GLASS: Plain glazing for 17th, 18th and 19th century windows. Repair and rebuilding of all early periods of glasswork. New handmade glass supplied to match originals. Ferramenta and casements repairs. Copies in wrought iron or stainless steel, leaded glazing in handmade glass only. Norgrove Studios is a small specialist glazing studio whose core business is stained glass, both restoration and newly commissioned work, with a specialist knowledge of vernacular buildings and glazing. The studio can help with any building restoration, from advice and consultancy to practical works. Handmade glass supplied for sash windows.

▶ RIVERSIDE STUDIO

9b Curzon Street, Hull, East Riding of Yorkshire HU3 5PH
Tel 01482 563742 Fax 01482 575900
E-mail riverside.studio@ic24.net

STAINED AND LEADED GLASS WINDOWS: Riverside Studio specialises in all aspects of the conservation, restoration and repair of stained and leaded glass windows. Wire guards and polycarbonate window protection are also manufactured and fitted. Working nationally, the studio undertakes work of the highest calibre for its clients who include ecclesiastical authorities, historic building owners and the top national conservation organisations. On its team Riverside has two UKIC accredited conservators. For free estimates and advice please contact Richard Green.

▶ THE STAINED GLASS SPECIALIST

10 Brackenhill Road, Wimborne, Dorset BH21 2LT
Tel/Fax 01202 882208
E-mail ssherriff@yahoo.com
Website www.lead-windows.co.uk

STAINED AND DECORATIVE GLASS: A fully comprehensive stained glass design studio offering repair, restoration and new design including *in situ* repairs of all aspects. This family run company offers sound advice on the repair and maintenance of all windows and their frames including the supply and fitting of all methods of window protection. Operating south of a line between Bristol and London, a reliable dependable service and top quality workmanship is always provided. Recent projects include new East windows for Wool, comprehensive restoration and renovation of listed school buildings leaded light windows and contract work upon the renovation of historical buildings.

WINDOW GLASS

▶ BURSLEDON BRICKWORKS CONSERVATION CENTRE

Bursledon Brickworks, Coal Park Lane, Swanwick, Southampton SO31 7GW
Tel/Fax 01489 576248
E-mail bursledon@ndirect.co.uk

TRADITIONAL DRAWN GLASS SHEET: *See also: profile entry in Mortars & Renders section, page 172.*

WINDOW GLASS

DOOR & WINDOW FITTINGS

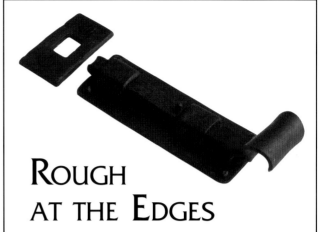

STRUCTURE & FABRIC
3.3 Metal, Glass & Wood

▶ BRAMAH

31 Oldbury Place, London W1U 5PT
Tel 08700 272624 Fax 020 7935 2779
E-mail lock.sales@bramah.co.uk

SASH WINDOW LOCKS AND HIGH SECURITY DOOR LOCKS: Bramah first made locks for historic buildings when they were being built in the early 1800s. Today they still make high security locks, and where they can, tailor them to suit the situation. This applies to their high security key registered deadlocks for secure areas, cabinet locks and cylinder locks. All Bramah locks may be mastered or keyed alike and are available in a variety of finishes, with black and bronze often being used. In addition Bramah manufacture the Rola range of window locks and bolts, where they have a niche expertise in the manufacture of sash window locks.

▶ CLAYTON MUNROE LIMITED

Kingston, Staverton, Totnes, Devon TQ9 6AR
Tel 01803 762626 Fax 01803 762584 E-mail mail@claytonmunroe.com
Website www.claytonmunroe.com

DOOR AND WINDOW FITTINGS AND ACCESSORIES: Formed in the early '80s in response to demand for authentic restoration hardware, Clayton Munroe Limited now offers an extensive selection of innovative door furniture, cabinet fittings and accessories, including a new range of stair rods. The 'Rough at the Edges' range, based on 18th century designs is hand forged and supplied with traditional fixings and addresses the search for authentic hardware to retain the true character of a period building. 'Rough at the Edges' products are used in historic buildings throughout the world as well as film sets for period dramas such as Jane Eyre. The company is also well known for its trade marked 'Patine' finish (similar to pewter in colour) and offers over 3,500 items featured in a new trade catalogue. *See also: display entry in this section, page 139.*

▶ DON FORBES SASH FITTINGS

Cotterton, Logiealmond, Perthshire PH1 3TJ
Tel/Fax 01738 880329

MANUFACTURING AND WHOLESALE IRONMONGERS: Specialist suppliers of all types of sash window fittings including: axle pulleys in a range of sizes and finishes, sash cords and chains, sash openers and easy clean fittings, a large selection of sash fasteners, lifts, stops, etc in a variety of styles and finishes. Don Forbes can refurbish existing fittings and supply replacement items to match existing. A comprehensive range of cast iron and lead sash weights and makeweights are available ex-stock, or made to specification. Kame lead and other lead products available. Brochure and price lists on request.

▶ MBL

55 High Street, Biggleswade, Bedfordshire SG18 0JH
Tel/Fax 01767 318695 E-mail info@mblai.co.uk
Website www.mblai.co.uk

SPECIALIST ARCHITECTURAL IRONMONGERS AND LOCKSMITHS: MBL is a specialist company involved in bespoke, unique and specialist architectural ironmongery and locking systems. Founded in 1990, MBL provides a quality service involving refurbishment, replication and restoration of historic and period fittings for all types of properties. From single commissions to full design, scheduling and supply of complete packages, MBL integrates modern security systems to complement historic buildings. Recent and on-going projects are Chicksands Priory, Bedfordshire; Waddesdon Manor, Buckinghamshire; Somerset House, London; St John's College, Cambridge; and various ecclesiastical projects and private houses.
See also: display entry on this page.

▶ QUALITY LOCK CO

Leve Lane, Willenhall, West Midlands WV13 1PS
Tel/Fax 01902 602942

HAND CRAFTED TRADITIONAL AND MODERN LOCKS: Quality Lock Company manufactures locks in the traditional way from original patterns and drawings and refurbishes and re-key locks dating from the early 1800s to the present time. Using your lock as a pattern, Quality Lock will produce new locks to exactly match the original. Their stock range of brass cased rimlocks harmonises well with historic buildings. They can produce master keyed locks and extensions to existing master key suites and are also able to produce short runs of specially designed locks to meet customers' individual locking needs.

WINDOW PROTECTION

▶ GOWAR GRILLES LTD

28–34 Rheidol Mews, Rheidol Terrace, Islington, London N1 8NU
Tel 020 7226 3644 Fax 020 7226 2969

HAND WOVEN BRASS AND STAINLESS RADIATOR GRILLES AND WINDOW GRILLES: A family firm, which was established in 1892 as traditional wire workers. Specialising in hand weaving decorative grilles for historic buildings nationwide. Also manufacturers of shaped stainless and galvanised church window grilles, protecting church windows while highlighting the glass shape in London and the Home Counties.

LOCKSMITHS

▶ A M EXPERIMENTAL

The Brew House (At the Lock Museum), 54 New Road, Willenhall, West Midlands WV13 2DA
Tel 07941 651193
E-mail a.m.experimental@btclick.com
Website http://home.btconnect.com/a.m.experimental

LOCKSMITHS: Established for 18 years. Specialist in the restoration of all types of historic locks and the making of replacement keys. New locks built on a one-off basis to suit your requirements, ie to an architectural style or for a specific level of security. A specialist engineering service for damaged household items offered with replacements for worn or missing parts. One-off springs, bezels and castings can be produced to put that treasured item back into service.

ARCHITECTURAL METALWORK

▶ ALBION MANUFACTURING

The Granary, Silfield Road, Wymondham, Norfolk NR18 9AU
Tel 01953 605983 Fax 01953 606764

ARCHITECTURAL METALWORK AND WIREWORK SPECIALISTS: Albion Manufacturing has over 60 years experience in new and restoration work from window guards to bronze doors. They have a comprehensive range of casting designs both old and new and in 2000 received an award of distinction for gates and railings supplied to Messrs Willis Coroon in Ipswich.

▶ ARCHITECTURAL METALWORK CONSERVATION

Unit 19, Hoddesdon Industrial Centre, Pindar Road, Hoddesdon, Herts EN11 0DD
Tel/Fax 01992 443132

BLACKSMITHS AND FERROUS METALWORK CONSERVATORS: Architectural Metalwork Conservation has wide experience in the traditional methods of working iron and a considered approach to repairs. Projects range from high quality reproductions and new designs in wrought iron, cast iron and steel to sympathetic restoration of historic ironwork. Consultancy services include the provision of training programmes, surveys and treatment recommendations. Contact Keith Blackney (National Council for Conservation-Restoration accredited conservator) for further details.

▶ B LEVY & CO (PATTERNS) LIMITED

37 Churton Street, Westminster, London SW1V 2LT
Tel 020 7834 1073 Fax 020 7630 8673

ARCHITECTURAL METALWORK: *See also: display entry on this page.*

▶ BARR & GROSVENOR LTD

Jenner Street, Wolverhampton, West Midlands WV2 2AE
Tel 01902 352390 Fax 01902 871342

ARCHITECTURAL CAST IRON: Barr & Grosvenor Ltd is one of the few remaining foundries with the skills and patience to produce complex items as used in Victorian times. The company is involved at all stages in restoration projects requiring castings, whether still existing or long since removed. Their services can help stabilise, restore or recreate castings for all purposes. Most of their work is made to order so that essential attention to detail is guaranteed. Specification advice, project management, pattern making, casting, machining, paint finishing and site fitting are all available. Group open days by appointment.

ARCHITECTURAL METALWORK

ARCHITECTURAL METALWORK

▶ BASSETT & FINDLEY LTD

Talbot Road North, Wellingborough, Northamptonshire NN8 1QS
Tel 01933 224898 Fax 01933 227731
E-mail info@bassettandfindley.ltd.uk
Website www.bassettandfindley.co.uk

ARCHITECTURAL METALWORK: 90 years experience in architectural metalwork using steel, stainless steel, brass, bronze, aluminium, copper, pewter and zinc. Full survey, design, manufacturing and fixing service available. The company employs time served craftsmen skilfully blending traditional methods using castings and hand forming metal, with modern techniques of CNC cutting, forming and folding to meet demanding standards at affordable prices The extensive finishing shop provides a polishing and patination service, ensuring products for listed buildings faithfully replicate original profiles, and colour blend unobtrusively with their environments. *See also: display entries in this section, page 141, and Windows & Doors section, page 129.*

▶ CAMBRIAN CASTINGS (WALES) LTD

Unit 19, Crofty Industrial Estate, Penclawdd, Swansea SA4 3RS
Tel/Fax 01792 850912 Mobile 07966 157139
E-mail jon@cambriancastings.co.uk
Website www.cambriancastings.co.uk

ARCHITECTURAL METALWORK: Specialising in iron, bronze and aluminium casting, Cambrian Castings undertakes all types of architectural metalwork creation, renovation and repair. A full design and fabrication service for stairs, railings, grilles and screens; both decorative and structural, can be provided. The company specialises in bronze and cast iron grillework and complex cast iron components for park gates and interior/exterior stairs. Visit Cambrian Castings' website for examples, or contact Jon Barnett for more information.

▶ CARREK LTD

Mason's Yard, Wells Cathedral, Wells, Somerset BA5 2PA
Tel 01749 689000 Fax 01749 689089
Website www.carrek.co.uk

HISTORIC BUILDING REPAIR COMPANY: *See also: profile entry in Building Contractors section, page 61.*

▶ CASTAWAY CAST PRODUCTS & WOODWARE

Brocklesby Station, Brocklesby Road, Ulceby, North Lincolnshire DN39 6ST
Tel/Fax 01469 588995

CAST METAL PRODUCTS INCLUDING BESPOKE ITEMS: Supplying nation-wide, Castaway undertakes any cast metalwork project using aluminium, bronze and gun metal, grey and nodular (SG) iron, carbon steels and stainless steels. Castings for historic buildings include gutter sections and other drainage ware, airbricks, wall retaining plates, brackets, gates and railings, signage, window frames and much more. Items can be made from drawings, photographs or from sight of originals - broken, corroded or intact. Items can be made from standard existing patterns or to your own designs in any quantities; from one upwards. In-house pattern making facilities are available if required. Contact John Wade to discuss your requirements in detail.

▶ CASTING REPAIRS LIMITED

Marine House, 18 Hipper Street South, Chesterfield, Derbyshire S40 1SS
Tel 01246 246700 Fax 01246 206519

RESTORATION OF IRONWORK: Casting Repairs Limited has over 40 years experience in the repair of decorative and structural ironwork. This covers work on buildings of major historical importance, large municipal buildings, bridges, public fountains, bandstands, statues, windows, gates and railings. The company can provide complete project management, where required, together with a combination of repair techniques including cold metal stitching, welding and brazing. With full fabrication facilities, new architectural ironwork can be designed and produced to specific requirements. Operating nationally, Casting Repairs Limited brings an expert and cost effective approach to every project.

▶ CHRIS TOPP & COMPANY WROUGHT IRONWORKS

Lyndhurst, Carlton Husthwaite, Thirsk YO7 2BJ
Tel 01845 501415 Fax 01845 501072
E-mail enquiry@christopp.co.uk
Website www.christopp.co.uk

BLACKSMITHS: Established in 1980, Chris Topp & Co has been one the pioneers in the return to the traditional art of the blacksmith. The company carries out a wide range of work throughout the UK in genuine wrought iron, cast iron and mild steel; including bespoke ironwork, restoration, renovation and repousee. A consultancy and design service is also available. Chris Topp & Co prides itself on using the highest craft standards and where appropriate the traditional material of the blacksmith, wrought iron. And as far as they know, through its sister company The Real Wrought Iron Company, Chris Topp & Co is the world's sole supplier of wrought iron. *See also: display entry in this section, page 144.*

▶ CROWNCAST LTD

The Foundry, Rushenden Road, Queenborough, Kent ME11 5HD
Tel 01795 662722 Fax 01795 666552

MANUFACTURERS OF IRON CASTINGS AND NON-FERROUS CASTINGS AND PATTERNS: Crowncast is a craft foundry which offers a bespoke service for special conservation and refurbishment projects. Crowncast's versatility means that they are often called on to cast replacement pieces using old castings as patterns. Projects have included lamp posts for Osborne House, gate posts for Kensington Palace Gardens and gates and fencing for Westminster and Greenwich. Recently they were heavily involved in the refurbishment of Brighton Railway Station.

▶ DOROTHEA RESTORATIONS LTD

INCORPORATING ERNEST HOLE (ENGINEERS) LTD
Northern office & works: New Road,
Whaley Bridge, High Peak, Derbyshire SK23 7JG
Tel 01663 733544 Fax 01663 734521 E-mail north@dorothearestorations.com
Website www.dorothearestorations.com
▶ Southern office & works: Riverside Business Park,
St Annes Road, St Annes Park, Bristol BS4 4ED
Tel 0117 971 5337 Fax 0117 977 1677 E-mail south@dorothearestorations.com
Website www.dorothearestorations.com

RESTORATION OF ARCHITECTURAL METALWORK, TRADITIONAL MACHINERY AND MILLS: With over 25 years as nation-wide leaders in traditional metalwork and engineering using authentic materials and methods, the company specalises in the repair and reinstatement of large scale wrought and cast iron structures. Free assistance offered to organisations preparing HLF submissions for municipal park restorations, public building and conservation area schemes. Recent projects include the 18th century main gates, Chirk Castle; Cookham Bridge, Maidenhead: Oswestry Park bandstand; Palace of Westminster gates; cast iron railings, British Museum; bronze leafwork, Albert Memorial; forged and leaded railings, Pavilion Gardens, Buxton. In-house services include condition surveys, historic records interpretation, repair of iron castings, engineering solutions to historic metalwork problems. *See also: display entry in this section, page 144.*

▶ EURA CONSERVATION LTD

Unit H10, Halesfield 19, Telford, Shropshire TF7 4QT
Tel 01952 680218 Fax 01952 585044
E-mail mail@eura.co.uk

ACCREDITED CONSERVATORS AND CONTRACTORS FOR ARTISTIC, ARCHITECTURAL AND MARINE METALWORK: Projects include the ornamental cladding of the Albert Memorial, Gravesend Pier, Crystal Palace Park Dinosaurs, the Hubert Fountain, the Canada, Australia and South Africa Gates and many municipal monuments. The company also deals with computerised conservation documentation systems, security fixings, packing and transporting museum objects and EU-supported research. Clients include English Heritage, the National Trust, the Department for Culture, Media and Sport, many local authorities, museums and private estates. The company acts as main, principal and sub-contractor. *See also: display entry in Statuary section, page 94.*

ARCHITECTURAL METALWORK

ARCHITECTURAL METALWORK

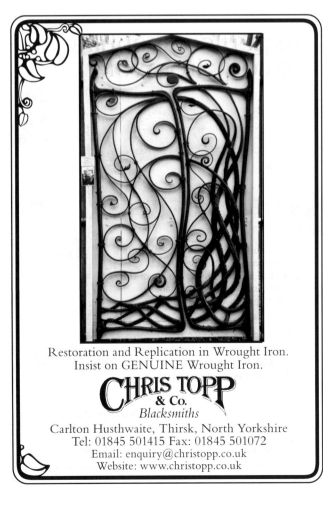

ARCHITECTURAL METALWORK

▶ GEORGE JAMES & SONS

22 Cransley Hill, Broughton, Kettering, Northants NN14 1NB
Tel/Fax 01536 790295

BLACKSMITHS: The family business was established in 1841 and continues to operate from the same forge and workshop in Northamptonshire. The firm has a reputation for producing specialist ironwork of a high quality. The company has considerable experience in the restoration of wrought and cast ironwork whether the need arises from neglect, old age or accidental damage. It also produces new forged ironwork to customer requirements. Recent commissions include the restoration of 18th and 19th century ironwork at Castle Ashby House, Northamptonshire and The Werner Gates at Luton Hoo, Bedfordshire.

▶ HADDONCRAFT FORGE

The Forge House, East Haddon, Northants NN6 8DB
Tel 01604 580559 Fax 01604 580541
E-mail info@haddoncraft.co.uk
Website www.haddoncraft.co.uk

SPECIALISTS IN ARCHITECTURAL WROUGHT IRONWORK: Haddoncraft Forge offers a comprehensive range of high quality decorative wrought ironwork, handcrafted using a coke-fired forge and anvil. A restoration service is available. *See also: display entry for Haddonstone in Cast Stone section, page 110.*

▶ HART BROTHERS ENGINEERING LTD

Albion Works, Cobden Street, Pendleton, Salford M6 6LY
▶ Also at: Soho Works, Soho Street, Oldham OL4 2AD
Tel 0161 737 6791 Fax 0161 633 5316

GREY CAST IRON WELDING: Specialising in high quality casting repairs, Hartbro successfully undertakes remedial works to castings that require a high degree of skill to repair. The result being a reliable casting returned as close as possible to its original specification. Full cross-sectional welding of material damage ensures structural integrity throughout. Castings repaired from broken, worn, or parts-missing condition, of any weight, size or quantity. Hartbro casting repairs may be machined, filed, drilled or tapped to return your component to its original condition. Parts may be finished to a condition where their repair is virtually impossible to detect or identify that the item has been damaged. *See also: display entry in this section, page 146.*

▶ HODGSONS FORGE DECORATIVE METALWORK & RESTORATION

2 Wesley Road, Terrington St Clement, Kings Lynn, Norfolk PE34 4NG
Tel/Fax 01553 828637

HAND FORGED WROUGHT IRON SPECIALISTS: Master blacksmiths for five generations, now a partnership of Colin and Ian Hodgson, this family-run business is proud of the quality of their work and the service they provide. Their range of products covers all estate, domestic and industrial applications. All site work is undertaken including the dismantling and installation of ironwork. They can design or reproduce the most complex features including fine tracery work and wrought iron reproduction using traditional blacksmithing methods. Their most recent commissions include refurbishing ironwork at Knowle House, Sevenoaks, Kent and Waddesdon Manor, Bucks.

▶ J H PORTER & SON LTD

13 Cranleigh Mews, Cabul Road, London SW11 2QL
Tel 020 7978 5576 Fax 020 7924 7081
E-mail info@jhporter.co.uk
Website www.jhporter.co.uk

BLACKSMITHS: After 60 years in Kensington J H Porter & Son moved their forge to larger premises. They manufacture, restore, reproduce and repair gates, railings, balustrades, balconies and staircases, together with security grilles, antiques, indoor/outdoor furniture etc. They work to their own or their clients' traditional or modern designs. Commissions include work for antique restorers, architects, artists, builders, churches, designers, English Heritage, estate managers, London Transport, the National Trust, photographers, shop-fitters and the general public. A work force of five blacksmiths produces all manner of ironwork to a very high standard.

HARTBRO
WELD ENGINEERS
Architectural & Structural Cast Iron Welding
Gas Fusion Specialists
No pattern & re-cast costs
Original castings in original settings
No material mis-match or visible damage reminder
Homogeneous repairs under controlled welding environment
Project management & consultancy service
See text entry details, page 145
Brochure on request
HARTBRO
Hart Brothers Engineering Ltd
Manchester: Tel: 0161 737 6791
Fax: 0161 633 5316

Mather & Smith are probably the oldest established architectural ironfounding and metalwork specialist in the UK, dating back to 1675, giving an unequalled depth of experience. Mather & Smith can offer a complete conservation and restoration service of all cast iron structures including railings, gates, stair balustrades, and balcony railings. We also offer a range of Victorian and modern cast iron wall railings and gates as well as a standard range of standard street furniture. We also offer a bespoke design and installation service.

A division of the MJ Allen Group of Companies

Hilton Road, Cobbswood Industrial Estate
Ashford, Kent TN23 1EW
Telephone 01233-622214 Fax 01233-632979
email sales@mjallen.co.uk Website www.mjallen.co.uk

Peter S Neale
BLACKSMITHS

SPECIALISTS IN THE RESTORATION AND CONSERVATION OF HISTORIC METALWORK

ARCHITECTURAL METALWORKERS
DRIVE AND ESTATE GATES • RAILINGS
STAIRCASES • CASEMENTS • CANOPIES etc

REPRODUCTION OF PERIOD DESIGNS –
BOTH HANDFORGED AND CAST IRONWORK

Scatterford Smithy, Newland, Nr. Coleford
Gloucestershire GL16 8NG
Phone/Fax (01594) 837309
Website: www.peter-s-neale.demon.co.uk
E-mail: peter@peter-s-neale.demon.co.uk

▶ MARSH BROS ENGINEERING SERVICES LTD

PO Box 3, Bakewell, Derbyshire DE45 1LT
Tel 01629 636532 Fax 01629 636003
E-mail info@marshbrothers.co.uk
Website www.marshbrothers.co.uk

STRUCTURAL AND DECORATIVE CAST IRON: Marsh Bros provide a specialist service to restore decorative and structural cast iron components. The work undertaken includes staircases, railings, gates and crestings and cast iron roof structures for lantern and atrium roofs. Also, restoration and repair of cast iron bridge parapets, facades and arch castings. Marsh Brothers provides a bespoke supply and design service to restore and manufacture identical replacement parts where loss or damage has occurred. The company covers all areas of the UK and principal customers include county councils, Crown Estates and major civil contractors.

▶ MATHER & SMITH LTD

Hilton Road, Cobbswood Industrial Estate, Ashford, Kent TN23 1EW
Tel 01233 622214 Fax 01233 631888
E-mail sales@mjallen.co.uk
Website www.mjallen.co.uk

IRON FOUNDER ARCHITECTURAL: Mather & Smith is probably the oldest established architectural iron founding and metalwork specialist in the UK, dating back to 1675, giving an unequalled depth of experience. Conservation and recent restoration works includes restoration of West Norwood Cemetery cast iron railings, and perimeter cast iron posts and railings at the Tower of London. Mather & Smith can offer a complete conservation and restoration service of all cast iron structures including railings, gates, stair balustrades, and balcony railings. They also offer a range of Victorian and modern cast iron wall railings and gates as well as a standard range of standard street furniture. A bespoke design and installation service is also offered. *See also: display entry on this page.*

ARCHITECTURAL METALWORK

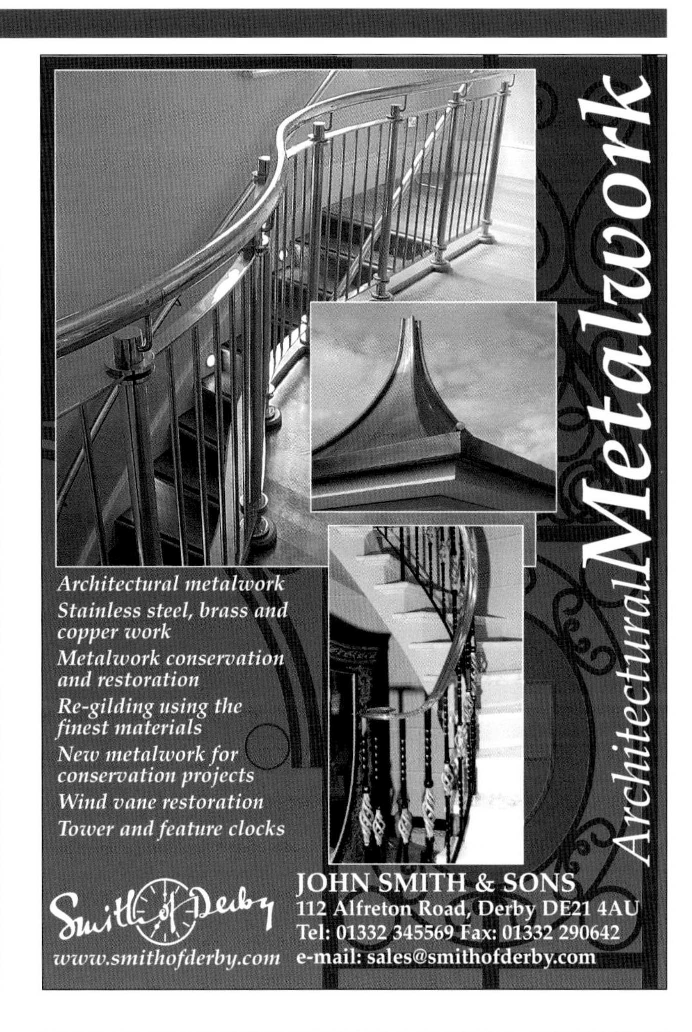

Architectural Metalwork
Architectural metalwork
Stainless steel, brass and copper work
Metalwork conservation and restoration
Re-gilding using the finest materials
New metalwork for conservation projects
Wind vane restoration
Tower and feature clocks
Smith of Derby
www.smithofderby.com
JOHN SMITH & SONS
112 Alfreton Road, Derby DE21 4AU
Tel: 01332 345569 Fax: 01332 290642
e-mail: sales@smithofderby.com

STRUCTURE & FABRIC
3.3 Metal, Glass & Wood

▶ **PLOWDEN & SMITH**
190 St Ann's Hill, London SW18 2RT
Tel 020 8874 4005 Fax 020 8874 7248
E-mail richard.rogers@plowden-smith.com
Website www.plowden-smith.com
CONSERVATION AND RESTORATION: *See also: display entry in Fine Art Conservation section, page 203.*

▶ **RUPERT HARRIS CONSERVATION**
Studio 5, No 1 Fawe Street, London E14 6PD
Tel 020 7515 2020 Fax 020 7987 7994
E-mail enquiries@rupertharris.com
Website www.rupertharris.com
ARCHITECTURAL METALWORK: *See also: display entry in Fine Art Conservation section, page 203.*

▶ **SHEPLEY ENGINEERS LIMITED**
Westlakes Science Park, Ingwell Hall, Moor Row, Cumbria CA24 3JZ
Tel 01946 599022 Fax 01946 591933
E-mail engineers@shepley.vhe.co.uk
ARCHITECTURAL METALWORK RESTORERS AND RENOVATORS: Shepley Engineers Limited has been responsible for engineering faithful and innovative solutions for many major architectural restoration projects. The company offers a comprehensive range of services acting as consultant, designer, contractor or as principal contractor depending on the size and specific requirements of the scheme. Major projects completed or in progress include: The Dorchester Hotel, London; Smithfield Market, London; The Curvilinear Range at the Botanical Gardens, Dublin; The Palm House at Sefton Park, Liverpool; The Paxton Pavilions, Sheffield Botanical Gardens; The Palm House, Botanical Gardens, Dublin and St Pancras Station, London. *See also: display entry on this page.*

SEL
SHEPLEY ENGINEERS LIMITED
Metalwork Specialists
Restoring & Conserving
Our Heritage
Westlakes Science Park
Ingwell Hall, Moor Row
Cumbria CA24 3JZ
Tel: +44 01946 599022
Fax:+44 01946 591933
Email: engineers@shepley.vhe.co.uk
A VHE GROUP COMPANY

RESTORING BATTERSEA PARK

HILARY TAYLOR

Battersea Park (Grade II*) has been enjoyed by Londoners for almost 150 years. By the late 1990s much of the landscape was showing signs of the wear and tear one might expect after such a long period of intensive use. As at so many other public parks, this was made worse by a lack of investment in upkeep and renewal. Furthermore, the original 19th century design had been overlaid with various additions which took little or no account of what had once been a very fine landscape. The most notable of these later developments was the Pleasure Gardens, introduced for the Festival of Britain in 1951. In their turn, these Gardens too had been partially dismantled and what remained was shapeless and ugly. In 1999, with the help of a major grant from the Heritage Lottery Fund (HLF), Wandsworth Borough Council (WBC) began a programme of work to restore and enhance the park. The landscape team engaged on this project is led by Hilary Taylor Landscape Associates Ltd (HTLA) and the work of restoration and renewal will continue until 2004.

BEGINNINGS OF THE PARK

Battersea Park had first been envisaged in 1845 when an outline plan was drawn up by James Pennethorne, architect to the Office of Works. The scheme included land on the periphery which was intended for the building of large mansions; the new park would inflate the value of the land and the ensuing sales and ground rents would help to support the park. This was a technique pioneered by Paxton at Prince's Park, Liverpool and then at Birkenhead Park, and was one of the ways the early public parks were financed. At Battersea, the plan inspired confidence and, in 1846, an Act of Parliament was passed which enabled the Commissioners of Woods and Forests to lay out the new park.

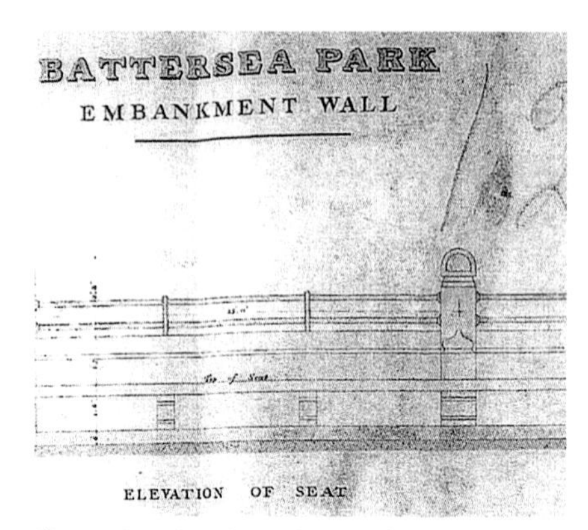

Figure 1 *James Pennethorne's drawings of the river wall*

RIVERSIDE PROMENADE

The site intended for the park extended along the River Thames across an area described as a 'dismal marsh', being regularly flooded. Hence, over the following decades, the embankment was gradually built up using material excavated from the Surrey docks and from other large engineering projects.

James Pennethorne had intended that there should be a broad and stylish promenade along the Thames, accompanied by an embankment wall. Hilary Taylor discovered Pennethorne's original, exquisite, watercolour drawings for the wall (*Figure 1*), which was to have been built from polished, pink granite. It would have provided a magnificent accompaniment to the promenade and would certainly have enhanced the delights of both the river and the park. In the event, Pennethorne did not live to see either promenade or wall. Along the river, the levels of the park were still being raised, bit by bit, for almost 30 years; even as late as 1877, a 30 foot length of the embankment was washed away at high tide.

Figure 2 *The concrete GLC wall, before the restoration*

More than 100 years later there was both a promenade and an embankment wall, but these were anything but grand or stylish. The walk was narrow, muddy, uneven, and unattractive; the utilitarian wall which overlooked it was built of concrete by the Greater London Council (*Figure 2*). Despite the popularity of Battersea Park, and the wonderful views available across the river, it was surprising how few people used this Thames-side walk. It was clear that the renewal of the promenade and the embankment wall should be a focus of the restoration.

In collaboration with WBC, the HLF and English Heritage, it was decided that Pennethorne's wall – though never realised – should be the starting point for a new one. Not surprisingly, however, the restoration project did not allow for one kilometre of pink

Figure 5 *Battersea Park embankment wall and promenade, 2003*

Figure 7 The Sub-Tropical Garden, c1860s

Figure 8 Engraving of the Sub-Tropical Garden, Battersea, 1877

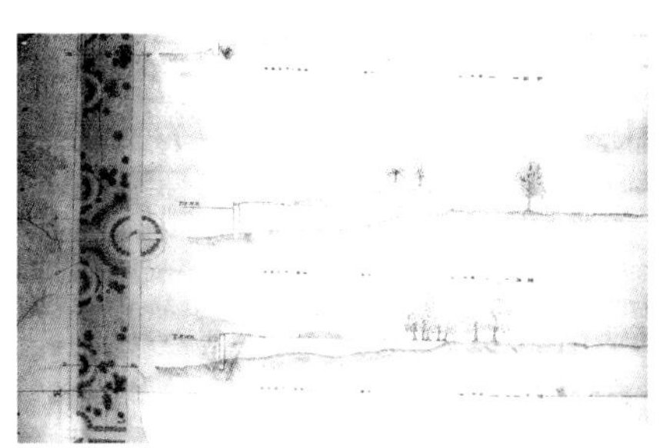
Figure 6 John Gibson's drawing, with plan and sections, showing the composition of trees and shrubs

Figure 9 The Russell Page Parterre and the Refreshment Tent, 1951

granite; HTLA had to find an alternative. So the practice's chief designer, Peter Vickers, embarked on an intense period of research. First of all, it was agreed that the existing wall should be clad, with a substance that acknowledged the granite originally proposed, but that also had a character and identity of its own. Pre-cast concrete was chosen for its robust constitution and its sculptural capacity, and the specification was discussed with numerous fabricators, eventually finding, in Evans Concrete Ltd, a company that would share the designer's own persistent and determined efforts to obtain a suitable colour and finish.

After several experiments, the desired result was achieved. It was discovered that the stone envisaged for the original wall would have come from one of two Scottish quarries. Hence, red granite aggregate was brought in from the only one of these quarries still operating and became a vital ingredient in the final mix, which also comprised various different sands, mixed with white cement.

The next task was to develop a series of units which could be fitted together and fixed over the GLC wall to achieve a result that reflected both the monumentality and the sensuousness of the historic design. Ultimately, numerous moulds were made, from which 1,872 separate units were cast, each designed to fit together as in a jigsaw (*Figure 3*). After being taken from the mould, every individual piece was carefully brushed with acid so that the surface texture of the granite aggregate was exposed (*Figure 4*). The final wall was completed at the end of 2002, some sections incorporating the seat, as in Pennethorne's drawings, and other sections somewhat lower, to ensure that no-one is cut off from views of the river (*Figure 5*). The wall is accompanied by a broad promenade.

The success of this part of the scheme is witnessed by thousands of people who visit daily to jog, stroll or meander their way along the embankment promenade, enjoying the park and revelling in the magnificent views.

SOFT LANDSCAPING

It was not until 1854 that the carriage drives, lake and mounds, which still form the defining elements of the park today, were designed and built by John Gibson, the first park superintendent. Gibson had been a pupil of Joseph Paxton, the remarkable head gardener to the Duke of Devonshire at Chatsworth, and also designer of one of the first public parks at Birkenhead. Paxton certainly taught Gibson to understand design and planting and one of the most distinctive characteristics of Battersea Park in its early days was the tremendous diversity of its shrubs, trees and flowers. A key aim for the restoration team was to recover something of this richness and variety, lost over the years to municipal practices of planting and maintenance.

Again, archival research proved rewarding. HTLA acquired some of Gibson's original plant orders. These lists enabled the team to identify much of the vast range of plants brought into the park. In fact, the list took some unravelling; a combination of hasty handwriting, slipshod Latin, and names of cultivars of which there is now no trace, made things difficult. The combined resources of Loudon's 1838, eight-volume *Arboretum et Fruticetum Britannicum,* Bean's *Trees and Shrubs Hardy in the British Isles* and an up-to-date *Manual of Trees and Shrubs* by Hillier proved invaluable and about 90 per cent of the original lists was deciphered.

Figure 10 In 2002, the recreated Parterre and new Pergola

Figure 13 *John Piper's Grand Vista concluded by an image of the 'Crystal Palace'*

Figure 12 *The Tea Terrace, Battersea Park, 2003*

Working with one or two nurseries, many of these plants have been sourced and will be brought into the Park once more.

Hilary Taylor even discovered watercolour plans and sections drawn by Gibson (*Figure 6*), showing exactly the way he would present and compose a collection of plants to enhance their impact. First, the ground was formed into mounds and hollows; the malleable earth treated as if it were a sculpture. The dramatic impact of this arrangement was then enhanced with groups of flowering and evergreen plants, comprising weeping, rounded and fastigiated (upright) forms. Such attention to the detail of a composition is rare in public parks today, but is being reintroduced at Battersea.

SUB-TROPICAL GARDEN

Exploiting his delight in sculptural form and unusual plants, in the early 1860s Gibson laid out his most famous feature of Battersea Park, the Sub-tropical Garden (*Figure 7*). Remarkably, it was again Paxton whom Gibson had to thank for his knowledge of exotic material. In 1835 he had been sent on a plant-hunting mission to India, via Madeira and South Africa. The letters he sent home from the trip are full of excitement at the plants he had encountered and, without doubt, this trip inspired in Gibson a vision which he realised in his Sub-tropical Garden.

Visitors came from miles around to see the fantastic garden. Contemporary descriptions and photographs reveal that Gibson experimented each year with different plants (*Figure 8*). He over-wintered much under glass. He was also able to take risks; he had cleverly created a sheltered bowl for his garden, framed on one side by a planted bank and on the other by the lake. Thus he discovered that some of the larger plants – including the banana (*Musa ensete*), and the dwarf fan palm (*Chamaerops humilis*) – would survive outside throughout the year.

Gibson launched a fashion. Within a short time sub-tropical gardens were springing up in public and private gardens across the land, accompanied by books and articles. One of these, William Robinson's 1871 *The Subtropical Garden*, has proved most useful to HTLA in planning the restoration of the Sub-tropical Garden at Battersea. Robinson wrote for people who did not have the opportunity to provide winter protection under glass for their more delicate plants. Instead, he described how the lush, exotic character of a subtropical garden could be created by planting and managing hardy material to achieve 'large and handsome leaves, noble habit, or graceful port'. Included

Figure 14 *The 'Crystal Palace Fountain', 2003*

amongst the plants he recommended are ornamental grasses and hardy bamboos, hardy perennials such as *Acanthus*, *Gunnera* and *Rheum*, and shrubs and trees like *Ailanthus*, *Rhus* and *Paulownia* which, by being cut back on a regular basis, will produce lavishly impressive leaves. The Sub-tropical Garden at Battersea will be restored in 2004 and will include plants selected by Gibson himself, as well as some of this hardier material, managed for exotic effect.

FESTIVAL PLEASURE GARDENS

In 1951, a swathe of land across the middle of Battersea Park was transformed for the Festival of Britain by the imposition of the Festival Pleasure Gardens. Amongst the principal designers responsible were James Gardner, Russell Page and John Piper.

Before the war, Gardner had been a designer for Cartier's, the jewellers, and had then worked in industrial display and design. In the army, he had been responsible for camouflage. Altogether, there could hardly have been a better preparation for directing the development of gardens where brilliant effects were created with the limited and poor quality materials then available.

Russell Page's contribution was focused on the flower gardens which lay in the centre of the site. His early training had been at the Slade School of Art, which placed emphasis on the importance of colour and drawing. His passion for plants led him to landscape architecture and, before the war, he had already built up a reputation as a fine designer who managed the colour, forms and textures of his plants with a novel brilliance.

John Piper, of course, was well-known as an artist, designer and illustrator. It was he who was principally responsible for the theatrical 'Grand Vista' – a long view down the length of two rectangular pools, concluded by a massive screen, fabricated in the form of the 1851 Crystal Palace, thus invoking a clever link between Britain at the height of Empire and at the end of a successful war.

By the late 1990s what was left of Gardner, Page and Piper's Pleasure Gardens was damaged and dilapidated. The restoration team determined to conserve these remnants and recover some of the most important elements of the Pleasure Gardens, thus reintroducing an area dedicated to fun, colour and celebration.

One of the most fascinating discoveries to emerge from HTLA's study of the Park was the fact that – despite their appearance of expressive, *ad hoc* dash – the Pleasure Gardens had actually been composed in an almost classical way. In plan and in elevation, the strict mathematical proportions of the Golden Section had dictated the layout. It was, therefore, important to recover as much of this structural clarity as possible.

A section of the Russell Page Flower Garden had survived, albeit truncated; now it was re-established on its proper footprint. In 1951, the southern edge of his parterre had been enclosed by a gargantuan refreshment tent, exotic in silhouette and detail (*Figure 9*). This had long-gone, so a bright, gold and white pergola was created in its place (*Figure 10*), designed to echo the original in the extravagant theatricality of its outline and scale (*Figure 11*). Called the 'Tea Terrace', this will also house tables, chairs and a refreshment kiosk, thus enhancing the facilities in the park and encouraging more people to sit and enjoy the colourful planting within the restrained geometry of the layout (*Figure 12*).

To the west of Page's Flower Garden lay the pools and fountains of John Piper's Grand Vista. The team felt that it was essential to restore this magnificent composition, even to the recovery of Piper's 'Crystal Palace', his poetic conclusion to the Vista (*Figure 13*). Hence, instead of an openwork screen, fabricated from cane and other transient materials as in 1951, a new 'Crystal Palace' was created, this time in the form of an extravagant fountain. Here, a bank of 19 jets of water describe the outline of the Crystal Palace in the air (*Figure 14*).

Now, in early 2003, families can be found, each morning, gathering at the end of the Grand Vista, awaiting the appearance of plumes of water and shimmering rainbows. The 'Crystal Palace Fountain' serves both to provide a poetic evocation of the original Pleasure Gardens, and to celebrate the fact that, once again, Battersea Park is being enjoyed by thousands of Londoners each day and is now looking forward to a reinvigorated future.

Dr HILARY TAYLOR is Director of HTLA Ltd, the lead landscape consultants on the restoration of Birkenhead Park and other historic landscapes throughout the country. An affiliate of the Institute of Historic Building Conservation, she is also a keen and knowledgeable plantswoman and an art historian with a PhD in art history.

EXTERNAL WORKS			
A D Calvert Architectural Stone Supplies Ltd	ss		100
A F Jones (Stonemasons)	ss		100
Abbey Heritage Ltd	pv ss		101
Abbey Masonry and Restoration Limited	ss		100
Acanthus Associated Architectural Practices		la ur	19
Albion Manufacturing Ltd	gt		141
Graciela Ainsworth	ss		95
Anderson and Glenn		la	15
Anthony Blacklay & Associates		la	20
Antique Bronze Ltd	ss		94
Antique Buildings Limited	pv		157
Architectural Metalwork Conservation	gt sf si		141
Arrol & Snell Ltd		la	20
B Levy & Co (Patterns) Limited	gt		141
Balmoral Stone Ltd	ss		101
The Bath Stone Group	pv		101
Boden & Ward	ss		101
Britannia Architectural Metalwork Ltd	gt sf si		143
Brock Carmichael Architects		la ur	21
Building Design Partnership		ur	21
Cambrian Castings (Wales) Limited	gt sf		142
Carthy Conservation Ltd	ss		102
The Cast Iron Company	gt sf si		143
Castaway Cast Products and Woodware	gt si		142
Casting Repairs Limited	gt sf si		143
Cathedral Works Organisation (Chichester) Limited	ss		102
Chilstone	sf ss		110
Chris Topp & Co Limited	gt sf		144
Clague		la	22
The Cleft Wood Co	fe		213
Cliveden Conservation Workshop Ltd	ss		102
The Conservation Studio		ur	35
Crowncast Ltd	gt sf si		142
The Cumbria Clock Company	cl		153
David Ashton Hill Architects		la	22
Don Barker Ltd	gt sf si		144
Donald Insall Associates Ltd		ur	24
Dorothea Restorations Ltd	gt sf si		144
The Downs Stone Company Ltd	ss pv		104
E T Clay Products Limited	pv		118
Edmund Kirby Architects		ur	24
English Woodlands Timber Ltd	fe		122
Eura Conservation Ltd	ss		94
Fairhaven of Anglesey Abbey	ss		103
Filkins Stone Company	ss sf		104
Fine Iron	gt sf si		145
George James & Sons, Blacksmiths	gt sf		145
Gilmore Hankey Kirke Ltd		la	24
Haddonstone Limited	ss sf		110
Hanson Brick	pv		118
Hargreaves Foundry Ltd	gt sf si		91
Hart Brothers Engineering Ltd	gt sf si		146
Hearns (Specialised) Joinery Ltd	cn sf		127
Hilary Taylor Landscape Associates Ltd		la	15
Hodgsons Forge	gt		145
Holden Conservation Ltd	ss		95
J H Porter & Son Ltd	gt sf si		145
Julian Harrap Architects		la	26
King Sturge Heritage		ur	45
Latham Architects		la ur	27
Malbrook Conservatories	cn		152
Mansfield Thomas & Partners		la	28
Marsh Brothers Engineering Services Ltd	gt sf si		146
Mather & Smith Ltd	gt sf si		146
Mowlem Rattee & Kett	ss		106
Niall Phillips Architects Ltd		ur	28
Nicolas Boyes Stone Conservation	ss		106
Nimbus Conservation Limited	ss		107
PAYE Stonework	ss		107
Peter S Neale Blacksmiths	gt sf		146
Plowden & Smith Ltd	cl ss		203
The Real Wrought Iron Company	gt sf		147
Rupert Harris Conservation	gt ss		203
Shepley Engineers Limited	gt		147
Smith of Derby Clockmakers	cl gt sf		153
Spencer & Richman	ss		189
The Standard Patent Glazing Co Limited	cn		89
Stoneguard	ss		66
Stonewest Limited	ss		67
Taylor Pearce Restoration Services Limited	ss		94
Thwaites & Reed Ltd	cl		153
Vale Garden Houses Ltd	cn		152
Victorian Classic Style	gt sf si		152
The York Handmade Brick Company Limited	pv		119

See also Architectural Salvage, pages 155–157

KEY

cl	clocks	pv	flagstones and paving
cn	conservatories	sf	street and garden furniture
fe	timber fencing	si	signage
gt	gates and railings	ss	statuary and stone carving
la	landscape architects	ur	urban designers

CONSERVATORIES

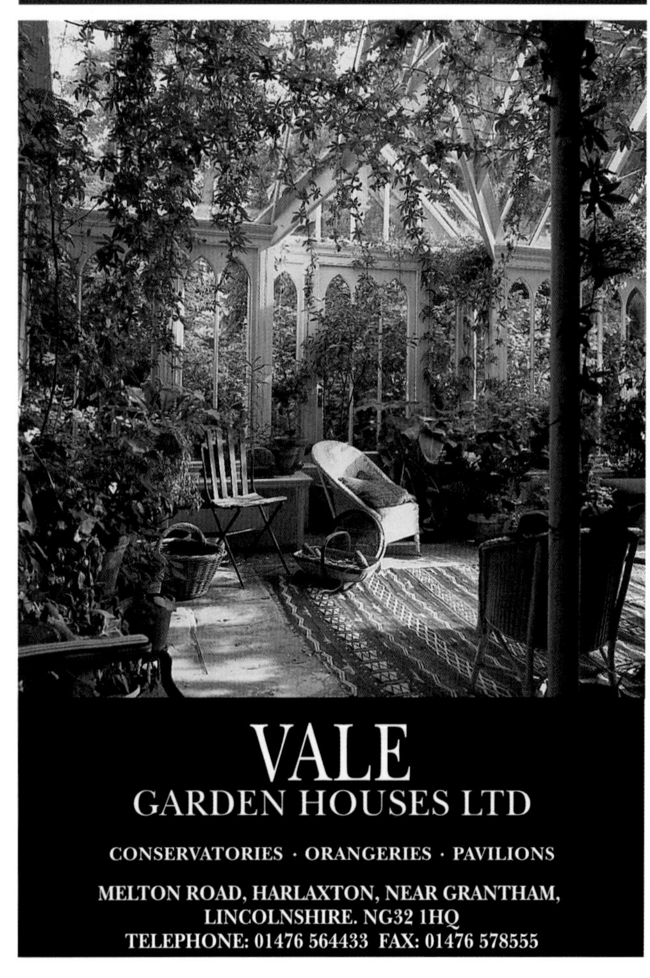

VALE
GARDEN HOUSES LTD
CONSERVATORIES · ORANGERIES · PAVILIONS
MELTON ROAD, HARLAXTON, NEAR GRANTHAM,
LINCOLNSHIRE. NG32 1HQ
TELEPHONE: 01476 564433 FAX: 01476 578555

▶ MALBROOK CONSERVATORIES LTD

2 Crescent Stables, Upper Richmond Road, Putney, London SW15 2TN
Tel 020 8780 5522 Fax 020 8780 3344 Website www.malbrook.co.uk
BESPOKE CONSERVATORIES: Malbrook specialises in the design, production and installation of fine buildings in timber and glass. Particular attention is paid to the architectural details of the property, and the traditional joinery methods of construction are well suited to providing quality double-glazed structures for restoration and refurbishment projects. The experienced staff at Malbrook offer a full design service to assist clients and their architects in order to ensure the best possible solution. This attention to detail is maintained by close supervision throughout the installation process including all the finishing touches on site.

GARDEN BUILDINGS

▶ CHILSTONE

Victoria Park, Fordcombe Road, Langton Green, Tunbridge Wells, Kent TN3 0RE
Tel 01892 740866 Fax 01892 740249
GARDEN STRUCTURES: Reconstructed stone temples, follies, porticos. *See also: profile entry in Cast Stone section, page 110.*

GATES & RAILINGS

▶ VICTORIAN CLASSIC STYLE

Unit 8d, Hillborough Business Park, Sweechbridge Road, Hillborough, Herne Bay, Kent CT6 6TE
Tel/Fax 01227 860705
CAST IRON GATES AND RAILINGS: Victorian Classic Style is a small company dedicated to recreating cast iron gates and railings from the Victorian era for use in domestic situations. All the company's patterns are based on original English designs and have successfully been installed in local government conservation schemes involving English Heritage and the National Lottery. Delivery can be arranged nation-wide. Victorian Classic Style also has pattern work facilities available.

GARDEN & STREET FURNITURE

▶ BRITANNIA ARCHITECTURAL METALWORK LTD

The Old Coach House, Draymans Way, Alton GU34 1AY
Tel 01420 84427 Fax 01420 89056
Website www.britannia.uk.com
See also: display entry in Architectural Metalwork, page 143.

▶ CHILSTONE

Victoria Park, Fordcombe Road, Langton Green, Tunbridge Wells, Kent TN3 0RE
Tel 01892 740866 Fax 01892 740249
GARDEN AND STREET FEATURES: Reconstructed stone benches, bollards, etc. *See also: profile entry in Cast Stone section, page 110.*

▶ DOROTHEA RESTORATIONS LTD

INCORPORATING ERNEST HOLE (ENGINEERS) LTD
Northern office & works: New Road, Whaley Bridge, High Peak, Derbyshire SK23 7JG
Tel 01663 733544 Fax 01663 734521
E-mail north@dorothearestorations.com
Website www.dorothearestorations.com
▶ Southern office & works: Riverside Business Park, St Annes Road, St Annes Park, Bristol BS4 4ED
Tel 0117 971 5337 Fax 0117 977 1677
E-mail south@dorothearestorations.com
Website www.dorothearestorations.com
RESTORATION OF ARCHITECTURAL METALWORK, TRADITIONAL MACHINERY AND MILLS: *See also: display entry in Architectural Metalwork, page 144.*

▶ EURA CONSERVATION LTD

Unit H10, Halesfield 19, Telford, Shropshire TF7 4QT
Tel 01952 680218 Fax 01952 585044 E-mail mail@eura.co.uk
ACCREDITED CONSERVATORS AND CONTRACTORS FOR ARTISTIC, ARCHITECTURAL AND MARINE METALWORK: *See also: display entry in Statuary section, page 94 and profile entry in Architectural Metalwork section, page 142.*

▶ FINE IRON

Buidling 1, Gilfach Uchaf, Aberbran, Brecon, Powys LD3 9NL
Tel 01874 636966 Fax 01874 638049
E-mail info@fineiron.co.uk
Website www.fineiron.co.uk
See also: display entry in Architectural Metalwork, page 145.

▶ HADDONSTONE LIMITED

The Forge House, East Haddon, Northampton NN6 8DB
Tel 01604 770711 Fax 01604 770027
E-mail info@haddonstone.co.uk
Website www.haddonstone.co.uk
STANDARD RANGE INCLUDES URNS, TROUGHS, STATUARY, SUNDIALS, FOUNTAINS, POOL SURROUNDS, SEATS, COPINGS, TEMPLES, PAVILIONS AND BALUSTRADING: Custom-made architectural stonework a speciality. *See also: display entry in Cast Stone section, page 110.*

▶ MATHER & SMITH LTD

Hilton Road, Cobbswood Industrial Estate, Ashford, Kent TN23 1EW
Tel 01233 622214 Fax 01233 631888
E-mail sales@mjallen.co.uk
Website www.mjallen.co.uk
IRON FOUNDER ARCHITECTURAL: *See also: display entry in Architectural Metalwork, page 146.*

▶ THE REAL WROUGHT IRON COMPANY

Carlton Husthwaite, Thirsk, North Yorkshire Yo7 2BJ
Phone 01845 501415 Fax 01845 501072
Email: enquiry@realwroughtiron.com
Website: www.realwroughtiron.com
GENUINE WROUGHT IRON
See also: display entry in Architectural Metalwork, page 147.

PAVING

✦ CHARMING AND FUNCTIONAL ✦
GENUINE COTSWOLD STONE FLAGS
HAND-FINISHED AND SUPPLIED DIRECT FROM OUR OWN QUARRY
OUR EXPERIENCED MASONS ARE AVAILABLE FOR FITTING IF REQUIRED
THE DOWNS STONE COMPANY LTD
SUPPLIER OF TRADITIONAL COTSWOLD STONE PRODUCTS
TELEPHONE: 01608 658357
FOR FURTHER INFORMATION AND BROCHURE

▶ RANSFORDS CONSERVATION AND RECLAIMED BUILDING SUPPLIES

Drayton Way, Drayton Fields, Daventry, Northants NN11 5XW
Tel 01327 705310 Fax 01327 706831
Website www.ransfords.com

SUPPLIERS OF YORKSTONE, SANDSTONE AND LIMESTONE FLAGS, AND SLATE: *See also: display entry and profile entry in Architectural Salvage section, page 155.*

▶ STONEGUARD

St Martins House, The Runway, Ruislip, Middlesex HA4 6SG
Tel 0870 241 6366 Fax 020 8839 9988
Website www.stoneguard.co.uk
▶ with offices also in Bath, Birmingham, Manchester and Stirling

BUILDING CONTRACTOR, PAVING AND HARD LANDSCAPING: *See also: profile entry in Building Contractors section, page 66.*

CLOCKS

Conserving Time
Tower and feature clocks
Clock conservation & restoration
Re-gilding using the finest materials
Auto winding, approved for historical clocks
Wind vane restoration
Maintenance schemes to suit your budget
Architectural metalwork
Smith of Derby
JOHN SMITH & SONS
112 Alfreton Road, Derby DE21 4AU
Tel: 01332 345569 Fax: 01332 290642
www.smithofderby.com
e-mail: sales@smithofderby.com

▶ THE CUMBRIA CLOCK COMPANY

Dacre, Penrith, Cumbria CA11 0HH
Tel/Fax 01768 486933

TOWER CLOCK RESTORATION: Their services include full restoration of mechanical movements, carillon and tune playing machines; the manufacture and installation of automatic-winding systems, and; the full restoration of dials. New clock systems and dials can be manufactured to the customer's requirements, complete with striking and chiming mechanisms. Their customer list includes: Salisbury and Hereford Cathedrals, the National Trust, Hampton Court Palace and many churches and local authorities throughout the country. Free quotations are given within the UK.

▶ PLOWDEN & SMITH

190 St Ann's Hill, London SW18 2RT Tel 020 8874 4005 Fax 020 8874 7248
E-mail richard.rogers@plowden-smith.com Website www.plowden-smith.com

CONSERVATION AND RESTORATION*: See also: display entry in Fine Art Conservation section, page 203.*

▶ SMITH OF DERBY

112 Alfreton Road, Derby DE21 4AU
Tel 01332 345569 Fax 01332 290642 E-mail sales@smithofderby.com

CLOCKS AND ARCHITECTURAL FEATURES: Smith of Derby builds and maintains tower clocks on churches and public buildings. Regional centres serve the entire UK for installation and maintenance work. Annual servicing schemes are available tailor made to the individual clock. Clock options include solar powered dials, electronic bell sound, timed animated movement, and automatic winding of weight driven mechanisms. Architectural feature work comprises bespoke interior and exterior metalwork for new developments and refurbishment schemes. Conservation work includes restoration of turrets and cupolas using brass, copper, mild steel, stainless steel and wood. Additionally, any traditional building material texture can be replicated using advanced mould and colouring techniques in GRP. *See also: display entries on this page, and in Architectural Metalwork section, page 147.*

GENERAL SUPPLIERS

Name			Page
Antique Buildings Limited	as		157
Best Demolition	as		155
Bursledon Brickworks Conservation Centre		sp	172
Carpenter Oak & Woodland Ltd		sp	72
Chalk Down Lime		sp	154
Coyle Timber Products		sp	154
Cornish Lime Company		sp	174
Drummond's Architectural Antiques Limited	as		155
G & N Marshman		sp	216
Hirst Conservation Materials Ltd		sp	174
H J Chard & Sons Ltd		sp	174
LASSCO	as		155
The Lime Centre		sp	174
Mike Wye & Associates		sp	174
Milestone Lime		sp	175
Minchinhampton Architectural Salvage	as		155
Minter Reclamation	as		156
Mongers Architectural Salvage	as		156
Nostalgia	as		189
Priors Reclamation	as		156
Ransfords Reclaimed Building Supplies	as		155
The Real Wrought Iron Company	as	sp	147
Retrouvius Architectural Salvage	as		157
Rose of Jericho		sp	175
Ryedale Conservation Supplies (Chalk Hill)		sp	175
SALVO	as		155
Solopark Plc	as		156
Swiss Cottage Furniture	as		157
Timber Supplies	as		157
The Traditional Lime Co		sp	176
Ty-Mawr Lime Ltd		sp	176
Victorian Woodworks	as		213
Walcot Reclamation Ltd	as		157
Weald & Downland Open Air Museum		sp	124
Westland London	as		189
Womersley's Limited		sp	176

KEY

as architectural salvage
sp builders merchants and specialist suppliers

GENERAL BUILDING MATERIALS

▶ CHALK DOWN LIME LTD

102 Fairlight Road, Hastings, East Sussex TN35 5EL
Tel 01424 443301 Fax 01580 830096
Mobile 0771 873 8708
E-mail chalkdownlime@supanet.com

SPECIALIST SUPPLIERS OF LIME MORTARS AND RELATED PRODUCTS: *See also: profile entry in the Mortars & Renders section, page 172.*

▶ COYLE TIMBER PRODUCTS

Bassett Farm, Claverton, Bath BA2 7BJ
Tel 01225 427409 Fax 01225 789979
E-mail info@coyletimber.co.uk
Website www.coyletimber.co.uk
Contact Joe Coyle

MANUFACTURERS AND SUPPLIERS OF SPECIALIST BUILDING MATERIALS: Coyle Timber Products produces an extensive range of traditional building materials including shingles traditionally cleft from English oak, riven batten and tile pegs for historic roofs, oak, Scots pine and chestnut plasterers' lath, sawn laths, oak staves, a wide selection of hand made oak framing pegs and rosehead nails. Having over 10 years experience, the company can undertake any shingling contract to the highest standard. Specialist orders are welcome. Coyle Timber Products is a supplier of FSC registered products. This gives you the assurance that the timber used in all the company's products comes from well-managed forests independently certified in accordance with the rules of the Forest Stewardship Council. For further information please contact Joe Coyle.

▶ G & N MARSHMAN

1 Nell Ball, Plaistow, Billingshurst, West Sussex RH14 0QB
Tel 01798 342427

MANUFACTURERS OF RIVEN OAK, CHESTNUT PLASTERERS' LATHS AND TILE AND STONE SLATE BATTENS: *See also: profile entry in Plasterwork section, page 216.*

▶ HIRST CONSERVATION MATERIALS LTD

Laughton, Sleaford, Lincolnshire NG34 0HE
Tel 01529 497517 Fax 01529 497518
E-mail materials@hirst-conservation.com
Website www.hirst-conservation.com

CONSERVATION MATERIALS: *See also: display entry on the inside front cover.*

▶ ROSE OF JERICHO

Horchester Farm, Holywell, Dorchester DT2 0LL
Tel 01935 83676 Fax 01935 83903
E-mail info@rose-of-jericho.demon.co.uk
Website www.rose-of-jericho.demon.co.uk

MANUFACTURERS OF TRADITIONAL MORTARS AND PAINTS: *See also: display entry in this section, page 175.*

▶ RYEDALE CONSERVATION SUPPLIES

North Back Lane, York YO60 6NS
Tel 01653 648112 Fax 01653 648112
E-mail info@ryedaleconservation.com
Website www.ryedaleconservation.com

SUPPLY OF MORTARS AND RENDERS AND RELATED TRAINING: *See also: display entry for Chalk Hill Lime Products in Mortars & Renders section, page 175.*

▶ TY-MAWR LIME LTD

Ty-Mawr, Llangasty, Brecon, Powys, Wales LD3 7PJ
Tel 01874 658249 Fax 01874 658502
E-mail tymawr@lime.org.uk
Website www.lime.org.uk

A WELSH CENTRE FOR TRADITIONAL AND ECOLOGICAL BUILDING: *See also: profile entry in Mortars & Renders section, page 176.*

ARCHITECTURAL SALVAGE

▶ BEST DEMOLITION

Harcourt Lodge Buildings, Burwash Road, Heathfield, East Sussex TN21 8RA
Tel 01435 862381/866170 Fax 01435 867203
▶ 10a Hawthorn Road Industrial Estate, Lottbridge Drove, Eastbourne, East Sussex BN23 6QA
Tel/Fax 01323 416572

ARCHITECTURAL SALVAGE: Best Demolition was formed in 1959 and is now one of the leading demolition companies in the South East. The past couple of decades has seen a greater awareness and emphasis on the care of our environment and Best Demolition has been doing its part by salvaging historic building materials from buildings being demolished. The company has saved many old architectural objects and a wide range of original materials which are now a valuable resource for conservation and refurbishment projects. Best Demolition has two yards in East Sussex which supply reclaimed bricks, tiles, timber and oak beams and more. Please ring to check stock availability.

▶ DRUMMONDS ARCHITECTURAL ANTIQUES LIMITED

The Kirkpatrick Buildings, 25 London Road (A3), Hindhead, Surrey GU26 6AB
Tel 01428 609444 Fax 01428 609445
Website www.drummonds-arch.co.uk

ARCHITECTURAL ANTIQUES, RESTORED BATHROOMS, RECLAIMED FLOORING AND GARDEN FEATURES: Specialist suppliers of fully restored antique bathrooms and fittings; reclaimed wood, tile and stone flooring; garden furniture, troughs and fountains; wood, stone and iron fire surrounds; cast iron stoves and radiators; hardwood and pine doors; door and window furniture; gates and railings; antique furniture; lighting; windows; antique decorative items. Makers of the finest handmade cast iron baths and fittings; brass door and window furniture. Proper vitreous enamelling used on all baths. Open Mon-Fri 9.00am to 6.00pm, Sat 10.00am to 5.00pm. Closed Sundays.

▶ LASSCO

St Michael's Church, Mark Street, Shoreditch, London EC2A 4ER
Tel 020 7749 9944 Fax 020 7749 9941
▶ Ropewalk, Maltby Street, London SE1 3PA
Tel 020 7394 2100 Fax 020 7394 2136
Website www.lassco.co.uk

ARCHITECTURAL SALVAGE: LASSCO is London's pioneering and largest specialist in architectural salvage, sourcing items from some of London's most prestigious buildings. Divided between four companies, LASSCO sells a huge range of decorative features and details: chimneypieces, panelled rooms, ecclesiastical furniture, stained glass, mirrors, lighting and garden ornament: doors, pub and museum interiors; radiators, bathrooms and kitchens; reclaimed and new hardwood flooring, as well as high quality replicas. LASSCO are based at two sites: the Ropewalk in Bermondsey, open 10–6 Monday to Saturday, and St Michael's Church in Shoreditch, open 10–5.30 Monday–Friday, 10–5 Saturday.

▶ MINCHINHAMPTON ARCHITECTURAL SALVAGE CO

Cirencester Road, Nr Minchinhampton, Glos GL6 8PE
Tel 01285 760886 Fax 01285 760838
E-mail masco@catbrain.com
Website www.catbrain.com

ARCHITECTURAL ANTIQUES, GARDEN STATUARY, CHIMNEYPIECES AND TRADITIONAL FLOORING: Minchinhampton Architectural specialises in large architectural features and garden ornament. The company also carries extensive stocks of both hardwood and softwood flooring. Situated in the heart of the Cotswolds, the company offers a comprehensive sourcing service, friendly advice and a garden design/consultancy department. Opening hours are Monday to Friday 9am–5pm and Saturday 9am–3pm. Telephone or e-mail for further details, or visit their comprehensive website. Subscribers to the Salvo Code.

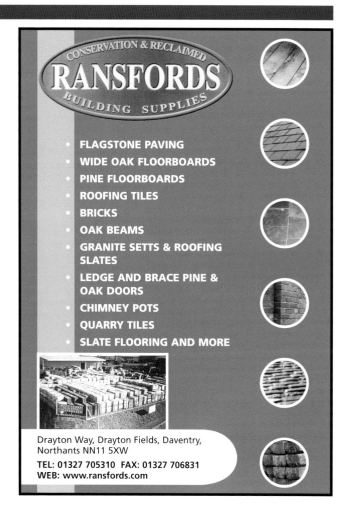

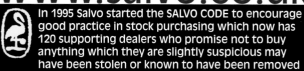

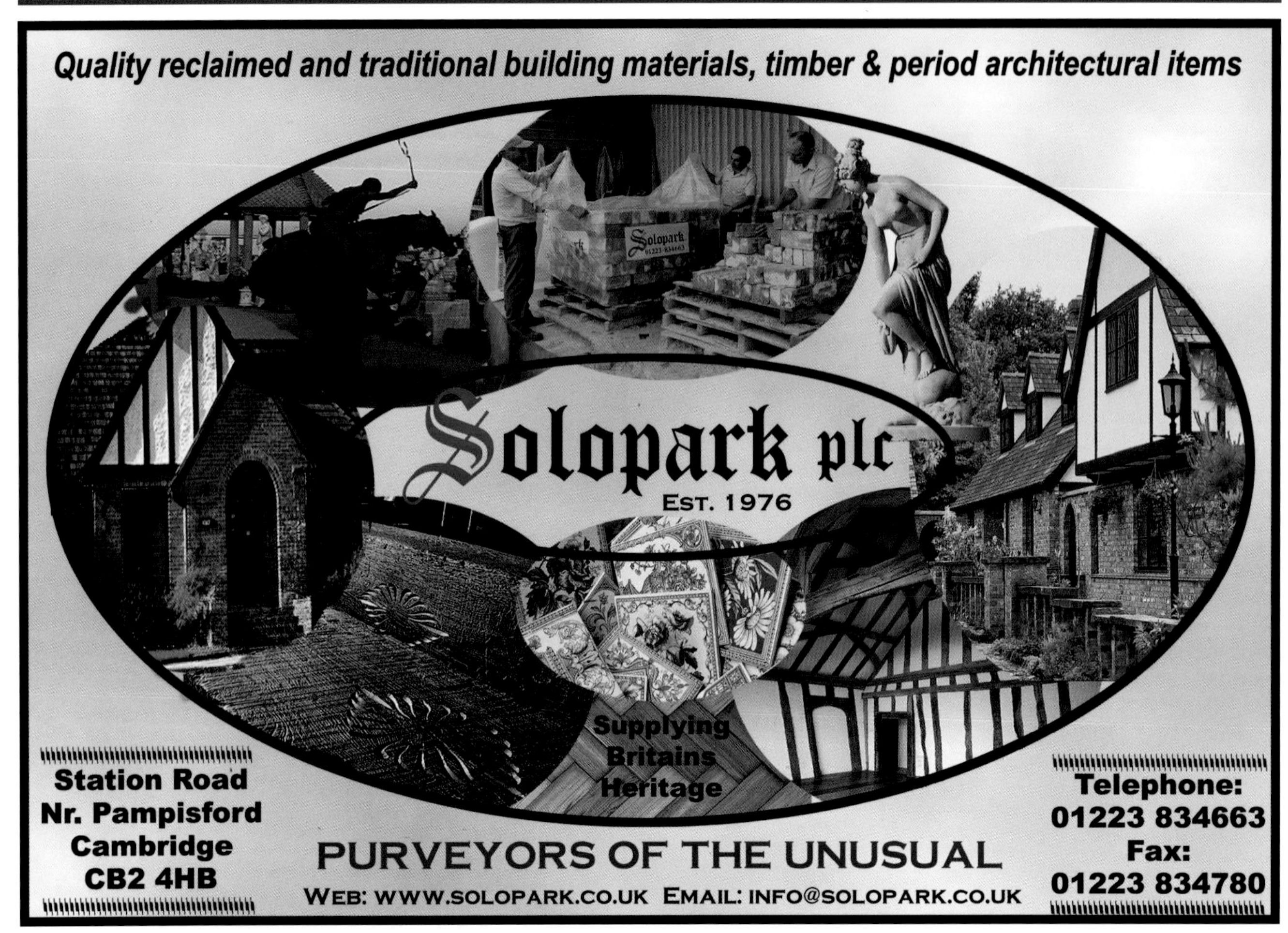

▶ MINTER RECLAMATION LIMITED

Lower Common Lane, Three Legged Cross, Dorset BH21 6RX
Tel 01202 828873 Fax 01202 828813
Website www.minterreclamation.co.uk

RECLAIMED PERIOD BUILDING MATERIALS AND ARCHITECTURAL ANTIQUES: Minter Reclamation Limited stocks a large selection of refurbished period bathrooms with reconditioned taps in brass or nickel, as well as bricks, slates, tiles, beams, pine floorboards, quarry tiles, fireplaces etc. The company specialises in the manufacture and installation of traditional kitchens made from reclaimed pine with maple worktops. Belfast sinks and refurbished mixer taps can be supplied. Clients include the National Trust and many listed buildings. Minter also operates a 'wants-and-offers' index and if items are not in stock they can usually be sourced.

▶ MONGERS ARCHITECTURAL SALVAGE

15 Market Place, Hingham, Norfolk NR9 4AF
Tel 01953 851868
E-mail mongers@mongersofhingham.co.uk
Website www.mongersofhingham.co.uk

ARCHITECTURAL ANTIQUES, SANITARYWARE, GARDEN ITEMS AND RE-CLAIMED FLOORING: Mongers use many local craftspeople to restore architectural antiques and objects in order that they can be re-used. Stocks include restored period bathrooms, fireplaces and chimneypieces, cast iron radiators, doors, door furniture, gates, garden benches, stone troughs, urns, reclaimed floorboards, tiles and architectural items of interest. A comprehensive list of items that they stock can be viewed on their website together with a photographic library of selective pieces. Their stock is sourced from demolition nationwide. Open Monday – Saturday. Salvo Code dealers.

▶ PRIORS RECLAMATION

Unit 2A, Ditton Priors Industrial Estate, Ditton Priors, Shropshire WV16 6SS
Tel 01746 712450
E-mail vicki@priorsrec.co.uk
Website www.priorsrec.co.uk

ARCHITECTURAL SALVAGE: Specialist suppliers of quality reclaimed flooring and doors. Large stocks of flooring, dry stored, trimmed, de-nailed and ready to lay. Stocks of original doors, supplemented by a made-to-measure service, making traditional doors from reclaimed pine or new oak. Personal service and attention to detail. Salvo Code dealer.

▶ RANSFORDS CONSERVATION AND RECLAIMED BUILDING SUPPLIES

Drayton Way, Drayton Fields, Daventry, Northants NN11 5XW
Tel 01327 705310 Fax 01327 706831
Website www.ransfords.com

ARCHITECTUAL SALVAGE: Ransfords has huge supplies of quality materials all used in conserving Britain's heritage: flagstones, wood flooring, bricks, roofing slates, tiles and fittings, quarry tiles, railway sleepers, doors, granite setts, cobbles, wallcoping, paviors, beams, fireplaces, chimney pots, flooring slate, pine flooring, oak flooring, Victorian lamps, windows, radiators and sinks. Opening hours: Monday – Friday 8.30-5.00pm, Saturdays 9.00-12.00pm. *See also: display entry in this section, page 155.*

ARCHITECTURAL SALVAGE

▶ RETROUVIUS ARCHITECTURAL RECLAMATION & DESIGN

2A Ravensworth Road, London NW10 5NR
Tel 020 8960 6060 Mobile 0378 210855 E-mail mail@retrouvius.com
Website www.salvo.co.uk/dealers/retrouvius
Contact Adam Hills or Maria Speake

ARCHITECTRUAL ANTIQUES AND BUILDING COMPONENTS: Retrouvius seeks out architectural antiques and building components nation-wide. They work with architects, builders, and demolition companies to purchase unwanted fittings and fixtures directly from site. Retrouvius abides by the SALVO Code. Both partners are architecturally trained and have worked in the conservation world. The company offers a unique and personal design service helping with the integration of salvaged materials into interior and landscape projects. The office and warehouse are now located in Kensal Green. Please phone for further information.

▶ SOLOPARK PLC

Tel 01223 834663 Fax 01223 834780
E-mail info@solopark.co.uk Website www.solopark.co.uk

ARCHITECTURAL SALVAGE: Solopark facilities provide a Mecca for all those interested in reclaimed and new building supplies including a vast selection of bricks, roofing tiles, slates and seasoned timbers, doors and panelling, together with a plethora of unique articles ranging from Victorian baths, porticos and clock towers. There is also a large showroom full of unique period architectural items including a wide selection of original and reproduction fireplaces, surrounds and ornamental accessories. Solopark provides the ideal venue for architects, builders and discerning homemakers from all over the UK and far beyond. *See also: display entry in this section, page 156.*

▶ SWISS COTTAGE FURNITURE

85 Westfield Crescent, Burley Road, Leeds LS3 1DJ
Telephone 01132 429994 E-mail info@swisscottageantiques.com
Website www.swisscottageantiques.com

ARCHITECTURAL SALVAGE: Located just behind Yorkshire TV in Leeds, West Yorkshire, Swiss Cottage Furniture deals in architectural and decorative items removed from houses and buildings in the West Yorkshire area. Large stock of doors, fireplaces, bathroom fittings and also church and school interiors. Suppliers of props for film and TV.

▶ TIMBER SUPPLIES

Ayres Quay, Farringdon Row, Sunderland, Tyne & Wear SL1 3QS
Tel 0191 567 0947 Fax 0191 567 7963 E-mail nelson@timber15.fsnet.co.uk

RECLAIMED TRADITIONAL BUILDING MATERIALS INCLUDING ROOFING SUPPLIES: Suppliers of roofing slates such as Welsh, Westmoreland, Burlington, Cornish and Voss. Tiles available include English and European machine and hand-mades. Timber supplied includes pitch pine, oak, yellow pine, and railway sleepers, new and reclaimed. Stone: facing, random, lintels, sills, water tablets. Contact Timber Supplies for bricks, machined and hand-made, granite setts, architectural salvage and much more.

▶ WALCOT RECLAMATION

108 Walcot Street, Bath BA1 5BG
Tel 01225 444404 Fax 01225 448163 E-mail rick@walcot.com/jane@reproshop.com
Website www.walcot.com/www.reproshop.com

ARCHITECTURAL ANTIQUES AND TRADITIONAL BUILDING MATERIALS: One of the country's leading dealers in affordable architectural antiques and salvage, Walcot has two large city centre sites. (1) The Walcot Street yard specialises in: chimney pieces, fireplaces, panelling, doors, wrought iron, bathroom fittings, brassware, etc. Stock is recorded on database with photographic support. A shipping and export service is available as is on site parking. Contact details above. (2) The Riverside Depot supplies all types of reclaimed flooring, roofing, paving, oak beams and stonework. Patinated oak and elm flooring a speciality.

▶ WESTLAND LONDON

St Michael's Church, Leonard Street, London EC2A 4ER
Tel 020 7739 8094 Fax 020 7729 3620
E-mail westland@westland.co.uk Website www.westland.co.uk

PERIOD AND PRESTIGIOUS CHIMNEYPIECES, ARCHITECTURAL ELEMENTS, PANELLING, FOUNTAINS, STATUARY, PAINTINGS AND FURNITURE: *See also: display entry in Fireplaces section, page 189.*

RECONSTRUCTED BUILDINGS

THE SALVO CODE

Architectural Salvage Dealers subscribing to the Salvo Code agree to:

1. Not buy any item if there is the slightest suspicion that it may be stolen.
2. Not to knowingly buy any item removed from listed buildings without listed building consent or from sites of scheduled monuments without scheduled monument consent.
3. To record the registration numbers of vehicles belonging to persons unknown who offer items for sale.
4. Where possible to keep a record of the provenance of an item, including the date of manufacture, where it was removed from, and any previous owners.
5. To agree to their business details being held on a list of businesses which subscribe to the Code.
6. To display a copy of this Code in a public position within their premises.

Reprinted with kind permission of Salvo Magazine.

A list of the dealers who have agreed to abide by the Code is available from Salvo, see page 155.

SIGNS AND SIGNAGE IN THE HISTORIC ENVIRONMENT

MICHAEL COPEMAN

There is a remarkable consensus about signage among conservation practitioners: too many signs clutter up our historic towns and obscure historic buildings and townscapes. Traffic signs, in particular, should be rationalised, sited carefully and with respect for their historic context. Advertisements should be controlled. Old features such as street nameplates should be kept wherever possible. Shopfronts should be designed and made in such a way that their proportions and materials respect the building and streetscape of which they form a part, maintaining their individuality, and standard corporate fascias should be avoided. Local distinctiveness, well designed signs, careful lettering, a diversity of imagery and individual expression should be encouraged to enrich the townscape.

These principles are sound but the practice is leaky. Compare almost any historic town centre with a photograph of the same place 30 years ago, and it will be clear that there has been an astonishing proliferation of signage. Standard traffic signs predominate, old details have been lost, and corporate logos prevail. To understand what might be going wrong, the assumptions in the consensus outlined above bear examination, both about the nature of the problem and how to resolve it. Although this article considers signs as objects bearing text or images in the historic environment, it should never be forgotten that signs are primarily messages, not things.

Not all signage in the urban scene is contentious. Some signs, such as traffic signage, shop signs, and advertising bill-boards, are felt to detract from the historic environment, while others, like street name plates, painted advertisements, or blue plaques, are treasured. This is not simply a matter of age or rarity. 'Clutter' is essentially a value judgement. Redundant or obsolete signage is 'clutter' because it is not old enough to be enjoyed. 'Clutter' may mean that there are too many signs, posts, boards and notices, literally getting in the way of our enjoyment of the historic scene, but equally true is the old cliché about communist Eastern Europe: that visiting Westerners found the streets without advertisements bleak and visually impoverished. Colour, lettering, individuality, and visual incongruity, are vital qualities in the urban scene.

Heritage clutter in Stoke-on-Trent

LEGISLATIVE BACKGROUND

The diversity of signage in the urban environment is extraordinary. The most prominent and numerous are traffic signage and advertising (including shopfronts), and it is these with which conservation is largely concerned. Planning authorities can exert a considerable control over signage in historic areas, but it is largely discretionary. PPG15[1] (for England) and the parallel guidance for Scotland and Wales set out a general approach. For example, PPG15 states that "it is reasonable to assume" that local authorities will apply more exacting standards to the control of advertisements in conservation areas, (paragraph 4.31). Similarly, "authorities should take advantage of" such flexibility on traffic signage as is allowed by the Department for Transport (DfT) [2] (paragraph 5.16). A wide range of practical advice is available (see footnotes and *Additional Reading*). Most local conservation area policies contain controls on advertising and shopfront design, and some guidance on traffic signage. Advertising and shopfronts can be controlled through the planning development control system, but traffic signs are a highways matter. In designated historic areas regional DfT offices will grant dispensations for non-standard street signs, although not enough authorities have fully effective mechanisms for securing them.

TRAFFIC SIGNAGE

Traffic signage in historic areas tends to be regarded as a necessary evil. The DfT[3] states that: "Modern usage of streets has demanded an increasing provision of street furniture including...traffic signs... sometimes at the expense of visual order." The guidance aims

Typical urban clutter in Clapham, London: note the old and new street name signs

to: "highlight…how traffic engineering and highway improvements can be designed sensitively in historic areas". In some situations this is a wholly appropriate approach. Traffic signs depend on a standardised code, quickly understood by the passing motorist. The English Historic Towns Forum[4] and English Heritage[5] guidance, for example, show how traffic signs can be reduced in number and size. One post can serve several functions, signs can be attached to buildings, reflective materials can obviate the need for additional lighting, and many traffic signs are unnecessary and ineffective. (One might ask why smaller, more discreet signs are workable in historic areas and not elsewhere.)

On the other hand, conservation standards have become an orthodoxy. Red phone boxes, miniature road signs, York stone, black and gold litter bins and finger-posts are visual shorthand for conservation area. These features may well reflect effective management of the historic environment and 'joined-up' thinking in the local council offices. Unfortunately, they are turning historic towns into indistinguishable, bland, sterilised imitations of themselves. This is not just a British problem. In the USA, "too many…re-done streets are over-designed. There is too much unified signage…too much good taste in general, or the pretension of it…too may designers have the same good taste and the result is bland conformity"[6]. Area conservation is predicated on the concept of local distinctiveness. If it results in the unthinking replacement of one standard with another, it has failed.

Historic towns are gathering and stopping places. Highways engineers want to move the traffic quickly and efficiently. We must decide to what extent we want through traffic to compromise other interests, but 'conservation' strategies need careful thought. Pedestrian priority schemes can lead to an explosion of new signage for pedestrian zones, parking restrictions, loading periods, park-and-ride schemes, heritage trails, interpretation and the rest.

Stick no bills – elegant if not eloquent

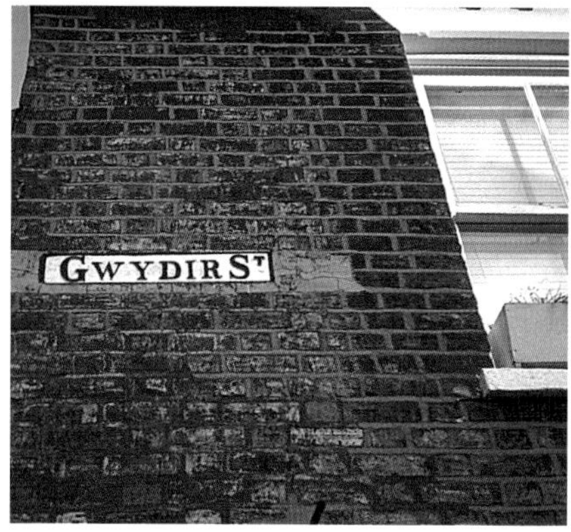

A fine old non-standard street name sign in Cambridge

An old facia sign above this book shop in Penzance has been retained by the new owners and left to grow old, gracefully

Ultimately, traffic is the problem, rather than road signs. However well designed, "traffic-oriented devices and signs reinforce the message that the space is primarily for traffic"[7]. There are radical possibilities. Removing direction signs in town centres, for example, could discourage through traffic in favour of local traffic. A recent experiment[8] in which traffic signs and signals were removed from the centres of several small Dutch towns resulted in better traffic flow. The scheme depended on low traffic speeds in relatively small urban centres, and while this may not work everywhere, it could be highly appropriate to some historic towns.

ADVERTISEMENTS

Advertising is ephemeral. Too much attention to design risks emphasising the permanence of something which will disappear of its own accord. Temporary or moveable signs may not only be cheaper but more appropriate than expensive designs, but if a business is required to make a detailed planning application for every shop sign, including temporary ones, there is no incentive to use them. If we are

Interpretation board in Barnsley: a well intentioned design but giving too much emphasis to the sign as an object

to maintain the character we value in old places, we must leave some things alone; let them fade, rust and decay, and make space for happy accidents and unintended juxtapositions. Not everything we value in historic townscapes is beautiful. Advertising can provide an opportunity for individuality and self-expression that need not be limited by preconceived notion of what is appropriate or tasteful.

Two well-known books from the 1960s make an instructive comparison. Gordon Cullen's *Townscape*[9] illustrates approvingly a corner shop haphazardly plastered with bills, posters and advertisements and calls its intricacy and colour 'delightful'. *Preservation and Change*[10], an official publication which articulated many of the principles underlying present day conservation practice, shows a virtually identical shop to illustrate 'before' in a 'before and after' example of the removal of clutter. Again, appropriate signage in historic areas is a matter of taste.

In this context, diversity is an absolute virtue, but the problem is whether it can be imposed. Diversity can be encouraged, not imposed, and it emerges from use, not from the imposition of superficial visual standards. Like traffic control, advertisement policies can have the opposite effect to that intended. The chaotic parade of signs along a down-market street of small shops has a wonderful variety which gentrification and conservation area policies will eradicate. Branded shop fascias and corporate logos represent an economic reality. Only a multiplicity of uses is likely to preserve visual diversity.

OLD SIGNS

The conservation of surviving old signs ought to be straightforward. If something is old and functional, it can safely be left alone. Regrettably, local authorities are still replacing old street nameplates with new 'branded' ones. This may be a well-intentioned attempt to create local identity where historically there was none, but the result is uniformity where difference was attractive and functional. There

The variety of signs, which is the product of the varied use, contributes their own character and interest to this street in Leith, Edinburgh

is no reason why nearby street nameplates should all conform to a particular style, as long as they are clear. A cast iron street nameplate proudly subtitled with the name of a long erased corporation is not only charming, but often more useful than a new one which bears the logo of the new and distant unitary authority. This is a larger issue than conservation. The new signs are advertisements for a local government system increasingly disconnected from real places and communities.

We value much the same qualities in old signs as we do in buildings, but we should not start listing them. Fine old lettering is a joy. The sight of an old faded advertisement painted on a gable end is one of the delights of urban life: its original message may now be obsolete but the sign acquires a new value in connecting us with the past. Old graffiti, such as the brown-shirt slogans that survive here and there, may provide an insidious reminder of our past, but they are essentially ephemera: by retaining it we may simply be encouraging further graffiti, leading to its eventual loss.

LEGIBILITY

The currently fashionable concept of "legible cities"[11] is underpinned by the idea that somehow we have lost the ability to understand places and that we need more signs to do so. The preoccupation with 'branding', about how cities need to "rethink how they present themselves…communicate more effectively with their users" is actually about marketing. The ubiquitous brown tourism signs and the 87 official and internationally incomprehensible icons which adorn them, are another form of advertising masquerading as information. Signs do not explain places; they direct our experience; and this is something far better done for yourself.

Traditional towns are legible; the relationship of church, street and market place, bridge, pub or guildhall is self evident. There is no need to explain that one building has higher status, or that the square is the centre and meeting place of the community. Even changing functions do not obscure this, and combined with simple nameplates few of us will get lost. The Barbican, an estate in the City of London, is notorious for its disorienting design and confusing topography. It lacks the familiar markers of a town or village so the visitor is dependent on signage.

This unwritten sort of legibility might help in the context of the Disability Discrimination Act[12]. An unthinking application of DDA standards to signage could result in ever more

A fine old cast iron fingerpost in Sussex

bigger, bolder signs. It is frequently segregation and special rules that produce the need for explanation. Perhaps these distinctions need thinking about, not the signage itself. Careful consideration of how a place works, based on the principle of equal and easy access for all, and a real understanding of its buildings, topography, development and uses ought to reduce the need for signs and make the place easier for everyone to use, understand and enjoy.

A great deal of signage is unnecessary, and reflects a failure to think about possible alternatives. Conservation can put too much emphasis on the sign as an object. Street signs are essentially modern, and often transient. There are no authentic historic traffic signs. The purpose of a street sign is to convey a message with as little interference from its physical form as possible. Signs may contribute to the townscape, but disguising a sign is as much contradiction of its purpose as trying to make it unobtrusive. In contrast, there is a sad lack of good lettering today, although it is the best means of strengthening the message without weighing down the object.

Too often a sign is a cheap substitute for decent services. Regular effective rubbish collection and street sweeping will do more to improve the environment than a sign asking people to use an overflowing bin. The ubiquitous clumps of empty signposts (although they are useful for locking up bikes) are a consequence of careless management and poor communication, not bad design or inadequate conservation policies. Removing redundant posts and using a single existing one is both cost effective and an environmental enhancement.

Towns and cities must contain a mass of conflicting interests, and conservation area management at its best is a model for working together to balance them: but why should high standards be restricted to historic environments? Ideas about community empowerment from sources like David Engwicht's book *Street Reclaiming* have roots in alternative politics but could be read as a paradigm for historic areas: human and not

Wick, Caithness; the sign is the architecture

mechanical, local not remote, distinctive not standardised.

Although signs are perceived as especially problematic in the historic environment, the underlying issues concern the public realm as a whole. Signage does not exist in isolation. It serves a purpose, and the best sign is the one that serves its purpose most effectively. The worst signs say one thing and mean another: nameplates or signs to tourist attractions which are really corporate advertising; information boards that are really bad sculptures, finger posts pointing to an imaginary past. Gigantic billboards are not, in themselves, works of the devil, but in the wrong place they are a blight. Signs do proliferate without control and obscure buildings, vistas and details, but the signs are only a symptom. If a town is bisected by trunk roads and its shops are all chain stores, that's what the road signs and shopfronts will say.

Recommended Reading

Conservation Area Management – A Practical Guide, EHTF Report No 38, 1998

Streets for All, English Heritage/Government Office for London; 2000

Planning in Small Towns, Planning Advice Note PAN 52, The Scottish Office, 1997

Llewelyn-Davies/Alan Baxter *Urban Design Compendium*, English Partnerships/The Housing Corporation London 2000

MICHAEL COPEMAN is an independent historic building consultant and writer based in London. He managed The Heritage Lottery Fund's Townscape Heritage Initiative grant scheme between 1997 and 2002, having worked previously as Historic Buildings Adviser for Essex County Council.

1 Planning Policy Guidance: *Planning and the Historic Environment* (PPG15), Department of the Environment/Department of National Heritage, 1994

2 *Street Signs Manual*, Department of Transport

3 *Traffic Management in Historic Areas*, Department for Transport Traffic Advisory Leaflet 01/96

4 *Traffic in Historic Towns*, EHTF, 1993

5 *Street Improvements in Historic Towns*, EH, 1994

6 Whyte, William City: *Rediscovering the Centre*, Anchor Press, NY 1990

7 Engwicht, David *Street Reclaiming*, New Society Publishing, Gabriola Island, British Columbia, Canada, 1999

8 The Observer, London, 30 June 2002

9 Cullen, Gordon, *Townscape*, Architectural Press, London, 1961

10 *Historic Towns: Preservation and Change*, Ministry of Housing and Local Government, HMSO, 1967

11 See (e.g.) *Building Legible Cities*, Kelly, Andrew, Adshel, 2001

12 Disability Discrimination Act 1995

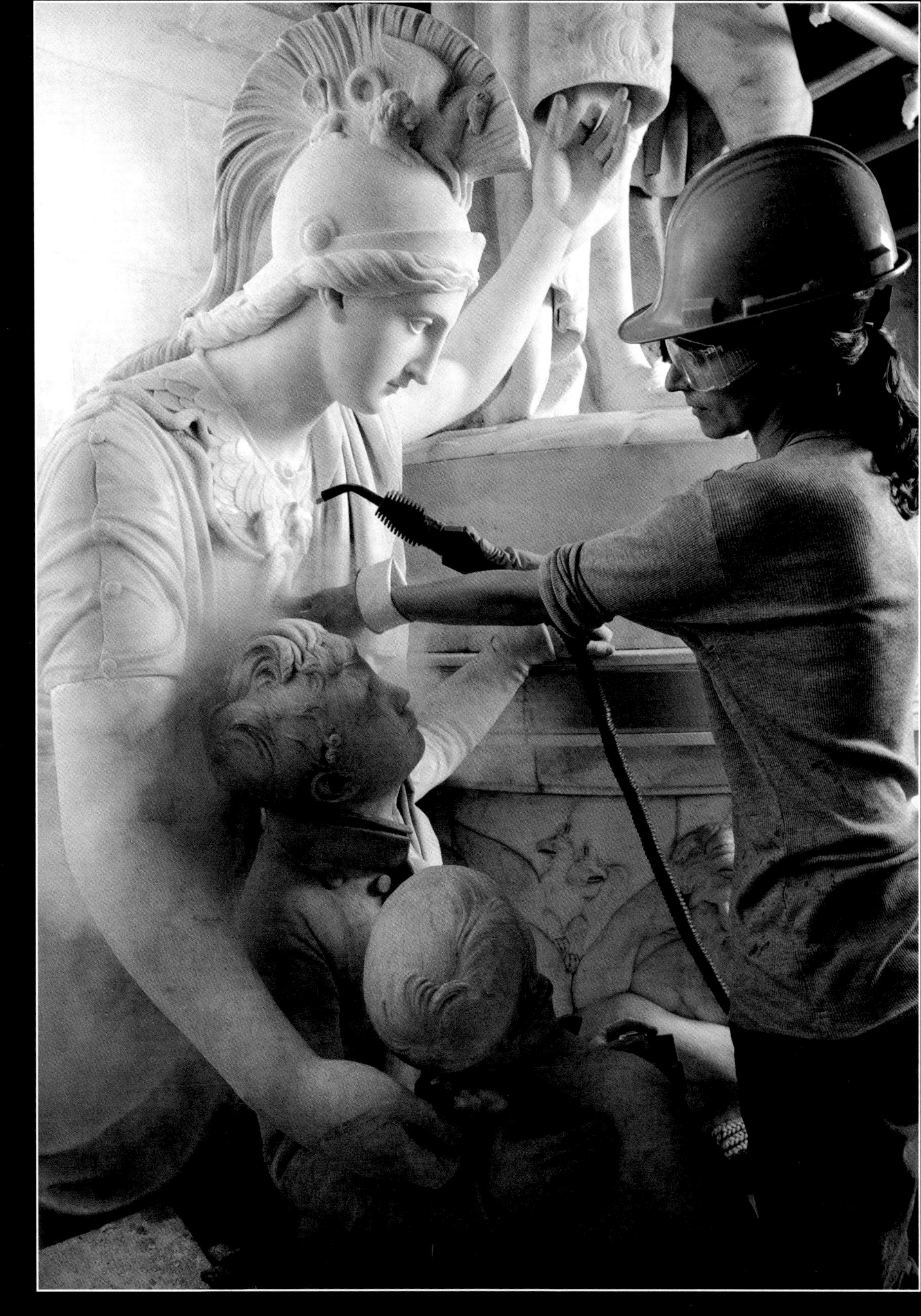

Patricia Maestre of Nimbus Conservation Limited cleaning the fine marble sculptures on the monument to Horatio Nelson at St Paul's Cathedral

Reproduced by kind permission of the Dean and Chapter of St Paul's. Photograph by Construction Photography

Chapter 4
SERVICES & TREATMENT

MASONRY CLEANING

MASONRY CLEANING			
A F Jones (Stonemasons)		ac cc gf pr sc wt	100
Abbey Heritage Ltd		ac cc gf ls pp pr sc wt	165
Anelays - William Anelay Ltd		ac cc gf pr sc wt	58
Arte Mundit	mp		167
Balmoral Stone		ac cc pr sc wt	101
Burleigh Stone Cleaning & Restoration Co Ltd		ac cc wt	166
Burnaby Stone Care Ltd		ac cc gf wt	166
C Ginn Building Restoration		ac	167
Carrek Limited		ac cc gf ls pr sc wt	61
Cathedral Works Organisation (Chichester) Limited		ac cc pr sc wt	102
Church Conservation Limited		ac cc sc wt	84
Cliveden Conservation Workshop Ltd		ac cc gf ls pr sc wt	102
Cyril John Decorators Ltd		gf	197
David Ball Restoration (London) Limited		ac cc gf pr sc wt	103
Flowplant Group Limited	mp	cc gf	168
Hare & Humphreys Ltd		pr	198
Hirst Conservation		ac cc ls pr sc wt	62
Holden Conservation		ac cc ls pr sc wt	95
J Joslin (Contractors) Ltd		ac pr sc wt	105
Keim Mineral Paints Ltd	mp		186
Lynton Lasers	mp	ls	168
Mather & Ellis Ltd		ac cc	105
Maysand Limited		ac cc gf sc wt	165
Minerva Stone Conservation		ac cc ls pr sc	106
Mott Graves Projects Limited		ac cc lw	106
Nicolas Boyes Stone Conservation		ls	106
Nimbus Conservation Limited		ac cc pp pr sc wt	166
O'Reilly Period Cornice Restoration & Cleaning		pr sc	216
PAYE Stonework		ac cc gf pr sc wt	107
Priest Restoration		ac cc pr sc wt	107
Rope Tech International Limited		ac	84
Russcott Conservation and Masonry		ac cc gf ls pr sc wt	108
Sarabian Limited		gf	168
StoneCo Limited		sc	108
Stonehealth Limited	mp	ac cc pp pr sc	169
Stonewest Ltd		ac cc gf pp pr sc wt	67
Strippers Paint Removers	mp		165
Suffolk Brick & Stone Cleaning Company Ltd		ac cc sc wt	169
Wells Cathedral Stonemasons		ac sc wt	109
William Sapcote & Sons Ltd		ac	67
William Taylor Stonemasons		sc wt	109

KEY

ac air/ water abrasive cleaning
cc chemical cleaning
gf graffiti protection and removal
ls laser cleaning
mp masonry cleaning products and materials
pp poultices
pr paint removal and poulticing
sc steam cleaning
wt low pressure water cleaning

MORTARS AND RENDERS

MORTARS AND RENDERS		
Advanced Chemical Specialties Ltd	cp	181
Anglia Lime	hy lm pz re	173
Balmoral Stone	pn	172
Beckwith Tuckpointing	pn	176
Bleaklow Industries	hy lm	173
The Bulmer Brick and Tile Company	hy lm pz	172
Bursledon Brickworks Conservation Centre	hy lm pz re	172
Carrek Limited	pn	61
Castle Cement Limited	hy	173
Cathedral Works Organisation (Chichester) Limited	hy lm pn pz re	102
Chalk Down Lime Ltd	lm re wd	172
Cliveden Conservation Workshop Ltd	pn	102
The Cornish Lime Company	hy lm pz re	174
E T Clay Products Limited	lm	118
Grall Plaster & Stone Specialist	pn	176
H J Chard & Sons	hy lm pz re	174
Hirst Conservation Materials Ltd	hy lm pz re	174
Keim Mineral Paints Ltd	cp	186
The Lime Centre	hy lm pz re	174
Limetec	hy	174
Mike Wye & Associates	hy lm pz re	174
Milestone Lime	hy lm	175
Rose of Jericho	hy lm pz re wd	175
Ryedale Conservation Supplies	lm pn	175
St Astier Natural Hydraulic Limes	hy	175
Strippers Paint Removers	hy lm	165
The Traditional Lime Co	hy lm re	176
Twyford Lime Products	lm	176
Ty-Mawr Lime Ltd	hy lm pz re	176
Wenlock Lime Ltd	lm	176
Womersley's Limited	hy lm pz re wd	176

KEY

cp stone consolidants
hy hydraulic lime
lm non-hydraulic lime (lime putty)
pn pointing with lime
pz pozzolanic additives
re hair and fibre reinforcement
wd wattle and daub

FIXINGS

FIXINGS	Cont	Supp	
Abbey Heritage Ltd	er		165
Avon Stainless Fasteners		fx	184
Bedford Timber Preservation Company	er		181
Britannia Architectural Metalwork Ltd		zw	143
Charterbuild Ltd	er zw		184
Falcon Repair Services	er zr zw		184
Glasgow Steel Nail Co Ltd		fx nl	184
Helifix Limited	er zr zw		184
Maysand Limited	zr zw		106
Mowlem Rattee & Kett	er		106
Priest Restoration	er		107
The Real Wrought Iron Company		nl	147
Robinsons Preservation Limited	er		182
Rotafix Ltd	er		183
Sarabian Limited	er		168
Tankerdale Ltd	er		128
Terminix Property Services	er		182

KEY

Cont contractors
Supp suppliers
er epoxy resin repairs
fx nuts, bolts, screws and nails
nl traditional nails
zr roof ties and plates
zw wall ties and plates

DAMP AND TIMBER DECAY

DAMP AND TIMBER DECAY			
Abbey Heritage Ltd		er	165
Advanced Chemical Specialties Ltd		dp	181
All Timber Infestation & Consultancy Services Ltd		dd ie	181
Allyson McDermott	ec		197
Balmoral Stone Ltd		dd	101
Bedford Timber Preservation Company		dd er ie nd	181
Charterbuild Ltd		er	184
D S Sheridan Building Services		dd	61
Delta Membrane Systems Ltd		dp	181
Demaus Building Diagnostics Ltd		dd tt	40
Falcon Repair Services		er	184
Gifford and Partners	ec		39
H-H-Heritage		nd	39
Helifix Limited		er	184
Hutton+Rostron Environmental Investigations Limited	ec	dd ie nd tt	182
International Fire Consultants Ltd		tt	188
Lightwright Associates	ec		190
Maysand Limited		dd ie tt	106
Minerva Stone Conservation		dd	106
Mowlem Rattee & Kett		er	106
Priest Restoration		er	107
ProBor		dp	181
Renlon Ltd		dd ie	182
Ridout Associates	ec	ie nd	182
Robinsons Preservation Limited		dd er ie nd tt	182
Rope Tech International Limited		nd	84
Rotafix Ltd		er dp	183
Ian Russell Engineering		nd	39
Ryder & Dutton		nd	47
Sarabian Limited		er	168
Sibtec Scientific	ec	dd tt dp	183
Tankerdale Ltd		er	128
Terminix Property Services		dd er ie	182
Thermo Lignum UK Limited		dd ie	183
Ward & Dale Smith Chartered Building Surveyors		dd ie	182

KEY

dd damp and decay treatment
dp damp and decay treatment products
ec environmental control
er epoxy resin repairs
ie insect and pest eradication
nd non destructive investigations
tt structural timber testing

BIRD CONTROL

BIRD CONTROL	
Balmoral Stone	183
Microbee Bird Control Limited	183
Rope Tech International Limited	84
Splitlath Ltd	67
Terminix Property Services	182

MASONRY CLEANING – NEBULOUS SPRAY

IAN CONSTANTINIDES and LYNNE HUMPHRIES

Conservation is generally not a dramatic process. It is frequently imperceptible and by its very nature, usually subtle. Consequently, cleaning can be one of the most satisfying processes of conservation because its results are immediately visible, and it appeals to building owners since their investment is readily seen. However, focusing on the aesthetic benefits of cleaning does risk overlooking the cause of the soiling and ignoring the history of the building. Cleaning has become one of the most controversial aspects of conservation, raising fundamental questions. Is it always necessary or even beneficial? Are we too ready to clean? Many buildings have been damaged by cleaning in the past, and even the most appropriate cleaning techniques can be harmful. Arguably the most beneficial aspects of cleaning are to reveal the condition of the building where the dirt may have concealed cracks or structural faults and to slow down deterioration by removing damaging materials.

Of the various methods available, nebulous spray cleaning is among the gentlest. This article looks at some of the many factors to be considered before selecting this cleaning method as the most appropriate one, as well as giving a very general account of what it involves.

TYPES OF SOILING

Dirt or 'soiling' can simply be defined as material which is in the wrong place[1]. The question is how to remove this material without causing irreversible damage to the material which is in the right place, either directly or by introducing new material. To clean a building or material successfully we need to start by understanding the nature of the dirt.

Dirt or soiling may take many forms: airborne particles, gaseous pollutants and organic aerosols from industrial or vehicular emissions; biological soiling by algae, fungi, bacteria and lichen; non-biological soiling by iron staining, paint or graffiti, for example; and the list goes on. In turn, these may all be affected by water, temperature and wind, and by the effects of microclimate.

It may be that the soiling causes stone deterioration or decay, or reduces the permeability of the substrate; or it may simply appear as an unsightly surface discoloration. Over time architectural surfaces build up a patina that is due in part to airborne particles, weathering cycles and the mineralogy of the stone itself. Unlike surface dirt, the patina

Flexible heads directing a carefully controlled fine spray of water onto stone mullions (photograph by Nimbus Conservation Limited)

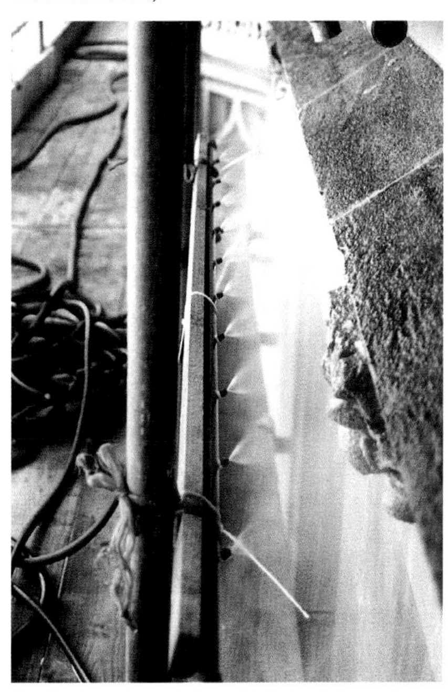

Fixed heads creating a nebulous mist effect on flat areas of masonry (photograph by Paye Stonework & Restoration Ltd)

does not simply lie across the surface of the stone but is combined to varying depths within the masonry, be it stone, brick or terracotta. Although not necessarily damaging in itself, removing this layer detracts from the historic interest of the original and may expose a weaker substrate to decay. Another consequence of removing the build-up of patinas or encrustations is the potential mobilisation of minerals beneath the stone surface, leading to discolouration.

Consideration should also be given to potential re-soiling of the stone. Industrial emissions and environmental factors have changed since many of our buildings were last cleaned, and it is unlikely that re-soiling will take the same form.

SELECTING THE CLEANING METHOD

To select a cleaning method or even to assess the need for cleaning, it is important to survey the building first. The aim is to establish the types of material, their condition, the architectural style, previous treatments and the nature, cause and pattern of the soiling for each area. All these criteria must be considered in the context of the building itself, its history, construction, location and proximity to other buildings etc.

Next, cleaning trials should be carried out on inconspicuous areas, preferably using the operator who will be doing the work finally, as skill is just as important as method. The trial will help to:

- further ensure that the correct method or methods are selected
- determine how clean the surface can become (the 'level of clean') without risk to the fabric
- highlight potential problems.

Trial areas should be selected on their ability to illustrate as far as possible the range of soiling types and fabric conditions, to establish levels of clean which are not just desirable but also achievable, with the least risk.

Bear in mind that a uniform surface is rarely achieved without excessive and highly damaging masonry cleaning. An uneven patchy finish is more likely as buildings are subjected to a variety of weathering patterns: regularly rain-washed areas often appear brighter than protected areas, particularly on limestone buildings: and flat facades may also have uneven soiling due to apparently similar stones varying in porosity, pore size, capillary action, or surface texture. The art of cleaning, on aesthetic grounds, is to find the balance between the extremes. Often it is better to under-clean.

APPROACHES

There is a multitude of different cleaning methods, which may be wet or dry, chemical- or water-based, abrasive and nonabrasive, many of which have a place in conservation. There are positive and negative points to all methods and there is rarely a single method suitable for all situations. The least harmful method or combination of methods should be selected for each case.

[1] *Science for Conservators Book 2, Cleaning, The Conservation Unit, 1983*

NEBULOUS SPRAY OR INTERMITTENT MIST SPRAY

Low-pressure water washing is probably the least aggressive form of cleaning. Its application is particularly useful where water-soluble dirt is present or water-soluble chemical compounds bind the dirt. Thicker encrustations of soiling which tend to form in protected areas of a building not regularly washed by rain may be softened by the water and subsequently mechanically removed. However, it cannot be used to remove soiling or staining which is insoluble in water.

Nebulous spray, also known as intermittent mist spray, is a development of low-pressure water washing. The aim is to apply the minimum amount of water for the minimum duration to soften the dirt, thereby enabling its removal by scrubbing or other relatively gentle treatment. Ordinary low pressure water washing, by comparison, risks saturating the masonry, causing damage to the wall by mobilising salts and causing fixings to corrode for example, as well as damaging other features fixed to the wall such as internal plasterwork, timber or decorations. It can also lead to dry rot.

Only once all the investigations have been carried out, questions answered, options considered and the conclusion drawn that nebulous water spray cleaning fulfils all the criteria, should cleaning be commenced by those trained and skilled in the use of this cleaning method and following the guidelines established during trials.

GENERAL PROCESS

The system of nebulous sprays is based on the principle of passing water through a very fine mesh or filter to create a mist that is then passed through fine nozzles. The mist spray system can be set up with nozzles at intervals along the building, concentrating on areas of greater need and reducing the level where less dirt is present. The level of water may be controlled electronically or by timers, allowing pulse or intermittent spraying, to avoid ever having water running down the face of the building. Before starting, the porosity of the stone can be assessed in order to balance the amount of water and duration required.

As the system produces such a fine mist it is important to place the nozzles close to the building's surface in order to ensure the water is directed correctly. Depending on the location and exposure of the elevation it is frequently necessary to erect a screen to reduce the risk of wind disturbance.

Nebulous spray systems can be designed to be incredibly flexible, directing the spray only where needed. Straight or flexible hoses may be employed depending on the requirements of the surface being treated and the nozzles from the hose may be grouped or spaced according to the severity of the dirt or encrustation being treated. Flat surfaces often require less water than a carved heavily soiled detail, which may require a cluster of nozzles positioned on an articulated hose to the profile of the carving.

ADVANTAGES

The most obvious advantages of cleaning with water are that water is cheap, readily available, safe and environmentally friendly. It is also particularly effective for cleaning limestone and marble.

The impact of the mist on the surface is negligible, reducing the risk of mechanical damage unless the surface is extremely friable. Consequently the risk of washing away weak pointing material or decaying stone is almost entirely eliminated.

Encrustations and dirt are softened progressively, reducing the risk of mechanical damage, and allowing greater control over removal and permitting more frequent monitoring of the surfaces. This ensures that the right levels of clean are achieved and reduces the risk of over cleaning. It also gives greater opportunity to re-evaluate the method or levels of cleaning than with many other cleaning methods.

Where the use of harsher methods of cleaning are unavoidable, prolonged use may be reduced by first cleaning with the nebulous spray system.

Removing softened material by brush between spraying cycles may accelerate the cleaning process and has the added advantage of enabling progress to be monitored.

A further advantage is the ability to control the quantity of water used. Excess run off, which this method avoids, is a particular problem with traditional water washing methods where weathered wash patterns formed by rainwater may channel the spray, avoiding adjacent areas of the masonry. As mist sprays use less water, a more even wash is achieved, avoiding the weathered wash channels and reducing the probability of saturation as the stone does not get so wet.

DISADVANTAGES AND RISKS

- Although the nebulous spray system reduces the risk of saturation enormously, this problem may still arise as a result of a failure in the timer, switch or in judging the porosity of the stone which can mean damage to internal finishes, hidden timber and ferrous fixings.
- Water cleaning methods may exacerbate deterioration when used on badly deteriorated masonry. The risk of water penetration through defective joints or fractures is still present with the nebulous spray system, illustrating the importance of carrying out a thorough survey externally, and continuous monitoring of the interior as cleaning progresses.
- As with all water treatments, the work should not be carried out when there is potential for frost damage.
- The network of hoses and bars situated close to the face of the building can restrict access and make monitoring or brushing down awkward.
- Efflorescence on the surface is possible where water treatments are carried out. Generally it is possible to estimate the risk of this prior to commencement.
- Water cleaning is less effective on siliceous stones such as granite and sandstone where the soiling is tightly bound to the silicate surface in insoluble compounds. Dirt on limestone is generally bound to relatively soluble chemical compounds.
- A frequent problem with many limestones and some sandstones is brown or orange staining caused by naturally occurring free iron within the stone being mobilised and carried to the surface. Consideration must also be given to the possibility of previous treatments, which may have been carried out, such as the application of a solution of copperas (ferrous sulphate) to Portland limestone in the 19th century in order to emulate the more fashionable Bathstone. Earlier conservation or cleaning treatments may also have a detrimental effect on the success of water cleaning.
- Finally the set-up and cleaning time required for the nebulous spray is greater than many other cleaning methods, however, this must be weighed against the increased control and gentleness of this type of spray.

RECENT DEVELOPMENTS IN TECHNOLOGY

Time controllers can be programmed to open a valve for a set period, the length and frequency of spray being determined by the nature of the material being treated.

Water flow meters are available to measure the quantity of delivered water and to calculate the output for sprays.

The use of articulated pipe allows greater control over the location of the nozzles.

FUTURE DEVELOPMENTS

The employment of moisture switches, which react to differing levels of moisture in the stone, may negate the need to predetermine the porosity of the stone.

Recommended Reading

Andrew, C, *Stone Cleaning A Guide for Practitioners*, Historic Scotland & The Robert Gordon University, 1994

Ashley-Smith, Jonathan, Scientific Editor, *Science for Conservators, Book 2, Cleaning*, Conservation Science Teaching Series, The Conservation Unit, 1983

Ashurst, John and Dimes, Francis G, *Conservation of Building and Decorative Stone*, Butterworth Heinemann, 1990

Mack, Robert C, Grimmer, A, *Assessing Cleaning and Water-Repellent Treatments for Historic Masonry Buildings*, Preservation Briefs 1, HPS, National Park Service, Technical Preservation Services

Matero, F, Bede, E, Tagle, A, *An approach to the evaluation of cleaning methods for unglazed architectural terracotta in the USA*, Architectural ceramics: Their history, manufacture and conservation, A joint symposium of English Heritage and the United Kingdom Institute for Conservation, 22–25 September 1994, James & James, 1996

Webster, Robin GM Editor, *Stone Cleaning and the nature, soiling and decay mechanisms of stone*, Proceedings of The International Conference held in Edinburgh, UK, 14–16 April 1992, Donhead

IAN CONSTANTINIDES is the the managing director of St Blaise Ltd (see page 64) and has worked in every aspect of historic building repair for over 20 years. Tel 01935 83662 E-mail info@stblaise.co.uk

LYNNE HUMPHRIES MA(RCA/V&A) is a conservator of both architecture and sculpture. She has worked and studied in both museums and historic buildings and formerly managed conservation for St Blaise Ltd. E-mail lynnehumphries@hotmail.com

PAINT REMOVAL

▶ HIRST CONSERVATION
Laughton, Sleaford, Lincolnshire NG34 0HE
Tel 01529 497449 Fax 01529 497518
REMOVAL OF SPECIFIC PAINT LAYERS TO INTERIOR AND EXTERIOR SURFACES: *See also: display entry on the inside front cover and profile entry in Building Contractors section, page 62.*

▶ SARABIAN LIMITED
Sarabian House, Kington St Michael, Chippenham, Wiltshire SN14 6JB
Tel 01249 750113 Fax 01249 750798
E-mail info@sarabian.co.uk
Website www.sarabian.co.uk
BUILT HERITAGE CLEANING AND RESTORATION: *See also: display entry in Masonry Cleaning section, page 168.*

▶ STONEHEALTH LIMITED
Bowers Court, Broadwell, Dursley, Glos GL11 4JE
Tel 01453 540600 Fax 01453 540609
E-mail info@stonehealth.com
Website www.stonehealth.com
PAINT REMOVAL AND ANTI GRAFFITI PRODUCTS: *See also: display and profile entries in Masonry Cleaning section, page 169.*

MASONRY CLEANING

▶ A F JONES (STONEMASONS)
33 Bedford Road, Reading, Berkshire RG1 7EX
Tel 0118 957 3537 Fax 0118 957 4334
E-mail af.jones@ukonline.co.uk
Website www.afjones.co.uk
MASTER STONEMASONS: *See also: display entry and profile entry in Stone section, page 100.*

▶ ABBEY HERITAGE LTD
Midstfields, Frome Road, Writhlington, Radstock, Near Bath, Somerset BA3 5UD
Tel 01761 420145 Fax 01761 437103
E-mail stone@abbeyh.co.uk
FACADE CLEANING, LASER CLEANING, RESTORATION AND MASONRY: Leaders in the field of laser cleaning for buildings, artefacts and artworks, Abbey specialises in facade cleaning – stone, terracotta, faience, granite, marble and brickwork – incorporating Jos/Torc and Doff systems, also anti-graffiti treatments. As part of their complete package they can implement masonry restoration and new stone projects. Abbey Heritage works closely with leading architects, surveyors, local authorities and consultancies throughout the UK. Dedication to detail and a high level of managerial input has earned them a reputation for integrity and reliability, generating consistent repeat business. As principal contractor Abbey provides associated leadwork, roofing, decoration, joinery, bronze and ironwork services. They offer full national and international consultancy services.

▶ C GINN BUILDING RESTORATION
89 Lunedale Road, Fleet Estate, Dartford, Kent DA2 6LW
Tel 01322 290505 Fax 01322 284839
E-mail c.ginnbuildingrestoration@btinternet.com
Website www.stonecleaning-restoration.com
RESTORATION AND CLEANING SPECIALISTS: *See also: display entry in this section, page 167 and profile entry in Stone section, page 102.*

▶ CLIVEDEN CONSERVATION WORKSHOP LTD
Head Office – The Tennis Courts, Cliveden Estate, Taplow, Maidenhead, Berkshire SL6 0JA
Tel 01628 604721 Fax 01628 660379
SCULPTURE, STONE AND WALL PAINTINGS CONSERVATION: *See also: profile entry in Stone section, page 102.*

▶ DAVID BALL RESTORATION (LONDON) LIMITED
104A Consort Road, London SE15 2PR
Tel 020 7277 7775 Fax 020 7635 0556
E-mail mail@dbr.uk.com
STONE CLEANING SPECIALISTS: *See also: display entry and profile entry in Stone section, page 103.*

▶ HIRST CONSERVATION
Laughton, Sleaford, Lincolnshire NG34 0HE
Tel 01529 497449 Fax 01529 497518
STONE CLEANING AND RELATED SERVICES: *See also: display entry on the inside front cover and profile entry in Building Contractors section, page 62.*

▶ MAYSAND LIMITED
109–111 Windsor Road, Oldham, Lancs OL8 1RH
Tel 0161 628 8888 Fax 0161 627 0996
E-mail sales@maysand.co.uk
MASONRY RESTORATION AND CONSERVATION: Maysand's expert team of surveyors and craftsmen provides the full range of masonry and building conservation services for historic, ecclesiastical and commercial projects. The company's range of services includes masonry stabilisation, stone and brickwork cleaning, repair and conservation/restoration, re-pointing, traditional stone masonry and new-build cladding. Members of Stone Federation Great Britain.
See also: display entry in Stone section, page 106.

SERVICES & TREATMENT
4.1 Remedial

MASONRY CLEANING

▶ NICOLAS BOYES STONE CONSERVATION

46 Balcarres Street, Edinburgh EH10 5JQ
Tel 0131 446 0277 Fax 0131 446 0283
Website www.nb-sc.co.uk

STONE CONSERVATION SERVICES: *See also: profile entry in Stone section, page 106.*

▶ NIMBUS CONSERVATION LIMITED

Eastgate, Christchurch Street East, Frome, Somerset BA11 1QD
Tel 01373 474646 Fax 01373 474648
E-mail enquiries@nimbusconservationlimited.com
Website www.nimbusconservation.com

STONE CONSERVATION AND MASONRY: Nimbus has been involved with the conservation and cleaning of historic buildings and monuments nationwide. Liaising with architects and surveyors Nimbus considers the specific merits of each project, and the systems used for cleaning are tailored to both the building materials and degree of soiling or decay. Internal and external elevations and detailed carved areas and sculpture can involve more than one method. Nimbus maintains that the skill of cleaning is provided by the careful operative rather than the technique. The company's teams are well trained, experienced and sensitive, and methods used include Jos, Doff, steam cleaning, nebulous sprays, poulticing, mechanical and the Arte Mundit system (as used at St Paul's Cathedral). *See also: display entry and profile entry in Stone section, pages 107 and 106.*

▶ PAYE STONEWORK

Stationmaster's House, Mottingham Station Approach, London SE9 4EL
Tel 020 8857 9111 Fax 020 8857 9222
E-mail kw@payestone.co.uk
Website www.payestone.co.uk

MASONRY CLEANING SPECIALISTS: *See also: profile and display entries in Stone section, page 107.*

C.GINN
BUILDING RESTORATION
"TAKING THE WEIGHT"
• MASONRY
• STUCCO
• PRE-CAST MOULDING
• STONE CLEANING
• BRICK CLEANING
UNIT 11, MILLSIDE INDUSTRIAL ESTATE, LAWSON ROAD, DARTFORD, KENT DA1 5BW
TEL 01322 290505 FAX: 01322 284839 EMAIL: c.ginnbuildingrestoration@btinternet.com
WEBSITE: www.stonecleaning-restoration.com

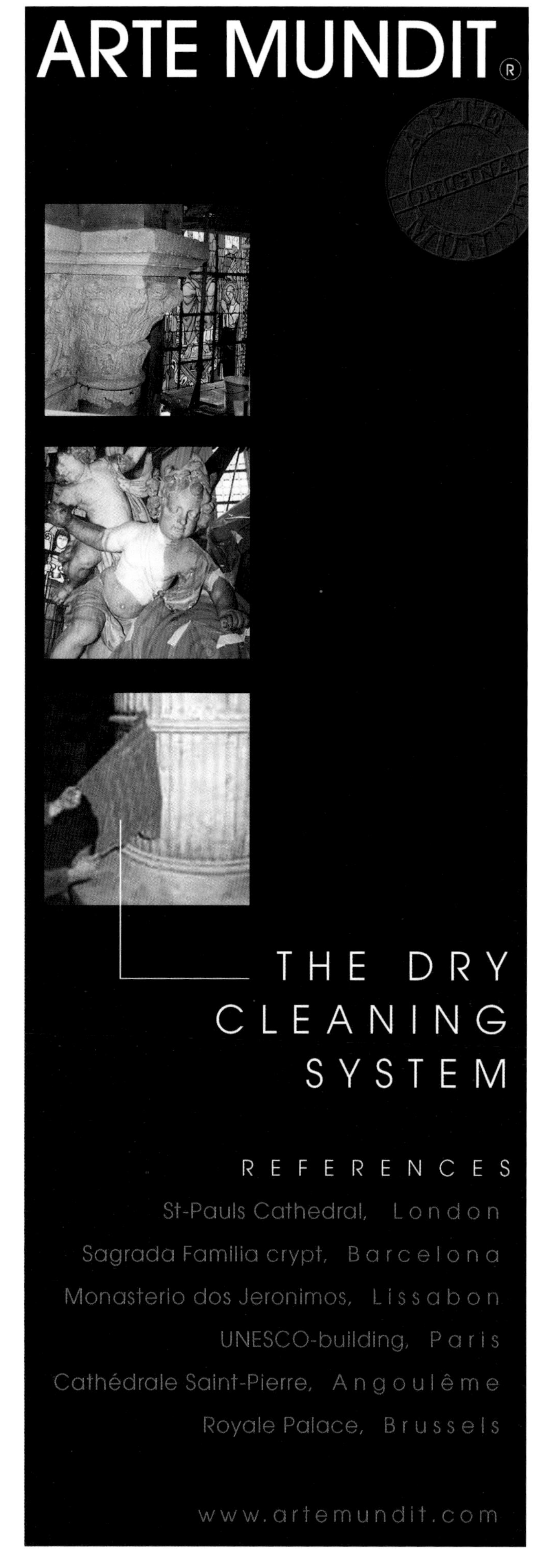
ARTE MUNDIT®
THE DRY CLEANING SYSTEM
REFERENCES
St-Pauls Cathedral, London
Sagrada Familia crypt, Barcelona
Monasterio dos Jeronimos, Lissabon
UNESCO-building, Paris
Cathédrale Saint-Pierre, Angoulême
Royale Palace, Brussels
www.artemundit.com

MASONRY CLEANING

▶ SARABIAN LIMITED

Sarabian House, Kington St Michael, Chippenham, Wiltshire SN14 6JB
Tel 01249 750113 Fax 01249 750798
E-mail info@sarabian.co.uk
Website www.sarabian.co.uk

BUILT HERITAGE CLEANING AND RESTORATION: As well as using approved conservation techniques for cleaning historic buildings internally and externally, Sarabian has developed significant expertise in removing paint from stone, including graffiti (anti-graffiti treatments available). A full repair, repointing and stone replacement service is also offered, including pinning where stone is suffering from severe cracking or instability. Lime plastering and rendering are new services offered. *See also: display entry in this section, page 168.*

▶ STONECO LIMITED

Woodland Close, Torquay TQ2 7BD
Tel 01803 616451 Fax 01803 616261
E-mail enquiries@stoneco.co.uk
Website www.stoneco.co.uk

MASONRY CLEANING: *See also: display entry in Stone section, page 108.*

▶ STONEGUARD

St Martins House, The Runway, Ruislip, Middlesex HA4 6SG
Tel 0870 241 6366 Fax 020 8839 9988
Website www.stoneguard.co.uk
▶ with offices also in Bath, Birmingham, Manchester and Stirling

MASONRY CLEANING AND PROTECTION: *See also: profile entry in Building Contractors section, page 66.*

▶ STONEHEALTH LIMITED

Bowers Court, Broadwell, Dursley, Glos GL11 4JE
Tel 01453 540600 Fax 01453 540609
E-mail info@stonehealth.com
Website www.stonehealth.com

STONE CLEANING SYSTEMS: Best known for the supply of leading specialist equipment, such as Jos/TORC and DOFF, the company also offers other cleaning methods for restoration and conservation. Stonehealth was responsible for establishing the 'Register' of trained operators, seen as important in helping to maintain standards, by allowing authenticity of a contractor to be checked prior to contract commencement. Stonehealth carries a wide up-to-date range of safer chemical products; from prevention, eg anti-graffiti, floor treatments, through to restoration, eg soot, metal oxide, graffiti removal etc. Capable of supplying 'specifically formulated' products and even extremely innovative ones, such as 'Clean-Film', applied as cream/paste, but peeled off 24 hours later as a latex film containing removed dirt and pollutants. *See also: display entry on this page.*

▶ STONEWEST LTD

Lamberts Place, St James's Road, Croydon CR9 2HX
Tel 020 8684 6646 Fax 020 8684 9323
E-mail stonewest@cwcom.net
Website www.stonewest.co.uk

BUILDING CONTRACTORS AND STONE MASONS: *See also: display entry in Building Contractors section, page 67.*

▶ SUFFOLK BRICK & STONE CLEANING COMPANY LIMITED

Dickens House, Old Stowmarket Road, Woolpit, Bury St Edmunds, Suffolk IP30 9QS
Tel 01359 242650 Fax 01359 241211

BRICK AND STONE CLEANING: Established in 1984, Suffolk Brick & Stone Cleaning Company has been involved in many major projects in the Eastern Counties using sympathetic methods to clean and restore brick and stone work. These projects include: Old Addenbrookes, Cambridge; Norwich Castle; Bedford Town Hall; Corpus Christi College, Cambridge and Eaton Park, Norwich. The company is a registered Jos system and Doff system contractor and uses other sympathetic cleaning methods as appropriate. Masonry repairs, pointing, waterproofing and paint removal are also undertaken as part of the overall service or as separate assignments. Please contact Greg Simonds, Director, for more information and helpful advice.

HYDRAULICITY

PAUL LIVESEY

Figure 1 The Eddystone lighthouse, now re-built on the Plymouth waterfront

The term 'hydraulicity' is derived from the French word 'hydraulique' which, at its simplest is defined as relating to water. It was adopted into construction usage to describe waterproof structures either to convey water or to withstand the ingress of water. In the early 18th century, French engineer Bernard Forest de Belidor used the term to describe construction techniques to resist the action of seawater. Early in the 19th century the eminent French engineer, Jean Louis Vicat, adapted the term to describe limes which would harden under water. They were thereby distinguished from the pure calcium limes, or 'air limes', which hardened by a different means resulting from the action of carbon dioxide when exposed to air. It was some time before the term came into common usage in Britain where an equivalent term 'water lime' had developed over a similar timescale. This initially arose with the work of Smeaton in developing his lime for seawater-resistant mortar for the Eddystone lighthouse *(illustrated on this page)* and continued through to the Portland cement precursor developed by Frost.

The term 'hydraulic' is now used internationally to describe cements and other binders which set and harden as a result of chemical reactions with water and continue to harden even if subsequently placed under water. There are a few exceptions but, in general, these chemical reactions involve calcium, silica and aluminium constituents which react with water to form a whole family of calcium silicate and calcium aluminate hydrates. *Table 1* sets out the chemical composition of typical hydraulic cements and binders.

The chemical composition in itself is only a limited guide to the hydraulic properties of the lime or cement. The key is in the way the components are combined into the various silicates and aluminates. *Figure 2* shows the relative strength obtained by the hydration of the common silicates and aluminates. In common with cement chemistry notation the four main compounds are described as:

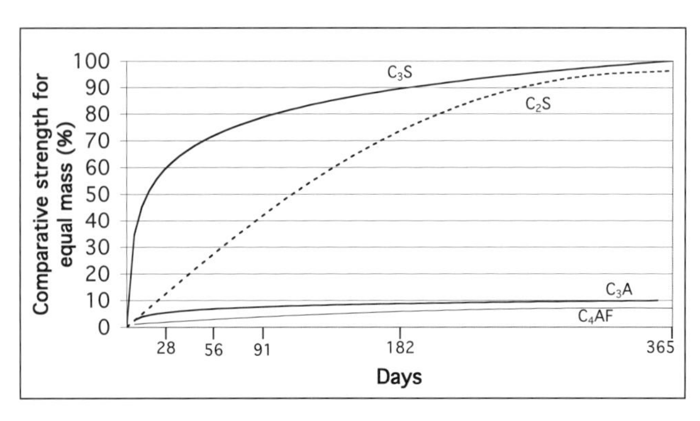

Figure 2 Comparative strength contribution of calcium silicates and aluminates

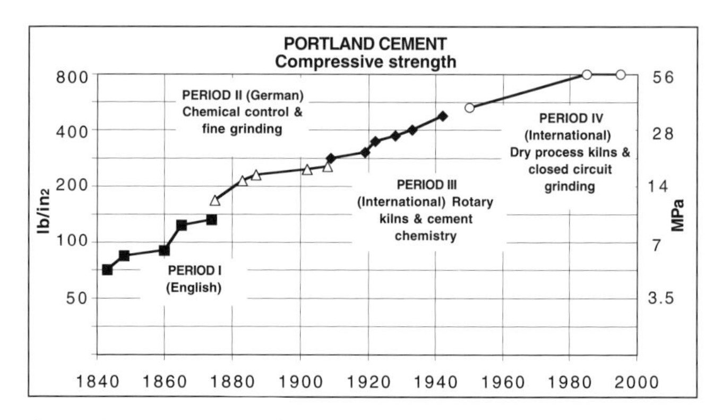

Figure 3 Changes in the properties of Portland cement

(* MPa = megapascals: the unit of resistance to compression which is equal to one Newton per square millimetre)

CATHEDRAL COMMUNICATIONS LIMITED

	Composition (%)				
	CaO	SiO_2	Al_2O_3	Fe_2O_3	MgO
Portland cement (CEM I)	64.0	21.0	5.0	2.5	2.0
High calcium lime (CL90)	73.5	0.8	0.1	0.1	0.9
Feebly hydraulic lime - A* (NHL 2)	62.8	12.2	1.3	0.4	1.2
Feebly hydraulic lime - B (NHL 2)	61.2	11.4	1.7	0.5	2.4
Moderately hydraulic lime - A (NHL 3.5)	60.3	10.4	3.8	1.5	1.5
Moderately hydraulic lime - B (NHL 3.5)	58.8	19.5	1.8	0.8	1.3
Moderately hydraulic lime - C (NHL 3.5)	63.6	12.6	1.3	0.6	2.1
Eminently hydraulic lime - A (NHL 5)	58.0	20.8	1.8	0.8	1.3
Eminently hydraulic lime - B (NHL 5)	38.0	23.7	10.9	3.3	2.8
Eminently hydraulic lime - C (NHL 5)	56.9	13.2	4.5	1.5	3.1
* letters A, B, C indicate different producers					

Table 1 *Main element oxide composition of hydraulic cements and limes*

	Composition (%)				
	CaO	SiO_2	Al_2O_3	Fe_2O_3	MgO
Italian pozzolana	2.2	58.9	16.6	3.4	0.8
Rhine trass	15.6	55.1	10.9	3.1	1.0
Yellow brick dust	12.0	66,2	10.0	3.4	3.6
Red brick dust	0.5	77.0	9.4	6.1	0.9
Metakaolin	0.1	55.4	39.9	0.6	0.3
Blastfurnace slag	40.8	36.0	13.7	0.7	7.5
Fly ash	4.8	52.2	28.1	7.4	2.2
Silica fume	0.3	95.5	0.8	1.3	0.2

Table 2 *Main element oxide composition of hydraulic pozzolans*

	Hydraulicity with age (days)				
	28	56	91	182	365
Feebly hydraulic lime (NHL 2)	0.8	0.8	0.7	0.7	0.5
Moderately hydraulic lime (NHL 3.5)	1.0	1.0	1.0	1.0	1.0
Eminently hydraulic lime (NHL 5)	1.3	1.5	1.4	1.4	1.4
Masonry cement (MC 22.5)	4.2	2.8	2.4	2.3	2.3
Portland cement (CEM I 42.5)	15.0	5.7	5.1	4.9	4.8
Italian pozzolana	0.3	0.9	1.2	1.4	1.8
Rhine trass	0.5	0.7	0.9	1.2	2.1
Yellow brick dust	0.3	0,5	0.7	0.9	1.0
Red brick dust	0.2	0.6	0.7	0.8	0.9
Metakaolin	13.1	10.4	9.8	8.6	6.6
Blastfurnace slag	7.4	5.7	4.4	4.3	4.1
Fly ash	0.4	1.7	2.0	3.3	3.6
Silica fume	10.4	8.1	6.4	6.1	6.1

Table 3 *Hydraulicity of cements and pozzolana compared with NHL 3.5 hydraulic lime*

C_3S – three parts calcium oxide combined with one part silicon oxide, also known as 'alite'

C_2S – two parts calcium oxide combined with one part silicon oxide, also known as 'belite'

C_3A – three parts calcium oxide combined with one part aluminium oxide

C_4AF – four parts calcium oxide combined with one part aluminium oxide and one part iron oxide, these latter two being generally termed 'aluminate phases'.

It can be seen that the calcium silicates are the main strength giving components in hydraulic limes and cements. The aluminate and ferrite compounds contribute little strength.

Although the overall bulk chemistry can be similar for different products, the burning process has a major bearing on their hydraulicity. Limes are burnt at temperatures below 1,000°C where the main reaction products are belite and free lime together with small amounts of low reactivity aluminates. Portland cements are burnt at temperatures up to 1,500°C where almost all lime is combined, forming alite, which is the main product, and reactive aluminates. The development of hydraulicity as measured by compressive strength resulting from changes in technology over the past two centuries is illustrated in *Figure 3*.

POZZOLANS

Despite their lack of either a detailed knowledge of chemistry or the technology to burn at high temperatures, Greek and Roman engineers succeeded in producing 'hydraulic' constructions. They achieved this by combining a calcium-bearing constituent (that is to say lime) with products providing the silica and/or aluminium constituents. Such products are referred to as 'latent hydraulic' materials: that is they are not hydraulic of themselves but become so when exposed to calcium-rich solutions. A number of materials fall into this category, both naturally occurring and manufactured. The best known traditional, naturally occurring example is the volcanic dust from the Mount Vesuvius region of Italy to the south of Naples. Termed pozzolana after the town of Pozzuoli around which the dust deposits are centred, the material was known by Roman builders to be capable of reacting with lime to produce superior mortars.

Other naturally 'pozzolanic' materials include the volcanic trass of the Rhine valley and various zeolites. Manufactured pozzolans are also commonly used today. These include metakaolin, a thermally treated china clay, and the waste by-products of several major industries such as blastfurnace slag, power-station fly ash and silica fume from the ferro-silicon industry. Thus the definition of a pozzolan could be a material that will react hydraulically with lime solution. Details of the composition of these pozzolans are set out in *Table 2*.

The degree to which such latent hydraulic materials react with lime is often termed their 'hydraulicity' or 'pozzolanicity' and a number of tests have been developed in order to measure this. The chemical test set out in BS EN 196-5 determines the quantity of lime absorbed from a standard lime solution by the pozzolan. In order to qualify for use in the standard pozzolanic cement of BS EN 197-1 a pozzolan must be capable of absorbing a specified amount of lime. Other more pragmatic tests have been developed to assess the strength-bearing properties in mortars and concretes. In the recent BS EN 206-1 concrete specification the term 'k factor' is used for latent hydraulic binders and is calculated as the amount of strength attributed to the pozzolan in concrete in comparison with a standard Portland cement. In the recent UK Foresight research into hydraulic limes a similar technique was used to determine the hydraulicity of pozzolans in mortar in comparison with an equal mass of the standard natural hydraulic lime used for the project. *Table 3* sets out the hydraulicity of these pozzolans in relation to the natural hydraulic lime and cement. The pozzolans vary in their speed of reaction so that a measure of hydraulicity depends, to some extent, on the age at which it is determined. In this experiment small amounts of the pozzolans were added to a 1:3 moderately hydraulic lime mortar (1:3 NHL3.5:sand). The proportion of pozzolan to lime did not exceed 30 per cent by volume.

In addition to supplementing the strength of mortars and concretes, while utilising otherwise waste materials, latent hydraulic

Figure 4 Concrete caisson breakwater at Brighton Marina

materials modify the hydration products formed and as a result modify the properties of the mortar or concrete. The values in *Table 3* illustrate the various reaction rates of different silicates, different proportions of the silicates and decay of strength gain as either the reactive silicate or the available lime becomes depleted. The initial low hydraulicity of some pozzolans is ascribed to the slow diffusion mechanism whereby the reactive silicates are released for reaction with the lime-rich solution.

In summary, the term 'hydraulicity' is the property of limes and cements to set and harden under water whether derived from a naturally hydraulic lime, cement or a pozzolan. 'Cements' in this context can be single products or combinations of calcium bearing cement or lime mixed with materials that contribute reactive silica and alumina. The hydraulic characteristic of such materials is derived from their reactive calcium and silicon phases which combine with water to form calcium silicate hydrates of various densities. The density of the hydrate phases provides the binding power of the cement or lime and determines the strength of the mortar or concrete produced from it. The relative strength of mortar or concrete is used to quantify the 'hydraulicity' of the material. Products of moderate hydraulicity are capable of being used for mortar in exposed situations such as that in *Figure 1*. Products of high hydraulicity are required as the main construction material in extreme exposure situations such as that in *Figure 4*.

Recommended Reading

Hewlett PC (ed), *Lea's Chemistry of Cement and Concrete*, Fourth edition, Arnold, London, 1998

Bye GC, *Portland Cement*, Second edition, Thomas Telford, London, 1999

Holmes S and Wingate M, *Building with Lime*, Intermediate Technology Publications, London, 1997

PAUL LIVESEY, Technical Manager, Castle Cement Limited, is a chemist with 40 years experience in the cement industry. He is responsible for the quality of Castle products, technical work in support of customer services, product applications, research and development, and company representation on industry, UK and European technical committees. He is lead UK expert to the British Standards and European Standards committees drafting the new generation of standards for cement, lime and methods of testing them, a director of Concrete Information Limited, Chairman of the British Cement Association Standards and Technical Committee and a member of the Building Limes Forum national committee.

MORTARS & RENDERS

▶ BALMORAL STONE LTD

Unit 1C, West Craigs Industrial Estate, Turnhouse Road, Edinburgh EH12 8NR
Tel 0131 339 5920 Fax 0131 317 9707

STONEMASONRY AND RESTORATION CONTRACTORS: *See also: display entry in Stone section, page 101.*

▶ THE BULMER BRICK & TILE CO LTD

Bulmer, Nr Sudbury, Suffolk CO10 7EF
Tel 01787 269232 Fax 01787 269040

LIME PRODUCTS: Stockists of lime putty, hydraulic lime and associated products. Also producers of ground brick pozzolans. *See also: profile entry in Brick Suppliers section, page 118.*

▶ BURSLEDON BRICKWORKS CONSERVATION CENTRE

Bursledon Brickworks, Coal Park Lane, Swanwick, Southampton SO31 7GW
Tel/Fax 01489 576248
E-mail bursledon@ndirect.co.uk

TRADITIONAL BUILDING PRODUCTS AND EDUCATIONAL COURSES: The Centre, administered by an educational trust, is based at the 19th century Bursledon Brickworks near Southampton. It has a museum and educational function and also supplies traditional building materials including lime putty, hydraulic lime, ready-mixed lime mortars, limewash, hair, stone dust, pozzolanic additives, ochres and pigments, pointing irons, riven laths, battens, air-dried oak and traditional drawn glass. Other products can be sourced. The Centre runs courses on traditional materials, techniques and the conservation of the built environment at Introductory, CPD and post graduate level. The Centre also provides consultancy advice on building conservation, and fully serviced facilities for seminars, exhibitions and conferences.

▶ CARREK LTD

Mason's Yard, Wells Cathedral, Wells, Somerset BA5 2PA
Tel 01749 689000 Fax 01749 689089
Website www.carrek.co.uk

HISTORIC BUILDING REPAIR COMPANY: *See also: profile entry in Building Contractors section, page 61.*

▶ CHALK DOWN LIME LTD

102 Fairlight Road, Hastings, East Sussex TN35 5EL
Tel 01424 443301 Fax 01580 830096
Mobile 0771 873 8708
E-mail chalkdownlime@supanet.com

SPECIALISTS IN MORTARS AND RENDERS: Chalk Down Lime stocks traditional building materials and offers a mortar analysis service. Suppliers of matured slaked lime putty, ready mixed mortars and renders, mortars made to individual requirements, limewash, natural pigments, laths, coal tar etc. Conservation projects undertaken by a team of dedicated craftsmen and professionals. They aim to provide a comprehensive service in the maintenance and repair of historic buildings.

▶ HARRY CURSHAM

Parks Farm, Cambridge, Gloucester GL2 7AR
Tel 01453 890297 Fax 01453 899121
Mobile 07986 185894
E-mail harry.cursham@lineone.net

CONSULTANT REPAIRS TO OLD, VERNACULAR AND TRADITIONAL BUILDINGS: *See also: profile entry in Heritage Consultants section, page 35.*

▶ GRALL PLASTER & STONE SPECIALISTS

Forest Corner, Purlieu Lane, Godshill, Fordingbridge, Hampshire SP6 2LW
Tel 01425 655430
Mobile 07973 832 846

MORTARS AND RENDERS *See also: display entry in Plasterwork section, page 214.*

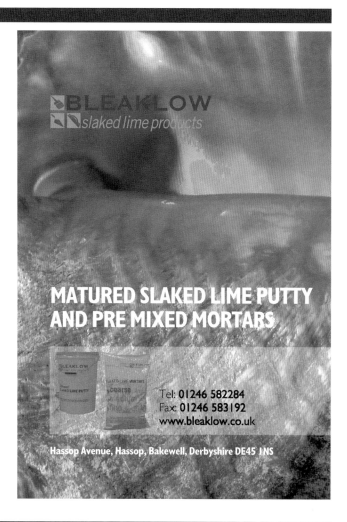

▶ H J CHARD & SONS

1 Cole Road, Bristol BS2 0UG
Builders' Merchants 01179 777681

LIME PUTTY AND RELATED PRODUCTS: H J Chard & Sons manufactures for sale direct to the trade and retail users lime putty from best quality Buxton non-hydraulic white quicklime. Lime putty is used in the production of a wide range of sand-lime mortars, including hair mortars. Other products available include pigments, tallow, linseed oil, PFA, HTI powder and French hydraulic lime.

▶ HIRST CONSERVATION MATERIALS LTD

Laughton, Sleaford, Lincolnshire NG34 0HE
Tel 01529 497517 Fax 01529 497518
E-mail materials@hirst-conservation.com Website www.hirst-conservation.com

CONSERVATION MATERIALS: The company produces lime putty, mortars, plasters, renders, daubs, grouts, limewashes, coatings and paints for historic building repair. A range of other conservation materials is also supplied. All products are supported by technical and practical guidance. Analysis and research services are available enabling historic materials to be closely matched where required. Supplies can be delivered anywhere in Europe. This company is a subsidiary of Hirst Conservation which can provide a team of specialists to apply the materials, giving a single source of responsibility. *See also: display entry on the inside front cover.*

▶ MIKE WYE & ASSOCIATES

Buckland Filleigh Sawmills, Buckland Filleigh, Beaworthy, Devon EX21 5RN
Tel/Fax 01409 281644
E-mail sales@mikewye.co.uk Website www.mikewye.co.uk

LIME PUTTY, LIME MORTARS, PLASTERS AND LIMEWASH: Mike Wye & Associates supplies the finest quality traditional lime products at unbeatable prices, together with a comprehensive range of natural building materials, paints, varnishes and waxes. Visit their website for latest offers, limewash colour charts, guidesheets and practical courses programmes.

MORTARS & RENDERS

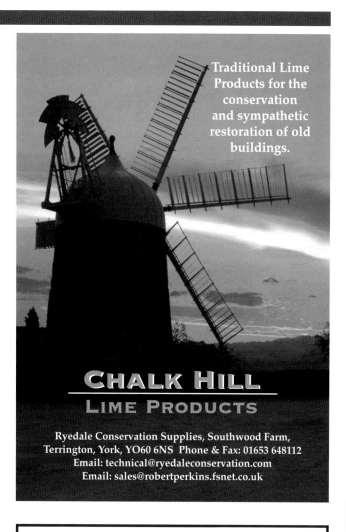

▶ MILESTONE LIME

(subsidiary of Whippletree Hardwoods)
Milestone Farm, Barley Road, Flint Cross, nr Royston, Herts SG8 7QD
Tel 01763 208966

TRADITIONAL LIME AND RELATED PRODUCTS: Suppliers of quality slaked lime putty and hydraulic limes for conservation and restoration of ancient buildings. Also stockists and producers of riven and sawn oak, sweet chestnut, hazel and larch lath. *See also: Whippletree Hardwoods display entry in Timber Suppliers section, page 124.*

▶ ROSE OF JERICHO

Horchester Farm, Holywell, Dorchester DT2 0LL
Tel 01935 83676 Fax 01935 83903
E-mail info@rose-of-jericho.demon.co.uk
Website www.rose-of-jericho.demon.co.uk

MANUFACTURERS OF TRADITIONAL MORTARS AND PAINTS: Rose of Jericho manufactures and supplies various lime putties and a wide range of pozzolanic additives and hydraulic lime. Ready-mixed mortars, plasters and renders are supplied together with lime based conservation materials. A wide range of sands, stone dusts and aggregates are stocked allowing accurate historic, geological and technically appropriate repair or matching mixes. Rose of Jericho manufactures limewashes, distempers, permeable emulsions, flat oils, and eggshells in historic and contemporary colours. Sophisticated mortar analysis uses DTA. *See also: display entry on this page.*

▶ RYEDALE CONSERVATION SUPPLIES

North Back Lane, York YO60 6NS
Tel 01653 648112 Fax 01653 648112
E-mail info@ryedaleconservation.com
Website www.ryedaleconservation.com

SUPPLY OF MORTARS AND RENDERS AND RELATED TRAINING: *See display entry for Chalk Hill Lime Products on this page.*

MORTARS & RENDERS

▶ SARABIAN LIMITED
Sarabian House, Kington St Michael, Chippenham, Wiltshire SN14 6JB
Tel 01249 750113 Fax 01249 750798
E-mail info@sarabian.co.uk
Website www.sarabian.co.uk
BUILT HERITAGE CLEANING AND RESTORATION: *See also: display entry in Masonry Cleaning section, page 168.*

▶ TWYFORD LIME PRODUCTS
1 Twyford Place, Tiverton, Devon EX16 6AP
Tel 01884 255407 Fax 01884 242446
E-mail arhunt@1twyford.fsnet.co.uk
MORTARS AND RENDERS: Manufacturer of lime putty, lime mortars, plasters and lime washes, skills for the repair and maintenance of historic buildings, conservation and repair of cob buildings.

▶ TY-MAWR LIME LTD
Ty-Mawr, Llangasty, Brecon, Powys, Wales LD3 7PJ
Tel 01874 658249 Fax 01874 658502
E-mail tymawr@lime.org.uk
Website www.lime.org.uk
A WELSH CENTRE FOR TRADITIONAL AND ECOLOGICAL BUILDING: Producers and suppliers of ecological and traditional building materials including pre-mixed lime mortars/plasters, hydraulic limes, laths, internal/external natural paints/finishes, sheepswool insulation. Next day deliveries throughout the UK. A full range of introductory and specialist courses are run on lime, plastering, natural paints and building technology as well as product-specific information seminars.

▶ WOMERSLEY'S LIMITED
Walkley Lane, Heckmondwike, West Yorkshire WF16 0PG
Tel 01924 400561 Fax 01924 403489
E-mail markwomersley@aol.com
Website www.womersleys.co.uk
LIME MORTARS AND ECO-FRIENDLY BUILDING MATERIALS

POINTING

▶ BECKWITH TUCKPOINTING
24 The Meadowway, Billericay, Essex CM11 2HL
Tel 01277 659949 Mobile 07958 727208
TUCKPOINTING, BRICK AND STONE RESTORATION: *See also: display entry on page 177.*

▶ GRALL PLASTER & STONE SPECIALISTS
Forest Corner, Purlieu Lane, Godshill, Fordingbridge, Hampshire SP6 2LW
Tel 01425 655430 Mobile 07973 832 846
MORTARS AND RENDERS *See also: display entry in Plasterwork section, page 214.*

▶ RYEDALE CONSERVATION SUPPLIES
North Back Lane, York YO60 6NS
Tel 01653 648112 Fax 01653 648112
E-mail info@ryedaleconservation.com
Website www.ryedaleconservation.com
SUPPLY OF MORTARS AND RENDERS AND RELATED TRAINING: *See also display entry for Chalk Hill Lime Products in Mortars & Renders section, page 175.*

EXTERIOR STUCCO

An introduction to the conservation and repair of forms of stucco commonly used to decorate the external façades of 19th century terraces in UK cities and towns

IAN CONSTANTINIDES and LYNNE HUMPHRIES

The term 'stucco' is of Germanic origin and its use has had many applications since the Middle Ages. These range from a coarse plaster or cement used chiefly for covering rough exterior surface of walls in imitation of stone, to a fine plaster, especially one composed of gypsum and pulverised marble, used for covering walls, ceilings and floors, and for making cornices, mouldings and other decorations. Consequently, the materials used in producing stucco vary considerably. Binders have included fully burnt gypsum; limes, sometimes with pozzolanic additives such as brick powder; cements; and linseed oil (mastic). Aggregates have included marble dust, crushed stone and sand. All have been used in varying mixes and hydraulic strengths with a wide range of colorants and other additives.

In Italy the term has had the widest use historically, generally referring to various soft materials capable of being shaped including mastic- or wax-bound glue for broken statues or lime/egg/brick powder for mosaic.[1]

In England, stucco was often defined by the technique used:

Common stucco an exterior render prepared from hydraulic lime, sand and hair

Rough stucco a fine plaster of sand and lime made from chalk or a very pure limestone which was used internally to imitate stonework

Bastard stucco a superior render prepared from fat lime putty (that is to say, non-hydraulic) and fine washed sand which was applied to a good backing coat, scoured and either polished or left floated

Trowelled stucco a fat lime render applied as a normal finishing coat, scoured, polished and painted.

In this article stucco is used to describe lime-based renders applied as a two- or three-part coating to external facades in the 19th century.

DESIGN AND USE OF STUCCO HISTORICALLY

The use of stucco or smooth render to simulate finely dressed stonework or rustication became popular in parts of Britain in the early 19th century. The material was often applied over brickwork but also sometimes over rubble stone. Not only was the appearance of finely jointed work or rustication achievable in stucco, it was also far more affordable than stone in many parts of the country.

Late 19th century stucco terraces in Holland Park, London

External stucco had been introduced into London in the later 18th century and was increasingly used to satisfy the Regency and early Victorian taste for smooth, evenly coloured house fronts, its cost amounting to about one quarter that of stone. Mid-Victorian fashions, however, as well as the fall in the price of stone, helped to phase out stucco very quickly after 1860. Later in the century, terracotta came into its own as a cheap and durable material for applied decoration and aggrandisement.

In the early Victorian period stucco was used in a variety of developments ranging from the highly prestigious to the less expensive. John Nash used stucco extensively in the early 19th century for his terraces in London, Brighton, Hastings, Southsea and Torquay as well as his Gothic and Italianate villas in Malvern, Leamington and Harrogate. Perhaps the finest example was his development of Regents Park, arguably his greatest work.

As the popularity for the material spread, highly elaborate stucco faced terraces and villas came to dominate the centres of several key towns and seaside resorts in England and Wales. It was generally used to cover the whole façade but not the sides and back. Fine examples of this can be seen on the Holland Park Estate in West London, built by William and Francis Radford between 1860 and 1879.

Although stucco remained popular in London for more conservatively designed houses until the 1870s, by the mid-Victorian period this form of embellishment was losing favour with many builders in London as the principles of Ruskin and Pugin filtered through to them and the Gothic Revival took hold, although it continued to remain popular in other parts of the country, particularly in seaside resorts, probably because stucco provides an excellent defence against salt-laden spray.

Stucco always remained a very regional material as it was rarely used if good stone was readily available, as in Bristol and Bath. Examples of stucco in Scotland are rare.

CONSTRUCTION

Stucco renders are of three basic types: a fat lime and sand mix sometimes with animal hair as reinforcement; a hydraulic mix containing either hydraulic lime and sand, or fat lime with a pozzolanic additive and sand; and various forms of mastic. Correct identification of the material used is essential when carrying out repairs to ensure both historical continuity and structural compatibility, as a different mortar mix may well be incompatible with the original.

LIME-BASED STUCCO

Generally, a pure or fat lime would have been prepared by slaking quicklime (calcium oxide) made from a pure limestone or chalk. Dry hydrated lime is sometimes used today for repair work, although mature fat lime is generally preferred.

A faster set could be achieved by using either a hydraulic lime or by adding a pozzolanic additive such as brick dust to a fat lime (see *Hydraulicity Explained* by Paul Livesey on page 170). Roman cement, used in stucco from the 1790s, is one form of hydraulic lime which was made from an argillaceous limestone (septarian nodules) and has a distinctive pinky-brown colour.

The sand and other aggregates used in a repair should match the existing in particle size, colour and type. For new stucco, washed and graded pit sand complying with the British Standard should be used. This should be hard, sharp, gritty and free from clay and organic impurities.

Hair should be long, strong, and free of dirt and grease, from the horse or ox and blended into coarse stuff in the ratio of 3–5kg per cubic metre of coarse stuff. It should be well teased into the mix, evenly distributed without tufts.

Lime plasters perform best in layers of uniform thickness. They are applied in two or three coats. The mix should be as dry as workably possible as this reduces the shrinkage and cracking on drying and, prior to the application of each coat, the surface should be sprayed down with clean water. This wetting helps to prevent moisture from being sucked out of each stucco application too rapidly, which results in cracking, loss of bond, and generally poor quality stuccowork.

The success of external lime-based stucco is in the tending. The longer it takes to dry out the better it will perform. Each coat of a fat lime stucco should be allowed to dry for between seven and 21 days, depending on the mix, season, weather and temperature, prior to applying subsequent coats. For a hydraulic lime mix the interval between coats can be as little as two or three days. It is most important to protect the work during and after application to prevent either accelerated or prolonged drying. In hot weather or situations where rapid drying is likely, the work should be protected with damp hessian. In addition, work must be completed long before the first frost.

MASTIC

In the late 18th and early 19th centuries different mastic recipes were patented by a succession of people trying to produce a high quality but inexpensive stucco. These included 'Adam's New Invented Patent Stucco' used by the Adam brothers, Robert and James (in fact based on stucco recipes patented by David Work in 1763 and Liadet in 1773), Christopher Dehl's mastic (1815), and Hamelin's Cement (1817).

The various forms of mastic generally consisted of a fine aggregate such as limestone, sands, crushed pottery and glass bound with linseed oil, often with litharge (lead monoxide) to aid drying. Dehl's mastic, for example, which is believed to have been used by Nash at Regent's Park and Carlton House Terrace, London, was made of 'linseed oil boiled with litharge and mixed with [fired] porcelain clay, finely powdered and coloured with ground brick or pottery, turpentine being used as the thinner'[2]. The background was liberally coated with linseed oil before applying the mastic.

Mastic can be recognised by the fact that it repels water. The material did not age well and tends to be very brittle. Where large areas have failed, John Ashurst recommends using a stucco made from hydraulic lime as an alternative to mastic. However, it can generally be said that repair of mastic stuccos is fraught with problems. On the whole the principal of 'like for like' may be followed.

ASHLAR EFFECTS AND FINISHES

The topcoat of stucco was often given a smooth, trowelled finish, and scored or lined in imitation of ashlar. This effect could be achieved in three ways: joint lines could be marked on while the top coat is still green using a tool called a jointer; the joints may be formed by sunken, slightly chamfered battens fixed to the second coat and then removed after the top coat stucco has set firm; or they may be run by a double horsed running mould. Special care must be taken to match to the existing work in position, spacing and style.

Generally, traditional renders of any period were painted, with the exception of some early experimental stucco recipes which were self-coloured, sometimes enhanced with a wash of copperas (iron sulphate) to give the

A detail of Nash's exuberant stucco at Carlton House Terrace (John Nash, 1827–32)

A stucco cornice at the Mansion House, Stowe, Buckinghamshire, in gauged lime, stone dust and hair on lath and wooden bracketing with applied cast enrichments

Forming the window entablature mitre in new lime stucco and, below, the finished result, at Wakefield Lodge, Northamptonshire

appearance of Bath stone, and the illusion of masonry joints was often created with a thin line of white lime putty, graphite, or some other pigment[3]. However, sooner or later painting became necessary to unify the finish, particularly as a result of patch repairs, and after these early self-coloured experiments, stucco was invariably painted in limewash or lead paint from the outset, often coloured with earth pigments in stone colours. If using pigments today, it is important to ensure they are compatible with lime.

COMMON MISTAKES AND CAUSES OF FAILURE

Arguably the most common reason for the failure of stucco is neglect. While stucco is a durable material, regular maintenance is required to prevent excessive water penetration and a breakdown of the surface. Failure of rainwater disposal systems, such as from blocked or damaged gutters is one of the most common causes of failure, which often leads to salt efflorescence, staining and biological growth. Excessive water penetration is liable to cause loss of adhesion either between the stucco coats or from the substrate, causing bulging and, ultimately, collapse. Trapped water may also freeze and expand causing dramatic failure, or it may move into the structural fabric of the building, possibly damaging the interior. Increased wetting and drying cycles contribute towards soluble salt activity as salts contained in the masonry are carried to the surface. Here they crystallise, resulting in an unsightly bloom of efflorescence or, worse still, a crumbling surface caused by 'cryptoflorescence' – the crystallisation of salts within the pores. Visible damp zones are frequently the first sign of a problem. Regular checks and maintenance of gutters, hoppers and downpipes are easily carried out and can prevent extensive failure and loss of original stucco surfaces.

All too often deterioration is caused by inappropriate repairs. The use of the wrong materials can exacerbate the rate of decay rather than slowing it down or preventing decay. Typical of this sort of problem is the use of hard cementitious materials incompatible with the stucco or lime based render. Cement based renders are more likely to crack than lime renders and consequently to let in the rain, as they move differentially to lime renders. Cracks will also form at the junction between the original and the modern repair.

Raising the ground level at the foot of elevations increases water retention within elevations and encourages rising damp. Prolonged moisture retention encourages biological activity such as moss growth which may cause local disaggregation. Ultimately larger plants may take hold, bringing with them further problems such as root penetration.

Surprisingly, modern synthetic paint systems which form impermeable barriers are still frequently applied over previously limewashed stucco, causing retention of water and associated problems as outlined above.[3] Modern impervious systems trap water behind the surface causing deterioration, which outweighs any possible advantage of imparting water repellence to the face.

REPAIRS

As with all conservation work, it is most important to record the area to be conserved prior to commencement as well as during work.

One of the first tasks may be to remove inappropriate materials and methods of repair and replace with the appropriate stucco/render. However, removal should only be carried out if this does not put greater risk on the original fabric. It may also be necessary to cut out defective areas of original that cannot be saved.

When conserving fine or delicate decorative details it may be advantageous to face-up original material around the perimeter of earlier repairs with acid-free tissue prior to removal of crude repairs to prevent any loss of detail.

Cracks greater than 2mm in a lime-based stucco should be carefully cut out to form a slight undercut which will act as a key, and thoroughly flushed out with water to remove dust and loose debris before being filled with fresh mortar based on trial results. Obviously a finer aggregate will be required where the crack is fine or hairline and it is often deemed unnecessary to undercut as the space is easily filled especially if limewash is to be applied.

Hollow areas and voids in a lime-based stucco should be flushed out likewise, although in this case it may be necessary to form a small hole at the base of the void to allow water to escape. Acetone may be used or added to the water to assist drying. A ten per cent solution of Primal WS24 may be injected into the void prior to grouting in order to increase the bond between grout and internal face, before injecting a fine grout based on lime putty or, in the case of a mastic, a similar mix based on analysis results. It is very important to observe the surface of the stucco while grouting to check for escape holes, surface bulging and consequent loss. Vulnerable areas should be supported until the grout has set. Finally the surface of the stucco is reinstated to its original profile, where possible without causing loss.

Salt efflorescence may be dry brushed and removed from all surfaces, as should all algal growth. A suitable biocide should be applied to affected areas only, to remove remaining algae and prevent re-growth.

Friable areas of a lime-based stucco may be consolidated with repeated applications of limewater. To avoid a white bloom it is most important not to let the limewater sit on the surface but to sponge it off with clean water.

Substituting modern materials for the original should always be avoided if at all possible. Wherever a high proportion of original stucco has survived a hundred years or more in the British climate, bear in mind that the original has been proved to work. This historic material, produced by craftsmen long ago, has its own intrinsic value like any antique and, with careful consolidation, suitable repairs and thorough maintenance, it should be possible to ensure that the original stucco work can still be seen by future generations.

Recommended Reading

Ashurst, John, *Mortars, Plasters and Renders in Conservation.* Ecclesiastical Architects' and Surveyors' Association, 1983

Beard, G, *Stucco and Decorative Plastering in Europe.* Thames and Hudson 1967

Grimmer, Anne, *The Preservation and Repair of Historic Stucco.* Preservation Briefs, Technical Preservation Services, National Park Service, Washington DC

Koller, Manfred; Paschinger, Hubert; Richard, Helmut (Author); Bromelle, NS; Smith, Perry (Editor); *Work in Austria on historic stucco – technique, colouring, preservation*, Case Studies in the Conservation of Stone and Wall Paintings, Preprints of the Contributions to the Bologna Congress, 21-26 September 1986

Millar, William, *Plastering, plain and decorative.* Batsford, 1899; Reprint, Donhead, Shaftesbury 1998

Simpson and Brown (Corporate Author), *Conservation of Plasterwork: a guide to the principles of conserving and repairing historic plasterwork*, Historic Scotland Technical Advice Note 2, Edinburgh, Historic Scotland 1994

Stagg, WD, Masters, RA, *Decorative Plasterwork: Its Repair and Restoration.* Second Edition, Attic Books 1986

IAN CONSTANTINIDES is the the managing director of St Blaise Ltd (see page 64) and has worked in every aspect of historic building repair for over 20 years. Tel 01935 83662 E-mail info@stblaise.co.uk

LYNNE HUMPHRIES MA(RCA/V&A) is a conservator of both architecture and sculpture. She has worked and studied in both museums and historic buildings and formerly managed conservation for St Blaise Ltd. E-mail lynnehumphries@hotmail.com

[1] Koller, Manfred; Paschinger, Hubert; Richard, Helmut (Author); *'Work in Austria on historic Stucco – technique, colouring, preservation',* Case Studies in the Conservation of Stone and Wall Paintings, Preprints of the Contributions to the Bologna Congress, 21–26 September 1986, Bromelle, NS; Smith, Perry (Editor)

[2] Davey, Norman, *A History of Building Materials.* Phoenix House, London 1961

[3] Grimmer, Anne, *'The Preservation and Repair of Historic Stucco', Preservation Briefs,* Technical Preservation Services, National Park Service

DAMP & TIMBER DECAY

▶ ADVANCED CHEMICAL SPECIALTIES LTD

9 Bofors Park, Artillery Road, Yeovil, Somerset BA22 8YH
Tel 01935 414012 Fax 01935 414022
E-mail billbeauford@ukonline.co.uk
Website www.acslimited.co.uk

CONSERVATION AND PROTECTION PRODUCTS FOR WOOD AND MASONRY: Suppliers of specialist conservation products including boron rods, a solid wood preservative for use in all timbers. Strategically inserted into damp timber they dissolve and distribute preservative in high risk areas. The Boracols are boron-based preservatives, which penetrate deeply into wet and dry wood of all types. ACS has considerable experience in façade treatments, offering a comprehensive range of water repellents for most substrates, including alkaline surfaces, and anti-graffiti coatings. ACS is a leading specialist for the Wacker stone consolidation system, which is used to preserve masonry and architecture.

▶ ALL TIMBER INFESTATION & CONSULTANCY SERVICES LTD

15 Baskerville Road, Wandsworth, London SW18 3RJ
Tel 020 8874 2013 Fax 020 8874 3021
E-mail mail@atics.co.uk
Website www.atics.co.uk
Contact CJD George MA PhD (biochemistry of decay) MSc (distinction, timber engineering) DIC AIWSc CSRT (credit)

DAMP AND TIMBER DECAY: Sound, practical, technically competent advice is provided on timber decay in buildings. Money is saved or spent efficiently. The managing director is uniquely qualified academically, and by experience, both as site operative and surveyor.

▶ BALMORAL STONE LTD

Unit 1C, West Craigs Industrial Estate,
Turnhouse Road, Edinburgh EH12 8NR
Tel 0131 339 5920 Fax 0131 317 9707

WOODWORM AND DRY ROT: *See also: display entry in Stone section, page 101.*

▶ DELTA MEMBRANE SYSTEMS LTD

Bassett Business Centre, Hurricane Way, North Weald, Essex CM16 6AA
Tel 08707 472181 Fax 08717 178060
E-mail info@deltamembranes.com
Website www.deltamembranes.com

DAMP TREATMENT: Delta Membrane Systems Ltd provides a range of products which are designed to provide lasting and cost effective solutions for damp contaminated and degraded buildings. From basements to vaults to floors and walls, Delta systems can completely transform almost any damp area. Delta has been used on many heritage type structures including historical churches, the Victoria Tower at the Houses of Parliament and Kensington Palace to name but a few. The systems are British Board of Agrément approved and also come with peace of mind in the form of a 30 year guarantee. A free site visit and assessment is available by contacting Delta on 08707 472181, alternatively visit the company website, or request a brochure.

▶ DEMAUS BUILDING DIAGNOSTICS LTD

Stagbatch Farm, Leominster, Herefordshire HR6 9DA
Tel 01568 615662
E-mail info@demaus.co.uk
Website www.demaus.co.uk

STRUCTURAL TIMBER TESTING AND BUILDING DIAGNOSTICS: Demaus Building Diagnostics specialises in the detection and assessment of decay, weakness and fire damage in structural timber using non-destructive techniques. *See also: profile entry in Non-destructive Investigations section, page 40.*

▶ HUTTON+ROSTRON ENVIRONMENTAL INVESTIGATIONS LIMITED

Netley House, Gomshall, Surrey GU5 9QA
Tel 01483 203221 Fax 01483 202911
E-mail ei@handr.co.uk
Website www.handr.co.uk
Contact Tim Hutton MA MSc MRCVS

CONSULTANTS ON TIMBER DECAY, BUILDING FAILURES AND ENVIRONMENTS: Simple solutions to common problems with expertise covering biodeterioration, insect attack, timber strength grading, damp, environmental health, non-destructive surveying, building monitoring systems and historic building consultancy. H+R carry out independent site and laboratory investigations providing specifications for remedial work or conservation. Expert witness work is also undertaken. They operate the Rothound® dry rot search dogs and install Curator® electronic moisture and structural monitoring systems. Resurgam®, a division of H+R, specialises in building conservation. Clients include The Royal Household, National Trust, English Heritage, national and local government, engineers, surveyors and property owners. *See also: Resurgam profile entry in Heritage Consultants section, page 35.*

▶ MAYSAND PRESERVATION CO LTD

109–111 Windsor Road, Oldham, Lancs OL8 1RH
Tel 0161 628 8888 Fax 0161 627 0996
E-mail sales@maysand.co.uk

TIMBER CONSERVATION CONTRACTORS: Maysand's preservation team includes highly skilled technical staff to offer solutions required to preserve and conserve the fabric of buildings. These include preventative treatment against dampness, fungal and insect infestations within buildings. Other services include chemical damp-proofing, underground waterproofing tanking systems, timber engineering, specialist timber preserving treatments and building contracting services for historic, ecclesiastical and commercial projects. Members of BWPDC. *See also: display entry in Stone section, page 106.*

▶ RENLON LTD

Richardson House, Boundary Business Court, Church Road, Mitcham, Surrey CR4 3TD
Tel 020 8687 4000 Fax 020 8687 4040
E-mail survey@renlon.com
Website www.renlon.com

DAMP AND TIMBER DECAY PROTECTION: For nearly 25 years Renlon has been providing a full range of services to protect buildings, including basement waterproofing. A qualified and experienced team of surveyors works together with the client's professional team to diagnose the cause of problems and design proposals to suit the property. The methods proposed offer the least disturbance to historic fabric and appearance with the minimum use of chemicals. Many of the South East's oldest and most valuable buildings are protected by Renlon guarantees. Insurance backing of their guarantee is available.

▶ RIDOUT ASSOCIATES

147a Worcester Road, Hagley, Stourbridge, West Midlands DY9 0NW
Tel 01562 885135 Fax 01562 885312
E-mail ridout-associates@lineone.net
Website www.ridoutassociates.co.uk

DAMP CONSULTANTS: Ridout Associates are international specialists in the scientific assessment of timber decay and other damp related problems in buildings who avoid the expensive damage caused by unnecessary or incautious remedial treatments, and reduce the quantities of hazardous chemicals introduced into the environment. The company also manufactures and markets moisture monitoring systems and devices to trap wood boring beetles. Clients include The Royal Household, National Trust and UNESCO. English Heritage and Historic Scotland commissioned Dr Ridout's recent book *Timber Decay in Buildings, The Conservation Approach to Treatment* which won an international award for 'the most outstanding book publication in the field of preservation technology during 1997-2000'.

▶ TERMINIX PROPERTY SERVICES

Heritage House, 234 High Street, Sutton, Surrey SM1 1NX
Tel 020 8661 6600 Fax 020 8642 0677 Branches 0800 789500

SPECIALISTS IN SOLVING DAMP AND TIMBER DECAY PROBLEMS: Founded over 50 years ago Terminix Property Services (formerly Peter Cox) has extensive experience in the repair of historic and listed buildings specialising in damp proofing, waterproofing, treatments for woodworm and fungal decay, epoxy resin repairs, wall stabilisation and condensation control. Specialist insurance to cover these problems is also available. Its unique Transfusion system for installing a remedial dpc has been proven in use in countless properties. A nation-wide service is provided and most work carries a long term guarantee. A member of the BWPDA, the company is also registered under ISO 9002.

▶ WARD & DALE SMITH, CHARTERED BUILDING SURVEYORS

The Walker Hall, Market Square, Evesham, Worcs WR11 4RW
Tel 01386 446623 Fax 01386 48215
E-mail wds@ricsonline.org
Contact Peter Rhodes FRICS DipBldgCons

INDEPENDENT AND INTEGRATED ADVICE FROM HISTORIC BUILDING CONSULTANTS. Identification and resolution of timber and damp related problems using measures that seek to minimize disturbance to buildings. Services include surveys, advice on causes/alternative approaches, specifications, obtaining tenders, contract administration etc. Expert witness work.

EPOXY RESIN REPAIRS

▶ ROBINSONS PRESERVATION LIMITED

38 Kansas Avenue, Salford M50 2GL
Tel 0161 872 3133 Fax 0161 872 6167
STRUCTURAL TIMBER REPAIRS: *See also: display entry in Damp & Timber Decay section, page 182.*

▶ ROTAFIX

Rotafix House, Abercraf, Swansea SA9 1UX
Tel 01639 730481 Fax 01639 730858
E-mail rotafixltd@aol.com
Website www.rotafix.co.uk
TIMBER ENGINEERING USING THE RESIWOOD™ SYSTEM: Rotafix is a leading European organisation in the development, formulation and manufacture of polymer systems for use in the repair, restoration and upgrading of ancient and modern timber structures – water mills, windmills, wind turbines, castles, cottages, museums, bridges, barns and boats. Contact Rotafix for literature and invaluable information on structural timber repair systems. Rotafix provides two one-day certificated training courses. The first covers the *Principles of Timber Engineering* with particular reference to restoration and repair and the second covers *Repair and Restoration of Brickwork, Stone, Masonry and Concrete.*

▶ TERMINIX PROPERTY SERVICES

Heritage House, 234 High Street, Sutton, Surrey SM1 1NX
Tel 020 8661 6600 Fax 020 8642 0677 Branches 0800 789500
EPOXY RESIN REPAIRS: *See also: profile entry in Damp & Timber Decay section, page 182.*

STRUCTURAL TIMBER TESTING

▶ DEMAUS BUILDING DIAGNOSTICS LTD

Stagbatch Farm, Leominster, Herefordshire HR6 9DA
Tel 01568 615662
E-mail info@demaus.co.uk
Website www.demaus.co.uk
STRUCTURAL TIMBER TESTING AND BUILDING DIAGNOSTICS: Demaus Building Diagnostics specialises in the detection and assessment of decay, weakness and fire damage in structural timber using non-destructive techniques. *See also: profile entry in Non-destructive Investigations section, page 40.*

▶ RIDOUT ASSOCIATES

147a Worcester Road, Hagley, Stourbridge, West Midlands DY9 0NW
Tel 01562 885135 Fax 01562 885312
E-mail ridout-associates@lineone.net
Website www.ridoutassociates.co.uk
NON-DESTRUCTIVE STRUCTURAL TIMBER SURVEYS: *See also: profile entry in Damp & Timber Decay section, page 182.*

▶ SIBTEC SCIENTIFIC, SIBERT TECHNOLOGY LTD

2a Merrow Business Centre, Merrow Lane, Guildford, Surrey GU4 7WA
Tel 01483 440724 Fax 01483 440727
E-mail ndt@sibtec.com
Website www.sibtec.com
MANUFACTURERS AND SUPPLIERS OF DECAY DETECTION EQUIPMENT

EXHUMATION OF HUMAN REMAINS

▶ CHERISHED LAND LIMITED

29 High Street, Crawley, West Sussex RH10 1BQ
Tel 01306 627321 Fax 01306 627321
E-mail information@cherishedland.com
Website www.cherishedland.com
EXHUMATION SPECIALISTS: Cherished Land provides a comprehensive exhumation service for professionals contemplating the disturbance of human remains, advising on every aspect of this complex and emotive subject including public relations, legal, ecclesiastical and health requirements, budget costings, site clearance and final disposition. Evaluation and advice provided without charge or obligation.

INSECT ERADICATION

BIRD CONTROL

▶ BALMORAL STONE LTD

Unit 1C, West Craigs Industrial Estate, Turnhouse Road, Edinburgh EH12 8NR
Tel 0131 339 5920 Fax 0131 317 9707
BIRD CONTROL: *See also: display entry in Stone section, page 101.*

▶ MICROBEE BIRD CONTROL LIMITED

Unit 1, Windsor Park, 50 Windsor Avenue, London SW19 2TJ
Tel 020 8540 9968 Fax 020 8540 7477
E-mail microbee.co.uk
Website www.microbee.co.uk
PEST BIRD CONTROL: Established in 1984 the company develops, manufactures and installs a wide range of devices to exclude and prevent feral pigeons from landing on buildings. These include Micropoint anti-perching pins, and Microwire stainless steel anti-perching wires as well as anti-pigeon enclosure netting to protect light wells. These products can be seen (with binoculars) on Canada House, Trafalgar Square, The Royal Opera House and thousands of other locations ranging from Royal Palaces to rubbish dumps. The company also fits exclusion netting to protect light wells and can sterilise and remove pigeon fouling.

ENVIRONMENTAL MONITORING

▶ RIDOUT ASSOCIATES

147a Worcester Road, Hagley, Stourbridge, West Midlands DY9 0NW
Tel 01562 885135 Fax 01562 885312
E-mail ridout-associates@lineone.net
Website www.ridoutassociates.co.uk
ENVIRONMENTAL MONITORING: Consultants on environmental monitoring and monitoring of dampness in buildings, including timber and masonry. Supply and installation of moisture monitoring systems, and data interpretation. *See also: profile entry in Damp & Timber Decay section, page 182.*

STRUCTURAL METAL TIES & FIXINGS

charterbuild
helifix
helibeam
cintec
grundtuben
duckbill
easipoint
hidden strengths...
an extensive range of sympathetic structural repair techniques for restoring integrity and strengthening listed or historic structures including brick/stone repair and reinforcement, trying and pinning systems, resin repair, grouting and repointing
to request our brochure, enquire online, or call for advice or to arrange a no-obligation meeting;
contact: peter christian
tel: 01245 425050
fax: 01245 325443
email: peter@charterbuild.co.uk
web: www.charterbuild.co.uk
address: 7 starboard view south woodham ferrers chelmsford CM3 5GR
restoring integrity

▶ CHARTERBUILD LTD

7 Starboard View, South Woodham Ferrers, Chelmsford CM3 5GR
Tel 01245 425050 Fax 01245 325443
E-mail peter@charterbuild.co.uk
Website www.charterbuild.co.uk

STRUCTURAL REPAIR SPECIALISTS: Charterbuild is a specialist remedial contractor with extensive experience in the sympathetic repair of failed or damaged masonry (brick and stone) and the restoration of structural integrity. Subsidence or fire damage, cracking, failure under load, bowed walls, delaminating masonry and the absence of structural tying are some of the more common problems which can be addressed by the use of a number of different specialist systems for which Charterbuild is an approved installer. *See also: display entry on this page.*

▶ HELIFIX LIMITED

21 Warple Way, London W3 0RX
Tel 020 8735 5222 Fax 020 8735 5223
E-mail info@helifix.co.uk
Website www.helifix.co.uk

STRUCTURAL REPAIR AND STABILISATION: For all remedial applications, Helifix has developed a comprehensive, cost-effective range of high performance, grade 304 and 316 stainless steel ties, fixings and non-disruptive masonry repair and reinforcement techniques. They are all independently tested, fully proven, widely used, manufactured to ISO9002 quality assured standards and supported by full technical data. These specially engineered stress free products are fully concealed for a sympathetic repair and provide a secure, reliable connection in bricks, blocks, stone, concrete and timber in buildings and masonry structures of all types and ages. Helifix provides a fully technical support service with advice, site visits, design specifications and quality installation by their national network of trained approved installers.

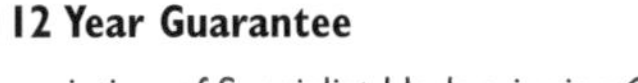

GENERAL FIXINGS & FASTENERS

▶ AVON STAINLESS FASTENERS

Avondale Business Centre, Woodland Way, Kingswood, Bristol BS15 1AW
Tel 0117 960 6665 Fax 0117 960 6668
E-mail sales@asfast.sagehost.co.uk

STAINLESS FASTENINGS: Full range of stainless steel fasteners (nuts, bolts, screws, woodscrews, coach screws, coach bolts, studding, plate washers, nails, tying wire, plain bar etc) for the timber building and conservation industries. Product guide available on request.

NAILS

▶ GLASGOW STEEL NAIL CO LTD

Lowmoss, Bishopbriggs, Glasgow G64 2HX
Tel 0141 762 3355 Fax 0141 762 0914
E-mail glasgowsteelnail@compuserve.com
Website www.glasgowsteelnail.com

MANUFACTURERS OF TRADITIONAL CUT NAILS IN MILD STEEL OR WROUGHT IRON

SECURITY FIXINGS

▶ EURA CONSERVATION LTD

Unit H10, Halesfield 19, Telford, Shropshire TF7 4QT
Tel 01952 680218 Fax 01952 585044
E-mail mail@eura.co.uk

SECURITY FIXINGS: *See also: display entry in Statuary section, page 94 and profile entry in Architectural Metalwork section, page 142.*

PAINTS & DECORATIVE FINISHES

Abbey Heritage Ltd	pd	165
Anglia Lime Company	lw	173
Advanced Chemical Specialties Ltd	pd	181
Bursledon Brickworks Conservation Centre	lw pg	172
Chalk Down Lime Ltd	lw	172
Classidur	pd	185
The Cornish Lime Company Ltd	lw pg	174
Craig & Rose	pd	186
Farrow & Ball	pd pg	186
HJ Chard & Sons	lw pg	174
Hirst Conservation Materials Ltd	lw pd pg	174
Keim Mineral Paints Ltd	pd	186
The Lime Centre	lw pg	174
Mike Wye & Associates	lw pd pg	174
Plastercraft	lw	216
Rose of Jericho	lw pd pg	175
Ryedale Conservation Supplies	pd	175
TheTraditional Lime Co	lw pg	176
Twyford Lime Products	lw	176
Ty-Mawr Lime Ltd	lw pd pg	186
Wenlock Lime Ltd	lw pg	176
Womersley's Limited	lw pg	176

KEY
lw limewash
pd paints and decorative finishes – general
pg pigments
See also Paint Analysis, page 200

PAINTS & FINISHES

▶ FARROW & BALL
Showroom, 249 Fulham Road, London
Tel 01202 876141 Fax 01202 873793
MANUFACTURERS OF TRADITIONAL PAPERS AND PAINT:
See also: display entry on page 186.

▶ HIRST CONSERVATION MATERIALS LTD
Laughton, Sleaford, Lincolnshire NG34 0HE
Tel 01529 497517 Fax 01529 497518
PAINTS AND RELATED PRODUCTS: *See also: display entry on the inside front cover and profile entry in Mortars and Renders section, page 174.*

▶ LISA OESTREICHER
Jubilee House, High Street, Tisbury, Wiltshire SP3 6HA
Tel 01747 871717 Fax 01747 871718
E-mail lisa.oestreicher@virgin.net
ARCHITECTURAL PAINT INVESTIGATIONS: Lisa Oestreicher provides a full range of analytical skills and techniques for the study of paint and decorative finishes within historic buildings. These include the identification of pigments and media as well as archival research. Full reports are prepared to provide a detailed insight into the historical development of interior and exterior schemes, for documentation purposes, conservation or accurate restoration. Assistance can also be given in the design and implementation of historically informed interior decoration schemes. Recent clients include the National Trust, Victoria & Albert Museum, architects, conservators and owners.

PAINTS & FINISHES

classidur®

Designed to be a real pain for stains

- LOW ODOUR
- ZERO TENSION
- WATER OR OIL BASED
- MATT OR SATIN FINISH
- PRIMER AND TOP COAT IN ONE
- HIGH VAPOUR PERMEABILITY
- EXCELLENT STAIN COVERING ABILITY

A primer and top coat in one, Classidur can be painted straight over stains in a wide variety of locations.

With amazing covering ability, Classidur is bad news for most stains including water damage, nicotine, grease and even fire damage.

For your FREE Classidur brochure and details of your nearest stockist, call today

(01275) 854911

solely imported by

Blackfriar Paints Limited
Blackfriar Road, Nailsea, Bristol BS48 4DJ
www.blackfriar.co.uk

▶ **RYEDALE CONSERVATION SUPPLIES**
North Back Lane, York YO60 6NS
Tel 01653 648112 Fax 01653 648112
E-mail info@ryedaleconservation.com
Website www.ryedaleconservation.com
SUPPLY OF MORTARS AND RENDERS AND RELATED TRAINING: *See display entry in Mortars & Renders section, page 175.*

▶ **TY-MAWR LIME LTD**
Ty-Mawr, Llangasty, Brecon, Powys, Wales LD3 7PJ
Tel 01874 658249 Fax 01874 658502
E-mail tymawr@lime.org.uk
Website www.lime.org.uk
A WELSH CENTRE FOR TRADITIONAL AND ECOLOGICAL BUILDING: *See also: profile entry in Mortars & Renders section, page 176*

HEAT & LIGHT	Products	Services	Page
The Cast Iron Company	li lf		143
Chelsom Limited	li	lc	190
Cico Chimney Linings Ltd		ch	188
Drummond's Architectural Antiques Limited	ra		155
FaberMaunsell Limited		lc sv	190
Gifford and Partners		sv	39
Hart Brothers Engineering Ltd		sv	145
Illumin Glass Studio	li		138
Julian Harrap Architects		sv	26
LASSCO	li		155
Light & Design Associates		lc	190
Lightwright Associates		ec lc sv	190
Mather & Smith Ltd	lf		146
Minchinhampton Architectural Salvage	ra		155
National Association of Chimney Engineers		ch	188
Norman & Dawbarn		sv	28
Original Features (Restorations) Ltd	ra		207
R Hamilton & Company Limited	li lf	lc sv	190
Robert Bloxham-Jones Associates		sv	190
Specialist Flue Service Ltd		ch	188
Walcot Reclamation Ltd	li ra		157

KEY

ch	chimney linings	lc	lighting consultants
ht	heating engineers	ra	radiators and stoves
li	light fittings: antique and decorative	sv	services engineers
lf	light fittings: display lighting		

FIRE PROTECTION & SECURITY	Products	Services	Page
A C Wallbridge & Co Ltd		lp	84
Advanced Chemical Specialties Ltd	fb		181
Applied Surveying & Design Group		fs	44
Carrek Limited		lp	61
Church Conservation Limited		lp	84
Donald Insall Associates Ltd		fs	24
Eura Conservation Ltd		se	142
FaberMaunsell Limited		fs	190
G&S Steeplejacks Ltd		lp	84
Gibbon, Lawson, McKee Ltd		fs	188
Gifford and Partners		fs se	39
International Fire Consultants Ltd		fp fs	188
Keim Mineral Paints Ltd	fb		186
Lightwright Associates		fs se	190
Mid Beds Locksmiths	se		140
Quality Lock Co	se		140
Specialist Flue Service Ltd	fb		188

KEY

fb	fire resistant coatings	lp	lightning protection
fp	fire protection systems	se	security products and services
fs	fire safety consultants		

WOOD & MASONRY PROTECTION

▶ ADVANCED CHEMICAL SPECIALTIES LTD

9 Bofors Park, Artillery Road, Yeovil, Somerset BA22 8YH
Tel 01935 414012 Fax 01935 414022
E-mail billbeauford@ukonline.co.uk Website www.acslimited.co.uk

CONSERVATION AND PROTECTION PRODUCTS FOR WOOD AND MASONRY: *See also: profile entry in Damp & Timber Decay section, page 181.*

INSURANCE

▶ LA PLAYA

The Stables, Manor Farm, Milton Road, Impington, Cambridge CB4 9NF
Tel 01223 522411 Fax 01223 237942
E-mail property@laplaya.co.uk
Website www.laplaya.co.uk
Contact Matthew Mullee

SPECIALIST INSURANCE SERVICES: Award-winning broker La Playa offers specialist insurance services and advice for period and listed properties, and for "non-standard" household contents including art, antiques, jewellery and ceramics. First class claims handling, highly personal service. Contact Private Client Director Matthew Mullee for advice or a second opinion on specialist policies, risk management/ security issues; insurance during extension, conversion and repair; use of craftsmen and specialist repairers/suppliers; liabilities for domestic staff and public, valuations for art and antiques. Commercial insurance also available.

INSURANCE

R·|·B·|·P·|·M
Scheme & Specialist Insurance

INSURANCE PROBLEMS? WE CAN HELP!

Listed and non-standard construction premises, archaeologists, all categories of commercial and personal insurances placed. With years of experience in placing all types of insurance, specialist brokers RBPM General Ltd have a solution for all your insurance requirements. Phone, fax or email your query to us, and we will do our best to assist you.

Volpoint House, Blakey Road, Salisbury SP1 2JG
Phone: (44) 1722 336998
Fax: (44) 1722 336407
Email: enquiries@rbpm.co.uk

MEMBER
General Insurance
STANDARDS COUNCIL

▶ LPOC INSURANCE SERVICES

45/47 High Street, Old Town, Hemel Hempstead HP1 3AF
Tel 0845 844 9300 Fax 01442 251568
E-mail lpocinsurance@ghbc.co.uk
Website www.lpocinsurance.co.uk
Contact Tracey Warren/Liz Allen

SPECIALISTS IN INSURANCE FOR LISTED AND PERIOD PROPERTIES: LPOC Insurance Services provides professional, sympathetic and individual advice for the insurance of listed and period properties as well as for contents, fine arts, jewellery and antiques. Specialist insurers are used, in many cases using appraisers, so that the risk of buildings underinsurance is removed. LPOC are the appointed insurance adviser to the Listed Property Owners Club. LPOC Insurance Services are a trading division of GHBC Ltd, who are members of the General Insurance Standards Council.

▶ RBPM GENERAL LTD

Volpoint House, Blakey Road, Salisbury, Wiltshire SP1 2JG
Tel 01722 336998 Fax 01722 336407
E-mail enquiries@rbpm.co.uk
Contact Bill Moss, John Brady or Tariq Mian

SCHEME AND SPECIALIST INSURERS: RBPM General Ltd has been involved in placing of insurance for listed buildings, period properties and buildings of non-standard construction since 1973. Placing insurances for both private and commercial clients, they have for many years worked closely with the archaeological and historical professions, placing all their insurance requirements, and have many markets to place all types of property and contents insurance, together with professional indemnity, liability, all risks, plant and hired-in equipment insurance. They are approved brokers to the Council for British Archaeology and administer their insurance scheme for archaeologists. *See also: display entry on this page.*

INSULATION

CHIMNEY CONSULTANTS

▶ THE NATIONAL ASSOCIATION OF CHIMNEY ENGINEERS (NACE)

PO Box 5666, Belper, Derbyshire DE56 0YX
Tel 01773 599095 Fax 01773 599195

ALL TYPES OF CHIMNEY CONSTRUCTION, REPAIR AND MAINTENANCE: Previously known as the National Association of Chimney Lining Engineers, (NACLE) the National Association of Chimney Engineers (NACE) is now an association of independent companies specialising in all types of chimney works. The association runs a register of competent companies detailing categories of specialisation for each company. In addition to this, NACE runs training schemes for operatives leading to an NVQ in chimney engineering, and information seminars for professionals (surveyors, architects, clerks of works etc) who require background information on chimneys, their design and correct construction.

▶ SPECIALIST FLUE SERVICE

Studio 4, Harpers Hill, Nayland, Colchester, Essex CO6 4NT
Tel 01206 262 469 Fax 01206 263193
E-mail neill@specialistflue.co.uk

CHIMNEY CONSULTANTS: The company, established in 1973, by the current technical director, Neill Fry, specialises in the design and installation of flue and chimney systems for almost any purpose from small fireplaces to large industrial freestanding chimneys. Their experience covers new design, renovation of existing buildings and fault finding. Flue types range from single wall, twin wall, flexible liners, fan dilution to fan assisted flues. The company completed contracts portfolio includes many Grade I and Grade II listed buildings, schools, universities, castles, palaces, hospitals, colleges, MOD establishments, airports etc. Facilities include computerized flue calculation programmes, full CAD capability, computer based health and safety and CDM system.

CHIMNEY FLUE LINERS

▶ CICO CHIMNEY LININGS LIMITED

Freepost, Westleton, Saxmundham, Suffolk IP17 3AG
Freephone 0500 833787 Fax 01728 648428
E-mail cico@chimney-problems.co.uk
Website www.chimney-problems.co.uk

NATIONAL CHIMNEY LINING SERVICE: Established for over 20 years, CICO Chimney Linings installs CICO's Refractory Cast-in-situ linings which have been independently approved by the BBA. CICO offer surveys without obligation nationwide via locally based branch offices.

▶ SPECIALIST FLUE SERVICE

Studio 4, Harpers Hill, Nayland, Colchester, Essex CO6 4NT
Tel 01206 262 469 Fax 01206 263193
E-mail neill@specialistflue.co.uk

FLUE LINERS

FIRE PROTECTION

▶ GIBBON, LAWSON, MCKEE LIMITED

41A Thistle Street Lane South West
Edinburgh EH2 1EW
Tel 0131 225 4235 Fax 0131 220 0499
E-mail dg@glmglm.co.uk

FIRE PROTECTION: Complying with the regulations can provide some assurance of safety but often at an unacceptable cost financially and in terms of alterations to historic building fabric. By approaching the problem from a fire engineering standpoint, combined with an understanding of historic buildings, GLM provides intelligent and cost-effective solutions to fire precautions in historic buildings. *See also: profile entries in Architects section, page 24 and Surveyors section, page 45.*

▶ INTERNATIONAL FIRE CONSULTANTS LIMITED

20 Park Street, Princes Risborough, Bucks HP27 9AH
Tel 01844 275500 Fax 01844 274002
E-mail ifc@intfire.com
Website www.intfire.com

INDEPENDENT FIRE SAFETY ENGINEERS: IFC is a team of 25 full time staff offering a timely and professional service to its clients. IFC has extensive knowledge and experience in dealing with fire safety and building conservation issues in a sympathetic manner. The firm's engineers have detailed structural knowledge and are able to offer clients full architectural standard outputs to your design team. Services include; building surveys, fire risk assessment, design and specification advice, co-ordination and design of fire tests, fire safety management, emergency plans and staff training. IFC has worked for English Heritage, The Historic Royal Palaces Agency, National Trust as well as many private owners of historic and listed properties.

FIREPLACES

▶ ALDERSHAW HANDMADE CLAY TILES LTD

Pokehold Wood, Kent Street, Sedlescombe, East Sussex TN33 0SD
Tel 01424 756777 Fax 01424 756888
E-mail tiles@aldershaw.co.uk
Website www.aldershaw.co.uk
FIREPLACES: *See also: profile entry in Clay Tiles & Roof Features section, page 85.*

▶ NOSTALGIA

Holland's Mill, Shaw Heath, Stockport, Cheshire SK3 8BH
Tel 0161 477 7706 Fax 0161 477 2267
Website www.nostalgia-uk.com
RECLAIMED ANTIQUE FIREPLACES: Nostalgia was formed 25 years ago and specialises in the supply of reclaimed antique fireplaces in wood, marble, cast iron, slate and stone. With over 1,000 items, dating from 1750–1910 always available, it has one of the largest collections of these items in the country. There is always a selection of restored fireplaces on display, but customers are welcome to choose a piece which they wish to have restored to order. Nostalgia takes particular pride in the restoration of their marble chimneypieces. A photo service is available.

▶ SOLOPARK PLC

Station Road, Nr Pampisford, Cambridgeshire CB2 4HB
Tel 01223 834663 Fax 01223 834780
E-mail info@solopark.co.uk
Website www.solopark.co.uk
ARCHITECTURAL ANTIQUES: *See also: display entry in Architectural Salvage section, page 156.*

▶ SPENCER & RICHMAN

Farrows, Aller, nr Langport, Somerset TA10 0QN
Tel 01458 253539 Fax 01458 253601
E-mail tracey@spencerandrichman.com
Website www.spencerandrichman.com
SPECIALIST CARVERS IN MARBLE AND STONE: Colour brochure on request.

▶ WESTLAND LONDON

St Michael's Church, Leonard Street, London EC2A 4ER
Tel 020 7739 8094 Fax 020 7729 3620
E-mail westland@westland.co.uk
Website www.westland.co.uk
PERIOD AND PRESTIGIOUS CHIMNEYPIECES, ARCHITECTURAL ELEMENTS, PANELLING, FOUNTAINS, STATUARY, PAINTINGS AND FURNITURE: Displayed in the vast interior of St Michael's Church, in the Westland, London warehouse, and on the company's extensive website www.westland.co.uk. Westland liaise, co-operate and supply for projects of architects, decorators, developers and individuals worldwide. The company will also buy any suitable items. There are in-house workshops covering most product areas; please view the website and contact Westland for your present and future requirements. Approach by car via Great Eastern Street or by tube, Old Street, exit four. Open Monday–Saturday and by appointment. *See also: display entry on this page.*

BUILDING SERVICES CONSULTING ENGINEERS

▶ FABERMAUNSELL LIMITED

23 Middle Street, London EC1A 7JD
Tel 020 7645 2000 Fax 020 7645 2099
Website www.fabermaunsell.com
Contact Stuart Tappin

CONSULTING ENGINEERS: International consulting engineers providing a complete range of historic building engineering skills including mechanical and electrical engineering; structural and civil engineering; lighting, acoustics; fire engineering and facilities management. Highly experienced engineers are able to provide a fully co-ordinated service to clients and are supported by continuous research, development and training. Their conservation engineers provide a sensitive and holistic approach to repairs, refurbishment and key issues such as the installation of services in historic buildings. They have particular expertise on 20th century historic buildings. Projects include: British Museum, Windsor Castle, Palace of Westminster, Bank of England, Somerset House, Royal Academy of Music, Royal Albert Bridge and Barbican Art Centre.

▶ GIFFORD AND PARTNERS

Carlton House, Ringwood Road, Woodlands, Southampton SO40 7HT
Tel 023 8081 7500 Fax 023 8081 7600

MECHANICAL AND ELECTRICAL SERVICES FOR HISTORICAL BUILDINGS: *See also: profile entry in Structural Engineers section, page 39.*

▶ LIGHTWRIGHT ASSOCIATES

11 Mill Lane, Heatley, Lymm, Cheshire WA13 9SD
Tel 01925 755359 Fax 01925 756925
E-mail lightwright@fsbdial.co.uk

BUILDING SERVICES CONSULTING ENGINEERS: Established 1973, this small specialist practice is experienced in dealing with the sensitive and sympathetic design and specification of lighting, electrical, fire, security, heating and ventilation installations in listed and historic buildings in England and Wales. Clients include the National Trust, local authorities, owners of historic houses, cathedrals and churches, museums and art galleries. The practice also conducts mechanical and electrical quinquennial inspections, compiles reports on existing systems, undertakes feasibility studies and the preparation of drawings and budget costs for grant applications.

▶ ROBERT BLOXHAM-JONES ASSOCIATES

4 Lancaster Drive, Upper Rissington, Cheltenham, Glos GL54 2QZ
Tel 01451 822820 Fax 01451 822821
E-mail r.bloxham-jones@talk21.com
Contact Robert Bloxham-Jones

BUILDING SERVICES CONSULTING ENGINEERS: Attention to detail is the key to successful integration of engineering services into historic buildings. Coupled with an understanding of their uses and contents is essential in ensuring minimal intervention into the fabric. Clients include Leeds Castle, Chequers, Treowen and The Shakespeare Birthplace Trust.

LIGHTING CONSULTANTS

▶ LIGHT & DESIGN ASSOCIATES

Unit 0615, Bell House, 49 Greenwich High Road, London SE10 8JL
Tel 020 8469 4000 Fax 020 8469 4005
E-mail design@lightanddesign.co.uk
Website www.lightinganddesign.co.uk

INTERIOR AND EXTERIOR LIGHTING SPECIALISTS: Architectural lighting design consultants, established in 1990, specialising in the interior and exterior lighting of historic buildings including churches, museums and palaces. Lighting design awards have been gained for new lighting systems at St Luke's, Battersea, and Hinde Street Methodist Church, Marylebone. Recent commissions include concept proposals for exterior lighting at Buckingham Palace, the re-lighting of the interior to The Chapel Royal, St James's Palace, façade lighting to the Royal Exchange building, London and the re-lighting of the interior to St Bartholomew the Great, Smithfields.

ANTIQUE & DECORATIVE LIGHTING

▶ CHELSOM LIMITED

Head Office: Heritage House, Clifton Road, Blackpool, Lancashire FY4 4QA
Tel 01253 831400 Fax 01253 698098

DECORATIVE LIGHTING AND REFURBISHMENT OF PERIOD LUMINAIRES: Chelsom has been supplying high quality lighting to the contract market for over 50 years. The range covers all types of luminaires in both traditional and modern styles. Chelsom specialises in the restoration of period light fittings and their conversion to low energy light sources when required. The company also designs and manufactures bespoke lighting to client specification. Restoration work takes place at the factory in Blackpool where there is also an experienced technical design department.

▶ R HAMILTON & COMPANY LIMITED

Quarry Industrial Estate, Mere, Wiltshire BA12 6LA
Tel 01747 860088 Fax 01747 861032

LIGHTING SPECIALISTS: Hamiltons manufactures to order wiring accessories and lighting control systems of quintessential quality. Traditional style and modern modular switches, sockets, dimmers, etc on brass, chrome, wood and polycarbonate plates. Designs include the familiar rope and bead edgings, sleek 'Linea' styles with concealed fixing, replicas of the original domed switch, System 45 with its continental influence. Bespoke configurations are a speciality. Lighting control systems for loads from 4 to 20 amp can be operated by control points mounted on plates matching their wiring accessories. Among the company's successes are National Trust and English Heritage sites plus prestige projects throughout the world.

A CONSERVATOR WORKING ON THE CEILING OF WORCESTER COLLEGE CHAPEL

Photograph by Cliveden Conservation Workshop Limited

INTERIOR CONSULTANTS & CONSERVATORS

Name			Page
A T Cronin Limited		ic	197
Abbey Heritage Ltd		fa	165
Graciela Ainsworth		fa	95
Allyson McDermott		ic	197
AMBO Architects	id		20
Anthony Beech Furniture Conservation and Restoration		an cb fr gi	205
Anthony Short and Partners	id		20
Antique Bronze Ltd		fa	94
A O C Archaeology Group		pa	34
Architecton	id		20
Arthur Brett		cb	126
Bricknell Conservation Limited		cb	59
Campbell Smith & Co Ltd		ic	200
Capital Crispin Veneer		ve	211
Carden & Godfrey Architects	id		22
Carthy Conservation Ltd		gi ic	102
Carvers & Gilders Limited		gi	124
Church Conservation Limited		gi	84
Clive Beardall		an cb fr gi ve	205
Cliveden Conservation Workshop Ltd		wa	102
Coltman Restorations		an fr	204
Corbel Conservation Ltd		ic	59
Cyril John Decorators Ltd		gi ic	197
David Brown and Partners	id		23
Devereux Decorators Ltd		ic wa	198
Donald Insall Associates Ltd	id		24
Elizabeth Holford Associates Ltd		fa ic pa wa	197
Farrow & Ball		wa	186
Feilden + Mawson	id		24
The Fine Iron Company		gi	145
Fisher Decorations Limited		gi ic	197
Francis Downing Painting Specialist		fa	203
French Polishing Contracts		fr	206
G R Pearce		fr	206
Gilmore Hankey Kirke Ltd	id		25
Giltwood Restorations Ltd		gi	124
H J Hatfield & Sons Ltd		an	205
Hamilton Weston Wallpapers Ltd		wa	203
Hare & Humphreys Ltd		gi ic pa	198
Hearns (Specialised) Joinery Limited		cb	127
Heritage Tile Conservation Ltd		ic	207
Hesp & Jones		gi ic	197
Hirst Conservation		fa gi ic pa wa	202
HOK Conservation and Cultural Heritage	id		26
Holden Conservation		ic	95
Holloway White Allom		ic	62
Howell & Bellion		gi ic	198
Huning Decorations		gi ic	200
International Fine Art Conservation Studios Ltd		fa ic wa	199
James Brotherhood & Associates Limited	id		26
Jameson Joinery		cb	73
Jonathan Rhind Architects	id		26
Julian Harrap Architects	id		26
KAW Design	id		200
Ken Biggs Contractors Limited		ic	63
The Leather Conservation Centre		an le	206
Linda Andrews		an fa fr gi pi	206
Lisa Oestreicher Architectural Paint Analysis		pa	200
Luard Conservation		ic	198
Mackenzie Wheeler	id		27
Mark Bridges Carver & Gilder Ltd		gi ic	125
Martin Ashley Architects	id		27
Martin A Horler		ic	200
Michenuels of London		an fr	206
Mott Graves Projects Ltd		cb ic	65
Mowlem Rattee & Kett		ic	106
N E J Stevenson		cb	206
Nevin of Edinburgh		gi ic pa	199
Nimbus Conservation Limited		ic	107
Norgrave Studios		ic	138
Norman + Dawbarn Architects	id		28
The Paint Practice		ic	199
The Perry Lithgow Partnership		fa gi ic wa	202
Peter Martindale		fa ic wa	202
Plowden & Smith Ltd		an fa fr gi ic le wa	203
Richard Crooks Partnership	id		28
Richard Griffiths Architects	id		29
Richard Ireland Period Restoration		wa	217
Robert Howie & Son		ic	199
Romark Specialist Cabinetmakers Limited		an cb fr	205
Rupert Harris Conservation		fa gi	203
Simpson & Brown Architects	id		30
Stuart Page Architects	id		30
Tankerdale Ltd		cb fr gi ve	128
Thermo Lignum		an	183
Tim Peek Woodcarving Services		cb fr gi id	125
Timothy Williams (Builders) Limited		ic	65
Treasure & Son Ltd		ic	65
United Kingdom Institute for Conservation of Historic and Artistic Works (UKIC)	id	ic	234
WCP (The Whitworth Co-Partnership with Boniface Associates)	id		213
Gillian M H Walker		fa	202
Michael Walton	id		31

KEY

- an antique and furniture restoration
- cb cabinet makers
- fa fine art conservation
- fr French polishers
- gi gilders
- ic decorators & conservators
- id interior designers
- lc leather conservation
- pa paint analysis
- pi picture framers
- ve veneers
- wa wall painting conservators

INTERIOR FITTINGS

Name			Page
Abbey Masonry and Restoration Limited		fi	100
Antique Buildings Limited		fi	157
Best Demolition		fi ki	155
Boden & Ward		fi	101
Carvers & Gilders Limited		fi	124
Chelsom Limited	li		190
Drummond's Architectural Antiques Limited		ba ra	155
Haddonstone Limited		fi	110
Hearns (Specialised) Joinery Limited		ki	127
Illumin Glass Studio	li		138
LASSCO	li	ba fi ki	155
Minchinhampton Architectural Salvage		ba fi ra	155
Nostalgia		ba fi	189
Original Features (Restorations) Ltd		fi ra	207
R Hamilton & Company Limited	li		190
Ransfords Reclaimed Building Supplies	li	fi	155
Retrouvius Architectural Salvage		ba fi ki	155
Solopark Plc		fi	157
Spencer & Richman		fi	189
The Bath Stone Group		fi	101
The Cast Iron Company	li		143
Thorverton Stone Company Ltd		fi	110
Tim Peek Woodcarving Services		fi	125
Treasure & Son Ltd		ki	65
Walcot Reclamation Ltd	li	ba fi ki ra	157
Weldon Stone Enterprises Ltd		fi	109
Wells Cathedral Stonemasons Ltd		fi	109
Westland London	li	fi	189

KEY

- ba baths, sinks and taps
- fi fireplaces
- ki kitchen and bathroom fittings
- li light fittings
- ra radiators and stoves

WALLPAPERS, PAINTS, TEXTILES & CARPETS			
A O C Archaeology Group		wl	34
Abbey Heritage Ltd	pd		165
Advanced Chemical Specialties Ltd	pd		181
Allyson McDermott		wl	197
Campbell Smith & Co Ltd		wp	200
Craig & Rose	pd		186
D & S Bamford (Makers) Ltd		ca	201
David Luckham Consultants Limited		ca ts	201
Devereux Decorators Ltd		wl	198
Farrow & Ball	pd	wp	186
Fiona Hutton Textile Conservation		ca tx	203
Hamilton Weston Wallpapers Ltd		wp	203
Hirst Conservation Materials Ltd	pd		174
Keim Mineral Paints Limited	pd		186
Linney Cooper		ca	201
Mike Wye & Associates	pd		174
Retrouvius Architectural Salvage		ts wp	156
Ryedale Conservation Supplies	pd		175
Ty-Mawr Lime Ltd	pd		176

KEY

- ca carpet and rug conservators
- pd paints and decorative finishes
- tx textile conservators
- ts textile suppliers
- wl wallpaper conservators
- wp wallpaper suppliers

CERAMICS, MOSAICS, MARBLE & GRANITE			
A F Jones Stonemasons	mg		100
Aldershaw Handmade Clay Tiles Ltd		ft	207
Burleigh Stone Cleaning & Restoration Co Ltd	mg		166
Craven Dunnill Jackfield Limited		ce ft mo	207
Carthy Conservation Ltd		mo	102
Daniel Platt Limited		ft	85
Heritage Tile Conservation Ltd		ce ft	207
Holden Conservation	mg		95
The Jackfield Conservation Studio		ce ft mo	207
London Mosaic Restoration		mo	208
National Federation of Terrazzo, Marble & Mosaic Specialists	mg	mo	208
Nicolas Boyes Stone Conservation		mo	106
Original Features (Restorations) Ltd		ce ft	207
The Original Flooring Company		ft	208
Original Oak		ft	211
Plowden & Smith Ltd		mo	208
Retrouvius Architectural Salvage	mg		156
Stonewest Ltd	mg		67
The Tile Gallery		ft	208
Trevor Caley Associates Limited		mo	208
Westland London	mg		189

KEY

- ce ceramics
- ft floor and wall tile suppliers
- mg marble and granite
- mo mosaic supplies

TIMBERS & PARQUET FLOORING		
A H Peck		211
Antique Bronze Ltd		211
Country Oak (Sussex) Ltd		211
Drummond's Architectural Antiques Limited		155
English Woodlands Timber Ltd		122
Jameson Joinery		211
Minchinhampton Architectural Salvage		155
Northwood Forestry Ltd		123
Original Oak		211
Priory Hardwoods Limited		211
Ransfords Reclaimed Building Supplies		211
The Round Wood Timber Co Ltd		155
Ternex Ltd		124
Victorian Woodworks		213
Walden Joinery		128
Weldon Flooring Limited		212
Whippletree Hardwoods		124

PLASTERWORK			
A T Mead Traditional Restoration		pl	58
A W Maguire, Original Plaster Mouldings		pl	213
Alba Plastercraft		pf pl	213
Anthony Hicks		pl	72
Babylon Tile Works		bt	85
Between Time Ltd		pl	58
Bosence & Co		pl	44
Bricknell Conservation Limited		pl	59
Bursledon Brickworks Conservation Centre	bt re		172
C R Crane & Son Ltd		pl	60
Carpenter Oak & Woodland Co Ltd		bt	72
Carthy Conservation Ltd		pl	102
Cathedral Works Organisation (Chichester) Limited	bt re		102
Chalk Down Lime Ltd	bt re	pl	172
Church Conservation Limited		pl	84
The Cleft Wood Co		bt	213
Cliveden Conservation Workshop		pl	102
Corbel Conservation Ltd		pl	59
The Cornish Lime Company Ltd	bt re		174
Coyle Timber Products	bt	pf pl	154
E G Swingler & Sons	bt		82
Farthing & Gannon		pf pl sa	214
Fine Art Mouldings		pf pl	214
G Cook & Sons Ltd		pf pl	214
Philip A Gaches		pl sa	214
George Jackson & Sons		pf pl	214
Grall Plaster & Stone Specialists		pf pl	214
H J Chard & Sons	re		174
Hayles & Howe Ltd Ornamental Plasterers		pf pl sa	214
Hirst Conservation		pl	202
Historic Buildings Conservation Limited		pl	61
Hodkin & Jones (Sheffield) Ltd		pf pl	214
Holden Conservation		pl	95
J H Upton & Son		pf pl	215
The Lime Centre	re		174
The Lime Plastering Company		pl	215
Locker & Riley Fibrous Plastering Ltd		pf	215
London Fine Art Plaster		pf pl	215
London Plastercraft Ltd		pf pl	215
Luard Conservation		pl	198
G & N Marshman	bt		216
James Martin		pl	216
Mike Wye & Associates	bt re		174
Minerva Stone Conservation		pl	106
Nimbus Conservation Limited		pl	107
O'Reilly Period Cornice Restoration & Cleaning		pf pl	216
Plastercraft		pf pl	216
Plowden & Smith Ltd		sa	203
Richard Ireland Period Restoration		pf pl sa	217
Standen Plastering Limited		pf pl	217
The Scagliola Company		sa	217
The Traditional Lime Co	bt re		176
Thomas and Wilson Limited		pf	217
Treasure & Son Ltd		pl	65
Trumpers		pf pl	216
Twyford Lime Products		pl	176
Ty-Mawr Lime Ltd	bt re		176
Weald & Downland Open Air Museum	bt		226
Whippletree Hardwoods	bt		124
Womersley's Limited		pl	176
Yorkshire Decorative Plasterers Ltd		pf pl	217

KEY

- bt battens and lath
- pf fibrous plasterwork
- pl lime plasterwork
- re hair and fibre reinforcement
- sa scagliola

LINLEY SAMBOURNE HOUSE

MANAGING TEAMWORK IN THE CONSERVATION OF AN HISTORIC INTERIOR

ALLYSON McDERMOTT

18 Stafford Terrace, Kensington, the home and studio of the Punch *cartoonist, Edward Linley Sambourne in the late 19th century – a fine, stucco-faced, mid-terrace London town house of the 1860s*

On Wednesday 11th April 2003 the Royal Borough of Kensington and Chelsea celebrated the completion of a successful two-year programme of structural repairs and conservation work at Linley Sambourne House, the remarkable Victorian terrace house at 18 Stafford Terrace, London.

The project illustrates a level of complexity which is typical of the larger historic interior conservation schemes. It was completed on time and within budget, providing an almost textbook example of the benefits of good planning, effective communication and efficient management, responding to the needs of this very special interior. This article examines how the management of the project contributed to its success.

HISTORICAL BACKGROUND

18 Stafford Terrace is a speculatively built mid-terrace London town house of the 1860s, arranged over four floors. Substantial structural alterations were made by its second owner, the Punch cartoonist Edward Linley Sambourne, between 1874 and 1900 to upgrade the interiors and provide studio facilities. The contemporary decorative finishes and fittings are typical of the aesthetic approach popular with the artistic community of the day and comprise significant examples of both William Morris wallpapers and embossed leather wallcoverings, as well as stained and painted glass, original services, painted and gilded finishes and encaustic tiles.

Continuous occupation by three generations of the same family, each with an appreciation of the history of the house, ensured that the interior was preserved with little alteration, and the last occupant, the Countess of Rosse, sold the property as a museum in 1980. Supported by an extensive archive, the interior provides an extraordinary example of a middle class household in the late Victorian and Edwardian periods.

Much of the house has been untouched since Sambourne's time, despite continuous occupation and the sympathetic introduction of modern conveniences. This invests the house with a further dimension as a landmark example of the modern conservation movement.

The Victorian Society was launched at the house in 1958, in opposition to the wholesale destruction of Victorian buildings and interiors which was taking place at the time.

By the end of the century the building was still in good order generally, but there were problems with recurring water ingress, particularly associated with an internal gutter which carried rainwater from the front, through the roof space to a down-pipe on the rear elevation of the building, thus avoiding spoiling the look of the classical façade. This gutter had become blocked, causing damage to the ceiling beneath. Dampness had also affected most of the walls on the rear elevation, damaging sensitive wallpapers and painted surfaces. Much of the interior had a significant layer of particulate dirt (the result of over a century of urban pollution) and many exposed surfaces had suffered the wear and tear of daily domestic use. Some areas which had been restored previously had discoloured and in some cases the repairs were lifting and flaking, damaging the original surfaces beneath. Whilst wear and tear could be considered historic, later restoration could not.

The bath which Linley Sambourne used for developing his pictures

The current programme of restoration commenced in Spring, 2001 with the aid of a grant from the Heritage lottery Fund. The main contract included the introduction of new services, some structural repairs and alterations and the development of the basement into an interpretation and education facility.

In preliminary discussions, the client voiced concerns that any interventive conservation could disrupt the delicate balance of so sensitive an interior. Similar projects have often proved disastrous, in the worst case combining inappropriate work with massive overruns of both time and budget, often caused by a breakdown in communication and the inevitable impact this has on the morale and performance of conservators.

Roy Sambourne's bedroom with wallpaper removed, revealing dampness behind.

The fireplace surrounded by layers of wallpaper

CONSERVATION STRATEGY

At Linley Sambourne House the conservation strategy needed to be underpinned by an appreciation of the history of the house, the quality of its original interior and its importance as a milestone for conservation within an historic building. Thus any approach to treatment needed to recognise the importance of all aspects of the interiors up to and including occupancy by the Countess of Rosse.

Details of the house are extensively documented and every opportunity was taken to further investigate and understand its archaeology in order to gain a comprehensive view of its appearance at various stages of development. This took the form of historical research supported by an extensive programme of scientific analysis. The results of these investigations enabled a better understanding of both the decorative chronology and the original materials and techniques used throughout the interior.

It seemed paramount that all work involving the conservation of the house and its interior should be carried out on the basis of minimum intervention, in which the primary aim was to prolong the life of all original materials, finishes and fittings by a process of informed and sympathetic conservation, rather than replacement or restoration.

Through its representatives, Bernard Burke and John Swindells, The Royal Borough of Kensington and Chelsea (RBK&C) proved to be an enlightened client and made a significant contribution to the success of the project, as did the knowledge and enthusiasm of the curatorial staff (Daniel Robbins and Reena Suleman) and the support of a sympathetic architect (Dante Vanoli of Purcell Miller Tritton) who had considerable experience in the conservation of historic interiors.

I was appointed as the conservation advisor at the outset and, as a member of the main design team, I was able to take a proactive role in all relevant aspects of both planning and implementation. This helped to ensure the smooth and efficient working of the conservation project and I was also able to provide useful expertise for other aspects such as environmental monitoring and control during the main contract works, protection, programming and other related tasks.

Embossed papers cover the ceiling and the upper wall surfaces: the vent in the centre of the ceiling rose was designed for the rather smokey gas lighting of the 1860s

The main contract was carried out under the supervision of the architect.

THE CONSERVATION SURVEY

Once the museum collections had been removed for safe storage off-site, it was possible to carry out a comprehensive condition survey of the interior, including its fixtures, fittings and finishes.

For ease of identification and numbering, each room was divided into its elements: floor, skirtings, dado, rail, wallface, upper wallface, cornice and ceiling. Other elements such as fitted furniture; fireplaces and ceiling roses were included wherever they were found. Each door and window was dealt with individually by the conservation team in conjunction with the architect, as he was responsible for any structural or external works.

The survey included details of all materials and techniques used as well as any preliminary historical research. Scientific analysis was also carried out where required either for the identification of materials or for an accurate interpretation of the decorative chronology. In some cases scientific analysis was also required to establish acceptable and effective methods of cleaning and consolidation, although at this preliminary stage such investigation was kept to the minimum required to gather a body of information.

Recommendations and options for treatment were then presented to the curatorial team for discussion and approval in line with the agreed conservation strategy and the following general principles.

THE BRIEF

1. General principles

Works were to be carried out only by suitably qualified and experienced conservators. Fittings, finishes and other components were only to be removed from the house for studio conservation under exceptional circumstances and only upon specific written permission, after first presenting a method statement for any conservation work to be carried out.

When conservation or investigation revealed previously hidden aspects, such as the location of 'lost' fixtures and fittings, and fragments of old wallpapers for example. These were to be brought to the attention of the curatorial staff and fully documented. Wherever possible, the programme of works was to be adapted to allow proper recording and cross referencing with documentation in the archive.

Should replacement be necessary, for example, of certain sections of wallpapers, full reference was to be made to the primary source material to obtain the closest possible match to the original item. It was also decided that, where any new materials were to be applied or inserted into original fabric, a barrier layer would be introduced, separating the original from the new to protect the original material and to facilitate reversibility. This approach is of particular significance in the treatment of painted or varnished surfaces or in the use of applied materials such as wallpaper or textiles.

In certain cases, agreed in advance, the conservation work was to be supported by the presentation of method statements and/or samples for inspection and agreement. Due to the significance and sensitivity of this interior, it was considered most important to gain the approval and agreement of both the conservation advisor and curatorial staff for these elements of work, and the contractor's brief allowed for considerable discussion and management attendance and supervision.

2. Schedule of work

Once individual approaches to treatment had been agreed, a detailed schedule of works was prepared and was subsequently divided into the appropriate specialist areas of expertise. This provided the foundation for a reasonably accurate budget projection and a planned programme of works which broke down into three phases. All contractors were required to tender on the same basis, with two main phases and some optional items of work which could be included if, for example, more funds became available.

Phase 1 was to be carried out on commencement of the interior conservation project and was to run for three weeks, with the exception of the architectural paint research and wallpaper investigation which would continue throughout the project. The scope of this phase included:

- removing certain items, such as water-damaged wallpaper, to facilitate effective repair and conservation and to permit access for inspection and repair to the wall beneath – these required careful packaging prior to removal from site for studio treatment

- removing and packing some stained and painted glass in preparation for removal from site for studio treatment
- obtaining approval and agreement from the conservation advisor for all objects to be removed before proceeding
- preparing photographic records and documentation (scope and format agreed by the conservation advisor)
- arranging safe packing and transportation, to the approval of the conservation advisor
- providing evidence of transit and studio insurance
- preparing receipts and documentation.

All off-site work was to be carried out by approved and experienced specialists, and the contractor was to allow for studio visits by the conservation advisor and the curatorial team, first to inspect the studios prior to accepting them and, later, to approve works being carried out.

Phase 2 involved the replacement of all the conserved items when related areas were sufficiently dry and stable, at a time agreed with the conservation advisor. This phase, which was expected to last ten weeks, would follow practical completion of the main contract programme (which was expected to last 36 weeks).

Optional – Where the client subsequently approved Optional items, the conservation advisor or architect would notify the contractor in writing. Such approved works would then have to be incorporated into the programme of works following Phase 1 procedures described above.

Specifications – Detailed specifications were also prepared. Generally these were as standard and included items such as fire protection, smoking, solvents, cleanliness, wet weather conditions, personal property, insurance, health and safety access, lighting, electrical equipment and protection. Specifications relating directly to the various specialisms were prepared individually.

THE CONSERVATION TEAM

The success of this project was largely due to the effectiveness of two-way communication, comprehensive planning and good management.

Throughout the planning stages, representatives from RBK&C Library and Arts, the curatorial team, the supervising architect and the conservation advisor worked closely together not only at the regular project team meetings but also informally by phone and e-mail. Other specialists also made a significant contribution to these discussions at various stages. These included the quantity surveyor, services engineer, structural engineer and main contractor. The result was a formidable multi-disciplinary team with the combined expertise to deal with all aspects of the programme, and in some cases even pre-empt potential difficulties.

Appliqué decorated panels on the dining room door, with another William Morris wallpaper on either side above the dado rail

The painted glass window in the drawing room

WHAT THE WORLD OF INTERIORS SAID...

"In the underwater gloom of 18 Stafford Terrace, Morris patterns run underfoot, across the furniture and up the walls, to disappear behind close hung picture frames. Traffic noises and thin London sunshine are shut out by tinted glass, Nottingham lace and the deep pelmets from which swags of coffee coloured velvet descend. Swimming up through the shadows comes a potent scent mingling tobacco, polish, draperies, dust – Victorian London."

The World of Interiors, May 2003

The extensive programme of conservation received little mention in this lavishly illustrated article marking the re-opening of Linley Sambourne House, and the 'scent of dust' it described was arguably a romantic notion, in view of the extensive programme of cleaning and conservation which had been completed. However, even as such, the article provides a measure of the sensitivity with which each treatment was carried out and a tribute to the painstaking work of all concerned.

As well as the support of the main project team, the conservation programme benefited from the knowledge and experience of many individual specialists. As the project moved from planning to implementation, the conservation advisor became conservation team leader, a novel method of management which enabled a team of dedicated conservators, many of them leaders in their field, to be employed directly, without having to use a large conservation contractor. This approach also eliminated a further tier of management, supervision and reporting and ensured unbroken communication with the main project team with significant benefits to both cost and effectiveness.

The interior conservation project was completed on time and well within budget, thereby allowing much of the 'optional' work to be carried out. All documentation is complete and is currently being transferred to an interactive database, which will record all the information acquired to date through historical research, scientific analysis and investigation. This will also provide a means of documenting the treatment carried out with recommendations for ongoing maintenance and care.

Once the project was complete, the curatorial team began the long and painstaking process of re-installing the contents in time for the re-opening of the museum in April 2003.

Recommended Reading

Banham Joanna, Porter Julia, and Macdonald Sally, *Victorian Interior Style.* Studio Editions, London 1995

Oman, Charles C and Hamilton, Jean, *Wallpapers: a history and illustrated catalogue of the collection of the Victoria & Albert Museum.* Sothebys & V&A, London 1982.

Guilding R, *'Full exposure.'* World of Interiors, May 2003.

Nicholson, Shirley, *A Victorian Household.* Sutton Publishing 1988

Saunders, Gill, *Wallpapers in Interior Decoration.* V&A Publications, London 2002

ALLYSON McDERMOTT established her own studios in 1980, specialising in the conservation of all aspects of the decorative arts and particularly those related to the Historic Interior. Whilst continuing as a practicing conservator, Allyson lectured in conservation at Northumbria University, was appointed Sotheby's Advisor on Conservation and subsequently became Advisor to the National Trust. She is currently undertaking research into the use of Chinese Wallpapers as part of the joint Victoria & Albert Museum / Royal College of Art conservation programme and is vice-chairman of the UKIC Historic Interiors Section. After a recent move to Gloucestershire, she intends to concentrate on her project management and consultancy work.

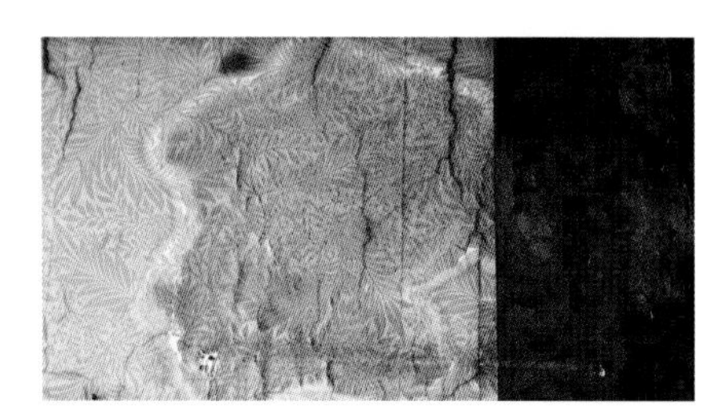

Detail of water damage to a William Morris wallpaper

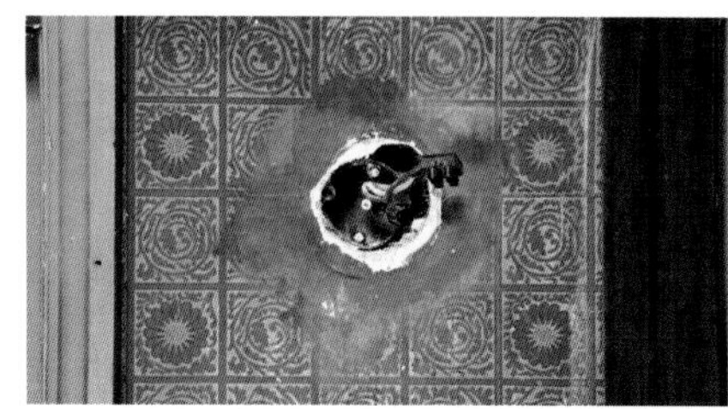

Morris 'Diaper' wallpaper around a light switch

Detail of embossed wallpapers in the studio

INTERIORS CONSULTANTS & CONSERVATORS

▶ A T CRONIN LTD

62A Valetta Road, London W3 7TN
Tel 020 8749 2995 Fax 020 8740 0508
E-mail ianblock@atcronin.co.uk Website www.atcronin.co.uk

INTERIORS CONSULTANTS AND CONSERVATORS: A T Cronin Ltd is a company specialising in curtain-making, upholstering and fabric walling. Their curtain-making workshop specialises in the use of traditional methods and materials. They can undertake period window treatments and the dressing of four-poster beds. An on-site team of fitters help to create all of the period details necessary to complete a job. Their upholstery workshop undertakes both the reconstruction of period pieces of furniture and the making of new pieces to your designs. In both cases, they use traditional methods. A fabric walling team undertakes all aspects of wall upholstery using traditional techniques.

▶ ELIZABETH HOLFORD ASSOCIATES LTD

The Tabaco Factory, Raleigh Road, Southville, Bristol BS3 1TF
Tel 0117 902 0269 Fax 0117 902 0243 Mobile 07977 997207
E-mail elizabeth.holford@blueyonder.co.uk

CONSERVATION AND RESTORATION OF PAINTINGS AND PAINTED SURFACES: The company has many years of experience specialising in the conservation and restoration of wall paintings, easel paintings and other painted surfaces. Work includes surveys, fully documented reports and analytical services; ultra-violet and infra-red reflectography, x-radiography, microscopy for cross-sections, pigment dispersions, and access to further extensive research facilities. Work has been undertaken for a range of heritage organisations, churches, museums, art galleries and private individuals both in the UK and abroad.

▶ FISHER DECORATIONS LIMITED

157 Marston Road, Stafford ST16 3BD
Tel 01785 251300 Fax 01785 251500
E-mail fisherdecs@aol.com

INTERIOR CONSULTANTS AND CONSERVATORS: Established in 1932, Fisher Decorations Limited was awarded the 2002 Rosebowl Award for Buxton Opera House. The team continually carries out well-researched projects to a high standard, its field of operations including specialist wallcoverings, gilding, marbling, graining, trompe l'oeil and murals. The firm also identifies and reproduces historic stencils and designs suitable alternatives. Consultancy work is carried out with a philosophical approach, always bearing in mind the customer's requirements. Technical reports incorporating the latest digital photography can be produced to show detail and colours, and advice on paints used can also be given. The restoration and conservation of churches and large public and private houses is the company's main area of work. The company's vast knowledge and expertise in ecclesiastical, period and heritage work greatly complements the company's current operations.

▶ HESP & JONES

The Cedars, Beningbrough, York YO30 1BY
Tel 01904 470256 Fax 01904 470937

SPECIALIST DECORATORS AND CONSERVATORS: Charles Hesp heads a small team of skilled craftsmen that carry out decoration and restoration work in stately homes, churches and large private houses throughout England and Europe. Over the last few years their contracts have included Harewood House, Warwick Castle, Castle Howard, the British Embassy in Paris and Prague, St Paul's Cathedral and numerous National Trust stately homes. They are skilled in all aspects of decoration with an emphasis on graining and gilding. They also advise and write reports for historic colour schemes.

▶ HOLDEN CONSERVATION LTD

6 Warple Mews, Warple Way, London W3 0RF
Tel 020 8740 1203 Fax 020 8749 8356, and
Dalshangan House, Dalry, Castle Douglas, Scotland DG7 3SZ
Tel/Fax 01644 460233

MARBLE, STONE, TERRACOTTA AND PLASTER: *See also: display entry in Statuary section, page 95.*

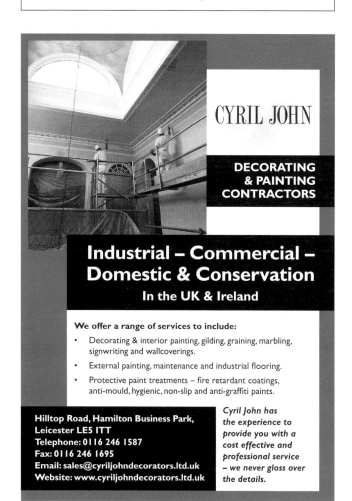

▶ HOWELL & BELLION

66A High Street, Saffron Walden, Essex CB10 1EE
Tel 01799 522 402 Fax 01799 525 696

CHURCH INTERIOR DECORATION, CONSERVATION AND RESTORATION: Howell & Bellion have many years experience of the decoration of churches and other fine buildings. Projects undertaken throughout the country have resulted in a prestigious client list, from local churches to buildings of national importance. Works include the cleaning and conservation of existing decorative schemes, restoration of lost or damaged decoration and the execution of new work such as gilding, stencilling, heraldry, hand painted ornament and the application of traditional materials. To provide a comprehensive service to clients projects often include associated small works such as repairs to carving, joinery, metalwork, stonework and the refurbishment of church metalware. Angels for riddel posts also supplied. Colour leaflets illustrating recent work are available upon request.

▶ INTERNATIONAL FINE ART CONSERVATION STUDIOS LTD

43–45 Park Street, Bristol BS1 5NL
Tel 0117 929 3480 Fax 0117 922 5511

CONSULTANTS, RESTORERS AND CONSERVATORS OF PAINTINGS AND MURALS: *See also: display entry in this section, page 199.*

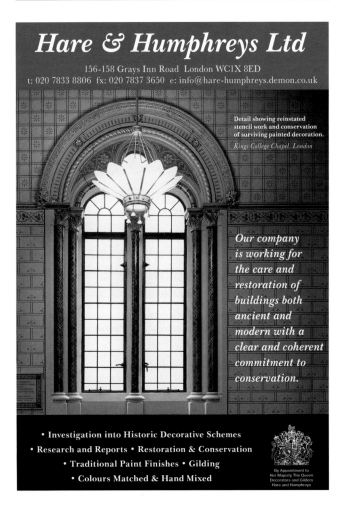

▶ LUARD CONSERVATION

23 Harlesden Gardens, London NW10 4EY
Tel 020 8961 7544 Fax 020 8961 7545 Mobile 07973 741117
E-mail david.luard@virgin.net

CONSERVATION AND RESTORATION OF WOOD, CARVINGS, LIME PLASTER, PANELLING, STRUCTURAL WOODWORK AND ASSOCIATED FINISHES: Luard Conservation provides a comprehensive service from initial enquiry to completion of contract; consultancy services are available as well as hands-on conservation. Contracts include conservation of Grinling Gibbons carvings at Windsor Castle, St James Palace, Burghley House, and Lyme Park; consultant conservator to the Grinling Gibbons Exhibition at the V&A, restoration of plaster ceilings at Longleat House, tender specification for St George's Lutheran Church in the City of London, and post-fire condition reports for the joinery in Peterborough Cathedral. David Luard is wood consultant to the London Diocesan Advisory Committee.

▶ NIMBUS CONSERVATION LIMITED

Eastgate, Christchurch Street East, Frome, Somerset BA11 1QD
Tel 01373 474646 Fax 01373 474648
E-mail enquiries@nimbusconservationlimited.com
Website www.nimbusconservation.com

STONE CONSERVATION AND MASONRY: *See also: display entry and profile entry in Stone section, pages 106 and 107.*

▶ NORGROVE STUDIOS

Bentley, Redditch, Worcs B97 5UH
Tel 01527 541545 Fax 01527 403692
Website www.norgrovestudios.co.uk

DECORATIVE AND STAINED GLASS: *See also: profile entry in Decorative & Stained Glass section, page 138.*

▶ PLOWDEN & SMITH

190 St Ann's Hill, London SW18 2RT
Tel 020 8874 4005 Fax 020 8874 7248
E-mail richard.rogers@plowden-smith.com
Website www.plowden-smith.com

CONSERVATION AND RESTORATION: *See also: display entry in Fine Art Conservation section, page 203.*

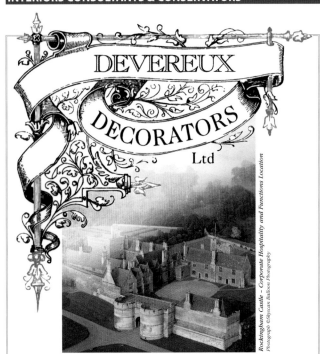

INTERIORS CONSULTANTS & CONSERVATORS

International Fine Art Conservation Studios Ltd

Conservators of Paintings, Murals & Historic Interiors

Inpainting of losses being carried out to an easel painting at our Bristol studio.

IFACS has been involved in important conservation projects throughout the UK and overseas for over 30 years. Recently these have included:

Conservation and redecoration of historic scheme in the Billiard and Drawing Room, Osborne House, Isle of Wight; conservation and technical studies of 19th century fresco paintings 'Neptune Resigning the Empire of the Seas to Britannia' by William Dyce, 1847 and 'Hercules & Omphale' by Anton von Gegenbaur, 1830 also both at Osborne House, Isle of Wight. Conservation of a wall painting on plaster, 'Charles II' by Antonio Verrio, 1707, Royal Hospital Chelsea; conservation of a canvas painting by William Holyoake entitled 'The Battle of Jersey' for the Jersey Museum, St Helier.

We have a fully qualified team of conservators and services can include technical analysis and research. Please visit our website at **www.ifacs.co.uk**

Richard Pelter, Director, is an accredited member of UKIC and fellow of ABPR.

43–45 PARK STREET, BRISTOL BS1 5NL Tel. 0117 929 3480 Fax. 0117 922 5511

National Painter of the Year Winners 1991, 1992 and 1993
Overall Supreme Scottish Decorators of the Year 1994

We offer a complete Interior Decorating service, specialising in old and historic settings.

Graining, Gilding, Marbling, Application of Historical Coatings, Fine Wallpapers, etc. to the highest standard.

Makers of White Lead paints and associated products.

A full Paint Analysis service including Pigment Identification under polarising microscope, all Cross Section work and exposing Stencil Work.

Trusted for the following prestigious assignments:

Kenwood House, London
Drum House, Edinburgh
Amhuinnsuidhe Castle, Harris
Paxton House, Berwickshire
3 Park Terrace, Glasgow National Galleries of Scotland
Weald & Downland Museum, Chichester

8 SWANFIELD, LEITH, EDINBURGH EH6 5RX
TEL: 0131 554 1711 FAX: 0131 555 0075
www.nevinofedinburgh.co.uk

Robert Howie & Son
Decorators
Colour Consultants Restoration Specialists
Est 1954
Working with
Historic Scotland, The National Trust & The Landmark Trust
for over 35 years
Projects have been carried out at the following properties:
Argyll's Lodgings, Stirling; Bowhill House, Selkirk;
Mary Queen of Scots' Chambers, Holyrood Palace, Edinburgh;
Turnberry Hotel; Culzean Castle, Ayrshire;
House of Dun, Montrose;
Auchendrane House; Blair House; Caprington Castle, Ayrshire;
Morlianich Long House, Killin, Perthshire;
Haddo Castle; Castle Fraser, Aberdeenshire;
Hill House, Helensburgh; The House of Binns, Linlithgow;
Mount Stuart House, Isle of Bute; Casa 'D'Italia, Glasgow;
Callander House, Falkirk...
...and many more
Tel: 01290 550373 Fax: 01290 553002
email: Roberthowiedecs@aol.com
Website: http://members.aol.com/roberthowiedecs/home/page.html

The Paint Practice

Your complete colour consultancy for period and present day properties

The Paint Practice provides a complete colour consultancy direct for clients, private and corporate, and as a support service for architects, surveyors, project managers, interior designers, consultants and restorers.

The Paint Practice also prepares technical reports on paints and substrate surfaces, and specifications on the way paints should be applied on different surfaces. Colours are researched as far back as the 12th century, and can be mixed by hand to re-create that special feel.

Castles, cottages, town houses, mansions, churches - all periods.

Next time don't wonder how or where, just phone The Paint Practice.

18 Hallam Chase, Sandygate, Sheffield,
South Yorkshire S10 5SW
Tel 0114 230 6828 Mobile 0781 434 8239

INTERIORS

INTERIOR CONSERVATORS & DECORATORS

▶ MARTIN A HORLER

The Old School House, Kilmersdon, Somerset BA3 5TB
Tel 01761 437372
E-mail plegstow@madasafish.com
Website www.martinhorler.20m.com
Contact Martin Horler FTC LCG Cert Ed

PAINTERS AND DECORATORS: Surveys, reports, specifications and supervision undertaken, also expert advice on all aspects of painting and decoration provided. Instructions to provide expert evidence reports. Any area of the UK covered. Martin Horler has 40 years experience as a specialist decorator, including 21 years as a lecturer in decoration. Past President of The Association of Painting Craft Teachers.

▶ HUNING DECORATIONS

18 Lower Oldfield Park, Bath BA2 3HL
Tel 01225 427459 Fax 01225 480846
E-mail huningdecor@aol.com
Website www.huningdecorations.co.uk

SPECIALIST DECORATORS: Joy and Wilm Huning have many years experience in the field of art and decorating, using traditional and modern materials, in small and large scale projects. Their skills cover: graining, marbling, gilding, stencilling, murals, trompe l'oeil, glaze effects and decorative painting. Specialisations also include copying from existing materials or finishes; hand mixing colours to match on a small or large scale; and creative work with murals, stencilling and colour schemes. Projects include work at Sir John Soane's Museum, Burlington House, The Royal Naval College in Greenwich, The Weald and Downland Open Air Museum, colleges and many small works for private clients. *See also: display entry on this page.*

INTERIOR DESIGNERS

▶ KAW DESIGN

38 Meadow Road, London SW8 1QB
Tel/Fax 020 7735 6088
E-mail kate@kawdesign.co.uk
Website www.kawdesign.co.uk
Contact Kate Ainslie-Williams, IIDA, BIDA, MSc Historic Conservation, associate member IHBC

INTERIOR ARCHITECTURE AND DESIGN: KAW DESIGN offers specialist advice and interior design in historic and traditional interiors, with experience stretching back 20 years over a wide range of building types. Work undertaken includes private residential property, hotel refurbishment, company headquarters and institutional buildings. Kate Ainslie-Williams has a particular interest in ensuring the continued viable use of old buildings and the sympathetic adaptation of their interiors to modern-day requirements. She combines design flair with extensive knowledge of interiors including the more practical aspects of a project and works closely with clients to achieve the optimum results.

HISTORIC PAINT ANALYSIS

▶ LISA OESTREICHER

Jubilee House, High Street, Tisbury, Wiltshire SP3 6HA
Tel 01747 871717 Fax 01747 871718
E-mail lisa.oestreicher@virgin.net

ARCHITECTURAL PAINT INVESTIGATIONS: Lisa Oestreicher provides a full range of analytical skills and techniques for the study of paint and decorative finishes within historic buildings. These include the identification of pigments and media as well as archival research. Full reports are prepared to provide a detailed insight into the historical development of interior and exterior schemes, for documentation purposes, conservation or accurate restoration. Assistance can also be given in the design and implementation of historically informed interior decoration schemes. Recent clients include the National Trust, Victoria & Albert Museum, architects, conservators and owners.

INTERIOR CONSERVATORS & DECORATORS

in the

Re-making, Designing, Conservation, Restoration and Cleaning

of

Carpets, Kilims and Textiles

Work to date includes:

- 18th Century Axminsters,
- Ushaks
- 19th Century Anglo-Persian
- Anglo-Indian carpets.

Large carpets up to 8m wide and runners up to 20m long can be woven.

For further information please contact
David or Sara Bamford

The Workhouse, Industrial Estate,
Presteigne, Powys LD8 2UF
Telephone: 01544 267849
Fax: 01544 260530

Linney Cooper
the carpet consultants

- Professional Service
- From Design to Installation
- Huge Choice of Carpets
- Prestige Products
- Quality Installation

BY APPOINTMENT
TO HRH THE PRINCE OF WALES
LINNEY COOPER LIMITED, NORTH WALES
CARPET SUPPLIER & INSTALLER

Jubilee House, Unit 1 Builder Street, Llandudno, North Wales, LL30 1DR
Tel: 01492 874000 Fax: 01492 873432 Email: carpets@linneycooper.co.uk

TEXTILE FLOOR COVERINGS

David Luckham Consultants Limited

Specialists in textile floor coverings

Supply, specification, design and development of carpets and other floorcoverings for historic properties.

7 Tuthill Park Wardington Banbury Oxfordshire OX17 1RR

Tel +44 (0) 1295 750022 • Fax +44 (0) 1295 758097
email enquiries@dl-consultants.com

INTERIORS

WALL PAINTING CONSERVATORS

▶ CLIVEDEN CONSERVATION WORKSHOP LTD

Head Office – The Tennis Courts, Cliveden Estate, Taplow, Maidenhead, Berkshire SL6 0JA
Tel 01628 604721 Fax 01628 660379
SCULPTURE, STONE AND WALL PAINTINGS CONSERVATION: *See also: profile entry in Stone section, page 102.*

▶ HIRST CONSERVATION

Laughton, Sleaford, Lincolnshire NG34 0HE
Tel 01529 497449 Fax 01529 497518
CONSULTANTS AND CONSERVATORS: *See also: display entry on the inside front cover and profile entry in Building Contractors section, page 62.*

▶ PETER MARTINDALE

1 Green Drove, Fovant, Salisbury, Wiltshire SP3 5JG
Tel/Fax 01722 714271
CONSERVATION OF WALL PAINTINGS AND ASSOCIATED DECORATIVE POLYCHROMY: Painted decoration within historic buildings is a unique and integral part of their fabric and history. To conserve these elements for future generations it is vital that appropriate, justifiable and sympathetic techniques and procedures are employed. This practice, founded in 1989, specialises in the conservation of wall paintings, decorative polychromy and paintings on panel and canvas. Clients include the National Trust, English Heritage and the Council for the Care of Churches. In addition to undertaking conservation projects, services include consultancy, condition surveys and technical analysis. Peter Martindale is an accredited member of the UKIC.

▶ THE PERRY LITHGOW PARTNERSHIP

1 Langston Lane, Station Road, Kingham, Oxon OX7 6UW
Tel 01608 658067 Fax 01608 659133
CONSERVATORS OF WALL PAINTINGS AND OTHER POLYCHROME DECORATION: Established 1983, the partnership operates throughout the UK and Eire specialising in wall paintings, panel paintings and paintings on canvas and stone. Clients include English Heritage, the National Trust, Council for the Care of Churches, architects and local governments. Extensive experience with 12th to 20th century schemes including recently: the 16th century High Great Chamber Frieze, Hardwick Hall; the 13th century Nave Ceiling, Peterborough Cathedral and; 13th to 17th century wall paintings, St Albans Cathedral. Conservation services include: project consultancy, condition surveys, monitoring, technical analysis, photographic and digitised graphic recording. Sympathetic methods and materials.

▶ PLOWDEN & SMITH

190 St Ann's Hill, London SW18 2RT
Tel 020 8874 4005 Fax 020 8874 7248
E-mail richard.rogers@plowden-smith.com Website www.plowden-smith.com
CONSERVATION AND RESTORATION: *See also: display entry in Fine Art Conservation section, page 203.*

▶ RICHARD IRELAND PERIOD RESTORATION

22 Avenue Road, Isleworth, Middlesex TW7 4JN
Tel 020 8568 5978 Fax 020 8568 5978
HISTORIC PLASTERWORK AND POLYCHROMATIC DECOR CONSERVATION: *See also: profile entry in Plasterwork section, page 217.*

PICTURE FRAMES

▶ LINDA ANDREWS

The Bakehouse, 1 Acreman Street, Sherborne, Dorset DT9 3NU
Tel 01935 817111 Fax 08707 058698 E-mail info@l-andrews.com
MUSEUM QUALITY CONSERVATION AND RESTORATION: *See also: profile entry in Gilders section, page 206.*

▶ PLOWDEN & SMITH

190 St Ann's Hill, London SW18 2RT
Tel 020 8874 4005 Fax 020 8874 7248
E-mail richard.rogers@plowden-smith.com
Website www.plowden-smith.com
CONSERVATION AND RESTORATION: *See also: display entry in Fine Art Conservation section, page 203.*

FINE ART CONSERVATORS

▶ LINDA ANDREWS

The Bakehouse, 1 Acreman Street, Sherborne, Dorset DT9 3NU
Tel 01935 817111 Fax 08707 058698 E-mail info@l-andrews.com
MUSEUM QUALITY CONSERVATION AND RESTORATION: *See also: profile entry in Gilders section, page 206.*

▶ FRANCIS W DOWNING

203 Wetherby Road, Harrogate, North Yorkshire HG2 7AE
Tel 01423 886962 E-mail francisdowning@msn.com
Website www.francisdowning.com
RESTORATION AND CONSERVATION OF PAINTINGS: Each painting is an individual work requiring specialised care and consideration when cleaning and restoring. The studio was established in 1976 to clean and restore paintings and paintwork in oil, tempera and acrylic, on canvas, panel and wood structures to the highest standards of conservation for private clients, stately homes, churches, museums and galleries. With photographs and detailed reports prepared during the work, ensuring that ethical guidelines are recorded at every stage. X-ray, infrared, ultra-violet and chemical analysis are used to aid examination. However, there is generally no charge for initial examinations or quotations. With full professional indemnity insurance, the practice covers the UK and Europe and is included on the register maintained by the UK Institute for Conservation of Historic and Artistic Works. Francis Downing also works as a forensic conservator and investigator for police, auctioneers and insurance companies. Over 30 years experience of art conservation, research and recovery. *See also: display entry on page 203.*

▶ HIRST CONSERVATION

Laughton, Sleaford, Lincolnshire NG34 0HE
Tel 01529 497449 Fax 01529 497518 E-mail hirst@hirst-conservation.com
Website www.hirst-conservation.com
SPECIALIST BUILDING AND ART CONSERVATORS: Consultancy and conservation work to painted and applied decoration on plaster, stone, canvas, wood and metal substrates. Restoration and recreation of historic decorative schemes. Also specialist building works including joinery, sculpture, marble, stonework, stone cleaning, stucco, pargetting, wall and floor plasters. Surveys, specifications and analysis services available. Hirst Conservation's policy is to provide a conservation service that is second to none. The company takes great pride in ensuring that it remains at the forefront of contemporary conservation ethics and thinking. The highly professional and dedicated team represents many different conservation skills and disciplines, and through its combined knowledge and experience is constantly striving to enhance current and develop future conservation practices. *See also: display entry on the inside front cover.*

▶ INTERNATIONAL FINE ART CONSERVATION STUDIOS LTD

43–45 Park Street, Bristol BS1 5NL
Tel 0117 929 3480 Fax 0117 922 5511
CONSULTANTS, RESTORERS AND CONSERVATORS OF PAINTINGS AND MURALS: *See also: display entry in Interiors Consultants & Conservators section, page 199.*

▶ RUPERT HARRIS CONSERVATION

Studio 5, No 1 Fawe Street, London E14 6PD
Tel 020 7515 2020 Fax 020 7987 7994 E-mail enquiries@rupertharris.com
Website www.rupertharris.com
FINE ART CONSERVATION: Conservators of fine metalwork and sculpture, including bronze, lead, zinc and electrotype; modern/contemporary art; fine ironwork; casting; replication; security fixing and maintenance. Advisor to the National Trust since 1982. Other clients include English Heritage and the Royal Collection, major museums, art galleries, city councils, architects and private collectors. *See also: display entry in this section, page 203.*

▶ GILLIAN M H WALKER

Unit 4, Whiting Street, Sheffield S8 9QR
Tel 0114 250 8161
CONSERVATION AND RESTORATION OF EASEL PAINTINGS: The studio specialises in the conservation of easel paintings from all periods, to museum standards. Clients include local authorities, stately homes, the National Trust and private owners. The head of the studio has UKIC accreditation.

FINE ART CONSERVATORS

CONSERVATION & RESTORATION

With over 35 years experience in conservation and restoration, Plowden & Smith is established as one of the world's leading conservators and restorers of fine art objects, furniture and paintings.

Each department is staffed by highly trained specialists. Departments include; furniture, paintings, ceramics, marble and stone, metalwork and decorative arts including ivories, waxes, gilding and ancient lacquer. The range of objects and paintings is diverse with early pieces from antiquity through to contemporary sculpture. Most objects are worked on at the company's 2 London premises but work is also carried out both on site and abroad.

In 1998 ***Plowden & Smith Exhibitions*** *was launched to build on the substantial experience in object mounting and exhibition displays.*

Plowden & Smith Projects *one of the leaders for on site conservation and restoration projects with a proven track record for completion on time and to budget. Projects are undertaken in all our specialist areas including painted ceilings, architectural and decorative metalwork, sculpture, gilding, stone and wood. Full project management facilities are available.*

PLOWDEN & SMITH LIMITED 190 ST ANN'S HILL LONDON SW18 2RT
Telephone: 020 8874 4005 Facsimile: 020 8874 7248
E-mail: richard.rogers@plowden-smith.com www.plowden-smith.com

RUPERT HARRIS CONSERVATION

conservation of fine metalwork
historic and modern sculpture

including
bronze, lead, zinc
and electrotype sculpture
modern and contemporary art
chandeliers and lanterns
gold and silver, jewellery
fine ironwork
arms and armour
ecclesiastical metalwork
casting and replication
gilding
consultancy and maintenance

Studio 5, No.1 Fawe Street
London E14 6PD
Tel. 020 7987 6231 / 020 7515 2020
Fax. 020 7987 7994
email:enquiries@ruperthatrris.com
www.ruperthatrris.com

Advisor to the National Trust
Diploma Metalwork Conservation V&A
Member of the IIC
Included on the UKIC Conservation Register

WALLPAPERS

▶ FARROW & BALL
Showroom, 249 Fulham Road, London
Tel 01202 876141 Fax 01202 873793
MANUFACTURERS OF TRADITIONAL PAPERS AND PAINT: Seven collections with over four hundred papers. *See also: display entry on page 186.*

▶ HAMILTON WESTON WALLPAPERS LTD
18 St Mary's Grove, Richmond, Surrey TW9 1UY
Tel 020 8940 4850 Fax 020 8332 0296 Mobile 07973 429568
E-mail info@hamiltonweston.com
Website www.hamiltonweston.com
HISTORIC WALLPAPERS: Hamilton Weston specialises in reproducing wallpapers for the restoration and refurbishment of period interiors. Hand and machine printed designs date from *c*1690. Documents for specific projects may be selected from the archive or reproduced from clients' designs. Handprints can be coloured to order in small quantities as required. Clients include the National Trust, leading architectural and design practices, museums, film and television companies. Robert Weston is an architectural historian and interior designer with over 25 years experience. An interior design consultancy service is available. Corporate members of BIDA and The Wallpaper History Society.

TEXTILES

▶FIONA HUTTON TEXTILE CONSERVATION
Ivy House Farm, Wolvershill Road, Banwell, Somerset BS29 6LB
Tel 01934 822449 E-mail fiona@textileconservation.co.uk
Website www.textileconservation.co.uk
TEXTILE CONSERVATION: Fiona Hutton Textile Conservation offers a full range of conservation treatments in its custom-designed studio, for antique or modern textiles damaged by accident or by poor storage or display conditions, including painted and printed textiles, woven tapestries, embroideries, upholstery and costume.

ANTIQUE AND FURNITURE RESTORERS

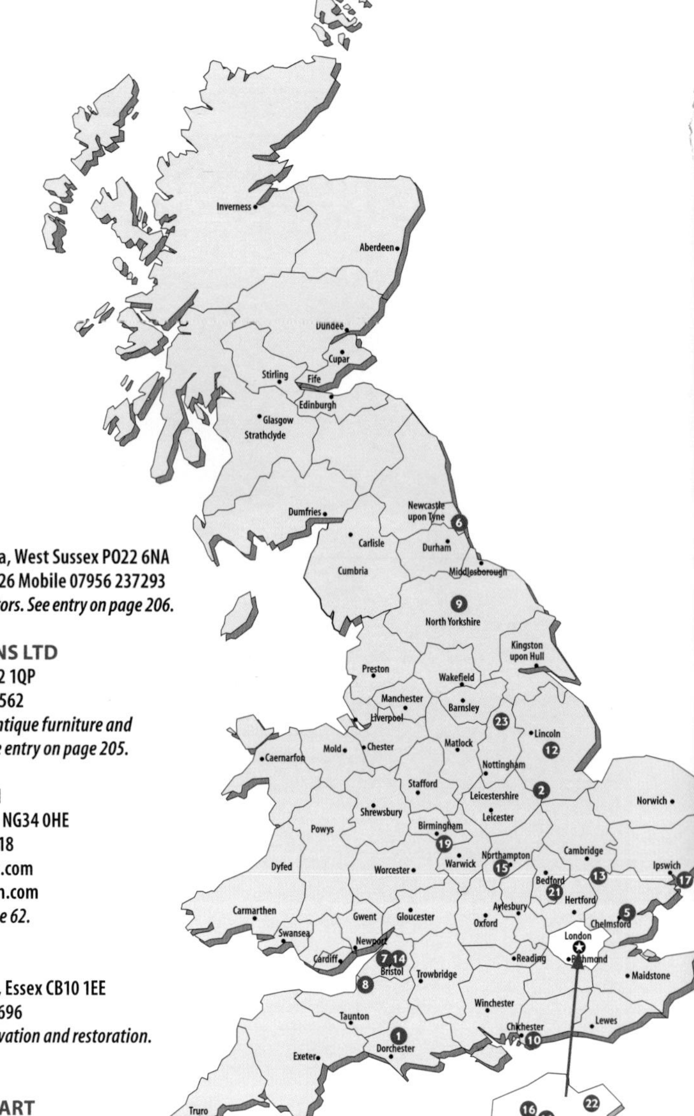

Regional designation is according to office location
Many firms operate nationally.

❶ LINDA ANDREWS
The Bakehouse, 1 Acreman Street, Sherborne, Dorset DT9 3NU
Tel 01935 817111 Fax 08707 058698
E-mail info@l-andrews.com
See entry on page 206.

❷ ANTHONY BEECH FURNITURE CONSERVATION AND RESTORATION
The Stable Courtyard, Burghley House, Stamford, Lincolnshire PE9 3JY
Tel 01780 481199
Specialist conservation of historic house furniture collections. See entry on page 205.

❸ ARTHUR BRETT
103 Pimlico Road, London SW1W 8PH
Tel 020 7730 7304 Fax 020 7730 7105
Website www.arthur-brett.com
Cabinet makers, antique furniture restorers and manufacturers of fine joinery. See entry on page 126.

❹ CARVERS & GILDERS
9 Charterhouse Works, Eltringham Street, London SW18 1TD
Tel 020 8870 7047 Fax 020 8874 0470
E-mail bcd@carversandgilders.com
Website www.carversandgilders.com
Restoration and conservation. Specialists in fine water gilding. See entry on page 124.

❺ CLIVE BEARDALL
Workshop and Showroom, 104b High Street, Maldon, Essex CM9 5ET
Tel 01621 857890 Fax 01621 850753
Website www.clivebeardall.co.uk
Antique furniture restoration and French polishing. See entries on page 205.

❻ COLTMAN RESTORATIONS
80 Meldon Terrace, Heaton, Newcastle NE6 5XP
Tel 0191 224 5209
Specialist restorers of antique furniture for four generations. Cabinet work, carving, wood turning, French polishing. Restoration of wall panels undertaken.

❼ ELIZABETH HOLFORD ASSOCIATES LTD
The Tabaco Factory, Raleigh Road, Southville, Bristol BS3 1TF
Tel 0117 902 0269 Fax 0117 902 0243 Mobile 07977 997207
E-mail elizabeth.holford@blueyonder.co.uk
Conservation and restoration of paintings and painted surfaces. See entry on page 197.

❽ FIONA HUTTON TEXTILE CONSERVATION
Ivy House Farm, Wolvershill Road, Banwell, Somerset BS29 6LB
Tel 01934 822449
E-mail fiona@textileconservation.co.uk
Website www.textileconservation.co.uk
Textile conservation. See entry on page 203.

❾ FRANCIS W DOWNING
203 Wetherby Road, Harrogate, North Yorkshire HG2 7AE
Tel 01423 886962 E-mail francisdowning@msn.com
Website www.francisdowning.com
Restoration and conservation of paintings. See entry on page 203.

❿ G R PEARCE
5, Ancton Drive, Middleton on Sea, West Sussex PO22 6NA
Tel 01243 583272 Fax 01243 584926 Mobile 07956 237293
Specialist French polishing contractors. See entry on page 206.

⓫ H J HATFIELD AND SONS LTD
42 St Michael's Street, London W2 1QP
Tel 020 7723 8265 Fax 020 7706 4562
Conservation and restoration of antique furniture and interior architectural features. See entry on page 205.

⓬ HIRST CONSERVATION
Laughton, Sleaford, Lincolnshire NG34 0HE
Tel 01529 497449 Fax 01529 497518
E-mail hirst@hirst-conservation.com
Website www.hirst-conservation.com
Art conservators. See entry on page 62.

⓭ HOWELL & BELLION
66A High Street, Saffron Walden, Essex CB10 1EE
Tel 01799 522 402 Fax 01799 525 696
Church interior decoration, conservation and restoration. See entry on page 198.

⓮ INTERNATIONAL FINE ART CONSERVATION STUDIOS LTD
43–45 Park Street, Bristol BS1 5NL
Tel 0117 929 3480 Fax 0117 922 5511
Consultants, restorers and conservators of paintings and murals. See entry on page 199.

⓯ THE LEATHER CONSERVATION CENTRE
University College Campus, Boughton Green Road, Moulton Park, Northampton NN2 7AN
Tel 01604 719766 Fax 01604 719649
E-mail lcc@northampton.ac.uk
Registered Charity No. 276485
Conservators of objects made wholly or partly of leather. See entry on page 206.

⓰ LUARD CONSERVATION
23 Harlesden Gardens, London NW10 4EY
Tel 020 8961 7544 Fax 020 8961 7545 Mobile 07973 741117
E-mail david.luard@virgin.net
Conservation and restoration of wood, carvings and other objects. See entry on page 198.

⓱ MARK BRIDGES CARVER & GILDER LTD
406 High Road, Trimley St Martin, Suffolk IP11 0SG
Tel 01394 273782 Fax 01394 277671
E-mail mbridges@woodcarving.fsnet.co.uk
Gilder and wood carver. See entry on page 125.

⓲ MICHENUELS OF LONDON
Unit 7, Titan Business Estate, Ffinch Street, London SE8 5QA
Tel 020 8694 9206 Fax 020 8694 9201
Antique and furniture restorers. See entry on page 206.

⓳ N E J STEVENSON
Church Lawford Business Centre, Limestone Hall Lane, Church Lawford, Coventry CV23 9HD
Tel 024 765 44662 Fax 024 765 45345
Designers and makers of distinctive commissioned furniture. See entry on page 206.

⓴ PLOWDEN & SMITH
190 St Ann's Hill, London SW18 2RT
Tel 020 8874 4005 Fax 020 8874 7248
E-mail richard.rogers@plowden-smith.com
Website www.plowden-smith.com
Conservation and restoration. See entry on page 203.

㉑ ROMARK SPECIALIST CABINETMAKERS LIMITED
Unit 3, Shaftesbury Industrial Centre, Icknield Way, Letchworth, Herts SG6 1HE
Tel 01462 684855 Fax 01462 684833
E-mail Romarkltd@aol.com
Cabinetmakers and antique furniture restorers. See entry on page 205.

㉒ RUPERT HARRIS CONSERVATION
Studio 5, No 1 Fawe Street, London E14 6PD
Tel 020 7515 2020 Fax 020 7987 7994
E-mail enquiries@rupertharris.com
Website www.rupertharris.com
Fine art conservation. See entry on page 203.

㉓ GILLIAN M H WALKER
Unit 4, Whiting Street, Sheffield S8 9QR
Tel 0114 250 8161
Conservation and restoration of easel paintings. See entry on page 202.

▶ ANTHONY BEECH FURNITURE CONSERVATION AND RESTORATION

The Stable Courtyard, Burghley House, Stamford, Lincolnshire PE9 3JY
Tel 01780 481199

SPECIALIST CONSERVATION OF HISTORIC HOUSE FURNITURE COLLECTIONS: Anthony Beech has been responsible for restoring and conserving many historically important collections and interiors. His remit includes conservation and restoration of furniture, bookcases and panelling. Furniture including boulle and giltwood decoration, carving and upholstery conservation. Clients include museums, country houses and churches.

▶ ARTHUR BRETT

103 Pimlico Road, London SW1W 8PH
Tel 020 7730 7304 Fax 020 7730 7105
Website www.arthur-brett.com

CABINET MAKERS, ANTIQUE FURNITURE RESTORERS AND MANUFACTURERS OF FINE JOINERY: A fifth generation family business, established in 1870, Arthur Brett manufactures the finest quality traditional furniture of English 18th and early 19th century styles. Much of the company's work involves special commissions, and in recent times it has expanded its capabilities into classic joinery work for leading architects. In addition Arthur Brett undertakes certain restoration and re-polishing work for both antique furniture and interior fitments, and can offer replica manufacturing facilities where an original item must be copied in every detail. *See also: display entry in Fine Joinery section, page 126.*

▶ CLIVE BEARDALL

Workshop and Showroom, 104b High Street, Maldon, Essex CM9 5ET
Tel 01621 857890 Fax 01621 850753
Website www.clivebeardall.co.uk

ANTIQUE FURNITURE RESTORATION AND FRENCH POLISHING: Established since 1982, Clive Beardall's attention to detail and authenticity of restoration and materials have earned him and his highly skilled team of craftsmen many prestigious commissions and a first class reputation. The team is experienced in all aspects of the antique furniture trade. Services include period furniture restoration, traditional hand French polishing, re-upholstery, marquetry, carving, gilding, leather desk lining, rush and cane seating, decorative finishes, valuations and made-to-order furniture, on site interior polishing, including panelling, staircases, floors etc. Clive Beardall is an accredited member of The British Antique Furniture Restorers Association and his workshops are included on the Conservation Register. *See also: display entry on this page.*

▶ H J HATFIELD AND SONS LTD

42 St Michael's Street, London W2 1QP
Tel 020 7723 8265 Fax 020 7706 4562

CONSERVATION AND RESTORATION OF ANTIQUE FURNITURE AND INTERIOR ARCHITECTURAL FEATURES: Founded in 1834, Hatfield and Sons has been responsible for the conservation and restoration of many collections of furniture and metalwork. In 1860 the company was commissioned to re-frame the miniatures in the Royal Collection. Hatfield and Sons will undertake conservation in all areas of the fine arts, specialising in furniture, especially boulle and marquetry, ormolu lighting, furniture mounts and interior architectural features. Advice is offered on conservation and restoration to museums, private and trade clients.

▶ PLOWDEN & SMITH

190 St Ann's Hill, London SW18 2RT
Tel 020 8874 4005 Fax 020 8874 7248
E-mail richard.rogers@plowden-smith.com
Website www.plowden-smith.com

CONSERVATION AND RESTORATION: *See also: display entry in Fine Art Conservation section, page 203.*

▶ ROMARK SPECIALIST CABINETMAKERS LIMITED

Unit 3, Shaftesbury Industrial Centre, Icknield Way, Letchworth, Herts SG6 1HE
Tel 01462 684855 Fax 01462 684833
E-mail Romarkltd@aol.com

CABINETMAKERS AND ANTIQUE FURNITURE RESTORERS: Romark Specialist Cabinetmakers Limited specialises in the restoration of fine antique furniture including marquetry, inlaid ivory, brass and tortoiseshell, wall paneling, wood carving, wax and French polishing, clock case restoration, desk-top leather lining, keys and locks. A complete re-upholstery service is also provided. Worm infested furniture treated by non-toxic methods and guaranteed clear. The firm also specialises in designing and making items of furniture to customers' specifications.

▶ RUPERT HARRIS CONSERVATION

Studio 5, No 1 Fawe Street, London E14 6PD
Tel 020 7515 2020 Fax 020 7987 7994
E-mail enquiries@ruperthaaris.com
Website www.rupertharris.com

ANTIQUE AND FURNITURE RESTORATION: *See also display entry in Fine Art Conservation section, page 203.*

CABINET MAKERS

▶ N E J STEVENSON

Church Lawford Business Centre, Limestone Hall Lane, Church Lawford, Coventry CV23 9HD
Tel 024 765 44662 Fax 024 765 45345

DESIGNERS AND MAKERS OF DISTINCTIVE COMMISSIONED FURNITURE: This company has established itself over many years as one of the finest furniture making companies in this country, and their skills are recognised as far afield as Korea, Russia and the USA. Their work encompasses free standing furniture through to fitted interiors, and has included major projects for the National Trust and English Heritage. They have undertaken many commissions for the Royal Household including the reproduction of the rosewood and gilt Pugin sideboard for Windsor Castle. Neil Stevenson and his workforce are committed to the very highest standards and delivering those standards on time.

GILDERS

▶ LINDA ANDREWS

The Bakehouse, 1 Acreman Street, Sherborne, Dorset DT9 3NU
Tel 01935 817111 Fax 08707 058698
E-mail info@l-andrews.com

GILDERS: Museum quality conservation and restoration of gilded, polychromed objects, oil paintings, icons, French mounts, washes and remounting of fine artwork. Traditional techniques include fine water gilding, patinas, gilding on glass, pastiglia, sgraffito, egg tempera, lacquerwork, decorative finishes and cleaning of fine art and decorative objects. Highly professional with extensive national and international portfolio.

▶ CARVERS & GILDERS

9 Charterhouse Works, Eltringham Street, London SW18 1TD
Tel 020 8870 7047 Fax 020 8874 0470
E-mail bcd@carversandgilders.com
Website www.carversandgilders.com

RESTORATION AND CONSERVATION: Specialists in fine water gilding. *See also: profile entry in Wood Carvers section, page 126.*

▶ MARK BRIDGES CARVER & GILDER LTD

406 High Road, Trimley St Martin, Suffolk IP11 0SG
Tel 01394 273782 Fax 01394 277671
E-mail mbridges@woodcarving.fsnet.co.uk

GILDER AND WOOD CARVER: *See also: display entry in Wood Carvers section, page 125.*

LEATHER CONSERVATION

▶ THE LEATHER CONSERVATION CENTRE

University College Campus, Boughton Green Road,
Moulton Park, Northampton NN2 7AN
Tel 01604 719766 Fax 01604 719649
E-mail lcc@northampton.ac.uk
Registered Charity No. 276485

CONSERVATORS OF OBJECTS MADE WHOLLY OR PARTLY OF LEATHER: Established in 1978, a non-profit-making organisation with charitable status offering world-wide service in leather conservation. The Centre undertakes practical conservation, training, research and analysis, consultancy work, and produces a wide-ranging series of publications on all aspects of leather conservation. Its principal activity of fee-earning practical conservation work covers a wide range of historical items, including decorated screens, wall-hangings, car and carriage upholstery, harness and saddlery, costume, and archaeological and ethnographic items. The LCC is included on the UKIC Conservation Register.

▶ PLOWDEN & SMITH

190 St Ann's Hill, London SW18 2RT
Tel 020 8874 4005 Fax 020 8874 7248
E-mail richard.rogers@plowden-smith.com
Website www.plowden-smith.com

CONSERVATION AND RESTORATION: *See also: display entry in Fine Art Conservation section, page 203.*

FRENCH POLISHERS

▶ G R PEARCE

5 Ancton Drive, Middleton on Sea, West Sussex PO22 6NA
Tel 01243 583272 Fax 01243 584926
Mobile 07956 237293

SPECIALIST FRENCH POLISHING CONTRACTORS: G R Pearce is a company engaged in high quality contracts in London, throughout the United Kingdom and overseas. Traditional hand French polishing and contemporary wood finishing is undertaken by experienced, reliable craftsmen who also specialise in colour matching new timbers to existing joinery. 42 years practical experience. Technical advice and colour/finish samples are provided by the company. Please contact Mr Geoffrey Pearce.

▶ MICHENUELS OF LONDON

Unit 7, Titan Business Estate, Ffinch Street, London SE8 5QA
Tel 020 8694 9206 Fax 020 8694 9201

FRENCH POLISHERS AND FURNITURE RESTORERS: The partnership team has firmly established itself as among the leading restorers in the UK. Widely known to the industry for their unusual and amazing techniques in colour matching and particularly recognised for their depth of knowledge and expertise on the conservation of listed buildings. Previous clients include Harrods, the MoD at Whitehall and the Royal Naval College, Greenwich.

▶ ALDERSHAW HANDMADE CLAY TILES LTD
Pokehold Wood, Kent Street, Sedlescombe, East Sussex TN33 0SD
Tel 01424 756777 Fax 01424 756888
E-mail tiles@aldershaw.co.uk
Website www.aldershaw.co.uk
HANDMADE CLAY TILES: *See also: profile entry in Clay Tiles & Roof Features section, page 85.*

▶ HERITAGE TILE CONSERVATION LTD
The Studio, 2 Harris's Green, Broseley, Shropshire TF12 5HJ
Tel/Fax 01952 881039
E-mail heritagetile@msn.com
Website www.heritagetile.co.uk
CONSULTANTS AND SPECIALISTS IN THE CONSERVATION AND RESTORATION OF ARCHITECTURAL CERAMICS AND GLASS: The company offers a comprehensive service for the restoration of pictorial tiled panels, geometric, encaustic and mosaic pavements, Opus Sectile, wall mosaic, tiling and faience facades. Restoration work is undertaken throughout the UK, or at their workshop, where they have pioneered the development of specialised techniques for the careful removal of architectural ceramics, wherever this proves necessary. Ceramic and glass material can be manufactured to match the original or provided from the company's stock of reclaimed materials. Clients include: Royal Museum of Scotland, Victoria & Albert Museum, Lichfield Cathedral, Foreign Office, London Transport Museum, Birmingham Museum & Art Gallery, Guy's and St Thomas' Hospital, Chatham Historic Dockyard Trust and the Highland Council.

▶ THE JACKFIELD CONSERVATION STUDIO
Jackfield Tile Museum, Ironbridge, Telford, Shropshire TF8 7AW
Tel/Fax 01952 883720
E-mail lesley@jackfield.fsbusiness.co.uk
CONSERVATION AND RESTORATION OF ARCHITECTURAL CERAMICS: A wide range of skills, expertise and experience in the conservation and restoration of architectural tile schemes from the medieval period through to the art deco period of the 1930s. The company can offer a complete tile service at Jackfield Tile Museum, including detailed reports and surveys covering the condition and treatment of historic schemes or collections; and manufacture of both glazed wall tiles and encaustic floor tiles. Included on the Conservation Register, member of IHBC and PACR accredited. Established in 1990, clients since 1996 include House of Commons, E H Medieval Tile Project, Royal Courts of Justice, St Albans Cathedral, Harrogate Turkish Baths, Warwick Castle, National Trust and Osgoode Hall, Toronto.

▶ ORIGINAL FEATURES (RESTORATIONS) LTD
155 Tottenham Lane, Crouch End, London N8 9BT
Tel 020 8348 5155 Fax 020 8341 4744
E-mail sales@originalfeatures.co.uk
Website www.originalfeatures.co.uk
CONSERVATION AND RESTORATION OF GEOMETRIC AND ENCAUSTIC TILED FLOORS AND PAVEMENTS: Original Features is a specialist company restoring Victorian and Edwardian tiled floors. In addition to restoration work, a consultancy service is available. Tiles can be supplied for restoration work by others. Original Features will also undertake restoration of period fireplaces (cast iron, stone and marble) and has a large stock of original fireplaces. Distributors of the 'Olde English' range of geometric and encaustic floor tiles.

FLOOR & WALL TILES

▶ THE ORIGINAL FLOORING COMPANY

230A Grange Road, Kings Heath, Birmingham B14 7RS
Tel 0121 605 8898 Fax 0121 605 8828
Mobile 07831 300628

DESIGN, INSTALLATION AND RESTORATION OF FLOOR AND WALL TILES: Brothers Michael and Mario Puopolo formed The Original Flooring Company after serving an apprenticeship with Italian craftsmen, learning traditional skills in tiling. Since the company's inception in 1987 the brothers' specialist skills have taken them all over the United Kingdom. They offer a complete service which includes the design, installation and restoration of Victorian and Edwardian floor tiles. Large stocks of original tiles are available. Recent work includes restoration to floors of The Birmingham College of Art and St James Church, Hartlebury, both Grade I listed.

▶ ORIGINAL OAK

Ashlands, Burwash, East Sussex TN19 7HS
Tel 01435 882228
Mobile 07771 890835

FLOOR TO WALL TERRACOTTA TILES: Quality handmade and reclaimed 6″/12″ terracotta floor tiles. Many sizes/shapes available. Very competitive prices. Also available marble/slate. *See also: profile entry in Timber Flooring section, page 211.*

▶ THE TILE GALLERY

1 Royal Parade, 247 Dawes Road, Fulham, London SW6 7RE
Tel 020 7385 8818 Fax 020 7381 1589
E-mail thetileg@lleryltd.freeserve.co.uk

FLOOR AND WALL TILES, CERAMIC AND NATURAL STONE

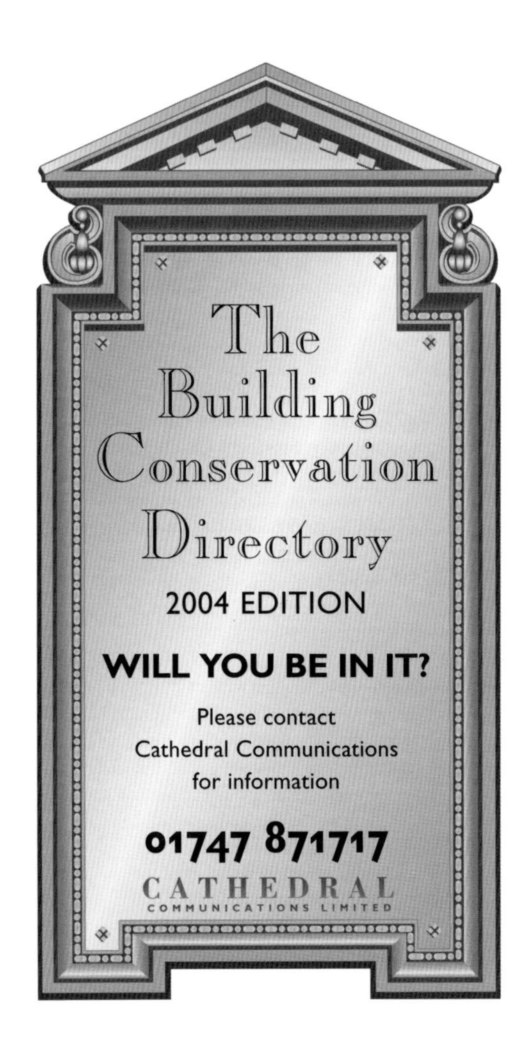

MOSAICS

▶ ANTIQUE BRONZE LIMITED (HISTORIC FLOORING DIVISION)

44 Hillway, Holly Lodge Estate, London N6 6EP Tel 020 8340 0931 Fax 020 8340 0743
E-mail info@antiquebronze.co.uk Website www.antiquebronze.co.uk

MOSAICS: Historic floor restoration by expert craftsmen with decades of experience specialising in the restoration, maintenance and repair of period flooring. All types of marble mosaic, tesserae, terrazzo, wood block flooring, cork and strip undertaken. Clients include Buckingham Palace, the Natural History Museum, University of London and many others.

▶ LONDON MOSAIC RESTORATION

2A Morrish Road, London SW2 4EH
Tel 07957 230873 E-mail julianhill@londonmosaicrestoration.co.uk
Website www.londonmosaicrestoration.co.uk

MOSAICS

▶ PLOWDEN & SMITH

190 St Ann's Hill, London SW18 2RT Tel 020 8874 4005 Fax 020 8874 7248
E-mail richard.rogers@plowden-smith.com Website www.plowden-smith.com

CONSERVATION AND RESTORATION: *See also: display entry in Fine Art Conservation section, page 203.*

▶ TREVOR CALEY ASSOCIATES LIMITED

Woodgreen, Fordingbridge, Hampshire SP6 2AU
Tel 01725 512320 Fax 01725 512420

MOSAIC CONSERVATION AND RESTORATION, DESIGN AND EXECUTION OF NEW WORKS: In addition to meticulous mosaic conservation work the company custom-designs and executes new mosaics in a wide range of traditional and contemporary styles. Winner of the RICS Building Conservation Award 1999 for mosaic work on The Albert Memorial, London, for English Heritage. Other conservation works include projects at Royal Courts of Justice, the Old Admiralty Building, on-going works at St Paul's Cathedral, London, and Charlotte Square for Scottish Heritage. The design and execution of new mosaics include projects for Westminster Cathedral, London, St Patrick's Cathedral, Melbourne, Australia, and at present for the Royal Albert Hall, London.

VICTORIAN AND EARLY 20th CENTURY ARCHITECTURAL TILE SCHEMES

PRINCIPLES OF CONSERVATION

LESLEY DURBIN

Tiles have been used as a medium for artistic and creative expression throughout the ages and the finest examples, such as murals and the work of William de Morgan at the end of the 19th century, have, rightly, always been regarded as individual works of ceramic art. However, tiled surfaces are common in old and historic buildings and the tiles themselves are, by their nature, so numerous that they are often regarded as being merely utilitarian and disposable, particularly the most recent examples. Furthermore, because tiles have always been mass-produced, generally using cheap and readily available materials, the individual object is often under-appreciated. The result, over the centuries, has been the loss of the greater mass of ceramic tile material representing successive ages of manufacture, often merely due to changes in fashion.

What therefore should be our guiding principle for conserving tiles today? Given the diversity of age, location and provenance found under the heading 'tile' it is not always helpful to fall back upon ethical conservation practices which dictate that all historical fabric is of equal importance and as a result must be preserved. Assessment of all facets of the tile or tile scheme in question must be prominent in our minds before any decisions leading to actions are made.

Age is of prime importance when we consider the practical options for the conservation of tiles dating from medieval times through to the end of the 18th century. We might almost say that any fragment of functional ceramic which has survived thus far deserves to be preserved whether it is aesthetically beautiful or not. Furthermore, we must consider whether particular tiles are too important or too fragile to remain in situ, when safe storage is available and desirable. Conversely, if the interpretation of the site is of primary importance, the integrity of the fabric may take precedence over its fragility, and there may be some justification for the tiles remaining in place, whatever the cost. If we use matching new material, must it be to enhance or extend the original only, or can it be used to replace the original if the original is best served by safe storage? Once the best

Detail of the Charing Cross Medieval Hunting Frieze, c1889 (by kind permission of the Ironbridge Gorge Museum Trust)

interests of the tile fabric, in terms of location, are determined, what is certain is that it is our duty to retain and preserve all examples of pre-industrial age ceramic tile whether fragmented or whole. The importance of examining, assessing and recording the underlying mortar should also be a factor for consideration in almost all circumstances.

In terms of any sort of conservation or preservation, 19th century tiles were largely ignored until the late 1970s except for a few small collections of tiles, housed in museums

Rock-a-bye Baby; one of the Ealing Hospital panels by Carters of Poole, c1930

with a particular connection to their provenance, which were classified as 'objects'. In particular, the Jackfield Tile Museum in Ironbridge, Shropshire, which was founded in 1983, is a museum of the history of tile manufacture covering the period between 1834 and 1960. In addition to conserving, restoring and displaying the very large collection of tiles and plaster patterns, the museum quickly became a focus for the conservation of tiles in the wider context of the built environment.

Initially the conservation emphasis was on rescuing and preserving large collections of decorative tile panels or ceramic art which were under immediate threat of destruction by demolition. Notable inclusions were ten pictorial hand-painted tile panels by WB Simpson from the Charing Cross Hospital, a further ten nursery rhyme panels by Carter's of Poole from Ealing Hospital and the ceramic mural by Gilbert Bayes, Pottery through the Ages from the Doulton House at Lambeth. In all three cases – and many since – when demolition and complete loss of decorative art panels has been threatened, ruthless, but effective methods employing large diamond cutting saws have been used to free the ceramics from their Portland cement settings. Such methods inevitably damage the tiles which subsequently require extensive restoration. Examples from each of these three buildings now enjoy permanent exhibition space at Jackfield Tile Museum, with the remainder on display at the new Charing Cross and Ealing hospitals and at the Victoria & Albert Museum respectively.

In the more enlightened time in which we now work, with the protection given to tiles by the listing of buildings, the emphasis is to allow important decorative pictorial panels to remain in place and to become positive elements of buildings which have been refurbished and put to new uses. Nevertheless, there are occasions when important pieces of tile art need to be moved. A recent example was a large tile mural, designed by Gordon Cullen, dating from the late 1950s and some 40 feet long, which was scheduled for re-siting due to re-development in Coventry city centre. In this case, a completely different approach was taken in order to significantly reduce the probability of damage to the ceramic. Rather than cutting and separating individual tiles from the cement substrate as in earlier examples of re-siting, the panel was cut into just eight large sections measuring approximately 2.0m x 1.5m each and moved, still adhered to some 350mm of pre-cast concrete substrate, using heavy lifting equipment. A great deal of careful preparation and precision work was required to cut, lift and re-seat each piece successfully, resulting in only minor damage to the ceramic fabric.

The emphasis during the above removal and re-siting projects, and many others like them, was on the time and effort expended to protect every individual tile from damage, and to retain all of the original tile material.

The same principles do not hold, however, as best practice for the restoration of many 19th and 20th century interior tile schemes in working buildings, where a more flexible attitude to repairs and alterations is required if the building is to remain in use. (A redundant building will rapidly deteriorate, losing all its historic fabric, not just its tiles.) Many historic interiors are now being restored to their former glory so that they can continue to function as working buildings well into the 21st century. Indeed, the original beauty and quality of schemes is often fully appreciated by the occupants of the building. That said, we also require high standards of cleanliness, functionality and quality for the fabric of our public buildings and places of work. In an important interior, such as one of the principal spaces of a hotel, chipped, fractured, missing, worn and dirty tiles are considered unacceptable, no matter how grand and splendid the interior was in its heyday. In these situations the conservation approach is usually applied in its broadest sense, and preservation of the scheme as a whole entity takes priority over individual tiles in much the same way that decayed stone and brick is replaced on external elevations. If a scheme can be retained and admired in the future by replacing the most badly damaged of its individual components and retaining all the rest, then that course is generally considered acceptable.

Fundamental to this approach is the absolute requirement for any replacement material to match the original exactly in terms of colour and form: anything less will be considered by the client to be poor work, and may actually jeopardise the survival of the work in the long term. Care should be taken to record the repairs so that it will be possible to tell which tiles are new and which are original in the future. Care should also be taken to ensure that original material is not removed and replaced unnecessarily when either its condition is still acceptable or a minor repair will restore its condition back to the realms of acceptability. On site repairs to individual tiles can, and should be carried out if the damage is minor, such as holes caused by obsolete fixtures and fittings. Repairs should also be carried out on site if a particularly elaborate one-off moulding is deemed too complex or expensive to replicate.

Victorian geometric and encaustic tile floors are fairly common in entrance halls of domestic and commercial buildings, as well as

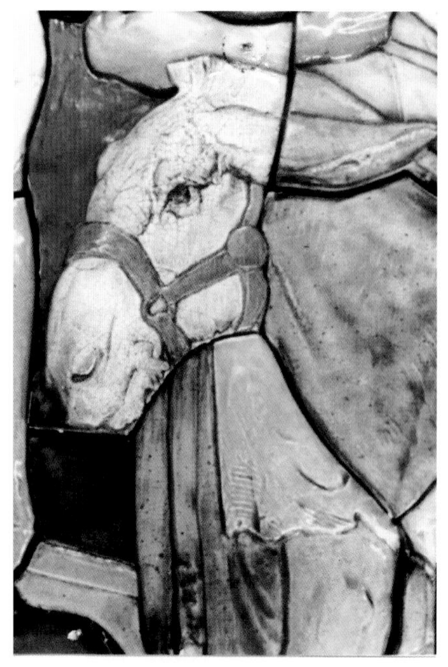

Detail of Pottery through the Ages *by Gilbert Bayes*

in churches. Whether complex and beautiful in design or simple and elegant, they all share a common history of many years of dirt and footfall, affecting their surface to a greater or lesser degree. In order to preserve these tile schemes for the future in situ, trip hazards must be removed if accidents are to be avoided. It is best to replace the material worst affected by severe surface wear and fracturing in high traffic areas. For the most important schemes, it may be possible to repair damaged tiles using a filler material such as coloured polyester resin (which is reversible), but usually floor tiles which are badly damaged or irreversibly stained are replaced with matching new tiles.

Although there is an onus on the owners of listed buildings to retain existing tile interiors within their refurbishment plans, quite often 19th century tile schemes simply do not fit with modern requirements for office layouts, disabled facilities, or public access for example. Doorways need to be let in, lifts installed and computer terminal networks established, particularly in hospitals, libraries or municipal buildings. Again the broad brush stroke of conservation is often applied by local authorities as guardians of their local heritage, and it has become increasingly acceptable, in all but our most important buildings, to retain what is the most effective

Detail of the art deco interior tile scheme in the D6 Building at Boots, Nottingham

part of the schemes, usually the ground floor, entrance halls and stairwells, whilst allowing upper floor or basement schemes to be used as sacrificial material for replacement tile. The Exchange Building in Birmingham is an example of this type of approach. Originally built in 1896, this was one of the first two telephone exchanges in the country, and it is of enormous historic importance. Nevertheless, the building has been upgraded considerably since then to remain commercially viable, and many alterations have been allowed to its tiled decoration.

All interior tile schemes should be judged on their individual merit and it is especially important that an experienced tile conservator is consulted at the early stages of any redevelopment to survey the material and assess its overall condition and significance, as well as outlining all the options available for preserving the fabric to the best advantage.

It can be argued that in some circumstances, particularly art deco schemes of the 1930s, which often rely on a mass of undecorated tiling enhanced by small but significant architectural detail, that it is the design and style of the interior which is worth preserving rather the mass produced fabric of which it is built, and that as long as the exact balance, colour and detail of the scheme are preserved in total, then it is acceptable to replace worn out and damaged component parts and in so doing securing its long term future. Maintaining a detailed record of all the alterations undertaken is a vital part of a conservation programme of this type. By contrast some schemes, such as the interior of Leighton House in South Kensington, are nationally important or have great artistic, historic or technological merit and must be preserved in their full originality. Whichever route towards preservation is taken, it should be our aim to maintain as many historic tile schemes in their original setting as possible.

Recommended Reading

Herbert T and Huggins K, *The Decorative Tile in Architecture and Interiors*, Phaidon

Van Lemmen H, *Tiles in Architecture*, Thames & Hudson

There are also many other books on the history of tiles but very little on their conservation. A useful contact is Buckland Books (Chris Blanchett), Littlehampton, West Sussex, Tel 01903 717648 E-mail cblanchett@lineone.net

LESLEY DURBIN is senior conservator in The Jackfield Conservation Studio, based at Jackfield Tile Museum, Ironbridge, Telford. She has worked in the conservation of architectural ceramics for 20 years, and is an accredited member of UKIC and a member of the IHBC. She has worked for some time with English Heritage on its maintenance programme for medieval tiles. Current projects also include the House of Lords and Commons at the Palace of Westminster and she has a lengthy agenda of work at Osgoode Hall, Toronto for the Supreme Court of Ontario.

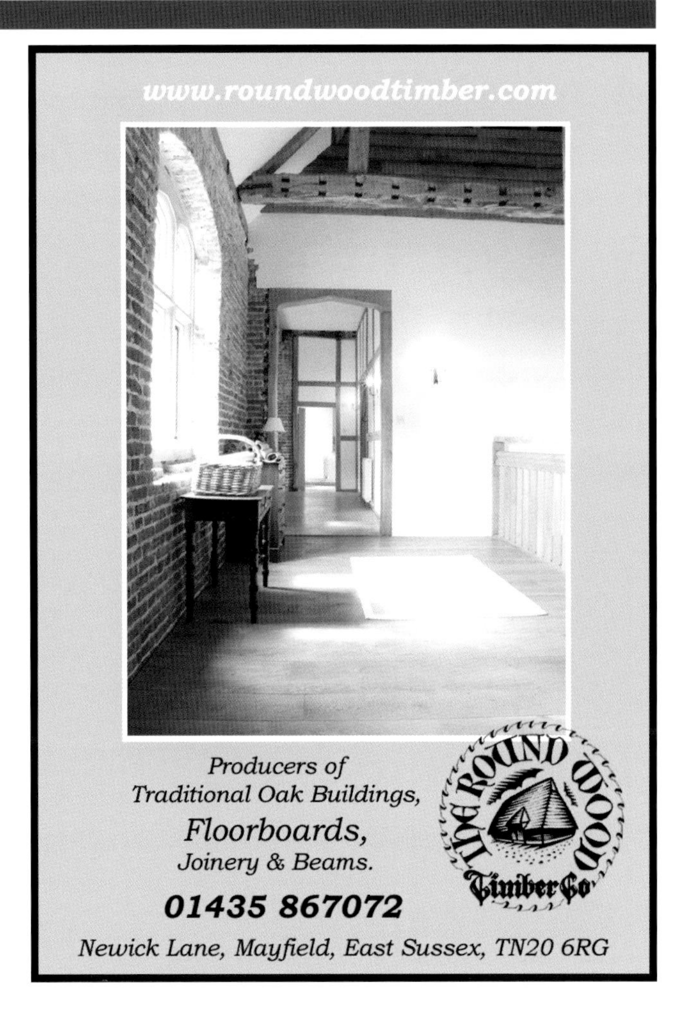

▶ ANTIQUE BRONZE LIMITED (HISTORIC FLOORING DIVISION)

44 Hillway, Holly Lodge Estate, London N6 6EP
Tel 020 8340 0931 Fax 020 8340 0743
E-mail info@antiquebronze.co.uk
Website www.antiquebronze.co.uk

TIMBER AND PARQUET FLOORING: Historic floor restoration by expert craftsmen with decades of experience specialising in the restoration, maintenance and repair of period flooring. All types of wood block, cork, strip, marble, mosaic, tesserae and terrazzo undertaken. Clients include Buckingham Palace, the Natural History Museum, University of London and many others.

▶ COUNTRY OAK (SUSSEX) LTD

Little Washbrook Farm, Brighton Road, Hurstpierpoint, West Sussex BN6 9EF
Tel 01273 833869 Fax 01273 831742
E-mail oakean@aol.com

SPECIALIST SUPPLIER OF EARLY OAK FLOORBOARDS AND BEAMS: From the 17th, 18th and 19th centuries, the boards generally come in widths of between 6″–14″. The company aim to maintain good stocks of around 1,000 square metres. They also carry large stocks of oak beams from the same period, including joists, rafters and beams of all sizes. Traditional plank doors and latches made to order in antique or new oak boards.

▶ JAMESON JOINERY

Hook Farm, West Chiltington Lane, Billingshurst, West Sussex RH14 9DP
Tel 01403 782868 Fax 01403 786766

SPECIALISTS IN TIMBER FLOORING, FURNITURE, STAIRCASES AND KITCHENS: *See also: display entry in Timber Frame Builders section, page 73.*

▶ ORIGINAL OAK

Ashlands, Burwash, East Sussex TN19 7HS
Tel 01435 882228 Mobile 07771 890835

BUILDING AND RESTORATION TIMBER SUPPLIERS: Original Oak supplies old/new oak, pine, elm and beech floorboards. These are often superb wide (12″) floorboards like those found in 17th century houses, and can be viewed in situ at their previous project sites, narrower widths also available. All flooring kiln dried with beautiful grain/character, samples on request. The company also supplies oak beams for restoration and construction which meet all stress grading criteria if necessary. Doors, both traditional plank and ledged and braced, and Victorian four panelled are also available. Quality handmade and reclaimed terracotta floor tiles/parquet flooring are also supplied. Marble and slate flooring, traditional oak timber framed buildings are constructed using traditional methods. Please ring Original Oak on 01435 882228/07771 890835 for prompt, personal attention.

▶ PRIORY HARDWOODS LIMITED

The Studio, High Wray, Haggs Lane, Cartmel, Cumbria LA11 6HD
Tel 01422 824900 Fax 01422 825507
E-mail info@prioryhardwoods.com
Website www.prioryhardwoods.com

TIMBER FLOORING

▶ RANSFORDS CONSERVATION AND RECLAIMED BUILDING SUPPLIES

Drayton Way, Drayton Fields, Daventry, Northants NN11 5XW
Tel 01327 705310 Fax 01327 706831
Website www.ransfords.com

SUPPLIERS OF RECLAIMED OAK AND PINE FLOORBOARDS: *See also: display entry and profile entry in Architectural Salvage section, page 155.*

TIMBER FLOORING

▶ TERNEX LTD

The Sawmill, 27 Ayot Green, Welwyn, Herts AL6 9BA
Tel 01707 324606 Fax 01707 334371

TRADITIONAL T&G FLOORING IN ENGLISH AND IMPORTED HARDWOODS AND SOFTWOODS: *See also: profile entry in Timber Suppliers section, page 124.*

▶ VICTORIAN WOODWORKS

54 River Road, Creekmouth, Barking, Essex IG11 0DW
Tel 020 8534 1000 Fax 020 8534 2000
E-mail sales@victorianwoodworks.co.uk
Website www.victorianwoodworks.co.uk

SPECIALIST HARDWOOD FLOORING MANUFACTURERS AND SUPPLIERS: Victorian Woodworks holds huge stocks of reclaimed and antique timbers for wide board, strip, woodblock and patterned flooring, and reclaimed beams for decorative or structural use. The company offers a range of reclaimed doors and bespoke one-off designs, and all products are complemented by a wide range of solid, new hardwoods. All Victorian Woodworks timber is stocked and kiln dried on site at the company's own sawmill, which also has its own bespoke joinery workshop, and where the showrooms and studio are located.

▶ WELDON FLOORING LIMITED

Glebe Farm Workshops, Caunton, Norwell, Nottingham NG23 6LB
Tel 01636 636962 Fax 01636 636961

PROFESSIONAL HARDWOOD FLOORING: The firm specialises in the design, specification, supply and fitting of elaborate marquetry and parquetry floors. They are a team of highly skilled craftsmen with many years of experience in both the restoration and creation of hardwood floors. They use only the finest materials and pay particular attention to detail, their emphasis is on superior quality and they are totally dedicated to their art. *See also: display entry in this section.*

PLASTERWORK

▶ A W MAGUIRE, ORIGINAL PLASTER MOULDINGS

25 The Street, Upper Halling, Rochester, Kent ME2 1HT
Tel/Fax 01634 240012 E-mail maguires@originalplastermouldings.co.uk
Website www.originalplastermouldings.co.uk
Contact Mr A W Maguire

PLASTER CRAFTSMEN: Original Plaster Mouldings is a family run business where skills have been handed down through generations. The company specialises in the design, manufacture and restoration of interior/exterior fibrous, lime, sand and cement mouldings, and its expertise also encompasses the removal of paint layers from historic mouldings *in situ*. The team of time-served craftsmen is trained in the traditional methods, yet has the skills to combine these methods with the latest technological developments. Recent projects include Fort Clarence 1812 of Rochester in conjunction with English Heritage. This project featured in the RIBA Annual Review 2002. For more information and technical advice please call 01634 240012.

▶ ALBA PLASTERCRAFT

12 Russell Road, Lee-on-the-Solent, Hampshire PO13 9HP
Tel 023 9255 3027 Fax 023 9279 9290 Contact Alan Bailey MCPG

FIBROUS PLASTERWORK, ALSO SAND/CEMENT MOULDINGS: Providing accurate, high quality craftsmanship, working on projects of all sizes across the UK. 30 years experience in restoration and specialist refurbishment work for commercial, public, private and listed buildings. Manufacturing and fixing mouldings, working from existing pieces, salvaged fragments, models, drawings or photographs. Involved with the main plastering contractors at Lloyds Register of Shipping, London, also at Windsor Castle and Uppark after the extensive fires. Offering a wide range of mouldings, some most unusual, creating classical, period and themed styles in new-build properties as well. Projects completed for: councils, the National Trust, Wimpey (UK) Ltd, Costain, Granada Hotels, Kier Southern, film industry, many smaller builders and private owners. Members of the Craft Plasterers Guild and League of Professional Craftsmen.

▶ BURSLEDON BRICKWORKS CONSERVATION CENTRE

Bursledon Brickworks, Coal Park Lane, Swanwick, Southampton SO31 7GW
Tel/Fax 01489 576248
E-mail bursledon@ndirect.co.uk

SUPPLIERS OF MATERIALS FOR TRADITIONAL LIME PLASTERWORK: *See also: profile entry in Mortars & Renders section, page 172.*

▶ THE CLEFT WOOD COMPANY

1 Littlecote Cottages, Littlecote, Dunton, Near Winslow, Bucks MK18 3LN
Tel 01525 240434 Fax 01525 240434
E-mail enquiries@cleftwood.com
Website www.cleftwood.com

RIVEN OAK LATHS, ROOFING BATTENS, TILE PEGS, HEWN BEAMS AND HAZEL WATTLE HURDLES: Traditional green woodsmen, the Cleft Wood Company sources its timber from well managed English woodlands, much cut by the company itself. They have been supplying the conservation trade for nine years. Products have gone to the National Trust, major museums, wholesalers and builders large and small around the country. The Cleft Wood Company aims to keep reasonable stocks but for large orders advance notice is desirable. Special commissions also undertaken.

▶ CLIVEDEN CONSERVATION WORKSHOP LTD

Head Office – The Tennis Courts, Cliveden Estate,
Taplow, Maidenhead, Berkshire SL6 0JA
Tel 01628 604721 Fax 01628 660379

SCULPTURE, STONE AND WALL PAINTINGS CONSERVATION: *See also: profile entry in Stone section, page 102.*

▶ CLIVEDEN CONSERVATION WORKSHOP LTD

Home Farm, Ammerdown Estate, Kilmersdon, Bath, Somerset BA3 5SN
Tel 01761 420300 Fax 01761 420400
E-mail info@clivedenconservation.co.uk

SCULPTURE, STONE, PLASTER, MOSAIC, WALL PAINTING CONSERVATION, STONE CARVING: *See also: profile entry in Stone section, page 102.*

INTERIORS

▶ COYLE TIMBER PRODUCTS

Bassett Farm, Claverton, Bath BA2 7BJ
Tel 01225 427409 Fax 01225 789979
E-mail info@coyletimber.co.uk
Website www.coyletimber.co.uk
Contact Joe Coyle
RIVEN OAK, CHESTNUT, SCOTS PINE LATHS, RIVEN BATTENS AND STAINLESS STEEL NAILS: *See also: profile entry in General Building Materials section, page 154.*

▶ FARTHING & GANNON

Filey Cottage, Mounts Lane, Newnham, Northants NN11 3ES
Tel/Fax 01327 310146
Website www.farthingandgannon.com
CONSERVATION AND RESTORATION OF SCAGLIOLA, HISTORIC PLASTERWORK AND APPLIED ORNAMENT: Experienced specialist in situ moulding, casting, mouldmaking and free hand modelling in lime, gypsum, papier-mâché and composition. New works and restoration of scagliola. Consolidation and stabilisation of ceiling and wall plaster. Consultancy with full material analysis and detailed reports.

▶ FINE ART MOULDINGS

Unit 6, Roebuck Road Trading Estate, 15–17 Roebuck Road, Hainult, Ilford, Essex IG6 3TU
Tel 020 8502 7602 Fax 020 8502 7603
SPECIALISTS IN FIBROUS PLASTERING: Established in 1982, Fine Art Mouldings has a vast amount of experience in the renovation and replication of all plaster features. From drawings and modelling, all work is carried out at their workshops where they produce fine plasterwork to be truly proud of. The company is very sympathetic to the wider uses of limes and putties and have a great deal of experience using lime products. Recent contracts include The Palace of Westminster and St James' Palace, both carried out in 1999. Please contact Andrew Barry to discuss your project or to request a catalogue showing the Fine Art Mouldings standard mouldings range.

▶ G COOK & SONS LTD

37 Montague Road, Cambridge CB4 1BU
Tel 01223 359511 Fax 01223 323966
SPECIALISTS IN FIBROUS PLASTERWORK AND TRADITIONAL LIME PLASTER: Founded in 1887, this family business has a reputation for high quality workmanship based on sound traditional principles gained over many years working in the Cambridge Colleges and surrounding country houses. Recent projects include the remodelling of Gonville and Caius College, Cambridge, Uppark and Ightham Mote for the National Trust, the new gallery for the Gilbert Collection at Somerset House and the Church of the Holy Sepulchre in Jerusalem. Continuity of experience has enabled the company to pass on the traditional skills of the craft and today it offers a highly trained work force capable of undertaking the most demanding work in plain or decorative plasterwork and external rendering.

▶ PHILIP A GACHES

Oddfellows Cottage, 42 Halfleet, Market Deeping, Peterborough PE6 8DB
Tel 01778 342188 Mobile 07932 403488
E-mail philipgaches@aol.com
PLASTERWORK: Conservation and restoration of all aspects of plasterwork also surveys, reports and specifications.

▶ GEORGE JACKSON & SONS

Unit 19, Mitcham Industrial Estate, Streatham Road, Mitcham CR4 2AP
Tel 020 8685 5000 Fax 020 8640 1986
Website www.clarkandfenn.skanska.co.uk
ORNATE PLASTERWORK AND COMPOSITION MOULDINGS: For over 200 years George Jackson & Sons has occupied a unique position 'by appointment' to provide much of the fine decoration, ornate plasterwork and composition mouldings in many notable buildings including Royal Palaces. The company is a division of Clark & Fenn and part of Skanska Construction. Services offered: interior conservation and restoration of historic buildings; composition enrichment, polished plaster, specialist glass reinforced gypsum products and fibrous plaster mouldings available from stock. For a catalogue call 020 8685 5020.

▶ HAYLES & HOWE LTD ORNAMENTAL PLASTERERS

25 Picton Street, Montpelier, Bristol BS6 5PZ
Tel 01179 246673 Fax 01179 243928
E-mail info@haylesandhowe.co.uk
Website www.haylesandhowe.co.uk
▶ HAYLES & HOWE INC
3500 Parkdale Avenue, Suite C1, Baltimore 21211–1408, Maryland USA
Tel (00 1) 410 462 0986 Fax (00 1) 410 462 0989
E-mail info@haylesandhowe.com
CRAFTSMEN IN CONSERVATION. HIGHLY EXPERIENCED THEATRE SAFETY TEAM: Hayles & Howe Ltd combines traditional skills with up to the minute technology to provide service which is second to none. The company specialises in scagliola, consolidation of early plasterwork, fire prevention and insurance advice, fine finishes and modern plasterwork from their standard range or to new design. For technical advice please fax 01179 243928. For an immediate response in all aspects of conservation please call 01179 246673 or 0800 652 1752. Traditional ceilings to order. Please call for their new brochure. Theatre enquiries please call 07768 233606.

▶ HODKIN & JONES (SHEFFIELD) LTD

Callywhite Lane, Dronfield, Sheffield, South Yorkshire S18 2XP
Tel 01246 290890 Fax 01246 290292
E-mail info@hodkin-jones.co.uk
Website www.hodkin-jones.co.uk
PLASTERWORK

▶ HOLDEN CONSERVATION LTD

6 Warple Mews, Warple Way, London W3 0RF
Tel 020 8740 1203 Fax 020 8749 8356, and
Dalshangan House, Dalry, Castle Douglas, Scotland DG7 3SZ
Tel/Fax 01644 460233
PLASTERWORK AND SCULPTURAL MATERIALS: *See also: display entry in Statuary section, page 95.*

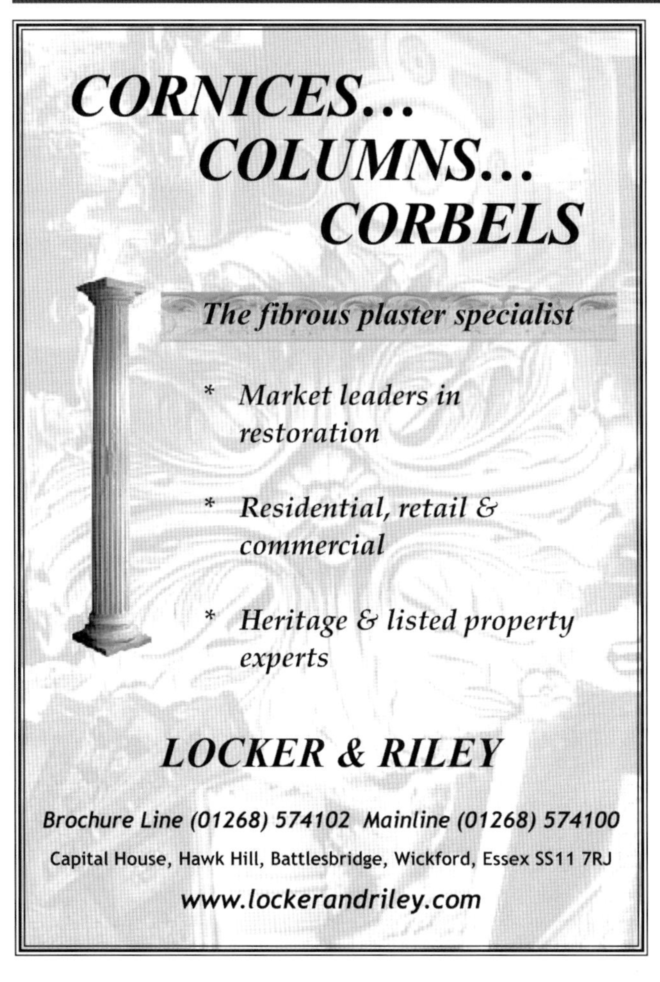

▶ J H UPTON & SON

189 London Road, Temple Ewell, Dover, Kent CT16 3DG
Tel 01304 823540/825456

DECORATIVE PLASTERWORK: J H Upton & Son specialises in all aspects of plain and ornamental plasterwork using traditional techniques and materials. The company can run mouldings *in situ* or in fibrous plaster, and more specialist aspects of the craft such as the *in situ* hand modeling of lime stucco and the repair and conservation of historic decorative plasterwork. J H Upton & Son can also restore or replace fibrous plasterwork to match its original state. All work involving historic plasterwork, plain or decorative is thoroughly researched regarding materials and methods originally employed. The company holds an extensive range of mouldings and can design authentic ceilings, cornices etc for any period or style.

▶ THE LIME PLASTERING COMPANY

No 3, 198 Whitecross Road, Hereford HR4 0DJ
Tel 01432 340054 Mobile 07932 951584
Contact Mark Holden

PLASTERWORK: Over 25 years of experience in the restoration and repair of both private and public listed buildings. Recent projects include St Michael's Church, Tenbury; repairs and re-rendering, internal and external, to Beaumaris Court House, Anglesey; 13th century Abbey Ford Farm, Leominster; and all external stuccowork to Middleton Hall in the National Botanical Gardens, Wales.

PLASTERWORK

▶ LOCKER & RILEY LTD

Capital House, Hawk Hill, Battlesbridge, Wickford, Essex SS11 7RJ
Tel 01268 574100 Fax 01268 574101
E-mail stevebrooks@locker&riley.com
Website www.lockerandriley.com

SPECIALIST FIBROUS PLASTERWORK AND GRG: An award winning fibrous plaster company with an impressive portfolio of works that include such notable projects as the Cabinet Offices, The Tate Gallery and The Royal Albert Hall, as well as high profile hotels and theatres such as The Dorchester, Grosvenor and London Palladium. Acknowledged experts in heritage and restoration work, Locker & Riley have a reputation for providing a level of service and workmanship that is simply second to none. For further information, to discuss your project or to order a brochure, please contact Steve Brooks on 01268 574100.

When contacting companies listed here, please let them know that you found them through *The Building Conservation Directory*

▶ LONDON PLASTERCRAFT LTD

314 Wandsworth Bridge Road, Fulham, London SW6 2UF
Tel 020 7736 5146 Fax 020 7736 7190
▶ 1 Poplar Court Parade, St Margaret's Road, Twickenham TW1 2DT
Tel 020 8744 2965
Website www.londonplastercraft.com

SPECIALIST PLASTERING: London Plastercraft Ltd is a specialist plastering company carrying out all types of interior and exterior moulded work including restoration and cleaning of cornices, ceilings and ornate plasterwork. Reproduction of any type of period moulding in either traditional materials such as lime mortar and lime putty or new modern equivalents. The company employs a team of highly skilled time-served craftsmen who carry out work of the highest quality throughout the UK and Europe. In addition to production from a large range of stock moulds, London Plastercraft Ltd offers a tailor made service, often working from photographs or architect drawings. Visit the newly refurbished Central London showroom to view the large range of decorative items such as cornices, ceiling roses, fire surrounds, columns and pilasters. Recent projects include Lambeth Palace, London; Royal College of Music, London; Kredit bank, Brussels; Bank of California and Museum of Mankind. *See also: display entry in this section, page 215.*

▶ G & N MARSHMAN

1 Nell Ball, Plaistow, Billingshurst, West Sussex RH14 0QB
Tel 01798 342427

MANUFACTURERS OF RIVEN OAK, CHESTNUT PLASTERERS' LATHS AND TILE AND STONE SLATE BATTENS: George & Nicholas Marshman have been supplying riven oak, chestnut plasterers' laths and tile and stone slate battens for over 20 years. They take pride in the quality of the materials they produce and treat every project as a challenge. Examples of projects are: Petworth House, Ightham Mote, Kings College, the Maritime Museum, Windsor Castle and the Globe Theatre. George & Nicholas Marshman supply all Great Britain and Ireland. Although they produce 320,000 feet annually, at times demand is greater than stock so please place your orders well in advance of the required delivery date.

▶ JAMES MARTIN

Blakemoor Cottage, Tenbury Road, Clows Top, Worcestershire DY14 9HE
Tel 01299 832017 Fax 01299 832017 Mobile 07989 879880

PLASTERWORK: Ornamental plaster repairs and mouldings carried out by experienced craftsmen in the restoration of Georgian, Victorian, Tudor, listed, domestic and commercial buildings. Internal and external: cornices and columns, porticos, capitals, decorative ceilings, external string moulds, friezes; all sympathetically restored using traditional methods. Cornices run *in situ*, lime mortars and plasters. Commissions also welcomed.

▶ O'REILLY PERIOD CORNICE RESTORATION & CLEANING

141 Pennine Drive, London NW2 1NG
Tel 020 8458 5917 Fax 020 8450 6450
E-mail oreilly@beeb.net

PERIOD CORNICE RESTORATION AND CLEANING: Mouldings cleaned using steam, repaired, replicated, fitted and expertly restored to original condition by craftsmen. Accumulated experience of father and son in fibrous and solid lath plasterwork using both traditional and compatible modern methods. Plain face cornice run *in situ*, enriched mouldings remodelled, replicated and matched to existing, multiple layers of paint removed *in situ*. The firm has many satisfied clients.

▶ PLASTERCRAFT

E I Flood & Sons t/a Plastercraft
Unit 13A, Douglas Road Industrial Park, Kingswood, Bristol BS15 8PD
Tel 0117 961 8040/Fax 0117 961 8060 Mobile 07802 795541
E-mail enq@flood-plastercraft.co.uk Website www.flood-plastercraft.co.uk

CONSERVATION AND ORNAMENTAL PLASTERING SPECIALISTS: This family business was established in 1959 and all staff are time-served craftsmen and have a reputation for high quality work. There are no facets of the plasterers' craft beyond the expertise of Plastercraft, including limework, external and internal *in situ* running, mouldings, rendering, polish plaster, GRC casting and fibrous plasterwork. For advice and consultancy nationwide on restoration and conservation please telephone Alan, Stan or Bill Flood, 0117 961 8040.

PLASTERWORK

▶ RICHARD IRELAND PERIOD RESTORATION

22 Avenue Road, Isleworth, Middlesex TW7 4JN
Tel 020 8568 5978 Fax 020 8568 5978

HISTORIC PLASTERWORK AND POLYCHROMATIC DECOR CONSERVATION: Consultant and practitioner specialising in the conservation, repair and restoration of decorative plaster, architectural paintwork and wall paintings. Applied ornament – lime, gypsum, scagliola, composition and papier-mâché. Decoration – recreation of historic paintwork, graining, marbling and gilding. Ceiling and wall paintings – conservation, cleaning, repair and restoration. A full pigment and materials analysis is available together with detailed conservation reports. Clients include English Heritage, the National Trust, Historic Royal Palaces Agency, ecclesiastical and private clients. Contracts throughout the UK and overseas from small scale projects to large commissions varying from the Church of the Holy Sepulchre, Jerusalem, to domestic interiors.

▶ STANDEN PLASTERING LIMITED

6b College Place, Brighton BN2 1HN
Tel 01273 680316 Fax 01273 676610

SPECIALIST AND TRADITIONAL PLASTERERS: Standen Plastering is a family run business having highly skilled craftsmen with over 40 years experience in traditional interior and exterior finishing. They have experience in *in situ* run cornice work and casting work. Recent projects have involved lath and plaster, wattle and daub work, restoration and conservation work as well as external stucco and internal enriched cornice work to various buildings and areas.

▶ THOMAS AND WILSON LIMITED

903 Fulham Road, London SW6 5HU
Tel 020 7384 0111 Fax 020 7384 0222
E-mail sales@thomasandwilson.com / info@thomasandwilson.com
Website www.thomasandwilson.com

SPECIALISTS IN FIBROUS PLASTERWORK: Established in 1919, Thomas and Wilson specialises in refurbishment work in period buildings. The company has been involved in a number of notable projects in buildings including Buckingham Palace, Windsor Castle, The British Museum, The Royal Opera House and The National Gallery. Manufacturing takes place in the company's workshops where mouldings are carefully reproduced from existing pieces or modelled from drawings and photographs. Thomas and Wilson has an extensive range of period mouldings and also specialises in the manufacture of external mouldings in GRP and GRC.

SCAGLIOLA

▶ THE SCAGLIOLA COMPANY

Chapeltown Business Centre, 231 Chapeltown Road, Leeds LS7 3DX
Tel 0113 262 6811 Fax 0113 262 5448
E-mail info@scagliolaco.u-net.com
Website www.scagliolaco.u-net.com

SCAGLIOLA AND HISTORIC PLASTERWORK: For the past 17 years The Scagliola Company has been producing scagliola finishes for the traditional and contemporary markets. Lapis, malachite, sienna, brocatelle, verde antique and many other precious surfaces are produced in the form of inlaid tops, pedestals, full size columns, statuary, chimneypieces, wall panels and floors. The Scagliola Company also runs a conservation studio and undertakes *in situ* restoration work of antique scagliola and early plasterwork and offers consultation and analysis of historic mixes.

CLAY PLASTERS

PHILIP ALLEN and NEIL MAY

Clay plasters have been used extensively in buildings in the UK and indeed all over the world, for thousands of years. Although it is not widely known, there are probably over a million buildings with clay materials in their structure in the UK, and a great many of these have clay plasters. Very often clay plasters are not recognised because they are painted or have a thin lime putty skim coat over them, both internally and externally. Many clay plasters are still performing well after many centuries, both in vernacular buildings and in higher status properties (including their use in mouldings, ornamental shields and so forth). The point of this article is to demonstrate not only that clay plasters are important historically, deserving proper repair and preservation, but also that clay plasters are a viable and high quality material for use in standard conservation and renovation work, the extension of historic buildings, and even within modern building contexts such as for new buildings in conservation areas and elsewhere.

Claytec range of plasters, blocks, boards and ancillaries

WHAT ARE CLAY PLASTERS AND HOW DO THEY COMPARE WITH OTHER PLASTER TYPES?

'Clay' refers to the binder in the plaster, just as lime, cement or gypsum refers to the binder in their respective plasters. Clay is therefore a better description than 'earth', as earth also contains aggregates and other materials which may be found in all types of plaster. The clay used in buildings is generally a graded mixture of particles in the clay and silt size ranges, which exhibit cohesion and plasticity. The clay-sized particles can be either finely ground rock or clay minerals.

The presence of the clay minerals exerts a considerable influence on the properties of the material, usually out of all proportion to the percentage content. There are various types of clay mineral formed from combinations of stacked crystalline sheets of silica and alumina. The most common minerals are kaolinite, illite, and montmorillonite. These have very different qualities due to the different arrangement of the stacked sheets and the different strength of the bond between the sheets. Basically, kaolinite has the strongest bonds (of hydrogen) and forms the most stable, least shrinkable clay, but it has weaker binding qualities and it is less hygroscopic (see below). Montmorillonite forms weaker bonds between its sheets (which are linked by water molecules as opposed to hydrogen molecules). Montmorillonite clays therefore have better binding qualities, exhibit considerable shrinkage and swelling and can absorb more water both through capillary and hygroscopic actions. The characteristics of illite clays fall somewhere between these two.

Clay undercoat plaster

Clay topcoat plaster

It is important to note that the clay is unfired, so it is not the same as the fired clay particles added to some lime plasters as pozzolans to make them set hydraulically. Fired clays have very different qualities.

Clay plasters (and mortars) also distinguish themselves from other plasters (and mortars) by the way that they bind and then harden or cure. For cement, gypsum and to some extent hydraulic limes, the process of curing is through a hydraulic reaction between the binders and water. For non-hydraulic limes, and to a lesser degree for hydraulic limes and cement, the process of curing is through carbonation, the conversion of calcium hydroxide to calcium carbonate by the absorption of carbon dioxide from the atmosphere. Clay plasters, however, become solid as excess water added during mixing is lost and the electrically charged surfaces of the clay particles move closer together. The relative strength of the forces of attraction and repulsion between clay particles is responsible for the amount of cohesion and binding force of clay in the plastic state and for the compressive and tensile strength in the solid state. The strength of the attraction between the clay particles can increase if the particles are brought closer together by external pressure (usually by a trowel or float, or if thrown) or if the water surrounding the particles has a higher concentration of positive ions (for example by the addition of additives such as urine). In addition to these electrostatic charges there are also capillary, electromagnetic, frictional and cementitious forces at a molecular level.

CATHEDRAL COMMUNICATIONS LIMITED

WHAT ARE THE QUALITIES OF CLAY PLASTERS COMPARED TO OTHER PLASTERS?

Clay plasters have very specific and, in many ways, unique qualities which are extremely well suited to historic building conditions for a number of reasons. These may be divided into 'breatheability', flexibility, reversibility and their aesthetic qualities.

Breatheability

In terms of their ability to 'breathe', clay plasters not only have excellent vapour permeability (a μ factor of around eight – that is to say, only eight times the equivalent thickness of air), but also extremely good hygroscopic qualities. What is significant here is not only the amount of moisture that can be absorbed from the air but also the rate of absorption. In most materials hygroscopic qualities relate to the capillary structure of the material, whereas in clay plasters, moisture can also be drawn in and held by ionic bonding with the clay particles themselves.

For this reason the type of clay is highly significant, as explained above. Kaolinite clays have less hygroscopic qualities, while montmorillonite clays have the ability to take in moisture very rapidly from the atmosphere when humidity rises. On the other hand, burnt clay and expanded clay have very poor hygroscopic qualities.

Clay plasters in an historic timber frame house

Because the rate at which clay plasters absorb moisture is much higher than that of other materials (timber, for example, takes in and releases large quantities of moisture but over a much longer period), clay plasters can act to protect vulnerable organic materials (and in particular timber) from high levels of relative humidity, when microbial and insect attack can be triggered. This can be an important strategy in the control of excess moisture in vulnerable buildings, threatened by increases in moisture level caused by showers, general indoor living, draught-proofing measures and the switch from open fires to central heating in particular.

In addition to the building benefits of clay, the hygroscopic qualities mean that moulds caused by condensation are minimised, and that a relative humidity of 50 to 60 per cent is maintained. This is the ideal level for mucous membranes of the human body, and also for the control of dust mites and other organisms which affect human health.

Clay plasters also have very good capillary qualities. They actually have less capillary draw than materials like lightweight brick, and even certain cement products, but more capillary draw than most types of timber. This means that within an exposed traditional timber frame building they will draw water droplets away from the timbers. By comparison, hard renders (including those containing eminently hydraulic lime or a high proportion of Portland cement) are notorious for drawing in moisture in at the crack where it abuts the timber and holding it there, trapped, causing the timber to decay.

Flexibility

Clay plasters are flexible in relation to their fibre content. In this sense they are similar to fat lime plasters. Their inherent soft and pliable qualities mean that fibres such as straw, flax, and hair are able to hold together the plaster without cracking in situations of minor or gradual movement, provided there is sufficient quantity of fibre. This is a significant quality in old buildings.

Reversibility

Clay plasters, to an even greater extent than fat lime plasters, are reversible. Unlike lime, cement and gypsum, clay is also reworkable, provided it is not contaminated (particularly by salts). Clay plasters are not only easy to remove, but they do not have the staining or caustic qualities of lime. Because of their 'breathing' qualities they can protect more vulnerable parts of structures, and absorb large amounts of moisture, salts and pollutants where these are a danger – in other words, they can be used sacrificially.

Aesthetic

Unpainted clay plasters have a very particular aesthetic. Due to the shrinkage of the clay particles on drying, the plasters always have an open texture even when polished. This means that light reflects and refracts on the surface in way in which there is always variation and never a gloss sheen. This is particularly noticeable in the modern self-coloured plasters that are becoming more commonly available.

CLAY PLASTERS IN PRACTICE

Earth materials were used extensively in historic buildings of all sorts prior to the 19th century. Their disappearance was partly an effect of the separation of agriculture from construction, as a result of the industrial revolution. The disappearance was not because the material was considered primitive or inadequate in any way, but because of the changing economic conditions of the time. This is an important fact to remember when looking at clay plasters in practice. These are not primitive materials, even if they are not synthetic or a result of high energy processes. Similarly the raw ingredients are not simple, even if they are commonly available.

Clay plasters have many complex qualities and require a proper understanding. Given that understanding, they can perform very effectively, and are durable and attractive.

Preparation

Clay plasters can either be made from local soil, or can be bought in a proprietary

Decorative clay plasterwork prior to lime washing

Spray plastering clay undercoats onto reed boards as part of a historic building renovation

A 15th century mixed timber frame and stone house in Garsington, Oxon, built originally with clay mortars and plasters throughout and renovated by Neil May Builders in 1999, using entirely clay undercoat and topcoat plasters.

form. If in a proprietary form, they come dry bagged and are mixed with water to a plastering consistency, prior to application. If the plasters are being made on site or from local material, a number of issues must be addressed, such as the type of clay, the size of aggregate, the proportion of clay, the even dispersion of clay (a problem if the clay is wet), the addition of fibres and many others. If this approach is to be taken, then soil analysis, good crushing and grinding machinery, possibly drying machinery and a lot of patience and knowledge are required.

All clay plasters can be stored dry indefinitely. If already mixed up with water the only material that will go off will be the fibre, and this will take several months.

Application

Clay plasters may be applied like any other plaster, by hand or by spray application. As with other plasters, background suction must be carefully controlled by wetting it, but without saturating it, and for top coat application the suction should be even across the surface. A very important point to note is that clay plasters should not be over-wetted during the plastering process. Too much water will increase the shrinkage on drying and then the plaster will crack or powder when it dries: lightly wetting just prior to application and then working wet on wet is the best strategy.

Clay plasters cure by the evaporation of water. If there is insufficient background suction, heat or air movement then they will not dry, but will stay soft. After some time moulds may also develop, particularly if there are natural fibres in the plasters. This is natural and can be reversed by carefully controlling the temperature and the amount of ventilation to improve the drying conditions. However, if there is too much suction, heat or air movement the plasters can dry too quickly and this can cause a separation from the substrate, and powdering of the surface. If this occurs then the product can be reworked as appropriate, either on the wall, or by being removed, knocked up again with water and reapplied.

If a mesh is required then hessian or glass fibre is best. Metal beading (galvanised or stainless) can also be used, but the heavy metal beading common for gypsum plastering is not suitable. On angle beads in particular a thin wire bead should be used and the clay plaster should cover the corner bead entirely.

For 'tight', less open grained plaster finishes a fine slightly damp sponge is used to consolidate the clay. For polished finishes specialist plastic trowels are used. If a steel trowel is used at this stage then the clay can be disfigured.

Limitations and protection requirements

Although clay plasters and renders are more vulnerable than lime renders if incorrectly detailed and protected, they can be used in all locations internally and externally without protection except where they are subject to direct spray.

When used externally they do need a limewash or a fine lime plaster skim to protect them against driving rain, depending on levels of exposure. Rain resistance can also be substantially improved by controlling the proportion of clays and fibres in the mix, and by additives such as casein.

Where exposed to direct spray internally (for example in showers, just above sinks and in other vulnerable areas), it is necessary to use tiling or some other impermeable material. It is also suggested that timber is preferable to clay plaster for lining the reveals around cold windows where condensation occurs frequently on the frames.

Where clay plasters get wet only occasionally, then a coat of casein or silicate paint, or limewash is usually sufficient protection.

Clay plasters can also be coated with matt emulsions and other more conventional paints, so long as these paints are reasonably vapour permeable (a μ factor of no more than 300). However, a build up of successive coats of emulsion will reduce their permeability, and rubbing down between coats can abrade the surface of the plaster. Non-vapour permeable coatings and harder coatings will cause problems in the long term for the plasters, and obviously effect their breathing qualities.

Clay plasters have good impact resistance, equivalent to that of many lime plasters, but they are not robust enough for all situations. Where a wall is subject to hard use or frequent impact (such as in an entrance lobby or at floor level for example) skirtings, dado rails, or panelling should be used.

Clay plasters are often overlooked or even purposely ignored in the repair and conservation of historic buildings. The reasons for this are partly a lack of understanding about the technical qualities of clay and partly a lack of available (clay) materials for repair. There is no reason, however, why either of these reasons should be valid in the future. It is vital for both the technical performance and the integrity and character of historic buildings that clay plasters and renders are respected, preserved appropriately and repaired where necessary with authentic (like for like) materials.

I would hope that not too far into the future clay technology will, once again, become a major part of the preservation of our historic building heritage.

Recommended Reading

Pearson, Gordon T *Conservation of Clay and Chalk Buildings*. Donhead Publishing, Shaftesbury, Dorset

Minke, Gernot *Earth Construction Handbook*. WIT Press

Guillard, Hubert and Houben, Hugo, *Earth Construction*. Intermediate Technology Publications

PHILIP ALLEN has a degree in civil engineering and is a Corporate Member of the Institution of Structural Engineers. He works as a freelance engineer and as a long-term consultant to NBT*. He worked for six months in Tanzania on earth construction, and has also worked physically on the construction of several earth buildings in the UK. He has extensive specialist knowledge of appropriate techniques for the repair of vernacular buildings and for the construction of new buildings using ecological principles.

NEIL MAY studied modern history at Oxford University and then an M Phil in sociology at Delhi University, India. During this time he co-directed an award winning documentary film about the coalfields in Bihar. At the end of this period he returned to England where he worked for four years as a general building labourer before becoming involved with ecological and traditional building. He founded Neil May Builders in 1994, an award winning ecological and conservation building company. He founded NBT in November 1999, a company that develops and sells ecological building materials for both historic and mainstream construction markets.

***Natural Building Technologies: www.natural-building.co.uk Tel 01844 338338**

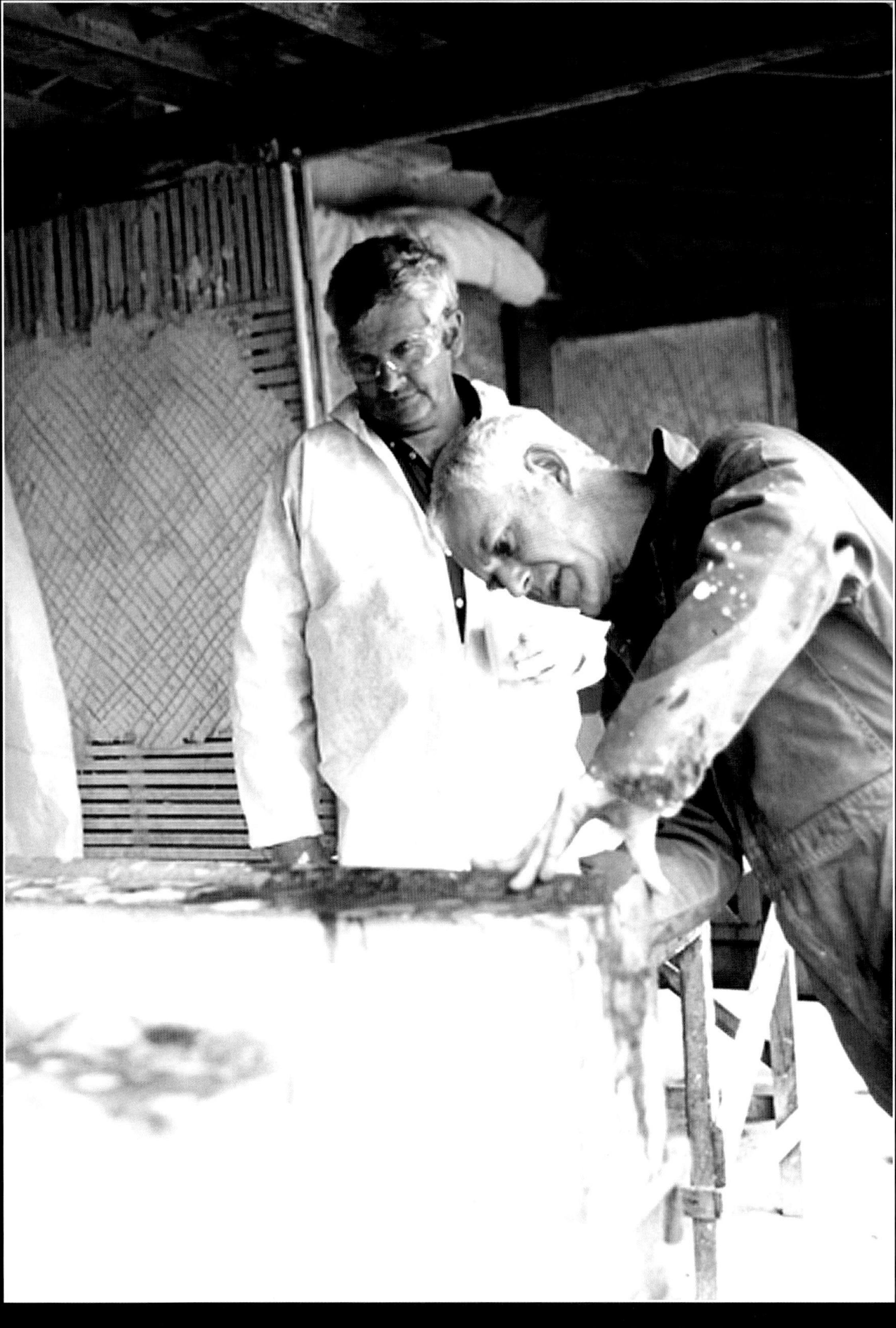

TRADITIONAL MASONRY TRAINING AT CHARLESTOWN WORKSHOPS
Photograph by Scottish Lime Centre Trust

Chapter 6
USEFUL INFORMATION

CONSERVATION COURSE LISTING 2003/2004

CRAFT TRAINING

CITY OF BATH COLLEGE 01225 816950	NVQ 2 Stonemasonry Craft incorporating Banker Mason and College Cert Architectural Stone carver NVQ 2 Stonemasonry Craft, Memorial Masonry NVQ 3 Stonemasonry Craft, Banker Mason NVQ 2&3/ICC Stone Craft	28 weeks over 2 years, block release 10-20 weeks over 2 years, block release 9 weeks, block release Flexible start and attendance
UNIVERSITY OF BOURNEMOUTH 01202 595444	HND Production and Conservation of Architectural Stonework	2 years FT
BUILDING CRAFTS COLLEGE, STRATFORD, LONDON 020 8522 1705	C&G Dip Advanced Stonemasonry C&G Dip Fine Woodwork - includes architectural joinery NVQ 2 & 3 Bench Joinery NVQ 2 & 3 Stonemasonry	1 year FT 2 years FT 6 months FT or 1-2 years PT 6 months FT or 1-2 years PT
CANNINGTON COLLEGE, SOMERSET 01278 655123	BTEC HND Horticulture (Garden Design) Foundation Degree in Landscape and Garden Design	2 years FT
CITY AND GUILDS OF LONDON ART SCHOOL 020 7735 2306/5210	C&G Dip Ornamental Wood Carving and Gilding C&G Dip Architectural Stone Carving	3 years FT 3 years FT
COLLEGE OF ESTATE MANAGEMENT, READING 0118 8986 1101	BSc Building Surveying (University of Reading)	4 years PT distance learning
COUNTRYSIDE AGENCY, GLOS 01242 521381	Courses in various rural trades, including; forgework, furniture making and restoration, thatching	
EDINBURGH'S TELFORD COLLEGE 0131 332 2491	HNC Architectural Conservation NC Building Conservation	1 year 1 year
UNIVERSITY OF GLAMORGAN 01443 482289	HND Architectural and Building Conservation	2 years FT
GLASGOW COLLEGE OF BUILDING AND PRINTING 0141 332 9969	NC Heritage	1 year
EXT 6523	HNC/D Furniture Restoration	1-2 years
EXT 4233	HNC/D Architectural Conservation	1-2 years
HEREFORDSHIRE COLLEGE OF TECHNOLOGY 01432 365391	BTEC ND Blacksmithing and Metalwork Forged Metals	1 year FT 2 years FT
LAMBETH COLLEGE, LONDON 020 7501 5478	Conservation Management (NVQ 3&4) C&G Diploma Mastercrafts - Conservation Crafts training in most fields (NVQ 1-3)	1 year per NVQ level 3 years PT (day) 1 year PT/FT per NVQ level
LEEDS COLLEGE OF ART AND DESIGN 0113 202 8000	C&G Furniture Restoration Specialist Techniques levels 2&3 BTEC (ND Design) Furniture Making and Restoration C&G Furniture Restoration Skills levels 2&3 OCN Woodcarving BTEC (HND) Furniture Restoration	PT 1 day per week 2 years PT 1 day per week PT 3 hours per week (evenings) 2 years FT
LONDON GUILDHALL UNIVERSITY 020 7320 1111	HND Furniture Restoration	2 years FT
PEMBROKESHIRE COLLEGE 01437 765247 X255	HNC/HND/Foundation Degree Architectural and Building Conservation	FT and PT options
UNIVERSITY OF PORTSMOUTH 023 9284 5031	BTEC HNC/HND Restoration and Decorative Studies	2 years FT
RYCOTEWOOD COLLEGE, OXON 01844 212501	HND Design (Furniture Restoration and Conservation)	2 years
SWANSEA INSTITUTE OF HIGHER EDUCATION 01792 481000	HND Building Conservation	2 years FT
UNIVERSITY OF LINCOLN 01522 895069	Access Certificate to HE in Conservation and Restoration (historic artefacts, not buildings)	1 year FT
WEYMOUTH COLLEGE, CONSTRUCTION SECTOR 01305 208946	HNC/D Applied Architectural Stonework ICC Architectural Carving and Banker Masonry NVQ 2/3 Architectural Stonemasonry and Carving	2 years FT 2 years FT 1 year FT per NVQ level
YORK COLLEGE 01904 770200	HNC/D Furniture Studies NVQ 1-3 Hand Crafted Furniture NVQ 3 Stone Masonry Conservation and Restoration C&G Progression Awards, Furniture Restoration/Conservation OCN Levels 1-3 Furniture Studies (including restoration/conservation)	36 weeks PT/FT 48 weeks PT 8 weeks, block release 36 weeks PT 36 weeks PT

UNDER GRADUATE COURSES

UNIVERSITY OF BOURNEMOUTH 01202 595444	BSc (Hons) Heritage Conservation	PT
UNIVERSITY OF BRIGHTON 01273 642398	BSc (Hons) Urban Conservation and Environmental Management	3 years FT 4 years sandwich 6 years PT
BUCKINGHAMSHIRE CHILTERNS UNIVERSITY COLLEGE 01494 522141	BA (Hons) Furniture Conservation and Restoration	3 years FT
CAMBERWELL COLLEGE OF ARTS 020 7514 6302	BA Conservation (works of art on paper)	3 years
UNIVERSITY OF CAMBRIDGE, BOARD OF CONTINUING EDUCATION 01954 280226	Certificate of Higher Education in Garden History Certificate of Higher Education in Historic Building Conservation	PT, varies PT - enrols every 2 years
CANNINGTON COLLEGE, SOMERSET 01278 655123	BSc Horticulture (Landscape & Garden Design) - course subject to approval	3 years FT
CANTERBURY CHRIST CHURCH UNIVERSITY COLLEGE 01227 782888	BA/BSc Built Heritage Conservation	3 years FT 6 years PT
UNIVERSITY OF CENTRAL LANCASHIRE 01772 893210	BA (Hons) Museum and Cultural Design BA (Hons) Heritage Management	3 years FT 4 years sandwich 5 years PT 3 years FT 4 years sandwich 5 years PT
CITY AND GUILDS OF LONDON ART SCHOOL 020 7735 2306/5210	BA (Hons) Conservation Studies Graduateship Diploma Ornamental Wood Carving and Gilding Graduateship Diploma Architectural Stone Carving	3 years FT 3 years FT 3 years FT
UNIVERSITY OF DERBY 01332 591736	BSc (Hons) Architectural Conservation	3 years FT 4 years sandwich 4-8 years PT
UNIVERSITY OF GLAMORGAN 01443 482289	BSc (Hons) Architectural and Building Conservation	3 years FT 4 years sandwich PT also available
UNIVERSITY OF LEICESTER 0116 252 2729	Certificate in Archaeology	9 months PT
UNIVERSITY OF LINCOLN 01522 895069	BA (Hons) Heritage Investigation	3 years
UNIVERSITY OF LINCOLN 01482 440550	BA Conservation and Restoration	3 years FT
LONDON GUILDHALL UNIVERSITY 020 7320 1111	BSc (Hons) Restoration and Conservation	3 years FT
MANCHESTER COLLEGE OF ARTS AND TECHNOLOGY 0161 953 4290	BA (Hons) Furniture Restoration/Conservation	3 years
UNIVERSITY OF NORTHUMBRIA AT NEWCASTLE 0191 227 4722	BSc (Hons) Architectural and Urban Conservation	3 years FT 4 years sandwich
NOTTINGHAM TRENT UNIVERSITY 0115 941 8418	BSc/BA (Hons) Landscape and Heritage	3 years FT 4 years sandwich
OXFORD UNIVERSITY 01865 280356	UGCert Vernacular Architecture	2 years PT
UNIVERSITY OF PORTSMOUTH 023 9284 5031	BA Restoration and Decorative Studies (top up from HND)	1 year FT
SWANSEA INSTITUTE OF HIGHER EDUCATION 01792 481000	BSc (Hons) Building Conservation Management	3 years FT
WRITTLE COLLEGE, ESSEX 01245 424200	BSc (Hons) Garden Design Restoration and Management	1 and 3 years FT

POST-GRADUATE COURSES

ANGLIA POLYTECHNIC UNIVERSITY 01245 493131 EXT 3410	MSc Conservation of Buildings	27 months PT, 15 months FT
ARCHITECTURAL ASSOCIATION, SCHOOL OF ARCHITECTURE 020 7636 0974	MA Conservation of Historic Landscapes, Parks and Gardens PGDip Conservation of Historic Landscapes, Parks and Gardens PGDip Conservation of Historic Buildings	2 years FT 2 years PT 2 years PT
UNIVERSITY OF BATH 01225 826908	MSc Conservation of Historic Buildings	1 year FT or 2 years PT
UNIVERSITY OF BOURNEMOUTH 01202 595444	MSc/PGDip Timber Building Conservation (at Weald & Downland Open Air Museum) MSc Architectural Materials Conservation MSc/PGDip Building Conservation	18 months PT plus dissertation 1 year FT (PT by arrangement) 18 months PT plus dissertation
UNIVERSITY OF BRISTOL 0117 954 6073	MA/PGDip Architectural Conservation	1-2 years PI
BUCKINGHAMSHIRE CHILTERNS UNIVERSITY COLLEGE 01494 603054	MA Furniture Restoration	1 year FT 3 years PT
CAMBERWELL COLLEGE OF ARTS 020 7514 6302	MA/PGDip Conservation	1 year/30 weeks
UNIVERSITY OF CENTRAL ENGLAND 0121 331 5130	MA Architecture: Conservation and Renewal	1 year FT 2 years PT
UNIVERSITY OF CENTRAL LANCASHIRE 01772 893210	MSc Building Heritage and Conservation MSc Architectural Materials Conservation MA International Building Heritage	1 year FT 2 years PT block release 1 year FT 2 years PT block release 1 year FT 2 years PT block release
CHESTER COLLEGE 01244 347341	BA History and Heritage Management	3 years FT
CITY AND GUILDS OF LONDON ART SCHOOL 020 7735 2306/5210	PGDip Conservation Studies PGDip Ornamental Wood Carving and Gilding PGDip Architectural Stone Carving	1 year FT 2 years PT 1 or 2 years FT 1 or 2 years FT
COLLEGE OF ESTATE MANAGEMENT 0118 8986 1101	RICS PGDip Building Conservation	2 years PT distance learning
COURTAULD INSTITUTE OF ART 020 7848 2777	MA Conservation of Paintings	3 years
DE MONTFORT UNIVERSITY 0116 257 7132	MSc Conservation Science MA Architectural Conservation	1 year FT 2 years PT 2 years PT
UNIVERSITY COLLEGE DUBLIN +353 1 716 2752	Master of Urban and Building Conservation (MUBC)	1 year FT 2-4 years PT
UNIVERSITY OF DUNDEE 01382 345236	MSc/PGDip European Urban Conservation	Variable (research based course)
UNIVERSITY OF DURHAM 0191 374 3627	MA Conservation of Historic Objects (Archaeology)	2 years FT
EDINBURGH COLLEGE OF ART 0131 221 6072	MSc/PGDip Architectural Conservation ***see also Heriot Watt University***	9/12 months FT 18/21 months PT
UNIVERSITY OF GREENWICH, SCHOOL OF ARCHITECTURE AND LANDSCAPE 020 8331 9100	MSc Building Rehabilitation MA Garden Design and History	1 year FT 2 years PT 1 year FT 2 years PT
HAMILTON KERR INSTITUTE, UNIVERSITY OF CAMBRIDGE 01223 832040	Dip Conservation of Easel Paintings	3-4 years
HERIOT WATT UNIVERSITY, EDINBURGH 0131 451 4641	MSc/PGDip Building Conservation (Technology and Management) ***see also Edinburgh College of Art***	2-7 years distance learning
IRONBRIDGE INSTITUTE 01952 432751	MA/PGDip Heritage Management	1-4 years
LAMBETH COLLEGE 020 7501 5478	MA Historic Building Conservation (jointly with South Bank University)	1 years FT 2 years PT
UNIVERSITY OF LEEDS 0113 233 5260	MA Country House Studies	1 year FT 2 years PT
UNIVERSITY OF NEWCASTLE 0191 222 6810	Certificate in Conservation and Planning MA/PGDip Town Planning (Urban Conservation) MA/PGDip Urban Conservation Cert Conservation Principles and Techniques	1 year PT 1-2 years FT 2 years PT 1 year FT 2 years PT 1 year PT
UNIVERSITY OF NORTHUMBRIA AT NEWCASTLE 0191 227 3250	MA Conservation of Fine Art	2 years FT
OXFORD BROOKES UNIVERSITY 01865 483458	MSc/PGDip Historic Conservation	9 months/1 year FT, 1/2 years PT
OXFORD UNIVERSITY 01865 280356	MSc/PGDip in Historic Conservation PGCert Architectural History	1/2 years PT 1 year PT
UNIVERSITY OF PLYMOUTH, CENTRE FOR EARTHEN ARCHITECTURE 01752 233608	MA/PGDip Architectural Conservation	2 years PT
UNIVERSITY OF PORTSMOUTH, INSTITUTE OF MARITIME AND HERITAGE STUDIES 023 9284 2421	MSc/PGDip/PGCert Historic Building Conservation	1 year FT 2 years PT
THE ROBERT GORDON UNIVERSITY 01224 263700	MSc/PGDip/PGCert Built Heritage Conservation	30 weeks
ROYAL COLLEGE OF ART AND THE VICTORIA AND ALBERT MUSEUM 020 7590 4532	MA conservation studentships in textiles/metals/furniture/paper conservation and other subjects	2-3 years
SHEFFIELD HALLAM UNIVERSITY 0114 225 4267	MA/PGDip/PGCert Heritage Management	1 year FT 2 years PT
SOUTH BANK UNIVERSITY 020 7815 7304	PGDip/MA Architecture - Historic Building Conservation	2 years PT
UNIVERSITY OF SOUTHAMPTON 023 8059 6718	MA Textile Conservation	2-5 years PT
UNIVERSITY COLLEGE LONDON	***see Institute of Archaeology***	
UNIVERSITY OF LEICESTER 0116 252 2729	PGCert Archaeology of Standing Buildings	
UNIVERSITY OF LINCOLN 01522 895069	MA Conservation of Historic Objects	1 year FT 2 years PT
UNIVERSITY OF THE WEST OF ENGLAND 0117 344 3210	MA/PGDip/PGCert Conservation of Buildings and their Environments MSc/PGDip/PGCert Countryside Conservation and Management	various 2 years +
UNIVERSITY OF YORK 01904 433963	MA in Conservation Studies (Historic Buildings) MA in Historic Landscape Studies	1 year FT 1 year FT

SHORT COURSES

ANGLIA LIME, SUFFOLK 01787 313974	Various short lime-related courses	1 day
CITY OF BATH COLLEGE 01225 816950	Beginners and advanced letter cutting	1 week each
UNIVERSITY OF BRISTOL 0117 954 6073	Maintenance of period houses	8 evenings
BRITISH WATERWAYS HERITAGE SKILLS TRAINING CENTRE, WARWICKSHIRE 01926 626100	Practical building conservation courses for homeowners, professionals and craftsmen	1 day
BRUCE LUCKHURST, KENT 01233 820589	Short courses in cabinet making, wood finishing and related skills	
BUILDING CRAFTS & CONSERVATION TRUST, KENT 01227 451795	Building conservation training	
BURSLEDON BRICKWORKS CONSERVATION CENTRE, SOUTHAMPTON 01489 576248	Various short courses in building conservation, traditional materials and trade techniques	

CONSERVATION COURSE LISTING 2003/2004

CITY AND GUILDS OF LONDON ART SCHOOL 020 7735 2306 /5210	Short courses in gilding and letter cutting	8 days
NMGM, LIVERPOOL 0151 478 4904	Introduction to laser cleaning in conservation	2 days
ESSEX COUNTY COUNCIL AT CRESSING TEMPLE BARNS 01245 437 672	Various short courses: lime plaster; pargetting with lime; leadwork; joinery repairs; rubbed and gauged brickwork; flint walling; wattle and daub; timber frame repairs; architectural metalwork; eco-friendly historic building; and barn conversions.	1-3 days
FLEUR KELLY, SOMERSET 01373 814651	Intensive short courses run on demand on a variety of fresco and panel painting techniques. Also occasional 2-week residential courses. Tailor made courses to suit level of student.	5 days or 2 weeks
HEREFORDSHIRE COLLEGE OF TECHNOLOGY 01432 352235	Blacksmithing	8-10 weeks PT
HOUSEMOUSE, SUFFOLK 01284 830492	Short courses for owners of timber-framed buildings	
ICCROM, ROME +39 06 58553 1	Various postgraduate short courses and training programmes in specific aspects of conservation principles and practice	Varies
LAMBETH COLLEGE, LONDON 020 7501 5478	Advanced craft skills courses: plastering, carpentry and joinery, decoration, masonry, stained glass	10 weeks
THE LEAD SHEET ASSOCIATION, KENT 01892 822773	Short courses in leadwork	Varies
THE LIME CENTRE, WINCHESTER 01962 713636	Bespoke courses on lime for contractors Introduction to the use of lime in traditional buildings (CITB approved) Advanced use of lime (CITB approved)	half to one day 1 day 1 day
LOW-IMPACT LIVING INITIATIVE, BUCKS 01296 714184	Natural paints and lime mortars and renders	3 days
MIKE WYE & ASSOCIATES, DEVON 01409 281644	Practical course in use of natural paints, oils waxes and varnishes Practical course in use of traditional lime and cob putty	1 day 1 day
OLD HOUSE STORE, OXFORDSHIRE 0118 969 7711	Practical courses in building conservation suitable for both building professionals and homeowners.	1 day
THE ORTON TRUST, NORTHAMPTONSHIRE 01536 761303	Practical courses in all aspects of stone masonry, including carving, lettering, conservation, modern stonework, tool sharpening and drawing. CITB approved.	3 days (Fri-Sun)
OXFORD BROOKES UNIVERSITY 01865 484872	Metric survey for historic buildings	3 day summer school
OXFORD BROOKES UNIVERSITY 01865 483458	Certificate in historic conservation	9 months PT
UNIVERSITY OF PLYMOUTH, CENTRE FOR EARTHEN ARCHITECTURE 01752 233608	Various short courses, including philosophies of conservation, conservations areas, conservation of historic gardens, and ecclesiastical conservation The care and conservation of cob buildings	
THE PRINCE'S FOUNDATION, LONDON 020 7613 8500	A series of seminars promoting the reuse of heritage industrial buildings	1 day
RYCOTEWOOD COLLEGE, OXON 01844 212501	Professional development award design (furniture restoration)	4 months FT/ 1 year PT
SCOTTISH LIME CENTRE TRUST, FIFE 01383 872722	Building conservation, ecological building and traditional building skills.	1-5 days
THE SOCIETY FOR THE PROTECTION OF ANCIENT BUILDINGS 020 7377 1644	Introduction to the repair of old buildings for non professionals SPAB Technical Days: subjects vary - ***see Events for details*** The repair of old buildings - a short course of lectures and site visits for contractors and building professionals	2 days (several times a year) 1 day 6 days (Spring and Autumn)
STURGE CONSERVATION STUDIO, NORTHAMPTON 01604 717929	Various courses in conservation techniques	1-3 days
THE STAINED GLASS SPECIALIST, DORSET 01202 882208	Short and long courses in stained glass for local authorities and colleges	Varies
TY-MAWR LIME LTD, POWYS 01874 658249	Various short courses in lime use, including: lime plastering; natural paints and finishes; ecological building	
UPKEEP 020 7815 7212	Upkeep delivers short courses on building construction services, repairs and maintenance	
VENICE EUROPEAN CENTRE FOR THE SKILLS OF ARCHITECTURAL HERITAGE CONSERVATION +39 (0)41 526 85 46	Historic building site manager programme - European Foundation for Heritage Skills (FEMP) Certificate Training courses on the conservation of architecture (stone, wood, iron and mortars) Intensive courses for learning specific crafts such as marmorino, fresco, wrought iron and marbled stucco techniques etc Mastro course for the conservation of the architectural heritage	7 weeks (3 in Venice) 3 months 2 weeks 500 hours
WEALD & DOWNLAND OPEN AIR MUSEUM, WEST SUSSEX 01243 811363	Various short courses in building conservation and traditional crafts and skills - ***see Events page for details***	1-7 days
WEST DEAN COLLEGE, WEST SUSSEX 01243 811301	A range of Building Conservation Masterclasses, in collaboration with English Heritage and Weald and Downland Open Air Museum - ***see Events page for details***	1-4 days
UNIVERSITY OF THE WEST OF ENGLAND, BRISTOL 0117 344 3210	Character appraisal of conservation areas Conservation areas: design control and enhancement	
WOODCHESTER MANSION TRUST, GLOUCESTERSHIRE 01453 750455	Various courses including; stonemasonry, lime plasters, stone repair, hurdle making	1or 2 days
UNIVERSITY OF YORK 01904 433963	Various short courses in the conservation of historic buildings and places, and historic landscapes and gardens	2-3 days

MISCELLANEOUS COURSES

BRUCE LUCKHURST, KENT 01233 820589	Conservation and restoration of antique furniture	1 year
THE CHIPPENDALE INTERNATIONAL SCHOOL OF FURNITURE, E LOTHIAN 01620 810680	Furniture design, making and restoration course	30 weeks
DAVID GRESHAM, BIRMINGHAM 0121 449 5666	Period house restoration and limewashing courses	
UNIVERSITY OF DUNDEE 01382 345236	Dundee Conservation Lectures - in association with the Architectural Heritage Society of Scotland	Evenings
LAMBETH COLLEGE, LONDON 020 7501 5478	Conservation seminars with expert speakers	
NORTH WEST KENT COLLEGE 01322 225471	Historic brickwork	Varies
OXFORD BROOKES UNIVERSITY 01491 832822	Introduction to Historic Building Legislation - in-house course for local councils, by arrangement Introduction to Conservation Law and Practice - in-house course for local councils, by arrangement	1 full day or half day full day, half day or evening only basis
PETER HOOD WORKSHOPS, CUMBRIA 01539 623662	Various conservation courses	
THE PRINCE'S FOUNDATION, LONDON 020 7613 8522	The Prince of Wales's Craft Scholarships - for practical training in traditional building/ conservation crafts	
RYEDALE CONSERVATION SUPPLIES, YORK 01653 648112	Practical use of lime mortars in traditional restoration	1 day
THE SOCIETY FOR THE PROTECTION OF ANCIENT BUILDINGS, LONDON 020 7377 1644	William Morris Craft Fellowship (craft training for qualified craftsmen or women) SPAB Scholarship (programme for young qualified architects, surveyors and others)	3 blocks, 2 months FT each 9 months FT
SOUTH KENT COLLEGE OF TECHNOLOGY 01303 858248	Traditional heavy carpentry, wattle and daub, timber frame repair	Varies
WOODCHESTER MANSION TRUST, GLOUCESTER 01453 750455	Residential courses for full-time mason students	2-3 weeks

COURSES & TRAINING

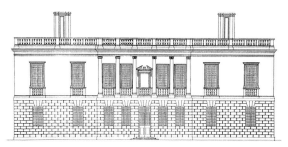

▶ BOURNEMOUTH UNIVERSITY

School of Conservation Sciences, Fern Barrow, Poole, Dorset BH12 5BB
Tel 01202 595176 Fax 01202 595255

COURSES IN BUILDING CONSERVATION: Bournemouth University offers the following programmes:
MSc Building Conservation (October 2004, three years part time, extended weekend study; CPD also available). For professionals working in historic building conservation who wish to improve their practical skills and theoretical background in all aspects of building conservation.
MSc Timber Building Conservation (October 2004, three years part time, extended weekend study). Delivered in partnership with the Weald & Downland Open Air Museum, in West Sussex, this programme enables building conservators to gain both practical and theoretical knowledge.
BSc Heritage Conservation (three years full time, part time available). One of the UK's most comprehensive heritage conservation degrees.
MA Architectural Materials Conservation (one year full time). Please call 01202 595812.

▶ THE BUILDING CRAFTS & CONSERVATION TRUST

Willow House, Lower Road, Woodchurch, Kent TN26 3SQ
Tel 01227 451795 Fax 01227 478797

BUILDING CONSERVATION TRAINING: The Trust's work is aimed at providing the tradesman and professional with the craft skills necessary to effect the accurate repair of historic buildings. The Trust manages building conservation training in lecture room, workshop and site environments. The Trust responds with courses on most conservation topics, when approached by groups of ten or more. The opportunity to experience conservation work on the re-building of a Georgian model farm is afforded by the Yonsea Farm Rebuilt project.

▶ BUILDING CRAFTS COLLEGE

Kennard Road, Stratford, London E15 1AH
Tel 020 8522 1705
Website www.thecarpenterscompany.co.uk

CITY AND GUILDS DIPLOMA COURSES IN WOODWORK OR STONEMASONRY: *See also: display entry in this section.*

▶ BURSLEDON BRICKWORKS CONSERVATION CENTRE

Bursledon Brickworks, Coal Park Lane, Swanwick, Southampton SO31 7GW
Tel/Fax 01489 576248 E-mail bursledon@ndirect.co.uk

SHORT COURSES IN BUILDING CONSERVATION, TRADITIONAL MATERIALS AND TRADE TECHNIQUES: *See also: profile entry in Mortars & Renders section, page 172 and course list, page 222.*

▶ CITY & GUILDS OF LONDON ART SCHOOL

124 Kennington Park Road, London SE11 4DJ
Tel 020 7735 2306/020 7735 5210 Fax 020 7582 5361
E-mail info@cityandguildsartschool.ac.uk
Website www.cityandguildsartschool.ac.uk

CONSERVATION COURSES: BA (Hons) and Post Graduate Diploma in Conservation Studies – The crafts, technical skills and theoretical knowledge needed for the conservation and re-establishment of works of art and artefacts made of wood, stone and related materials with polychromed and gilded surfaces.
Diploma and Post Graduate Diploma in Architectural Stone Carving – For specialists in replacement stone carving and the training of designer craftsmen.
Diploma and Post Graduate Diploma in Ornamental Wood Carving and Gilding – For specialists in replacement or reproduction carving and gilding and the training of designer craftsmen.
P/T Courses in Lettering and Letter Carving – The understanding of letter forms and the tools, implements and materials involved.

▶ THE COLLEGE OF ESTATE MANAGEMENT

Whiteknights, Reading RG6 6AW
Tel 0118 986 1101 Fax 0118 975 0188
E-mail prospectuses@cem.ac.uk
Website www.cem.ac.uk

RICS POSTGRADUATE DIPLOMA IN BUILDING CONSERVATION: *See also: display entry, page 225.*

▶ EDINBURGH COLLEGE OF ART

Dept of Architecture, Lauriston Place, Edinburgh EH3 9DF
Tel 0131 221 6072/6168 Fax 0131 221 6006/6157

DIPLOMA AND MASTERS DEGREE IN ARCHITECTURAL CONSERVATION: Conservation has been taught at Edinburgh College of Art since 1969. The Scottish Centre for Conservation Studies offers a one year Diploma/MSc course available for full, or part-time, study offering a thorough grounding in conservation principles and methods. Admittance is by a first degree in any area related to the built environment. Students proceed to the MSc (by dissertation) upon successful completion of the initial Diploma course. Core subjects (History, Theory, Building Analysis, Conservation Technology, Design Intervention, Planning Law, and Area Conservation) are supplemented by options including Conservation of Historic Parks and Gardens, Vernacular Architecture, and Modern Movement buildings.

▶ THE ORTON TRUST

20 Copelands Road, Desborough, Kettering, Northants NN14 2QF
Tel 01536 761303
Website www.ortontrust.org.uk

STONE MASONRY TRAINING COURSES

COURSES & TRAINING

▶ OXFORD BROOKES UNIVERSITY and UNIVERSITY OF OXFORD

c/o School of Planning, Oxford Brookes University, Oxford OX3 0BP
Tel 01865 483458 Fax 01865 483559

HISTORIC CONSERVATION COURSE: Established in 1990, the course is the product of collaboration between the two major institutions of higher education in Oxford. The course aims to be particularly relevant to the role of the conservation officer working within the planning system, but also aims to meet the needs of a wider range of professionals working within the public, private and voluntary sectors. The modular construction of the course enables full or part-time study for an MSc or Diploma in Historic Conservation (Oxford Brookes University), or part-time study for a Certificate in Architectural History (University of Oxford). Individual modules can be studied on a CPD basis and there are opportunities for MPhil or PhD research.

▶ RYEDALE CONSERVATION SUPPLIES

North Back Lane, York YO60 6NS
Tel 01653 648112 Fax 01653 648112
E-mail info@ryedaleconservation.com
Website www.ryedaleconservation.com

SUPPLY OF MORTARS AND RENDERS AND RELATED TRAINING: *See also: Chalk Hill Lime Products display entry in Mortars & Renders section, page 175.*

▶ SCOTTISH LIME CENTRE TRUST

The Schoolhouse, Rocks Road, Charlestown, Fife KY11 3EN
Tel 01383 872722 Fax 01383 872744
E-mail training@charlestownworkshops.org
Website www.charlestownworkshops.org

COURSES AND TRAINING IN LIME: *See also: display entry in this section, page 226, and profile entry in Heritage Consultants section, page 35.*

▶ SOCIETY FOR THE PROTECTION OF ANCIENT BUILDINGS

37 Spital Square, London E1 6DY
Tel 020 7377 1644 Fax 020 7247 5296
Website www.spab.org.uk

CONSERVATION COURSES:
The Repair of Old Buildings – a six day course of lectures and visits held each Spring and Autumn, for architects, surveyors, engineers, planners, builders and craftsmen.
SPAB Owners' Courses – weekend courses of lectures for owners of old houses, showing how to care for and sensitively repair old buildings, one in London, others regional.
The SPAB Scholarships – a nine-month specialist training for young qualified architects, building surveyors and structural engineers.
The William Morris Craft Fellowships – a six-month specialist training in three blocks for qualified building craftsmen or women of any trade.
Single day courses on a range of topics are offered throughout the year. Further details are available from SPAB.

▶ UNIVERSITY OF BATH

Department of Architecture and Civil Engineering, Bath BA2 7AY
Tel 01225 386908 Fax 01225 386691
E-mail E.S.J.Greeley@bath.ac.uk

MASTERS DEGREE IN THE CONSERVATION OF HISTORIC BUILDINGS (MSc): The course provides technical training within an academic framework including the teaching of classical architecture and the philosophy of conservation. Teaching units include: structural conservation; materials, construction and skills; history and theory; and the law relating to conservation and urban management. Taking place within the world-heritage city of Bath, the course may be taken over one year full time or two years part-time. Architects, engineers, surveyors and suitably qualified candidates from other fields with first degree or equivalent are eligible.

CHARLESTOWN WORKSHOPS

SCOTTISH **LIME** CENTRE

Charlestown Workshops is the training arm of the Scottish Lime Centre Trust, providing a programme of high quality practical training and cpd in traditional building technology and skills with 2, 3 and 5 day workshops for building professionals, contractors and homeowners.

The use of appropriate traditional materials and techniques is important if repairs to historic or traditional buildings are to be effective and we have a particular expertise in 'lime and stone' – the essential materials of much of our built heritage.

For 2003, we are running the following workshops:

- Lime mortars for traditional masonry
- Repair of traditional masonry
- Traditional flatwork plastering
- Traditional decorative plasterwork
- Introduction to traditional masonry
- Historic metalwork
- Environment conscious building
- Masonry building conservation
- National Units in Conservation Masonry accredited by the Scottish Qualifications Authority
- Bespoke training for specific conservation materials and techniques
- Natural stone pavier skills
- In association with the SPAB Homeowners Repair Weekend

For further details telephone us on **01383 872722** or e mail us at **info@charlestownworkshops.org**

Find out more by logging onto our website at
www.charlestownworkshops.org

Scottish Lime Centre Trust The Schoolhouse Rocks Road
Charlestown Fife KY11 3EN

▶ WEALD AND DOWNLAND OPEN AIR MUSEUM

Singleton, Chichester, West Sussex PO18 0EU
Tel 01243 811363 Fax 01243 811475
E-mail wealddown@mistral.co.uk
Website www.wealddown.co.uk

CONSERVATION TRAINING, SERVICES AND SUPPLIES. The Museum has an established reputation as a provider of specialist training/education in historic building conservation and the use of traditional building materials and processes, led by Director Richard Harris. The 48 historic buildings reconstructed on its site give the museum an unrivalled teaching resource in this specialised field. Courses for surveyors, architects, conservation officers and craftspeople are suitable for CPD. MSc in Timber Building Conservation with Bournemouth University. English Heritage Building Conservation Masterclasses in partnership with nearby West Dean College. Exciting and innovative new Downland Gridshell building providing workshop space and artefacts store. Research library designed for use by professionals. *See also: display entry on this page.*

▶ WEST DEAN COLLEGE

West Dean College, Chichester, West Sussex PO18 0QZ
Tel 01243 811301 Fax 01243 811343
E-mail enquiries@westdean.org.uk
Website www.westdean.org.uk

BUILDING CONSERVATION MASTERCLASSES: West Dean College in collaboration with English Heritage and the Weald and Downland Open Air Museum offers short intensive training courses in building conservation, a development of English Heritage's previous Masterclass programme. Courses offer 'hands-on-training and are designed for professionals and craftsmen to enhance their understanding of suitable materials, methods and techniques for sympathetic repairs. Most courses are residential, lasting 3–4 days. Professional development courses are also offered for conservators, as are Post Graduate Diploma programmes in conservation of historic artefacts: books and library materials, ceramics, clocks, furniture and fine metalwork.

THE UNIVERSITY *of York*

TAUGHT MASTERS COURSES

Conservation Studies (Historic Buildings)

Archaeological Heritage Management

Archaeology of Buildings

Historic Landscape Studies

The Department of Archaeology (incorporating the former IoAAS) is based at the historic city campus of King's Manor, with dedicated facilities for postgraduate study including student workspaces, computing facilities including CAD class-room, laboratories for the study of materials and environmental archaeology, and an expanded King's Manor library.

THE DEPARTMENT ALSO RUNS PROFESSIONAL DEVELOPMENT SHORT COURSES

Further information can be obtained on the World Wide Web at:
http://www.york.ac.uk/depts/arch/gsp
Or write to: Graduate Secretary, Department of Archaeology, University of York, The King's Manor, York YO1 7EP
Tel: 01904 433963 Fax: 433902 email: pab11@york.ac.uk

THE UNIVERSITY OF YORK –
DEDICATED TO EXCELLENCE IN TEACHING AND RESEARCH

WEALD & DOWNLAND OPEN AIR MUSEUM

learning from the past

A centre of excellence in building conservation and the use of traditional tools, materials and methods. The forty-eight historic buildings on the site give the Museum an unrivalled teaching resource in this specialised field. New subjects are being added each year.

- ❑ **CPD day schools and longer courses** for architects, surveyors, conservation officers and craftspeople, many with a hands-on element.
- ❑ **English Heritage Masterclasses in Building Conservation** in partnership with nearby West Dean College.
- ❑ **MSc in Timber Building Conservation** validated by Bournemouth University.

Many courses are based in the new innovative **Downland Gridshell** workshop. The Gridshell artefact and archive store and the library are accessible by appointment to research students.

Please contact Diana Rowsell, Course Development Officer at the Weald & Downland Open Air Museum,
Singleton, Chichester, West Sussex PO18 0EU
Tel: 01243 811464 **Fax:** 01243 811475
Email: courses@wealddown.co.uk www.wealddown.co.uk

ACCREDITATION AND TRAINING

RICHARD DAVIES

From December 2003, the lead professional advisor on all projects grant aided by English Heritage must be either an architect or a chartered building surveyor accredited in building conservation. This principle has applied in Scotland since April 2003. Historic Scotland intends to take a decision by this autumn on an appropriate date for the general introduction of accreditation as a condition of its historic buildings repair grants scheme. This article outlines recent developments for those commissioning professional services for building conservation. It may also be useful for practitioners seeking more information on current and likely future trends in accreditation and training.

Of the total output of the construction industry, 50 per cent is now concerned with repair and maintenance of the existing stock, and this amounts to an expenditure of £30 billion per annum. The skills required for these processes are essentially different from those of new build. It is therefore essential that clients have some signposts on what skills to ask for and how to obtain them.

Currently a number of the professional institutions that are concerned with building conservation have developed an accreditation system for their members who have the relevant skills in this field, and several others are considering doing so. Almost all such systems are concerned with professionals who are competent as general practitioners and who also have the necessary experience and knowledge required for maintenance of buildings of special historic value.

INTER-DISCIPLINARY ASSESSMENT

There is now a common method of assessment, agreed by most of the professional institutions, which should provide a potential client with reliable guidance to a professional's competence to advise on building conservation work. The aim is not to create an elite class of specialists in building conservation, rather to establish a reliable baseline. From this, an intelligent client will be expected to carry out further enquiries from a shortlist of registered candidates in order to select the one most suitable for the particular circumstances. An essential foundation for the assessment is a sufficiently rigorous method of testing to ensure consistent judgement of performance. This must not only apply to competencies within a discipline but it must also facilitate comparison between disciplines. (This is particularly important in the field of conservation, where professionals from more than one discipline may be equally qualified to advise on the conservation and repair of an historic building.)

Under the agreed common method of assessment for accreditation, a candidate for any of the specialist registers is required to submit evidence of completed work (the most usual is a summary of five recent projects), along with an analysis of the range and value of the experience that it demonstrates in relation to agreed international standards (as defined in the ICOMOS Guidelines for Education and Training). A reference is obtained from a candidate's client. Panels comprising professional peers and informed lay people are then required to judge the evidence provided, in a way that is systematic and auditable. The criteria for this assessment are made public knowledge and are thus available to the candidates in advance of their application to aid in the preparation of their submissions.

For some years now, there have been registers up and running for architects and surveyors. These have been operated by the Royal Incorporation of Architects in Scotland (RIAS), the Architects Accredited in Building Conservation (AABC) and the Royal Institution of Chartered Surveyors (RICS). They have evolved over time and have been recently reviewed with the aim of improving their consistency with the common method.

It has been a matter of some concern that the AABC register, which had originated within the RIBA, had not had the formal endorsement of that institution. Happily there is now an RIBA representative proposed for the AABC Board and with the support of the new President, George Ferguson, it can be predicted that the register will be returning to its natural home in the near future.

For structural and civil engineers the Conservation Accreditation Register for Engineers (CARE), drawing heavily on the common method, will be available by courtesy of the Institute of Civil Engineers from late summer 2003. Under consideration are registers for Mechanical and Electrical Engineers and Chartered Builders (Building Site Managers through CIOB). The Institute of Historic Building Conservation (IHBC) already operates a form of conservation accreditation covering all full members, although with very different aims, and the Institute may consider introducing an alternative or additional level of accreditation by the common method of assessment in the future.

To qualify for registration as an accredited professional, a person needs to understand the principles of conservation, the methods of investigation, and the wider social and financial issues (including legislation), as well as being able to implement and manage a conservation scheme. Thus there is a need to demonstrate a significant knowledge base to underpin the practical experience. Formal academic or vocational qualifications are considered as valuable components of any application.

Evidence of relevant continuing professional development (CPD) is essential for all applications, as it will be in due course for renewals. However, there is a bewildering variety of academic and practical training courses and events, and there are also a number of distance learning mechanisms under development. Therefore there is a need to provide practitioners and students with a map for judging their own training needs and for judging any training on offer. As a part of the agreement between the professions, a common framework of skills has been developed to assist on-going training. This evolved through the efforts of universities with a track record in building conservation, with the specific aim of developing training modules to complement the requirements of the registers. Addresses are provided below for access to the common framework and for some of the education and training currently on offer.

A significant factor in the drive for improvement in the quality of conservation work through registration has come from national departments and agencies. Historic Scotland initiated the development of the common framework in co-operation with English Heritage (EH). Both have given advanced warning of the need for work grant aided by them to be managed by competent (registered) professionals. Through their joint funding with EH of the Repair Grants to Places of Worship scheme, the Heritage Lottery Fund (HLF) has also endorsed the need for suitably qualified and experienced building professionals. EH has announced that the lead professional advisor on all new grant offers made after the beginning of December 2003 must be either an architect or a chartered building surveyor accredited in building conservation. Historic Scotland has made the employment of a conservation accredited architect, or building surveyor, a condition of the grants awarded from April 2003 under new arrangements made with the National Heritage Memorial Fund for places of worship in Scotland. Historic Scotland intends to take a decision by this autumn on an appropriate date for the general introduction of accreditation as a condition of its historic buildings repair grants scheme.

Below is a list of the institutions operating or working towards conservation registers that relate to the common agreement. Of course there are many others and it is not possible here to list or comment on these. Also shown are useful contact points for education and training.

CONTRACTORS, CONSERVATORS AND CRAFTS STANDARDS

This article has highlighted the recent progress by the professional institutions in the construction industry to improve their own standards in the important field of building conservation. However, this is not solely a matter for the traditional 'white collar' sector. There is a specialist qualification for building site managers which should become a benchmark for the proposed CIOB specialist register. As a driver for this and other specialist craft qualifications, there is now to be a condition of EH and HLF grants, that all contractors need to be registered with the Construction Industry Training Board (CITB). This means that each building contractor can be asked to supply a training plan evaluating the skills and training needs of all their employees. CITB is in the process of developing its own UK-wide National Heritage Training Group, aimed at serving the training needs of the industry as a whole but, initially at least, with special emphasis on craft training. This should be fully operational by December 2003.

Considerable progress has also been made in the development of performance standards for archaeologists by the Institute of Field Archaeology. The UK Institute of Conservation (UKIC) has long maintained a register of conservators and currently there are developments for aligning these with a wider range of specialisms in the field. The case has been made for comparing/reviewing progress in the various fields from time to time.

In conclusion, the professional institutions and CITB are working from inside the construction industry to improve their members' performance in conservation, refurbishment and reuse of buildings. But the message needs to get across to the customers large and small, that conservation is a specialist field, requiring specialist expertise. The greater the demand for registered professionals, contractors, conservators and craftsmen, the greater the pressure on the individual to attain and then maintain high standards.

Further Information

(All addresses and contact details see 'Useful Contacts', pages 234–236)

Chartered Institute of Building (CIOB) Heritage Group
COTAC (Conference on Training in Architectural Conservation)
English Heritage
Historic Scotland
Institute of Historic Building Conservation
Institution of Civil Engineers
Institution of Structural Engineers
National Council of Conservation-Restoration
Register of Architects Accredited in Building Conservation
Royal Institute of British Architects (RIBA)
Royal Institution of Chartered Surveyors (RICS)
Royal Incorporation of Architects in Scotland (RIAS)
United Kingdom Institute for Conservation of Historic and Artistic Works(UKIC)

RICHARD DAVIES is an architect with experience in the private and public sectors who has been closely involved with the care and development of the built environment throughout his career. He is a partner in the practice MRDA Architects & Conservation Consultants and is currently Director of COTAC, The Conference on Training in Architectural Conservation.

EVENTS

July 1-3 **Introduction to Lime in Traditional Buildings (P1)** A course for building professionals. **Venue:** Charlestown Workshops (see below)

July 2 **Stone Conservation** Short course with Mark Hancock. **Venue:** Woodchester Mansion, Gloucestershire **Contact:** Woodchester Mansion Trust Tel 01453 750455 Fax 01453 750457 E-mail office@woodchestermansion.org.uk

July 7 **Traditional Lime Plasters and Renders** A practically based day school covering the fundamentals of lime plastering from simple renders to ornamental work. **Venue:** WDOAM (see below)

July 8 **Town Visit – Maldon Contact:** English Historic Towns Forum Tel 0117 975 0459 Fax 0117 975 0460 E-mail ehtf@uwe.ac.uk

July 10 **Plastering with Lime** One-day course. **Venue:** Old House Store, nr Henley-on-Thames **Contact:** Barry Martynski Tel 0118 969 6949 Fax 0118 969 7771 E-mail info@oldhousestore.co.uk

July 14-18 **Traditional Roofing Methods** Five linked days exploring the traditions, methods and materials used in the roofing industries. Each day will include lectures, demonstrations and practical experience with the diverse materials. **Venue:** WDOAM (see below)

July 15-18 **The Conservation of Flint Buildings** This detailed study of flint buildings covers the origins and characteristics of flint, regional variations in method and style of building, and the strengths and weaknesses of flint construction in different contexts. **Venue:** West Dean College, nr Chichester **Contact:** Patricia Jackson, West Dean College, West Dean, Chichester PO18 0QZ E-mail pat.jackson@westdean.org.uk

July 15-18 **Conservation Masonry Accreditation** A course for apprentices or working masons. **Venue:** Charlestown Workshops (see below)

July 21-25 **Decorative Plasterwork** A course for those in the building trades. **Venue:** Charlestown Workshops (see below)

July 22 **Understanding Historic Buildings** Essex County Council Summer Seminar. **Venue:** Cressing Temple Barns, nr Witham, Essex **Contact:** Essex County Council (see below)

July 24 **Lime Mortars and Stone Repair** Short course with Max Knowles. **Venue:** Woodchester Mansion, Gloucestershire **Contact:** Woodchester Mansion Trust Tel 01453 750455 Fax 01453 750457 E-mail office@woodchestermansion.org.uk

August 5-7 **Introduction to Lime in Traditional Buildings (P1)** A course for building professionals. **Venue:** Charlestown Workshops (see below)

August 7 **Lime Plaster: Hands-on Traditional Building Skills Courses** One-day course showing how to repair lime plaster. **Venue:** Cressing Temple Barns, nr Witham, Essex **Contact:** Essex County Council (see below)

August 19 **'Access' to Historic Buildings** A look at how the Disability Discrimination Act affects historic buildings. **Venue:** Cressing Temple Barns, nr Witham, Essex **Contact:** Essex County Council (see below)

August 19-21 **Masonry Building Conservation – Principles and Techniques (P2)** A course for building professionals who have attained P1. **Venue:** Charlestown Workshops (see below)

September 2-4 **Introduction to Lime in Traditional Buildings (P1)** A course for building professionals. **Venue:** Charlestown Workshops (see below)

September 5-7 **Annual Conference of AAI&S** Association of archaeological Illustrators and Surveyors. **Venue:** London **Contact:** Margaret Mathews E-mail info@aais.org.uk

September 5 **Stone Roofing Technical Day/Symposium** A joint event of the Construction History Society and SPAB. **Venue:** Royal Agricultural College, Cirencester **Contact:** Michael Tutton, Construction History Society Tel 01344 630741 Fax 01344 630764 E-mail michael.tutton@virgin.net

September 8-9 **An Introduction to Laser Cleaning in Conservation** A two day course. **Venue:** Whitechapel, Liverpool **Contact:** Martin Cooper Tel 0151 478 4904 Fax 0151 478 4990 E-mail martin.cooper@nmgm.org

September 10 **Historic Carpentry** A course suitable for the professional, craftsman and homeowner. **Venue:** British Waterways Heritage Skills Training Centre, Hatton, Warwickshire (see below)

September 10 **Brick Repairs and Re-pointing** A course suitable for the professional, craftsman and homeowner. **Venue:** British Waterways Heritage Skills Training Centre, Hatton, Warwickshire (see below)

September 10 **Lime Rendering** A course suitable for the professional, craftsman and homeowner. **Venue:** British Waterways Heritage Skills Training Centre, Hatton, Warwickshire (see below)

September 11 **The Builders' Lime Day** SPAB technical day. **Venue:** Suffolk **Contact:** SPAB, 37 Spital Square, London E1 6DY Tel 020 7377 1644 Fax 020 7247 5296 E-mail info@spab.org.uk

September 12-15 **Civic Trust Heritage Open Days 2003** 2,000 properties, ranging from castles to follies, open free of charge all over England. Details of opening times on website. **Venue:** Nationwide **Contact:** Nina Frentrop, Heritage Open Days Programme Manager Tel 020 7389 1394 Fax 020 7321 0180 E-mail hods@civictrust.org.uk

September 12 **The Building Limes Forum** 12th Conference and Gathering. **Venue:** The Athenaeum, Bury St Edmonds, Suffolk **Contact:** BLF Administrator Fax 0131 553 7158 E-mail admin@buildinglimesforum.org.uk

September 15-17 **Arboricultural Association** Annual Conference. **Venue:** Nottingham Tel 01794 368717 E-mail admin@trees.org.uk

September 16 **Paints and Historic Buildings** A day of illustrated talks on paints suitable for use on historic buildings. **Venue:** Cressing Temple Barns, nr Witham, Essex **Contact:** Essex County Council (see below)

September 16-18 **NDT 2003** Conference of the British Institute of Non-Destructive Testing. **Venue:** Worcester **Contact:** P A Kolbe Tel 01604 630124 E-mail info@bindt.org

EVENTS

September 20-21
Hurdle Making Short course with Thomas Doherty. **Venue:** Woodchester Mansion, Gloucestershire **Contact:** Woodchester Mansion Trust Tel 01453 750455 Fax 01453 750457 E-mail office@woodchestermansion.org.uk

September 23-26
Conservation Masonry Accreditation A course for apprentices or working masons. **Venue:** Charlestown Workshops (see below)

September 25
Lime Course A practical day of slaking lime, analysing old mortars, plasters, renders and preparing matching mixes and limewashes. **Venue:** Sudbury, Suffolk **Contact:** Rory Sumerling, General Manager Tel 01787 313974 E-mail info@anglialime.com

September 26
Eastern Delights: Stained Glass Weekend at Ely Venue: The Stained Glass Museum **Contact:** Susan Mathews Tel 01353 660347 Fax 01353 665025 E-mail admin@stainedglassmuseum.com

Sept - 4 Oct **The Repair of Old Buildings: A Course of Lectures and Visits** The Society's Repair Course, which has been run annually since 1950. It is intended for architects, surveyors, structural engineers, planners and conservation officers, builders and craftsmen. **Venue:** London E1 **Contact:** The SPAB Education Officer Tel 020 7377 1644 Fax 020 7247 5296 E-mail info@spab.org.uk

29 Sept - 5 Oct
Timber Framing From Scratch Timber framing, starting with the tree. Participants carry out hand conversion by hewing and sewing, then lay out and jointing the frame using only traditional tools and methods. £425. **Venue:** WDOAM (see below)

October 1 **Lime Mortars** A course suitable for the professional, craftsman and homeowner. **Venue:** British Waterways Heritage Skills Training Centre, Hatton, Warwickshire (see below)

October 1 **Basic Masonry Repairs** A course suitable for the professional, craftsman and homeowner. **Venue:** British Waterways Heritage Skills Training Centre, Hatton, Warwickshire (see below)

October 1 **Historic Metalwork** A course suitable for the professional, craftsman and homeowner. **Venue:** British Waterways Heritage Skills Training Centre, Hatton, Warwickshire (see below)

October 2 **Cleaning and Paint Removal** A course suitable for the professional, craftsman and homeowner. **Venue:** British Waterways Heritage Skills Training Centre, Hatton, Warwickshire (see below)

October 2 **Dry Stone Walling** A course suitable for the professional, craftsman and homeowner. **Venue:** British Waterways Heritage Skills Training Centre, Hatton, Warwickshire (see below)

October 7-8
Lime Mortars for Traditional Masonry (C1) A course for those in the building trades. **Venue:** Charlestown Workshops (see below)

October 9 **Construction and Repair of Timber Frame Buildings** Exploration of the background to timber framed buildings, dating techniques, structural problems and sympathetic remedial methods. **Venue:** Weald & Downland Open Air Museum (see below)

October 10-11
The 21st Century Agenda - Making your Trust Fit the Bill The UK APT conference. **Venue:** Celtica Conference Centre, Machynlleth, Wales **Contact:** Nicola Richardson, UK Association of Preservation Trusts, Clareville House, 26-27 Oxendon Street, London Tel 020 7930 1629 Fax 020 7930 0295 E-mail apt@ahfund.org.uk

October 15-17
Annual Conference and AGM - Bath Contact: English Historic Towns Forum Tel 0117 975 0459 Fax 0117 975 0460 E-mail ehtf@uwe.ac.uk

October 21-23
Repair of Traditional Masonry C2 A course for those in the building trades who have attained C1. **Venue:** Charlestown Workshops (see below)

October 22-23
Introduction to Stone and Stonemasonry Techniques Short course with Mark Hancock. **Venue:** Woodchester Mansion, Gloucestershire **Contact:** Woodchester Mansion Trust Tel 01453 750455 Fax 01453 750457 E-mail office@woodchestermansion.org.uk

October 24 **Our Uplands Heritage and the Future of Wales** Conference. **Venue:** National Museum of Wales, Cardiff **Contact:** David Browne, RCAHMW Tel 01970 621200 E-mail david.browne@rcahmw.org.uk

October 24-26
Modern Earth Buildings 2003 International Conference and Exhibition. **Venue:** Berlin **Contact:** Peter Steingass Tel +49 30 61 77 62 43 Fax +49 30 264 76 229 E-mail info@moderner-lehmbau.com

October 24-25
Lime Plastering A hands on two-day course for plasterers on the use of traditional lime plaster. **Venue:** Northamptonshire **Contact:** SPAB, 37 Spital Square, London E1 6DY Tel 020 7377 1644 Fax 020 7247 5296 E-mail info@spab.org.uk

October 25 **Superstition, Myth and Magic: 15th - 19th Century** The ways in which people sought to protect their homes, trades and families from witchcraft. Particular emphasis given to the 17th century. **Venue:** Weald & Downland Open Air Museum (see below)

November **Botanical Society of the British Isles** Annual Exhibition Meeting. **Venue:** (details to be finalised) **Contact:** Peter Fry Tel 020 7942 5002

November **Town Visit – Durham Contact:** English Historic Towns Forum Tel 0117 975 0459 Fax 0117 975 0460 E-mail ehtf@uwe.ac.uk

November 5-7
Association of Building Engineers Annual Conference. **Venue:** Torquay **Contact:** Gillian McKenzie Tel 01604 404121 Fax 01604 784220 E-mail cpd@abe.org.uk

November 5-8
7th Historic Structures International Scientific Conference Historic Structures and Disasters. **Venue:** Transylvania Trust, Cluj-N, Romania **Contact:** Imola Kirizsan, Transylvania Trust Tel +40 264 435 489 Fax +40 264 436 805 E-mail emre@trust.dntcj.ro

November 13-14
Environmental Monitoring of our Cultural Heritage Sustainable Conservation Solutions. **Venue:** Edinburgh **Contact:** Angela Carolan, Events Co-ordinator, RIAS Tel 0131 229 754 Fax 0131 228 2188

November 18
The Historic Buildings, Parks and Gardens Event 2003 A key note address; a range of Specialist Seminars; all day exhibition. **Venue:** The Queen Elizabeth II Conference Centre, London SW1 **Contact:** Hall-McCartney Ltd Tel 01462 896688 Fax 01462 896677 E-mail hbpge@hall-mccartney.co.uk

December 5
Urban Design Conference The creation of quality places is central to town planning. This conference will focus on achieving developments, which deliver quality for all users of public space. **Venue:** London **Contact:** Claire Hooker, Event Executive

December 5
The Painted House: 15th –19th Centuries How and when was colour applied to historic buildings externally and internally? What are the clues to look for? Main focus on brick and timber framed buildings. Suitable for specialists and home owners. **Venue:** Weald & Downland Open Air Museum (see below)

2004

April 16-18 **British Sundial Society** International Conference. **Venue:** St Anne's College, Oxford **Contact:** The Secretary, British Sundial Society Tel 01344 772303 E-mail douglas.bateman@btinternet.com

July 8-9 **Conservation 2004** A conference to explore how conservators have responded to the challenges of the current approaches to heritage policy and funding. **Venue:** Merseyside Maritime Museum, Liverpool **Contact:** Sue Frye, ICE Conferences, Institute of Civil Engineers, One Great George Street, London Tel 020 7233 1743 Fax 020 7665 2315 E-mail conservation2004@ice.org.uk

British Waterways Events
British Waterways Heritage Skills Training Centre, Hatton, Warwickshire Tel 01926 626100 Fax 01926 626101 E-mail hstc@britishwaterways.co.uk

Charlestown Workshops
Scottish Lime Centre Trust, The Schoolhouse, 4 Rocks Road, Charlestown, Fife Tel 01383 872772 Fax 01383 872744 E-mail info@scotlime.org

Essex County Council
Pauline Hudspith, Tel 01245 437672 Fax 01245 258353 E-mail pauline.hudspith@essexcc.gov.uk

Weald & Downland Open Air Museum
Diana Rowsell, Training Co-ordinator, Weald & Downland Open Air Museum, Singleton, Chichester, West Sussex Tel 01243 811348 E-mail courses@wealddown.co.uk

PUBLICATIONS

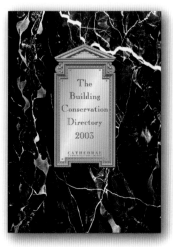

PUBLICATIONS

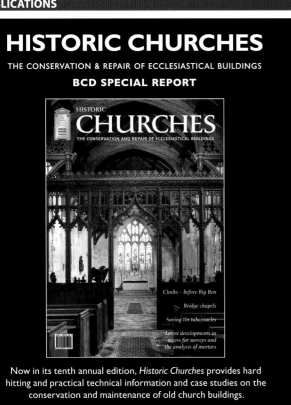

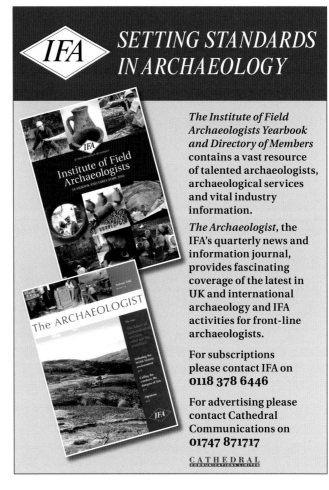

USEFUL INFORMATION

PUBLICATIONS

▶ THE COLLEGE OF ESTATE MANAGEMENT

Whiteknights, Reading RG6 6AW
Tel 0118 986 1101 Fax 0118 975 0188
E-mail prospectuses@cem.ac.uk
Website www.cem.ac.uk

CPD STUDY PACKS: Structural Failures in Traditionally Built Domestic Buildings; Health and Safety; Party Walls. COURSES: The College provides the two year RICS Postgraduate Diploma in Building Conservation by distance learning. *See also: display entry in Courses & Training section, page 225.*

WWW.BUILDINGCONSERVATION.COM

The Building Conservation Directory's website *www.buildingconservation.com*, in association with Blackwell's on-line bookshop, offers a unique selection of recommended book titles for historic building conservation, refurbishment and maintenance. If you want to learn more about the history and protection of our architectural heritage then click in to *www.buildingconservation.com*. Avoid those time consuming trips to specialist bookshops by selecting and ordering your books direct via the Internet using Blackwell's secure credit card purchasing system. *See also: inside back cover.*

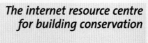

THE INSTITUTE OF HISTORIC BUILDING CONSERVATION

Malcolm Airs, President

In his *Lectures on Architecture and Paintings* John Ruskin wrote

> *There is time enough for everything else, time enough for teaching,*
> *Time enough for criticising, time enough for inventing.*
> *But time little enough for saving.*
> *Hereafter we can create, but*
> *it is only now that we can preserve.*

These words have a particular resonance today as we wrestle to reconcile the particular needs of our built inheritance with the legitimate demands of a modernising society which too often sees the historic environment as an obstacle to economic progress. This sentiment, as anyone reading this *Directory* will know, is a caricature which has no basis in the reality of the modern world. Nevertheless, the simplistic belief that architectural invention and conservation are mutually exclusive is still current in a surprising number of influential circles. If conservation is to play its proper part in managing necessary change for the benefit of us all then the various strands of the sector must unite to demonstrate that this particular canard is out of date and unsustainable.

The Institute for Historic Building Conservation came into being in 1997 partly to meet this challenge by providing a focus for all conservation professionals in the United Kingdom and, in 2000, for Ireland. Membership is open to all eligible professionals whose principal skill, expertise, training and employment is in the provision of specialist advice for the conservation of the historic environment. Our current members are drawn from a wide field of disciplines including architecture, town planning, surveying, engineering, estate management, archaeology and architectural history. They come from both the public and the private sectors. In order to use the letters IHBC after their names they have all had to demonstrate that they have satisfied the ten core areas of professional competence which the Institute has set as the appropriate standard for practice.

Our key activities include improving education and training in conservation, not only in the United Kingdom and Ireland but also in Europe where our Romanian initiative has been an outstanding success over the past four years. We act as a consultee and active participant on conservation matters in partnership with central and local government and other professional institutes and amenity bodies. We are dedicated to promoting the role of the historic environment in economic and social regeneration and we support excellence in all aspects of conservation including new design in historic settings. In addition to the commitment of all our members, our great strength is the autonomy of our 14 branches who organise their own training events and our Annual School as well as providing a forum for regional issues.

IHBC members on a tour of regeneration sites in Great Yarmouth, led by Stephen Earl. The Royal Naval Hospital, shown here, was recently rescued from dereliction and converted to flats.

We have set ourselves an enormous agenda and to succeed we need the support of all those actively engaged in the field. I would urge everybody who reads this to visit our website at www.ihbc.org.uk to see the full scale of our work and to consider joining us in creating a new role for conservation in the 21st century. The invaluable *Building Conservation Directory* has a circulation in excess of 10,000 amongst individuals and organisations. Fewer than 15 per cent of you belong to the Institute.

USEFUL CONTACTS

The Advisory Board for Redundant Churches
Cowley House, 9 Little College Street, London SW1P 3SH
Tel 020 7898 1871 / 020 7898 1872 Fax 020 7898 1870

Almshouse Association
Billingbear Lodge, Carters Hill, Wokingham,
Berkshire RG40 5RU Tel 01344 452922
Fax 01344 862062 www.almshouses.org

Amberley Working Museum
Amberley, Arundel, West Sussex BN18 9LT
Tel 01798 831370 Fax 01798 831831
www.amberleymuseum.co.uk

Ancient Monuments Board for Wales
National Assembly for Wales, Crown Building,
Cathays Park, Cardiff CF10 3NQ Tel 029 2082 6376

Ancient Monuments Society
St Ann's Vestry Hall, 2 Church Entry, London EC4V 5HB
Tel 020 7236 3934 Fax 020 7329 3677
www.ancientmonumentssociety.org.uk

The ARCH Foundation (Art Restoration for Cultural Heritage)
Keltenallee 3, Anif 5081, Austria Tel +43 662 833340
Fax +43 662 822867 www.arch.at

Architectural Heritage Fund
Clareville House, 26/27 Oxendon Street, London SW1Y 4EL
Tel 020 7925 0199 Fax 020 7930 0295 www.ahfund.org.uk

Architectural Heritage Society of Scotland
The Glasite Meeting House, 33 Barony Street,
Edinburgh EH3 6NX Tel 0131 557 0019
Fax 0131 557 0049 www.ahss.org.uk

Architectural Salvage Index
c/o Hutton & Rostron, Netley House, Gomshall,
Surrey GU5 9QA Tel 01483 203221 Fax 01483 202911
www.handr.co.uk/salvage_home.html

The Art Loss Register
Suite 101, Luton House, 164-180 Union Street, London
SE1 0LH Tel 020 7928 0100 Fax 020 7928 7600
www.artloss.com

Association of Conservation Officers –
see Institute of Historic Building Conservation

The Association for Industrial Archaeology
AIA Office, School of Archaeological Studies, University of
Leicester, Leicester LE1 7RH Tel 0116 252 5337
Fax 0116 252 5005 www.industrial-archaeology.org.uk

Association for Studies in the Conservation of Historic Buildings
Institute of Archaeology, 31-34 Gordon Square,
London WC1H 0PY Tel 020 7973 3326 Fax 020 7973 3090

Association of Diocesan and Cathedral Archaeologists
Chester Archaeology, 27 Grosvenor Street, Chester CH1 2DD
Tel 01244 402026 www.britarch.ac.uk/adca

Association of Preservation Trusts
Clareville House, 26-27 Oxendon Street,
London SW1Y 4EL Tel 020 7930 1629 Fax 020 7930 0295
www.heritage.co.uk/apt

Association of Small Historic Towns and Villages
7-9 Gerrard Street, Warwick CV34 4HD Tel 01926 400717
Fax 01926 400717 www.ashtav.org

Association of Town Centre Management
1 Queen Anne's Gate, Westminster, London SW1H 9BT
Tel 020 7222 0120 Fax 020 7222 4440 www.atcm.org

Avoncroft Museum of Historic Buildings
Stoke Heath, Bromsgrove, Worcestershire B60 4JR
Tel 01527 831363 Fax 01527 876934 www.avoncroft.org.uk

Beamish, The North of England Open Air Museum
Beamish, County Durham DH9 0RG Tel 0191 370 4000
Fax 0191 370 4001 www.beamish.org.uk

Brick Development Association Ltd
Woodside House, Winkfield, Windsor, Berkshire SL4 2DX
Tel 01344 885651 Fax 01344 890129 www.brick.org.uk

British Architectural Library
RIBA, 66 Portland Place, London W1B 1AD Tel 020 7580 5533
Fax 020 7631 1802 www.riba-library.com

British Artist Blacksmiths Association
Yew Tree Cottage, Bredenbury, Bromyard, Herefordshire
HR7 4TJ Tel 01885 482572 www.baba.org.uk

British Geological Survey
Keyworth, Nottingham NG12 5GG Tel 0115 936 3100
Fax 0115 936 3593 www.bgs.ac.uk

British Institute of Architectural Technologists
397 City Road, London EC1V 1NH Tel 020 7278 2206
Fax 020 7837 3194 www.biat.org.uk

British Institute of Non-Destructive Testing
1 Spencer Parade, Northampton NN1 5AA
Tel 01604 630124 Fax 01604 231489 www.bindt.org

British Interior Design Association
1/4 Chelsea Harbour Design Centre, London SW10 0XE
Tel 020 7349 0800 Fax 020 7349 0500 www.bida.org

The British Metal Casting Association
Boardesley Hall, The Holloway, Alvechurch, Birmingham,
West Midlands B48 7QB Tel 01527 585222
Fax 01527 590990 www.bmca.org

British Museum, Department of Medieval & Later Antiquities
Department of Medieval and Modern Europe,
Great Russell Street, London WC1B 3DG
Tel 020 7323 8741 Fax 020 7323 8496
www.thebritishmuseum.ac.uk

British Slate Association
Construction House, 56-64 Leonard Street, London EC2A 4JX
Tel 020 7608 5094 Fax 020 7608 5081
www.stone-federationgb.org.uk

British Society of Master Glass Painters
c/o Steven Clare, Holy Well Glass Ltd, Glaziers Yard,
Lovers' Walk, Wells BA5 2QL Tel 01749 671061
www.bsmgp.org.uk

British Standards Institution
389 Chiswick High Road, London W4 4AL
Tel 020 8996 9001 Fax 020 8996 7001 www.bsi.org.uk

British Sundial Society
4 New Wokingham Road, Crowthorne, Berkshire
RG45 7NR Tel 01344 772303 Fax 01344 772303
www.sundialsoc.org.uk

British Urban Regeneration Association
63-66 Hatton Garden, London EC1N 8LE
Tel 020 7539 4030 Fax 020 7404 9614 www.bura.org.uk

British Waterways
The Locks, Hillmorton, Rugby, Warwickshire CV21 4PP
Tel 01788 566030 Fax 01788 541076
www.britishwaterways.co.uk

British Wood Preserving and Damp-proofing Association
1 Gleneagles House, Vernon Gate, South Street,
Derby DE1 1UP Tel 020 8519 2588 Fax 020 8519 3444
www.bwpda.co.uk

The Brooking Collection
University of Greenwich, Oakfield Lane, Dartford,
Kent DA1 2SZ Tel 020 8331 9897 Fax 01380 816565
www.dartfordarchive.org.uk/technology/art_brooking.shtml

The Building Crafts & Conservation Trust
27 Orchard Street, Canterbury, Kent CT2 8AP
Tel 01227 451795 Fax 01233 478797

The Building Conservation Directory
Cathedral Communications Limited, High Street, Tisbury,
Wiltshire SP3 6HA Tel 01747 871717 Fax 01747 871718
www.buildingconservation.com

The Building Limes Forum
The Glasite Meeting House, 33 Barony Street,
Edinburgh EH3 6NX Fax 0131 553 7158
www.buildinglimesforum.org.uk

The Building of Bath Museum
Countess of Huntingdon's Chapel, The Vineyards,
The Paragon, Bath BA1 5NA Tel 01225 333895
Fax 01225 445473 www.bath-preservation-trust.org.uk

Building Research Establishment
Garston, Watford, Hertfordshire WD25 9XX
Tel 01923 664000 Fax 01923 664010 www.bre.co.uk

Cadw: Welsh Historic Monuments
Crown Building, Cathays Park, Cardiff CF10 3NQ
Tel 029 2050 0200 Fax 029 2050 0300
www.cadw.wales.gov.uk

Capel - Chapels Heritage Society
c/o RCAHMW, Crown Building, Plas Crug, Aberystwyth,
Ceredigion SY23 1NJ Tel 01970 621210
Fax 01970 627701 www.rcahmw.org.uk/capel

Cast Metals Federation
National Metalforming Centre, 47 Birmingham Road,
West Bromwich, West Midlands B70 6PY
Tel 0121 601 6390 Fax 0121 601 6391
www.castmetalsfederation.com

Cathedral Architects' Association
46 St Marys Street, Ely, Cambridgeshire CB7 4EY
Tel 01353 660660 Fax 01353 660661

Centre for Earthen Architecture
University of Plymouth, School of Architecture,
The Hoe Centre, Notte Street, Plymouth PL1 2AR
Tel 01752 233608 Fax 01752 233634
www.tech.plym.ac.uk/soa/arch/index.htm

The Chapels Society
1 Newcastle Avenue, Beeston, Nottinghamshire NG9 1BT
Tel 0115 922 4930
www.britarch.ac.uk/chapelsoc/index.html

Chartered Institute of Building (CIOB) Heritage Group
Englemere, Kings Ride, Ascot, Berkshire SL5 7TB
Tel 01344 630700 Fax 01344 630777 www.ciob.org.uk

The Chartered Institution of Building Services Engineers
222 Balham High Road, London SW12 9BS
Tel 020 8675 5211 Fax 020 8675 5449 www.cibse.org

Chiltern Open Air Museum
Newland Park, Gorelands Lane, Chalfont St Giles,
Buckinghamshire HP8 4AB Tel 01494 871117
Fax 01494 872163 www.coam.org.uk

The Church Monuments Society
34 Bridge Street, Shepshed, Leicestershire LE12 9AD
Tel 01509 650637 Fax 01837 851483
www.churchmonumentssociety.org

The Churches Conservation Trust
89 Fleet Street, London EC4Y 1DH Tel 020 7936 2285
Fax 020 7936 2284 www.visitchurches.org.uk

The Cinema Theatre Association
45 Arnold Road, London E3 4NU Tel 020 8981 7844
Fax 020 8981 7844 www.cinema-theatre.org.uk

Civic Trust
17 Carlton House Terrace, London SW1Y 5AW
Tel 020 7930 0914 Fax 020 7321 0180 www.civictrust.org.uk

Civic Trust for Wales
3rd Floor, Empire House, Mount Stuart Square, The Docks,
Cardiff CF10 5FN Tel 029 2048 4606 Fax 029 2048 2086
www.civictrustwales.demon.co.uk

Commission for Architecture and the Built Environment (CABE)
The Tower Building, 11 York Road, London SE1 7NX
Tel 020 7960 2400 Fax 020 7960 2444 www.cabe.org.uk

The Concrete Repair Association
99 West Street, Farnham, Surrey GU9 7EN Tel 01252 739145
Fax 01252 739140 www.concreterepair.org.uk

Conference on Training in Architectural Conservation (COTAC)
The Building Craft College, Kennard Road, Stratford,
London E15 1AH Tel 020 8221 1150 Fax 020 8221 2708
www.cotac.org.uk

The Conservation of Historic Thatch Committee
The Red Barn, Compton Durville, South Petherton,
Somerset TA13 5ET Tel 01460 240027

The Conservation Register
UKIC, 109 The Chandlery, 50 Westminster Bridge Road,
London SE1 7QY Tel 020 7721 8246 Fax 020 7721 8722

Construction Confederation
Construction House, 56-64 Leonard Street,
London EC2A 4JX Tel 020 7608 5000 Fax 020 7608 5101
www.thecc.org.uk

Construction History Society
c/o Library and Information Service, The Chartered
Institute of Building, Englemere, Kings Ride, Ascot,
Berkshire SL5 7TB Tel 01344 630741 Fax 01344 630764
www.constructionhistory.co.uk

Copper Development Association
5 Grovelands Business Centre, Boundary Way,
Hemel Hempstead HP2 7TE Tel 01442 275700
Fax 01442 275716 www.cda.org.uk

Corpus Vitrearum Medii Aevi Archive
National Monuments Record Centre, Kemble Drive,
Swindon, Wiltshire SN2 2GZ Tel 01793 414600
Fax 01793 414606 www.english-heritage.org.uk

The Corrosion Prevention Association
99 West Street, Farnham, Surrey GU9 7EN
Tel 01252 321302 Fax 01252 333901
www.corrosionprevention.org.uk

Council for British Archaeology
Bowes Morrell House, 111 Walmgate, York YO1 9WA
Tel 01904 671417 Fax 01904 671384 www.britarch.ac.uk

The Council for British Archaeology - Wales
c/o CPAT, 20 High Street, Welshpool,
Powys SY21 7JP Tel 01938 552035 Fax 01938 552179
pages.britishlibrary.net/cba.wales

The Council for Scottish Archaeology
c/o National Museums of Scotland, Chambers Street,
Edinburgh EH1 1JF Tel 0131 247 4119 Fax 0131 247 4126
www.britarch.ac.uk/csa

Council for the Care of Churches
Church House, Great Smith Street,
London SW1P 3NZ Tel 020 7898 1866 Fax 020 7898 1881
www.churchcare.co.uk

CoPAT – Council for the Prevention of Art Theft
c/o Studio 25, 1 Adler Street, Aldgate, London E1 1EG

Department of Culture, Media and Sport
2–4 Cockspur Street, London SW1Y 5DH Tel 020 7211 6200
Fax 020 7211 6210 www.culture.gov.uk

USEFUL CONTACTS

Department for Transport, Local Government and the Regions
Eland House, Bressenden Place, London SW1E 3EB
Tel 020 7944 3000 www.dtlr.gov.uk

The Disability Rights Commission
Freepost MID 02164, Stratford-upon-Avon CV37 9HY
Tel 0845 762 2633 www.drc-gb.org

DOCOMOMO – UK (International Working Party for Documentation and Conservation of Buildings, Sites and Neighbourhoods of the Modern Movement)
70 Cowcross Street, London EC1M 6EJ Tel 020 7490 7243
Fax 01223 311166

Dúchas – the Heritage Service
Department of Arts, Heritage Gaeltacht and Islands,
7 Ely Place, Dublin 2, Republic of Ireland
Tel +353 1 647 2300 Fax +353 1 662 0283
www.heritageireland.ie

The East Anglian Earth Buildings Group
Ivy Green, London Road, Wymondham,
Norfolk NR18 9JD Tel 01953 601701 Fax 01953 601701
www.eartha.20m.com

Ecclesiastical Architects & Surveyors Association
c/o Elden Minns & Co Ltd, Chartered Architects, Surveyors,
453 Glossop Road, Sheffield S10 2PT Tel 0114 266 2458
Fax 0114 266 2459

English Heritage
23 Savile Row, London W1S 2ET Tel 020 7973 3144
Fax 020 7973 3111/020 7973 3130
www.english-heritage.org.uk

English Heritage - East Midlands Region
44 Derngate, Northampton NN1 1UH
Tel 01604 735400 Fax 01604 735401

English Heritage - East of England Region
62-74 Burleigh Street, Cambridge CB1 1DJ
Tel 01223 582700 Fax 01223 582701

English Heritage - London Region
23 Savile Row, London W1S 2ET Tel 020 7973 3000
Fax 020 7973 3792 www.english-heritage.org.uk

English Heritage - National Monuments Record
23 Saville Row, London W1S 2ET Tel 020 7973 3000
www.english-heritage.org.uk

English Heritage - North East Region
Bessie Surtees House, 41 Sandhill, Newcastle upon
Tyne NE1 8JF Tel 0191 261 1585 Fax 0191 261 1130

English Heritage - North West Region
Canada House, 3 Chepstow Street,
Manchester M1 5FW Tel 0161 242 1400
Fax 0161 242 1401

English Heritage - South East Region
Eastgate Court, 195-205 High Street, Guildford,
Surrey GU1 3EH Tel 01483 252000 Fax 01483 252001

English Heritage - South West Region
29 Queen Square, Bristol BS1 4ND Tel 0117 975 0700
Fax 0117 975 0701

English Heritage - West Midlands Region
112 Colmore Row, Birmingham B3 3AG
Tel 0121 625 6820 Fax 0121 625 6821
www.english-heritage.org.uk

English Heritage - Yorkshire and Northern Lincolnshire
37 Tanner Row, York YO1 6WP Tel 01904 601901
Fax 01904 601999 www.english-heritage.org.uk

English Historic Towns Forum
PO Box 22, Bristol BS16 1RZ Tel 0117 975 0459
Fax 0117 975 0460 www.chtf.org.uk

Entrust
Acre House, 2 Town Square, Sale, Cheshire M33 7WZ
Tel 0161 972 0044 Fax 0161 972 0055
www.entrust.org.uk

Environment & Heritage Service, DoE NI
Built Heritage, 5-33 Hill Street, Belfast BT1 2LA
Tel 028 9023 5000 Fax 028 9054 3111 www.ehsni.gov.uk

Europa Nostra
Lange Voorhout 35, 2514 EC The Hague, The Netherlands
Tel +31 70 3024051/55 Fax +31 70 3617865
www.europanostra.org

European Confederation of Conservator-Restorers' Organisations (ECCO)
Secretariat, Diepestraat 18, B-3061 Leefdaal, Belgium
Tel +32 2 767 9780 palimpsest.stanford.edu/byorg/ecco

European Foundation for Heritage Skills
c/o Palais de l'Europe, F-67075 Strasbourg Cedex,
France Tel +33 (0)3 90 21 45 37 Fax +33 (0)3 88 41 2755
www.culture.coe.fr/skills

The Fire Protection Association
Bastille Court, 2 Paris Garden, London SE1 8ND
Tel 020 7902 5300 Fax 020 7902 5301 www.thefpa.co.uk

The Folly Fellowship
7 Inch's Yard, Market Street, Newbury, Berkshire
RG14 5DP Tel 01635 42864 Fax 01635 552366
www.heritage.co.uk/follies

The Fountain Society
26 Binney Street, London W1K 5BL Tel 020 7355 2002
Fax 020 7355 2002 www.fountainsoc.org.uk

Friends of Friendless Churches
St Ann's Vestry Hall, 2 Church Entry, London EC4V 5HB
Tel 020 7236 3934 Fax 020 7329 3677
www.friendsoffriendlesschurches.org.uk

Friends of War Memorials
4 Lower Belgrave Street, London SW1W 0LA
Tel 020 7259 0403 Fax 020 7259 0296
www.war-memorials.com

The Furniture History Society
1 Mercedes Cottages, St John's Road, Haywards Heath,
West Sussex RH16 4EH
Tel 01444 413845 Fax 01444 413845

The Garden History Society
70 Cowcross Street, London EC1M 6BP Tel 020 7608 2409
Fax 020 7490 2974 www.gardenhistorysociety.org

The Garden History Society in Scotland
The Glasite Meeting House, 33 Barony Street,
Edinburgh EH3 6NX
Tel 0131 557 5717 Fax 0131 557 5717
www.gardenhistorysociety.org

The Georgian Group
6 Fitzroy Square, London W1T 5DX Tel 020 7529 8920
Fax 020 7529 8939 www.georgiangroup.org.uk

Heritage Building Contractors Group (UK)
c/o Quonians, Lichfield, Staffordshire WS13 7LB
Tel 01543 414234 Fax 01543 410065

Heritage Conservation Trust
2 Chester Street, London SW1X 7BB Tel 020 7259 5688
Fax 020 7259 5590 www.hha.org.uk

The Heritage Council
Rothe House, Kilkenny, Irish Republic Tel +353 56 70777
Fax +353 56 70788 www.heritagecouncil.ie

Heritage Education Trust
Boughton House, Kettering, Northamptonshire NN14 1BJ
Tel 01536 515731 Fax 01536 417255
www.heritageeducationtrust.org.uk

Heritage Lottery Fund
7 Holbein Place, London SW1W 8NR Tel 020 7591 6000
Fax 020 7591 6001 www.hlf.org.uk

Historic Buildings Council for Wales
National Assembly for Wales, Crown Building,
Cathays Park, Cardiff CF10 3NQ Tel 029 2082 6311
Fax 029 2082 6375

Historic Chapels Trust
29 Thurloe Street, Kensington, London SW7 2LQ
Tel 020 7584 6072 Fax 020 7225 0607 www.hct.org.uk

Historic Churches Preservation Trust
Fulham Palace, London SW6 6EA Tel 020 7736 3054
Fax 020 7736 3880 www.historicchurches.org.uk

Historic Farm Buildings Group, Museum of English Rural Life
University of Reading, Whiteknights, PO Box 229,
Reading, Berkshire RG6 6AG Tel 0118 931 8663
Fax 0118 975 1264

Historic Gardens Foundation
34 River Court, Upper Ground, London SE1 9PE
Tel 020 7633 9165 Fax 020 7401 7072
www.historicgardens.freeserve.co.uk

Historic Houses Association
2 Chester Street, London SW1X 7BB Tel 020 7259 5688
Fax 020 7259 5590 www.hha.org.uk

Historic Royal Palaces
Surveyor of the Fabric Department,
Hampton Court Palace, Surrey KT8 9AU
Tel 020 8781 9829 Fax 020 8781 9809

Historic Scotland
Longmore House, Salisbury Place, Edinburgh EH9 1SH
Tel 0131 668 8600 Fax 0131 668 8788
www.historic-scotland.gov.uk

ICCROM (International Centre for the Study of the Preservation and Restoration of Cultural Property, Rome)
via de San Michele, 13, I-00153 Rome, Italy
Tel +39 06 58553 1 Fax +39 06 58553 349
www.iccrom.org

ICOMOS UK (International Council on Monuments & Sites UK)
10 Barley Mow Passage, Chiswick, London W4 4PH
Tel 020 8994 6477 Fax 020 8747 8464
www.icomos.org/uk

Institute of Field Archaeologists
The University of Reading, 2 Earley Gate, PO Box 239,
Reading, Berkshire RG6 6AU Tel 0118 378 6446
Fax 0118 378 6448 www.archaeologists.net

Institute of Historic Building Conservation
Jubilee House, High Street, Tisbury, Wiltshire SP3 6HA
Tel 01747 873133 Fax 01747 871718 www.ihbc.org.uk

Institute of Maintenance and Building Management
Keets House, 30 East Street, Farnham, Surrey GU9 7SW
Tel 01252 734062 Fax 01252 737741 www.imbm.org.uk

Institution of Civil Engineers; Panel for Historical Engineering Works (PHEW)
1 Great George Street, London SW1P 3AA
Tel 020 7665 2250 Fax 020 7976 7610 www.ice.org.uk

Institution of Structural Engineers
11 Upper Belgrave Street, London SW1X 8BH
Tel 020 7235 4535 Fax 020 7235 4294
www.istructe.org.uk

The International Institute for Conservation of Historic and Artistic Works
6 Buckingham Street, London WC2N 6BA
Tel 020 7839 5975 Fax 020 7976 1564
www.iiconservation.org

International Society for the Built Environment
30 Kirby Road, Dunstable, Bedfordshire LU6 3JH
Tel 01582 690187 Fax 01582 690188
www.ebssurvey.co.uk

Irish Georgian Society
74 Merrion Square, Dublin 2, Irish Republic
Tel +353 (0)1 676 7053 Fax +353 (0)1 662 0290
www.irish-architecture.com/igs

Ironbridge Gorge Museum
Ironbridge, Telford, Shropshire TF8 7AW Tel 01952 433522
Fax 01952 432204 www.ironbridge.org.uk

Ironbridge Institute
Ironbridge Gorge Museum, Ironbridge, Telford,
Shropshire TF8 7AW Tel 01952 432751 Fax 01952 432237
www.ironbridge.bham.ac.uk

Landfill Tax Credit Scheme – *see Entrust*

The Landmark Trust
Shottesbrooke, Maidenhead, Berkshire SL6 3SW
Tel 01628 825920 Fax 01628 825417
www.landmarktrust.co.uk

The Landscape Institute
6-8 Barnard Mews, London SW11 1QU Tel 020 7350 5200
Fax 020 7350 5201 www.l-i.org.uk

The Lead Sheet Association
Hawkwell Business Centre, Maidstone Road, Pembury,
Tunbridge Wells, Kent TN2 4AH Tel 01892 822773
Fax 01892 823003 www.leadsheetassociation.org.uk

The Lighthouse Society of Great Britain
Gravesend Cottage, Torpoint, Cornwall PL11 2LX
www.lsgb.co.uk

The London Parks and Gardens Trust
Duck Island Cottage, c/o The Store Yard, St James's Park,
London SW1A 2BJ Tel 020 7839 3969 Fax 020 7839 3969
www.londongardenstrust.org

The London Stained Glass Repository
Glaziers Hall, 9 Montague Close, London Bridge,
London SE1 9DD Tel 020 7403 3300 Fax 020 7407 6036
www.worshipfulglaziers.com

Maintain our Heritage
Weymouth House, Beechen Cliff Road, Bath BA2 4QS
Tel 01225 482228 Fax 0870 137 3805
www.maintainourheritage.co.uk

Master Carvers Association
Unit 20, 21 Wren Street, London WC1X 0HF
Tel 020 7278 8759 Fax 020 7278 8759
www.mastercarvers.co.uk

Metal Roofing Contractors Association
PO Box 101, Newmarket, Cambridgeshire CB8 8NE
Tel 0870 842 8422 www.mrca.org.uk

The Millennium Commission
26th Floor, Portland House, Stag Place,
London SW1E 5EZ Tel 020 7880 2001 Fax 020 7880 2000
www.millennium.gov.uk

Monumental Brass Society
Lowe Hill House, Stratford St Mary, Suffolk CO7 6JX
www.mbs-brasses.co.uk

Museum of Domestic Design and Architecture
Middlesex University, Cat Hill, Barnet EN4 8HT
Tel 020 8411 5244 Fax 020 8411 6639
www.moda.mdx.ac.uk

Museum of Garden History
Lambeth Palace Road, London SE1 7LB Tel 020 7401 8865
Fax 020 7401 8869 www.museumgardenhistory.org

USEFUL CONTACTS

National Archives of Scotland
HM General Register House, 2 Princes Street, Edinburgh EH1 3YY Tel 0131 535 1314 Fax 0131 535 1360 www.nas.gov.uk

The National Assembly for Wales
Crown Building, Cathays Park, Cardiff CF10 3NQ Tel 029 2050 0200 Fax 029 2082 6375 www.wales.gov.uk

National Association of Decorative Fine Arts Societies
NADFAS House, 8 Guilford Street, London WC1N 1DA Tel 020 7430 0730 Fax 020 7242 0686

National Council for the Conservation of Plants and Gardens
The Stable Courtyard, Wisley Gardens, Wisley, Woking, Surrey GU23 6QP Tel 01483 211465 Fax 01483 212404 www.nccpg.com

National Council of Conservation-Restoration
Warwickshire Record Office, Priory Park, Cape Road, Warwick CV34 4JS Tel 020 7721 8721 Fax 020 7721 8722 www.ukic.org.uk

National Council of Master Thatchers Associations
Foxhill, Hillside, South Brent, Devon TQ10 9AU Tel 07000 781909

The National Federation of Builders
Construction House, 56-64 Leonard Street, London EC2A 4JX Tel 020 7608 5150 Fax 020 7608 5151 www.builders.org.uk

NFMSLCE (National Federation of Master Steeplejacks and Lightning Conductor Engineers)
4d St Mary Place, The Lace Market, Nottingham NG1 1PH Tel 0115 955 8818 Fax 0115 941 2238 www.nfmslce.co.uk

National Federation of Roofing Contractors
24 Weymouth Street, London W1N 4LX Tel 020 7436 0387 Fax 020 7637 5215 www.nfrc.co.uk

National Heritage Memorial Fund
7 Holbein Place, London SW1W 8NR Tel 020 7591 6000 Fax 020 7591 6001

National Monuments Record of Scotland
John Sinclair House, 16 Bernard Terrace, Edinburgh EH8 9NX Tel 0131 662 1456 Fax 0131 662 1499 www.rcahms.gov.uk

National Monuments Record of Wales
Crown Building, Plas Crug, Aberystwyth, Ceredigion SY23 1NJ Tel 01970 621233 Fax 01970 627701 www.rcahmw.org.uk

National Physical Laboratory
Queens Road, Teddington, Middlesex TW11 0LW Tel 020 8977 3222 Fax 020 8943 6458 www.npl.co.uk

The National Piers Society
4 Tyrrell Road, South Benfleet, Essex SS7 5DH Tel 01268 757291 Fax 020 7483 1902 www.piers.co.uk

National Preservation Office
The British Library, 96 Euston Road, London NW1 2DB Tel 020 7412 7612 Fax 020 7412 7796 www.bl.uk/npo

National Society of Master Thatchers
20 The Laurels, Tetsworth, Thame, Oxfordshire OX9 7BH Tel 01844 281568 Fax 01844 281568 www.nsmt.co.uk

The National Trust for Scotland
28 Charlotte Square, Edinburgh EH2 4ET Tel 0131 243 9300 Fax 0131 243 9301 www.nts.org.uk

The National Trust
Head Office, 36 Queen Anne's Gate, Westminster, London SW1H 9AS Tel 020 7222 9251 Fax 020 7222 5097 www.nationaltrust.org.uk

The National Trust
Estates Department, 33 Sheep Street, Cirencester, Gloucestershire GL7 1RQ Tel 01285 651818 Fax 01285 657935 www.nationaltrust.org.uk

The Newcomen Society
The Science Museum, London SW7 2DD Tel 020 7371 4445 Fax 020 7371 4445 www.newcomen.com

Northern Ireland Assembly
Information Office, Room B2, Parliament Buildings, Stormont, Belfast BT4 3XX Tel 028 9052 1133 Fax 028 9052 1961 www.niassembly.gov.uk

The Office of the Deputy Prime Minister
Eland House, Bressenden Place, London SW1E 5DU Tel 020 7944 4400 Fax 0207 944 6589 www.odpm.gsi.gov.uk

The Orton Trust
20 Copelands Road, Desborough, Kettering, Northamptonshire NN14 2QF Tel 01536 761303 www.ortontrust.org.uk

Passive Fire Protection Federation
99 West Street, Farnham, Surrey GU9 7EN Tel 01252 739152 Fax 01252 739140 www.associationhouse.org.uk/pfpf.html

The Prince's Foundation
19-22 Charlotte Road, London EC2A 3SG Tel 020 7613 8500 Fax 020 7613 8599 www.princes-foundation.org

Professional Accreditation of Conservator-Restorers
NCC-R, 109 The Chandlery, 50 Westminster Bridge Road, London SE1 7QY Tel 020 7721 8721 Fax 020 7721 8722 www.ukic.org.uk/pacr

Public Monuments and Sculpture Association
72 Lissenden Mansions, Lissenden Gardens, London NW5 1PR Tel 020 7485 0566 Fax 020 7267 1742 www.unn.ac.uk/~hcv3

Public Record Office
Kew, Richmond, Surrey TW9 4DU Tel 020 8392 5200 Fax 020 8878 8905 www.pro.gov.uk

Railway Heritage Trust
PO Box 686, Melton House, 65-67 Clarendon Road, Watford, Hertfordshire WD17 1XZ Tel 01923 240250 Fax 01923 207079

Regeneration Through Heritage (Business in the Community)
The Prince's Foundation, 19-22 Charlotte Road, London EC2A 3SG Tel 020 7613 8518 Fax 020 7613 8599 www.princes-foundation.org

Regional Development Agencies - Advantage West Midlands
3 Priestley Wharf, Holt Street, Aston Science Park, Birmingham B7 4BN Tel 0121 380 3500 Fax 0121 380 3501 www.advantagewm.co.uk

Regional Development Agencies - East Midlands Development Agency
Apex Court, City Link, Nottingham NG2 4LA Tel 0115 988 8300 Fax 0115 853 3666 www.emda.org.uk

Regional Development Agencies - East of England
The Business Centre, Station Road, Histon, Cambridge CB4 9LQ Tel 01223 713900 Fax 01223 713940 www.eeda.org.uk

Regional Development Agencies - London
LDA Headquarters, Devon House, 58-60 St Katharine's Way, London E1W 1JW Tel 020 7680 2000 Fax 020 7680 2014/2040 www.lda.gov.uk

Regional Development Agencies - North West Development Agency
PO Box 37, Renaissance House, Centre Park, Warrington, Cheshire WA1 1XB Tel 01925 400100 Fax 01925 400400 www.nwda.co.uk

Regional Development Agencies - SEEDA (South East)
Cross Lanes, Guildford, Surrey GU1 1YA Tel 01483 484200 Fax 01483 484247 www.seeda.co.uk

Regional Development Agencies - South West of England
Sterling House, Dix's Field, Exeter, Devon EX1 1QA Tel 01329 214747 Fax 01329 214848 www.southwestrda.org.uk

Regional Development Agencies - Yorkshire Forward
Victoria House, 2 Victoria Place, Leeds LS11 5AE Tel 0113 394 9600 www.yorkshire-forward.com

Regional Development Agencies - One North East
Great North House, Sandyford Road, Newcastle Upon Tyne NE1 8ND Tel 0191 261 2000 Fax 0191 232 9069 www.onenortheast.co.uk

Register of Architects Accredited in Building Conservation
33 Macclesfield Road, Wilmslow, Cheshire SK9 2AF Tel 01625 523784 Fax 01625 548328 www.aabc-register.co.uk

Royal Archaeological Institute
c/o Society of Antiquaries of London, Burlington House, Piccadilly, London W1J 0BE Tel 020 7479 7092 www.royalarchaeolinst.org

Royal Commission on the Ancient and Historical Monuments of Scotland (RCAHMS)
John Sinclair House, 16 Bernard Terrace, Edinburgh EH8 9NX Tel 0131 662 1456 Fax 0131 662 1477 www.rcahms.gov.uk

Royal Commission on the Ancient and Historical Monuments of Wales
Crown Building, Plas Crug, Aberystwyth, Ceredigion SY23 1NJ Tel 01970 621200 Fax 01970 627701 www.rcahmw.org.uk

Royal Commission of the Historical Monuments of England
National Monuments Record Centre, Kemble Drive, Swindon SN2 2GZ

Royal Fine Art Commission for Scotland
Bakehouse Close, 146 Canongate, Edinburgh EH8 8DD Tel 0131 556 6699 Fax 0131 556 6633 www.royfinartcomforsco.gov.uk

Royal Horticultural Society
80 Vincent Square, London SW1P 2PE Tel 020 7834 4333 Fax 020 7630 6060 www.rhs.org.uk

Royal Incorporation of Architects in Scotland
15 Rutland Square, Edinburgh EH1 2BE Tel 0131 229 7545 Fax 0131 228 2188 www.rias.org.uk

Royal Institute of British Architects
66 Portland Place, London W1B 1AD Tel 020 7580 5533 Fax 020 7255 1541 www.architecture.com

The Royal Institution of Chartered Surveyors Building Conservation Group
12 Great George Street, Parliament Square, London SW1P 3AD Tel 020 7222 7000 Fax 020 7222 9430 www.rics.org.uk

Royal Society of Architects in Wales
Bute Building, King Edward VII Avenue, Cathays Park, Cardiff CF10 3NB Tel 029 2087 4753 Fax 029 2087 4926 www.architecture-wales.com

Royal Society of Ulster Architects
2 Mount Charles, Belfast BT7 1NZ Tel 028 9032 3760 Fax 028 9023 7313 www.rsua.org.uk

Royal Town Planning Institute
41 Botolph Lane, London EC3R 8DL Tel 020 7929 9494 Fax 020 7929 9490 www.rtpi.org.uk

The Rug Restorers Association
Lower Clatcombe House, Sherborne, Dorset DT9 4RH Tel 01935 813274

SALVO
PO Box 333, Cornhill on Tweed, Northumberland TD12 4YJ Tel 01890 820333 Fax 01890 820499 www.salvo.co.uk

SAVE Britain's Heritage
70 Cowcross Street, London EC1M 6EJ Tel 020 7253 3500 Fax 020 7253 3400 www.savebritainsheritage.org

The Scottish Civic Trust
The Tobacco Merchant's House, 42 Miller Street, Glasgow G1 1DT Tel 0141 221 1466 Fax 0141 248 6952 www.scotnet.co.uk/sct

Scottish Conservation Bureau
Historic Scotland, Longmore House, Salisbury Place, Edinburgh EH9 1SH Tel 0131 668 8668/83 Fax 0131 668 8669 www.historic-scotland.gov.uk

Scottish Executive Development Department
Planning Division, Victoria Quay, Edinburgh EH6 6QQ Tel 0131 244 7076 Fax 0131 244 7083 www.scotland.gov.uk

The Scottish Historic Buildings Trust
33 High Street, Cockenzie, East Lothian EH32 0HP Tel 01875 813608 Fax 01875 813608

Scottish Lime Centre Trust
The Schoolhouse, Rocks Road, Charlestown, Fife KY11 3EN Tel 01383 872722 Fax 01383 872744 www.charlestownworkshops.org

Scottish Record Office - *see National Archives of Scotland*
www.sro.gov.uk

The Scottish Redundant Churches Trust
4 Queen's Gardens, St Andrews, Fife KY16 9TA Tel 01334 472032 Fax 01334 470767

Scottish Society for Conservation and Restoration
Chaunston, Tartraven, Bathgate Hills, West Lothian EH48 4NP Tel 01506 811777 Fax 01506 811888 www.sscr.demon.co.uk

Sir John Soane's Museum
13 Lincoln's Inn Fields, London WC2A 3BP Tel 020 7405 2107 Fax 020 7831 3957 www.soane.org

The Society for the Protection of Ancient Buildings
37 Spital Square, Spitalfields, London E1 6DY Tel 020 7377 1644 Fax 020 7247 5296 www.spab.org.uk

The Society for the Protection of Ancient Buildings in Scotland
The Glasite Meeting House, 33 Barony Street, Edinburgh EH3 6NX Tel 0131 557 1551 Fax 0131 557 1551 www.spab.org.uk/scotland/index.html

Society of Architectural Historians of Great Britain
6 Fitzroy Square, London W1T 5DX Tel 020 7387 1720 Fax 020 7387 1721 www.sahgb.org.uk

Stained Glass Museum
The Chapter House, Ely Cathedral, Ely, Cambridgeshire CB7 4DN Tel 01353 660347 Fax 01353 665025 www.stainedglassmuseum.com

The Stationery Office
PO Box 29, Norwich NR3 1GN Tel 0870 6005522 Fax 0870 600 5533 www.hmso.gov.uk

Steel Window Association
The Building Centre, 26 Store Street, London WC1E 7BT Tel 020 7637 3571 Fax 020 7637 3572 www.steel-window-association.co.uk

Stone Federation Great Britain
Construction House, 56-64 Leonard Street, London EC2A 4JX Tel 020 7608 5094 Fax 020 7608 5081 www.stone-federationgb.org.uk

Stone Roofing Association
Ceunant, Caernarfon, Gwynedd LL55 4SA Tel 01286 650402 Fax 01286 650402 www.stoneroof.org.uk

The Theatres Trust
22 Charing Cross Road, London WC2H 0QL Tel 020 7836 8591 Fax 020 7836 3302 www.theatrestrust.org.uk

Tiles and Architectural Ceramics Society
Oakhurst, Cocknage Road, Rough Close, Stoke on Trent, Staffordshire ST3 7NN Tel 01782 397996 Fax 01782 395599 www.tilesoc.org.uk

Timber Research and Development Association
Stocking Lane, Hughenden Valley, High Wycombe, Buckinghamshire HP14 4ND Tel 01494 569600 Fax 01494 565487 www.trada.co.uk

The Timber Trade Federation
4th Floor Clareville House, 26-27 Oxendon Street, London SW1Y 4EL Tel 020 7839 1891 Fax 020 7930 0094 www.ttf.co.uk

The Tool and Trades History Society
22 Windley Crescent, Darley Abbey, Derby DE22 1BZ Tel 01225 837031 Fax 01225 837031

Town & Country Planning Association
17 Carlton House Terrace, London SW1Y 5AS Tel 020 7930 8903 Fax 020 7930 3280 www.tcpa.org.uk

The Traditional Paint Forum
c/o The National Trust for Scotland, 28 Charlotte Square, Edinburgh EH2 4ET Tel 0131 243 9449 Fax 0131 243 9599

The Twentieth Century Society
70 Cowcross Street, London EC1M 6EJ Tel 020 7250 3857 Fax 020 7251 8985 www.c20society.org.uk

Ulster Architectural Heritage Society
66 Donegall Pass, Belfast BT7 1BU Tel 028 9055 0213 Fax 028 9055 0214 www.uahs.co.uk

United Kingdom Institute for Conservation of Historic and Artistic Works (UKIC)
109 The Chandlery, 50 Westminster Bridge Road, London SE1 7QY Tel 020 7721 8721 Fax 020 7721 8722 www.ukic.org.uk

Upkeep
Room 203, South Bank University, 202 Wandsworth Road, London SW8 2JZ Tel 020 7815 7212 Fax 020 7815 7213 www.sbu.ac.uk/upkeep

Urban Design Alliance (UDAL)
70 Cowcross Street, London EC1M 6DG Tel 020 7251 5529 Fax 020 7251 5529 www.udal.org.uk

Urban Design Group
70 Cowcross Street, London EC1M 6EJ Tel 020 7250 0872 www.udg.org.uk

Urban Parks Forum
Caversham Court, Church Road, Caversham, Berkshire RG4 7AD Tel 0118 946 9060 Fax 0118 946 9061 www.urbanparksforum.co.uk

Vernacular Architecture Group
c/o Mrs Brenda Watkin, Ashley, Willows Green, Great Leighs, Chelmsford, Essex CM3 1QD Tel 01245 361408 www.vag.org.uk

The Victorian Society
1 Priory Gardens, Bedford Park, London W4 1TT Tel 020 8994 1019 Fax 020 8747 5899 www.victorian-society.org.uk

The Vivat Trust
61 Pall Mall, London SW1Y 5HZ Tel 0845 090 0174 Fax 0845 090 0174 www.vivat.org.uk

Wallpaper History Society
c/o 49 Glenpark Drive, Southport, Merseyside PR9 9FA Tel 01704 225429

Weald & Downland Open Air Museum
Singleton, Chichester, West Sussex PO18 0EU Tel 01243 811363/811464 Fax 01243 811475 www.wealddown.co.uk

World Monuments Fund in Britain
2 Grosvenor Gardens, London SW1W 0DH Tel 020 7730 5344 Fax 020 7730 5355 www.worldmonuments.org

The Worshipful Company of Glaziers and Painters of Glass
Glaziers Hall, 9 Montague Close, London Bridge, London SE1 9DD Tel 020 7403 6652 Fax 020 7403 6652 www.worshipfulglaziers.com

The Wrought Iron Advisory Service
Carlton Husthwaite, Thirsk, North Yorkshire Tel 01845 501415 Fax 01845 501072 www.realwroughtiron.com/wiac.htm

For more up to date contacts, please check our website www.buildingconservation.com

SPECIALISTS INDEX

SPECIALISTS INDEX

SPECIALISTS INDEX

When contacting companies listed here, please let them know that you found them through *The Building Conservation Directory*

PRODUCTS & SERVICES INDEX

CW00972510

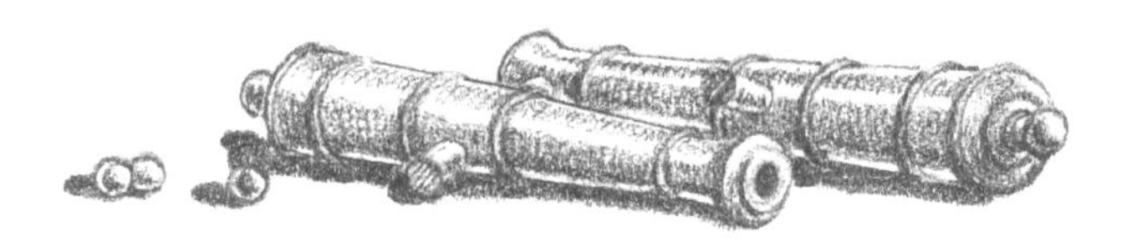

THE SHIPS OF JOHN PAUL JONES

Dedicated to Jean Boudriot

THE SHIPS OF JOHN PAUL JONES

BY WILLIAM GILKERSON

PUBLISHED BY THE UNITED STATES NAVAL ACADEMY MUSEUM, THE BEVERLEY R. ROBINSON COLLECTION, AND THE NAVAL INSTITUTE PRESS

M C M L X X X V I I

Other books by William Gilkerson:
Gilkerson on War
The Scrimshander
Maritime Arts by William Gilkerson
An Arctic Whaling Sketchbook
American Whalers in the Western Arctic (with John Bockstoce)

CONTENTS

Library of Congress Catalog Card Number: 87-50530
ISBN 0-87021-619-8
Printed in the United States of America

Foreword

Just prior to 1980, I met Wm. Gilkerson when he contacted me with some questions on matters of history. Having corresponded with him at length on the subject of French naval small arms of the American War of the Revolution, our professional relationship deepened, and his activites as a marine painter, particularly of 19th century whaling ships, were of great interest to me. The development of our association was further inspired by questions pertaining to the BONHOMME RICHARD.

Confronted with the problems posed by the task of recreating this famous vessel in aquarelle, he requested my help. I hesitated at becoming too involved, for the undertaking was a difficult one. However, I was won over by the mutual affection we had developed for one another, and so I dedicated myself to the long and difficult task of reconstructing John Paul Jones's ship. My own reward has been the succession of aquarelles that he has created for this book, prepared for the United States Naval Academy Museum.

Now, I am asked to write a foreword to this handsome collection, a request by which I am most honored and somewhat embarrassed. Not being a man of letters, I am more comfortable at my drawing table or buried in the archives than at expressing myself on paper.

Gilkerson's art demonstrates a perfect understanding and knowledge of the naval world of the last decades of the 18th century. Here he has shown enormous progress, having interested himself previously more with the 19th century. Since he is a demanding individual, dedicated to his studies and therefore to the understanding of the time, I have helped as best I could. This knowledge and understanding are indispensable; I would go so far as to call it the foundation of an artist's work. Next, there is the observation of the subject, in this case the ship's attitudes, those admirable spectacles which can only be envisioned through attentive examination of contemporary works. This involves the study of the difficult equilibrium between the wind and the sails, the sea and the ship.

For a marine painter to be worthy of his subject, what a formidable quantity of knowledge is required before he is intelligently able to address even the most insignificant representation of an historic ship! This is just the point. Aptitude and talent are the qualities that separate true artists from the imposters, so numerous in the domain of art.

Now that this artist is so at ease with his subject that he can view it globally as well as from the perspective of its detail, he may let his talent flow, calling forth his inventiveness, sensitivity and tact. Since he has this level of expertise at his command, we will watch his future work to see how well he protects himself from his own skill, not permitting it to become the enemy of sincerity. Indeed, I idealize the marine painter, hoping to recognize him in Gilkerson, whose work is so precise that even amateurs recognize it. The sea is his profession. His ships sail naturally; the wind blows through his pictures, and in them the ocean is restless. Gilkerson's character surfaces in his paintings. He is likeable, and the same can be said for his vision of the sea, ships and sailors. Now I will watch for a rugged look to develop in his work.

Having followed his career for several years, I see in this collection the result of long hours and attentive work. To you he brings days gone by to life in a manner that gives rise to emotion and reverie. He has indubitably established himself as a painter dedicated to this maritime world of old.

Jean Boudriot
Paris, 1987

INTRODUCTION

The pictures in this book represent research and development spanning 20 years, by the artist William Gilkerson. In 1968, he was commissioned to portray Jones's first naval command, the sloop PROVIDENCE, and also his most famous command, BONHOMME RICHARD. Gilkerson soon discovered to his surprise that little accurate evidence survived as to how these specific, 18th century ships appeared, and the sources that did exist — models and paintings — were at best imaginative, at worst fictions.

So, he began reading documents of the period, including the logs, reports, correspondence and narrative accounts, gathering their information, incorporating them into the evolving series of compositions and vignettes presented here. A close examination of this work will show the viewer the progress the artist has made in pursuit of historical accuracy. Included are many of the early drawings, and all of the most recent ones. As Gilkerson's knowledge evolved, so did the reconstructions of the vessels.

The viewer with little or no interest in 18th century naval architecture may appreciate the compositions on their artistic merit. For the careful student, the list of plates in the back of this book is annotated with an itemized list of errata, the artist's own technical critique of some of the earlier pictures, where subsequent study has revealed any inaccuracy.

The most enigmatic of all of Jones's ships proved to be BONHOMME RICHARD. Having been built as a French East Indies Company ship, then converted to a man-of-war, she sank 30 hours after her epic battle with HMS SERAPIS, and no plans of her have ever come to light. Even contemporary paintings of the action show only representations of the ship, with no two alike.

In the course of his ongoing study, Gilkerson contacted many scholars, among them Anglo-American historian Peter Reaveley, who had spent as many years as the artist in pulling together whatever he could find on the ship, the battle and its participants. An exchange of information led ultimately to a full fledged collaboration into which was drawn the leading authority on French 18th century naval architecture, France's Jean Boudriot. Noted for his remarkable skill as a draughtsman in recording every detail of hull and rigging, Boudriot had established a world reputation in 1973 with publication of *Le Vaisseau de 74 Canons,* subsequently translated into English: *The Seventy-Four Gun Ship.*

With the combined material in Reaveley's and Gilkerson's collections of English language documentation, plus Boudriot's encyclopedic knowledge of BONHOMME RICHARD's type, an educated reconstruction became possible.

Gilkerson had made several trips to Paris, returning with brief-cases of data, courtesy of Boudriot. When translated, there was enough material to permit the Annapolis based historic ship architect Thomas C. Gillmer to create preliminary orthographic drawings of RICHARD's basic hull. These were forwarded to Paris for critique, and Boudriot was favorably impressed.

At this point in the story of an American marine artist trying to do a creditable rendering of one of Paul Jones's ships, the French connection, Jean Boudriot, became intrigued enough with the project to set aside for a year his own life work (drafting all of the major types of French naval vessels of the late 18th century), in order to undertake detailed plans of BONHOMME RICHARD for future artists, modelists and historians. The results of this work are being published under the title *John Paul Jones and the Bonhomme Richard*, and it is in very many ways a companion book to this one. Besides the reconstruction, it contains the definitive account of BONHOMME RICHARD's battle with SERAPIS by Peter Reaveley, an analysis of the ship's hydrodynamics by Tom Gillmer, more of

William Gilkerson's pictures, and this writer's essay on the uniforms of John Paul Jones.

As a result of all of these efforts orchestrated by Gilkerson, BONHOMME RICHARD, once the most poorly understood of all of Jones's ships, has metamorphosed into the best known of them all. This academic side to the artist's work has led to other revelations of value to historians. Notable are Gilkerson's authoritative studies of the 18th century sailors; their activities, apparel, weapons, even musical instruments.

Besides its academic dimension, the artist's work contains others at least as important. Gilkerson begins with quick sketches, taken from life, and produces many vignettes in the process of developing a major composition in watercolor or oil. Among the details most appreciated in his work are the crews. By immersing himself in the working of the vessels he paints, he is able to convey something of the trials and the triumphs of all the men who sailed them, ordinary seamen along with famous captains. A compassionate understanding is discernible in his writing as well as in his paintings.

Gilkerson is a sailor. Born in 1936, he learned to sail as a child on a Wisconsin lake, and he went to sea at age 14 as a mess-boy aboard a Norwegian freighter. By 1961, he was master of his own yacht, sailing the North Sea, the English Channel and many of the waters raided by Jones nearly two centuries earlier. Returning to the United States, he moved to San Francisco, and there lived aboard his 40-foot ketch while working as a feature writer and editor for the *San Francisco Chronicle* until 1970, when he quit in order to pursue his artwork full time. Since then, he has published a number of books, his work has been anthologized, it has appeared in countless magazines and periodicals, and has been collected and shown by various public institutions.

In 1975, Gilkerson sent a small pencil drawing of BONHOMME RICHARD to the U.S. Naval Academy Museum, thus beginning a long association. Today, the Museum owns 15 of his works, and is pleased to host the exhibition of many more of them. This book is not so much a catalogue of an exhibition as an extension of the art and scholarly research of this artist. *The Ships of John Paul Jones* will endure for many years as a resource on its subject.

James W. Cheevers
Senior Curator
U.S. Naval Academy Musuem

BONHOMME RICHARD, 1973

Beginnings

John Paul Jones was christened John Paul in the Year of Our Lord 1747. He was born the son of a gardener at Arbigland, on Scotland's mercantile southwest coast near Kirkcudbright, and some undetermined number of years later (five or six, at a guess) he clambered aboard his first command, no doubt a rowboat. Though the humblest of vessels, that tiny craft has nevertheless provided the first and last taste of real independence to many a great captain.

The boy John Paul's first experience of shipping on the Solway Firth is suggested by the words of his biographer, Samuel Eliot Morison:

> *Sloops, schooners and brigs passed before his eyes daily, inward bound to Dumfries or outward bound to the Isle of Man, Belfast, Liverpool, and even the colonies overseas. If the tide did not serve sailing craft to mount the River Nith to Dumfries, they would call at the little port of Carsethorn, a mile and a half from Arbigland, and lie on the sand until floated by the flood. (William) Craik's son recalled, after Paul Jones became famous, that the child would run to Carsethorn whenever his father would let him off, talk to the sailors and clamber over the ships; and that he taught his playmates to maneuver their little rowboats to mimic a naval battle, while he, taking his stand on the tiny cliff overlooking the roadstead, shouted shrill commands at his imaginary fleet.*

The little rowboats of John Paul's not totally imaginary fleet were lapstraked, edge-fastened skiffs, built of yellow pine. Like most of the other craft indigenous to the Solway Firth, they were descended from Viking ancestry. Perhaps John Paul's first taste of sailing was aboard one of the local fishing vessels, a herring boat or a trawling smack. The trawlers fished for flounder in the winter with two or three men, but in the spring and summer went out for shrimp with a man and boy for crew. They were double-ended, clinker built smacks of some 25 to 30 feet in length, with a good sheer and full deck, both of which were needed against the dangerous chop that gets up in the Solway when the wind turns against the powerful tide. He learned to pilot those waters well enough to navigate them again some two decades later, when he returned to put raiding parties ashore. In 1760, at age 13, John Paul received his parents' blessing to sign for a seven-year indentured apprenticeship aboard the brig FRIENDSHIP of Whitehaven, the vessel on which his seagoing career began. Not much specific information about FRIENDSHIP has survived, although much is known of her type, for the trading brigs of that time were used throughout the British Isles and Europe, with only regional design differences. They probably moved more of the western world's cargo than just about any other type of vessel.

A brig by definition has two masts, with square sails and fore-and-aft sails to both. The brig and its variations (including snows and brigantines) generally ranged from 60 to 90 feet, measured on deck. This made them economical to operate and small enough to sail in most waters. They were capable of working in and out of the tightest harbors, but also were large enough to cross oceans, which many did routinely.

According to documents of 1760, FRIENDSHIP was a brig of 179 tons. Four years later she somewhat enigmatically appears in *Lloyd's Register of Shipping* as a 200-ton vessel, a discrepancy perhaps attributable to the inexactitude of thc admeasurer's science at that time. In either case, FRIENDSHIP was a fair sized brig for her time, measuring some 80 to 85 feet. Her beam, or width at her widest point, would have been about 22 feet, and her draft would have measured perhaps 11 feet. In today's terms, she would be seen as a very small vessel indeed, barely qualifying for the "tall ship" category.

The opportunity to obtain his deep water initiation aboard FRIENDSHIP no doubt provided John Paul with some insights that would be valuable to him when he unexpectedly found himself a naval officer 16 years later. FRIENDSHIP carried 18 deck guns, and a large crew for a merchantman — 28 men to handle guns and sails in case of need. England and France were in the waning days of the Seven Years' War, and it was usual for the merchantmen of both nations to be well armed. As far as is known, FRIENDSHIP was not called upon to use her batteries, but her guns would have been exercised regularly. FRIENDSHIP's routine was to make one round-trip voyage each year, from Whitehaven to Barbados, where she would dispose of her cargo of consumer goods and take on another of rum and sugar, which she would carry to Virginia. This was the "triangle trade", and it gave John Paul the annual opportunity to visit his elder brother, William, who had emigrated to America and become a tailor in Fredericksburg. On these visits William introduced John Paul to life in the American colonies. After buying and loading a new cargo of tobacco, pig iron and barrel staves, FRIENDSHIP would then sail home to Whitehaven.

FRIENDSHIP

John Paul spent some three years in her, learning the ropes, ship handling, navigation and the skills of the merchant — buying and selling cargo and keeping books. In 1764, FRIENDSHIP was sold by her owners. Peace had come, bringing a period of post-war recession to the shipping industry. With the disposal of the brig, the apprentice was released from his indenture, leaving him a free agent.

Unquestionably John Paul sought to further his career with a deck officer's berth of some sort, but found few attractive opportunities for a 16-year-old apprentice with no particular family influence in a sluggish industry. The opening which did come his way was a third mate's berth aboard another Whitehaven vessel about which we know nothing today, except its name — KING GEORGE — and that it was a slaver, or "black-birder" in the parlance of the time.

The traffic in slaves between Africa and the Americas was not yet illegal, but was in bad odor figuratively as well as literally. The awful conditions that prevailed aboard those ships have been sufficiently discussed elsewhere to need no repetition here. Always a humanitarian in his later career, John Paul no doubt attempted to introduce some measure of compassion into the KING GEORGE. In any case, he remained in her for some two years before engaging as first mate aboard another slaver in Kingston, Jamaica.

More is known of his third vessel. The TWO FRIENDS is recorded on the Naval Office lists of London as having arrived at Kingston on April 18, 1767, a 30-ton brig (or brigantine) built at Philadelphia in 1763, with a crew of six officers and hands, carrying a cargo of "77 Negros from Africa." TWO FRIENDS could not have been more than 50 feet in length, and the conditions aboard her are hardly to be imagined. One voyage in her was enough for John Paul. He left her and the slave traffic forever, expressing revulsion with what he later termed "that abominable trade." His resignation left him on the beach, without prospects, a teenager far from home.

Brigantine TWO FRIENDS

Having been paid off in Kingston, he shortly thereafter met the master of the newly arrived brig, JOHN, registered in Liverpool but sailing out of his home town of Kirkcudbright, under the command of Samuel McAdam, who was coincidentally a fellow Scot from Jones's birthplace. At McAdam's invitation, John Paul entered the brig as passenger, bound for home. During the voyage both McAdam and the mate died of fever. As Jones was the only qualified ship's officer and navigator left aboard, he assumed command and returned her in good order to her Scottish owners, who were pleased enough with his performance to offer him command of the vessel.

JOHN, at 60 tons, would have been about 55 or 60 feet in length, with a beam of 18′ or so, and a draft of about seven or eight feet. She is listed by one source as a brig, and by another as a brigantine, which is probably what she was. Although a brigantine was and is technically a brig with square sails only on the foremast, the term "brig" was frequently used rather loosely on shipping documents, and JOHN's extremely small size suggests she carried the lighter rig of a brigantine.

Today it seems remarkable to contemplate the lengthy voyages of these tiny craft in the hands of men such as John Paul. The seamanship of these men was of an order of awareness that we today can appreciate only dimly (if at all), because the necessity for such skill has been largely erased by sophisticated mechanical assistance in many forms. When John Paul took up his first command, he had no way to determine longitude other than by elaborate guesswork. The charts he used were frequently inaccurate, and the only constructed navigational markers were lighthouses which were few, far between and frequently unreliable. His sense of his ship's position, his knowledge of her limits of tolerance under stress, his weather forecasting system — these and other perceptions all depended on his senses and the seat of his pants. The result was attainment of a sensitivity — a level of awareness at which today's high tech sailor can only guess, although to the 20-year-old shipmaster John Paul it was unremarkable.

As to the vessel with which he faced the vicissitudes of the sea (pirates, hurricanes, lee shores and other perils), JOHN would not have sailed closer to the wind's eye than 60 degrees, if that; she might have averaged seven knots in fine conditions, and would today be considered so unsafe by the various maritime authorities that she would be unable to obtain the simplest license for commercial operation. In the context of her own age, however, she was a highly evolved, complex and efficient machine. In her, with a crew of seven men including himself, John Paul departed Scotland with a cargo of salt provisions and firkins of butter. He stood south — through the Bristol Channel, past Biscay and Spain, until finding the northeast trade winds. There, JOHN was pointed west and nudged to the latitude of her destination, Kingston, her latitude being the only line of position which she could keep with any degree of certainty. Upon arrival, it was incumbent upon the young master to undertake whatever haggling was necessary to favorably dispose of her goods, and purchase an equally favorable return cargo.

According to record, JOHN departed Kingston on June 23, 1769, bound for Kirkcudbright with 49 hogsheads and 6 casks of sugar, 156 puncheons of rum, 44 bags of pimento, 6 bags of cotton, 75 mahogany planks, and two and a half tons of Campechean dyewood. His return route possibly took him through the Windward Passage, thence along Cuba's north coast, or he might have rounded Cuba to the west. In either event, he would have sailed north through the Florida Strait in search of the westerlies that would waft him home.

John Paul made two round trip voyages in JOHN, earning the unqualified approval of the brig's owners. Their recommendation was helpful to him in obtaining his second command, a full-rigged ship, which he secured in October of 1772.

BETSY of London was definitively a ship; that is, she had three masts, carrying square sails on all three. In that age, it was more or less the maximum rig, although as with other classes of craft, tonnage

Ship BETSY

could not be deduced from rig, and BETSY's tonnage is not recorded.[1]

Also unrecorded are BETSY's length, beam and draft, so anything other than her rig is guesswork. Most likely, she was a ship of some 200 or 300 tons, and most likely her new master bought a share in her, judging by the handsome profits she earned him. In her, he again traded to the West Indies, from London via Madeira to Tobago. BETSY was a lucky ship for John Paul financially; however, she would prove unlucky in other ways.

Beginning her second voyage, in January of 1773, BETSY leaked so dangerously that she did not clear the English Channel before having to put into Cork for survey. No fewer than 30 of her mid-section frames were discovered fractured, and the repairs delayed the voyage for half a year. When BETSY finally arrived at Tobago, John Paul found himself embroiled in a dispute with his crew over pay, which resulted in a minor mutiny. He was attacked on BETSY's deck by the ringleader, a "prodigious brute of thrice my strength," who swung at him with a club of some sort, whereon John Paul defended himself with his sword, running the fellow through and killing him on the spot.

Immediately following the incident, Master Paul abruptly departed his ship, putting it into the hands of his agent, and then fled Tobago on another vessel, eventually finding his way to Virginia. There he changed his name — first to John Jones, and later to John Paul Jones.

Mystery surrounds this incident. The man that Jones killed was a native of Tobago, with friends and relatives in that small community. Jones's biographers are persuaded that he feared being lynched, without benefit of the hearing which would have exonerated him. Having once fled to save his life, he might well have feared being convicted of murder *in absentia*, his flight having been taken as a token of guilt, in conjunction with testimony from the only witnesses, all of whom were cronies of the deceased. The fear of being haunted by a murder conviction would certainly be reason for changing his name. Flight was contrary to his nature, as would become evident.

Whatever the circumstances, the consequences placed John Paul, now John Paul Jones, with his brother in the Colony of Virginia in early 1774. He was 27 years old, perhaps a fugitive, and again on the beach. He would this time remain there long enough to get acquainted with America, some of America's influential sons (who would open doors to his naval career), and one or two of her daughters, too. Meanwhile, distant events shaped his destiny . . . the Boston Tea Party, the gathering of the Continental Congress, the battles at Lexington and Concord and the siege of Boston. During the summer of 1775, he offered his services to the newly-formed Continental Navy, was accepted, and appointed Lieutenant. His next vessel would be a ship of war.

1. A merchant ship would range anywhere from 150 to upwards of 1200 tons, according to the measurement system of the time which expressed a vessel's tonnage on her theoretical capacity. In other words, at the time of the Revolution, a vessel's tonnage was reckoned by applying a mathematical formula to the length of keel and the midship breadth. These admeasurers' "tons" are not to be confused with *displacement* tons, which today are sometimes used to express a vessel's actual weight. Hence, in the old system, a ship of 100′ length with a beam of some 27′ would be registered at perhaps 275 tons. As a rule of thumb, a vessel could actually carry a weight of cargo roughly one-third greater than her measured tonnage. All references in these pages are to "old" tons.

PROVIDENCE (EX-KATY)

On December 3, 1775, at Philadelphia, John Paul Jones ran up the Grand Union flag (predecessor of the Stars and Stripes) aboard ALFRED, flagship of the Continental fleet. Four days later, he received confirmation of his commission as the first of the American Navy's First Lieutenants. Although ALFRED was the first warship to which Jones was assigned, it was not the first in the chronology of his naval commands. That designation belonged to the sloop PROVIDENCE, ex-KATY, which is where we should begin, with a vessel whose military career predated Jones's.

The curtain had gone up on the naval drama of the Revolution back in 1772, when an embattled group of Rhode Island colonists seized and burned His Majesty's customs schooner GASPEE, having first lured her aground in Narragansett Bay. The perpetrators were never brought to justice by the Crown, though not for want of effort, and so began a kind of cat-and-mouse relationship between British patrol vessels and American shipmasters, whose cargos and activities fell under ever closer scrutiny as political tensions worsened. With increasing frequency, a reinforced British naval presence disrupted maritime traffic, halting, searching, and sometimes seizing vessels.

By 1774, with war clouds darkening, the same quorum of Rhode Islanders that had burned GASPEE convened again to liberate a number of cannon from Fort Island, off Newport, ferrying them away aboard the sloop KATY, arguably the first American warship of the Revolution. KATY belonged to the Providence shipowner John Brown, who had been arrested (the word most frequently used at the time was "kidnapped") by a British naval patrol at the same time that two packet boats were seized for carrying flour to rebel groups. The incident stirred up more than local wrath. Eventually released, Brown armed his KATY with a battery of four-pounder cannon,

probably from the supply taken from Fort Island, and gave command of the sloop to Captain Abraham Whipple, with whom he had led the GASPEE raid. Brown then arranged to charter KATY (for $90 per month) to the Rhode Island General Assembly, whose resolution of June 12, 1775, in part reads:

> *It is voted and resolved that the committee of safety be, and they are hereby directed to charter two suitable vessels, for the use of the colony, and fit out the same in the best manner to protect the trade of this colony . . . That the largest of the said vessels be manned with eighty men, exclusive of officers; and be equipped with ten guns, four-pounders; fourteen swivel guns, a sufficient number of small arms, and all necessary warlike stores . . .*

So, armed with weapons and a warrant, KATY went prowling the Bay for the seized packets, which the British had meanwhile armed and incorporated into their squadron as tenders. One was soon located — the DIANA — and attacked. The ensuing cannonade was brisk and brief, ending with the DIANA driven ashore, where she was recaptured by the rebels and refloated.

The British commander at Newport was unable to bring KATY to account, although his squadron repeatedly was reinforced. His frigates were too deep for Narragansett Bay's inner waters, and his shallow-draft vessels were too weak, so KATY busied herself for most of 1775 haunting the shallows, sometimes darting out, harassing the British in a variety of ways — ferrying rebel passengers and supplies, disrupting enemy communications and protecting a wide area from British marauding parties, while posing a constant threat to the squadron's tenders.

In November, Rhode Island's Esek Hopkins was appointed as the Continental Navy's first Commander in Chief, and KATY was ordered to Philadelphia to join the rest of the fleet. This consisted of four merchantmen which had been purchased and were being converted into warships: the ship ALFRED, initially armed with 20 nine-pounders and 10 six-pounders; the ship COLUMBUS, with 18 nine-pounders and 10 sixes; the brig ANDREW DORIA, with 16 sixes; and the brig CABOT, with 14 sixes. It was this flotilla over which KATY and Captain Whipple found John Paul Jones raising the flag on the 3rd of December.

KATY joined the fleet and was formally commissioned into the Continental Navy, at which time her name was changed to honor her home: PROVIDENCE. She was given another pair of carriage guns, bringing her armament to 12 four-pounders, plus a similar number of swivels. As the fleet fitted out, she underwent repairs which cost £20, 12 s.

When Captain Whipple was promoted to COLUMBUS, John Paul Jones was offered command of PROVIDENCE. It was an offer he could refuse, which is what he did, electing to remain second officer on ALFRED. Why did Jones, the ambitious maverick, choose a secondary role aboard a converted merchantman when he could have commanded a trim, small cruiser of proven ability? He offered as his reason his complete lack of training or experience in the skills and protocols of naval service; he wanted some on-the-job training.

No doubt he was also prudently wary of PROVIDENCE. The big colonial sloops were fast and handy to be sure, but they also had a well-deserved reputation for being dangerous and difficult to handle.

PROVIDENCE's single tall mast carried a complete set of fore-and-aft sails: staysail, jib, and outer jib tacked to the end of an outrageously long jib boom; a huge mainsail clewed loose-footed to a boom of approximately the same length as the vessel itself, and perhaps a leg-of-mutton topsail sheeted to the gaff. In the square sail department, she is known to have carried a square topsail and studding sails, over which she probably hoisted an unbraced topgallant. Below, she was likely able to stretch a huge, square course in light weather.

This inventory of sails gave PROVIDENCE tremendous versatility, allowing her to be handled as either a square-rigger or a fore-and-aft rigger, or both at once: to sail well under the merest scrap of canvas in a gale, or ghost happily in barely discernible air under a towering corporation of sails.

Constant vigilance was necessary to prevent her being struck by an unheralded squall, for even a sudden wind shift could lay her flat in the water, and the power of her full mainsail in an uncontrolled jibe would have proven catastrophic in even a moderate breeze. PROVIDENCE spread much larger individual sails than other vessels of her size — schooners, brigs, ketches, and so on. The split-rigged craft with more manageable sails were handled more easily in an emergency, and by fewer men. It is a great mistake to think of PROVIDENCE as a diminutive vessel because of her single-masted rig. She is known to have had the capacity to carry and sustain a complement of 80 men, not an excessive number to handle her sails and guns simultaneously in action, and provide prize crews and officers to her captures. In fact, she was seldom able to get anywhere near a full crew.

Some clues to PROVIDENCE's appearance survive, including the observations of a British spy at Philadelphia, who described her as: "A sloop, all black, long, and low . . . with crane irons over the quarters for oars." No plans or measurements of PROVIDENCE have survived, but extrapolation based on her known capacities and characteristics reveals her as one of the larger examples of her genre.

Probably she measured some 70' in length, with a beam of around 22' and a draft of 10' or 12', depending on how much drag she was given in John Brown's Providence shipyard, where she is thought to have been built (possibly prior to 1763, as a privateer during the French and Indian Wars). The earliest of her voyages for which there is record took her to Surinam (on South America's north coast) in 1769 on a trading venture. When GASPEE was burned, KATY was off on a whaling cruise.

The sloop type had been developed in America over several generations, after originating as the "Jamaica" sloop, which has come down as the "Bermudian" sloop. Fishing boat, whaler, cargo carrier for coastal or offshore voyages, smuggler, slaver, privateer — the sloop was the answer to most of the colonists' everyday maritime needs. It was the preferred craft of the pirates in their heyday. As a warship, the sloop possessed the desirable quality of being able usually to escape from that which it could not fight, and to catch that which it could. Although the British Navy did not formally adopt sloops and cutters until 1760, it had long employed them for a variety of jobs — not least of which was chasing their counterparts.

The next time Jones was offered PROVIDENCE, he would accept her. For the moment, the command went to John Hazard, who was characterized by one of his fellow captains as ". . . a Stout Man Very Vain and Ignorant — as much low cunning as Capacity," an assessment with which history has agreed.

The story of the navy's first fleet is well known. Delayed by ice in the Delaware River, it was joined by FLY, another small sloop which was also offered to Jones, and also declined. By the time the squadron broke free, everyone on the vessels was well chilled, and instead of pursuing its original orders — scouring the Chesapeake (where small British warships were playing havoc with the vital water commerce) — the fleet headed directly for Nassau in the Bahamas, where it landed Marines and captured a vital store of ordnance and 24 barrels of gunpowder.

Well ballasted with captured cannon, the squadron returned north intact to a rendezvous off Block Island. There, it encountered the light British frigate GLASGOW, 24, to everyone's subsequent regret. In Jones's first engagement, his teachers gave him little inspiration. From his vantage point on ALFRED's main gun deck, he watched the frigate and squadron approach one another. It was a beautifully moonlit night. Commodore Hopkins issued no orders whatever, and formed no line or plan of battle. CABOT was the first vessel to encounter, beginning the engagement without orders. In return, she was disabled by the well-aimed broadside of the frigate's heavier guns.

ALFRED was next to engage, but had her tiller ropes shot away almost immediately, so that she drifted out of control and was raked ferociously by the enemy's fire. By the time ANDREW DORIA avoided collision — first with CABOT, and then with the drifting ALFRED — and was at last able to engage, she found herself alone because COLUMBUS had been wind blanketed by the other vessels and was unable to help. PROVIDENCE stayed clear entirely. By the time COLUMBUS managed to enter the action, her gunnery was wild and completely ineffectual. By the way of finale, the plucky GLASGOW beat a retreat, firing stern chasers at her pursuers as she went, and effected her escape.

In the action, Jones had from the enemy a lesson in cool professionalism, and from his superiors a lesson in total incompetence. He sat on the subsequent courts-martial. Captain Hazard was cashiered for his non-role in the action with GLASGOW (among other things), and Jones was again offered PROVIDENCE, with the temporary rank of Captain. This time he made no hesitation. No doubt

Sloop PROVIDENCE pursuing a Halifax schooner over the Grand Banks

feeling his naval instruction had been sufficient, he accepted command on May 10, 1776.

After a brief refit, PROVIDENCE was assigned an assortment of errands which included escorting a convoy of colliers to Philadelphia. All of these duties were performed under the nose of the enemy, and provided Jones the experience he needed in handling his command. This time he was more fortunate in his tutor, PROVIDENCE's seasoned second lieutenant John Peck Rathbun, who was thoroughly familiar with the quirks and capabilities of his sloop. Both Jones and Rathbun were within a year of 30, and apparently they became friends and mentors to one another. Both were archetypal warriors, cast from the same mold; both had the personal zeal and panache beloved of history.

PROVIDENCE was lucky in her entire crew, which Jones later described as the best with which he had ever sailed. Despite the problem of recruiting, which was exacerbated by competition from a gaggle of new privateers offering generally more seductive terms to seamen, PROVIDENCE signed on enough hands to bring her complement to 73 officers and men, which was about 90% of her capacity. Included were 25 Marines.

When we pause to remember that each man brought onto the ship his own gear, and that a number of specialists aboard kept shop with elaborate inventories of specialized equipment (cook, carpenter, sailmaker, "surgeon," gunner, boatswain, steward, and perhaps an armorer or smithy as well), we may dimly perceive that in the small cruisers of the sailing navy the fine art of packing men, supplies and material into small spaces reached an epitome of compaction, making the interior of any modern warship look as spacious as a polo field by comparison.

Then, besides the myriad gear, the vessel carried at least enough water and food for a month and a half; more if possible, and cordwood for cooking. At the time Jones took up her command, PROVIDENCE's ration list stipulated that each seaman's diet was to include a pound of bread daily, and also a pound of beef or pork every day except Wednesday, when the meat would be replaced by a half-pint of rice with two ounces of butter and four ounces of cheese. On Mondays, Thursdays and Saturdays there was a half-pint of dried peas, four ounces of cheese on Monday and Saturday, and a pound of turnips, onions and potatoes each week. There was "pudding" (probably a duff concocted from pork fat, molasses or brown sugar and dough) twice a week, and a half-pint of vinegar and the same quantity of rum was allowed daily, with a discretionary extra rum ration allowed on appropriate occasions.

Occupying the area under PROVIDENCE's long quarterdeck, the officers were supplied from separate stores of somewhat more fastidious fare. On deck, the 16-or 18-foot longboat on its chocks was filled with crates of chickens and pigs, along with their feed and perhaps a few tethered sheep.

At Philadelphia, on August 6, Jones received his sailing orders. Confirmation of his captain's commission came two days later. His orders directed him to "proceed immediately on a Cruise against our Enemies & we think in & about the Latitude of Bermuda may prove the most favourable ground for your purpose," with the admonishment to ". . . Seize, take, Sink, Burn or destroy . . ." enemy shipping.

Within a fortnight PROVIDENCE had provisioned and sailed, capturing a whaling brigantine soon after clearing the Delaware Capes. Heading south to the latitude of Bermuda, she found a convoy, investigated, and was in turn investigated by the escort, H.M. Frigate SOLEBAY, 28 guns, which gave chase. By Jones's account, there was a "strong cross sea," kicked up by lots of wind. Wind was what the frigate wanted. Being heavier, SOLEBAY was stiffer and steadier in blustery conditions. She was favored also by her greater length of waterline, which in sufficient wind translated into more hull speed than the smaller vessel could attain. For once the conditions favored the enemy, and PROVIDENCE was in a dangerous situation.

Caught in such circumstances, Jones's only recourse was to exploit the advantage of PROVIDENCE's fore-and-aft sails, which permitted him to hold closer to the wind's eye than the square-rigger could, thereby prolonging the chase. His best hope was that the frigate would content itself with driving him over the horizon, and then return to its flock. SOLEBAY smelled blood, however, and persisted in thc chase, gradually closing the distance that separated them. It was a foaming, close-hauled pursuit which went on hour by hour. PROVIDENCE's advantage of weatherliness was largely negated by the steep, short seas through which she was forced to bash, impeding her more than the frigate. SOLEBAY inexorably approached to leeward toward the point where its broadside guns would bear. Surrender or catastrophe seemed the only alternatives.

Jones and Rathbun in council had all day to contrive a different denouement, however, cooking up a surprise maneuver which called for careful timing. By 5:30 p.m. SOLEBAY had worked up to "within musket shot" of PROVIDENCE's lee quarter, and had opened fire with its forwardmost gun. Abruptly, Jones ordered his helm put *up*, and all of the square sails including studding sails were broken out in a rehearsed maneuver, sending PROVIDENCE shooting downwind directly across the bows of her pursuer. The British captain was no doubt poised for a forlorn last evasive attempt at tacking to windward by the sloop, and was stunned to find his prey suddenly under his leeward guns, which were unprepared. By the time he was able to respond, PROVIDENCE was out of range downwind, travelling at hull speed and opening her lead by the moment under a blossom of square sails as SOLEBAY was forced to ponderously retrim her more complex rig to a downwind situation.

Sloop PROVIDENCE eluding H.M. Frigate SOLEBAY and firing a swivel cannon

This is the moment depicted in the painting on page 16. SOLEBAY's staysails are backwinding and her topgallants are being trimmed as PROVIDENCE escapes into the gathering twilight. A swivel gunner is getting off an impertinent shot with the only ordnance that will bear aft. The chances of the little cannon doing any damage in the circumstances truly would have been a long shot, but a lucky round might just have severed a halliard or a brace, further inconveniencing SOLEBAY's pursuit. Jones is known to have more than once used impudent shooting as much for the amusement and morale of his crew as for the discomfiture of his enemy.

In early September, PROVIDENCE took two good prizes, and then headed north for Nova Scotia, where another British frigate, H.M.S. MILFORD, was encountered, this time under less threatening circumstances. In lighter airs, Jones put PROVIDENCE through her steps, first cracking on sail, then shortening down ". . . to give him a wild goose chase, and tempt him to throw away powder and shot . . . He excited my contempt so much by his continual firing at more than twice the proper distance, that when he rounded to give his broadside, I ordered my Marine officer to return the salute with only a single musquit."

On September 22, Jones navigated PROVIDENCE to Canso harbor to take on water and stovewood. He found there three fishing schooners; one he sank, another he burned, and the third he made a prize. Having recruited several volunteers from among the local fishermen, he headed for the east side of Canso Bay, where, according to an informant, lay an entire fleet of British fishing vessels. Jones found them distributed between two anchorages, sent a raiding party into each, and patrolled between them in PROVIDENCE while the whole assemblage of nine vessels — schooners, brigs and ships — were captured along with some 300 fishermen. Having destroyed the shore installations, Jones returned three of his prizes to his prisoners, gave them provisions, and sent them home.

Jones's cruise effectively was terminated on September 25 by a violent gale which drove two of the prizes ashore and separated another with 25 men and PROVIDENCE's small boat. This left Jones too shorthanded to continue, and he headed for home. On the way, he made a prize of a whaling sloop, and dropped anchor in Narragansett Bay on October 7.

In all, Jones had taken 16 prizes, of which four brigantines, a sloop and a ship reached safe harbor. He also had destroyed a valuable increment of His Majesty's fisheries, thereby further straining the resources of an already overextended British Navy in its efforts to control the coasts of the entire northeastern American continent.

ALFRED

Jones had little rest. As soon as he made his reports, he received a new assignment — to continue depredations of Canada's maritime provinces, with ALFRED, PROVIDENCE, and an armed brig, HAMPDEN. Captain Saltonstall of ALFRED had recently been posted to one of Connecticut's two frigates newly off the ways, so Jones moved his gear back aboard his former ship, this time as Captain and Commodore.

No doubt ALFRED felt very familiar to Jones. As a square-rigged merchantman, she was essentially the type of vessel upon which he had received his whole professional training and experience prior to embarking on his naval career. In PROVIDENCE, he had tasted the ambrosia of a fast ship, and he never wanted any other kind after that. It was something ALFRED would never be as a cruiser, for although formidably armed,[1] and a more prepossessing craft than PROVIDENCE, ALFRED's sailing qualities had been severely compromised by her conversion to a warship. Foremost, she had been conceived and built to carry cargo for a living, not guns, and so she had been taken into the service through necessity. At the time of ALFRED's purchase, the Congress had realized the need for proper cruisers, and authorized the accelerated construction of 13 frigates, two of which were now manning in Providence, competing with Jones's vessels for seamen.

Jones complained that ALFRED was "crank," too easily dipping her leeward gunports, making her battery on that side unserviceable. Commodore Hopkins shared that view, reporting her as "tender sided" and "The most unfit vessel in the fleet." Hopkins noted also that she still carried ". . . a nine-pound shot through her mast," a souvenir of her brush with GLASGOW the previous year.

ALFRED was built at Philadelphia just before the War. There are no plans or reliable pictures of her, but some valuable clues to her appearance have survived, including an exceptionally detailed description by a British agent reporting his observations to the Admiralty:

> *"The Alfred was formerly a Merchantman (and) is about 275 or 300 tuns at most. (She is) pierced for 20 guns 9 pounders & carries 6 four pounders on the Quarter Deck & forward. the Capt. (is) and American. She they say has 160 men, & . . . sails dully. her Appearance when her Guns are house & ports lowered down had very little of the ship of war. she is square sterned, without quarter Gallery or Badges, has a figurehead painted yellow with a remarkable large Plume of feathers on his helmet painted white. The ship is painted plain black & yellow with a white bottom . . . is very taunt but not square*

1. Not so formidable as originally, at Philadelphia, when she was given 30 guns — 20 nines and 10 sixes. Apparently Jones removed most or all of her six-pounder upper deck guns into the bilge, which would have substantially improved — though not restored — her sailing qualities. ALFRED's armament was altered several times over the course of her career.

rigged . . . has a Gaft Mizen . . . Top Armour & Quarter Cloths Blue with 13 white Stars the same as in the upper Corner of their Colors . . ."

This describes features characteristic to Philadelphia-built ships, including the square stern, no quarter galleries, and "taunt but not square rigged," meaning she had very tall masts, but not overly-long yards. She also would have had a flush deck with no waist, and only a slightly raised quarterdeck. A 1775 customs declaration confirms most of these features. Extrapolating from a deduced tonnage of 300, ALFRED probably had an overall length of about 100′ and a beam of 27′. She had been launched in 1774 or 1775, and originally named BLACK PRINCE. As such, she had been given a flamboyant figurehead showing her namesake drawing his sword. It worked just as well for King Alfred, whom the Naval Committee honored as the founder of the British Navy. Her "plain black & yellow" meant she was black with a yellow stripe *above* her gunports.

It is significant that the highly detailed intelligence report does discuss ALFRED's gunports, but not as having appeared abnormally low on her hull. Such a feature would have been remarkable; important both in identifying her and in evaluating her potential performance. We may therefore infer that ALFRED's indubitable tendency to dip her lee ports was at least as much a consequence of her slab-sided shape and her extreme tenderness as low port placement. The heavy guns she was never intended to carry raised her center of gravity from bilge to main deck, destabilizing her.

Whatever ALFRED's physical handicaps, the most pressing problem at the outset of the cruise was a critical shortage of men for Jones's vessels. His flagship had only 40 hands out of a complement initially rated as 220, and the other two vessels of his squadron were not much better off. Nor was there any apparent way of finding the men he needed. The available manpower pool had been drained, certainly by the manning of the two new frigates, but primarily by the ubiquitous privateers. ALFRED was outclassed on all fronts in the competition for volunteers. After various frustrations, Jones eventually distributed the crew of HAMPDEN into his other two craft, and sailed with only ALFRED and PROVIDENCE, on October 27, 1776.

Sailing up Vineyard Sound, he investigated Tarpaulin Cove off Naushon Island, an anchorage long favored by pirates and privateers. Indeed, Jones found there one of the Rhode Island privateers to which he owed thanks for his crew shortage. Lieutenant Rathbun and an armed party were sent aboard the privateer, where four deserters from the Navy were found hiding behind a false bulkhead. Their services were promptly reclaimed, along with those of 20 prime seamen, appreciably augmenting Jones's crew.[2] He sailed north.

On November 11, Jones took a brigantine bound for Halifax, and the following morning made a tremendously important capture — the 350-ton armed transport MELLISH with 10,000 complete winter uniforms "from hats to shoes" intended for the British army in Canada, as well as some weapons, cavalry equipment, stocks of medicine and general cargo. Wanting to keep her securely with him, Jones added MELLISH to his squadron, sending aboard her 25 seamen and 10 cannon, presumably ALFRED's entire six-pounder battery of which he was happy to be justifiably rid.

A week later, another small prize was taken in a hard chase during which PROVIDENCE strained her seams. The squadron had experienced punishing weather all along. Finally the sloop's pumps

2. Impressment was against the law in the Continental Navy, and although there were some ambiguities to this incident, Jones and Rathbun were later sued over it.

could only marginally keep her afloat. Her captain, Hoystead Hacker, turned for home during a snowsquall.

Again short-handed, Jones continued his cruise with ALFRED and MELLISH, taking three colliers. These, too, were manned and ordered to stay close. Jones turned for home, encountering and taking still another ship, the 10-gun letter-of-marque[3] JOHN. Still another prize crew somehow was scraped together, and the convoy proceeded in unremitting bad weather until December 8, when a British frigate was encountered. It turned out to be Jones's old friend, H.M.S. MILFORD, full of men and more than a match for his combined but weak force.

When the wary MILFORD moved to investigate this unknown but formidable flotilla, Jones stolidly held his course, making a flurry of signals and no doubt breaking out British colors. With night coming on and a northwesterly gale blowing, MILFORD went on to a parallel course with the apparently friendly convoy, accompanying it in a commanding position. Having instructed MELLISH and the three colliers to hold their course for home and not follow ALFRED and JOHN, Jones ordered lights shown in both those vessels, and about midnight tacked to the northeast. It was a move calculated to draw the frigate to investigate. It worked. At dawn, MILFORD was almost upon them, and the other vessels were out of sight. It was just the kind of improvisational theater at which Jones was so talented. Having accomplished his ruse, he fired a broadside into MILFORD and headed again for home. Instead of pursuing ALFRED and her small tender, the frigate turned on JOHN, which was recaptured.

Jones's primary concern had been for the safety of MELLISH and her valuable cargo. When he sailed into Boston on December 15, he learned to his relief that she had come into New Bedford, where her priceless supply of winter clothing was already being unloaded for shipment to General Washington's frozen few in Pennsylvania. It would arrive just in time to boost the army's morale prior to its Delaware crossing and victory at Trenton. Conversely, the loss of that cargo severely inconvenienced Burgoyne's army in Canada, so that the timeliness of the capture made MELLISH one of the more valuable prizes of the war.

PROVIDENCE had made it safely into Narragansett Bay, only to be bottled up there in early December by the appearance of a huge British invasion fleet which blockaded the Bay and landed an army to occupy Newport. Its warships and transports were so numerous "... there was nothing to be seen but ships," as Marine Lieutenant Trevett noted in his journal after viewing the armada from the deck of PROVIDENCE.

Although blockaded, PROVIDENCE was assigned again to Jones in mid January, 1777, and ALFRED was reassigned, which did not please Jones. The bad news was followed quickly by the even worse news that the Continental Navy's original Captains List had been released, and on it he was only number 18. Nevertheless, because of his successful cruises, and no doubt partially because of the howl he raised, the Marine Committee decided to give Jones command of a proper squadron, including ALFRED again, as well as COLUMBUS, CABOT, HAMPDEN and PROVIDENCE. It was a squadron only on paper, however, and never destined to form, for the actual ships were scattered, blockaded, unmanned, or under repair.

In March, Jones and his colleague Rathbun travelled overland to Philadelphia to see what they might accomplish with a personal visit to the halls of government. As a result, Rathbun was promoted Captain and given command of PROVIDENCE, while Jones was eventually assigned to RANGER, a new 20-gun ship fitting out at Portsmouth, New Hampshire.

3. A merchantman licensed to capture prizes.

The squadron had experienced punishing weather all along

RANGER

In mid July 1777, Jones arrived in Portsmouth, New Hampshire, to oversee the completion and outfitting of his new command, and he wrote of her: "The RANGER taken altogether will in my judgement prove the best Cruizer in America . . . She will always be able to Fight her Guns under a most excellant cover and nothing can be better calculated for sailing Fast or making good Weather . . ." By

excellant cover, he meant her entire battery was under the quarter-deck, the forecastle or the waist gangways, and well-sheltered from enemy small arms fire. She was a proper warship.

RANGER had been authorized by Congress (along with two other vessels of her class) in the late fall of 1776, and had been built at Hackett's shipyard on Rising Castle Island, adjacent to Portsmouth, on the Piscataqua River. Launched in May, 1777, and originally named HAMPSHIRE, she had been renamed RANGER to honor the role of American riflemen. It is thought that RANGER had been intended to be a brig, but at the suggestion of her designer, William Hackett, she was built as a copy of the ship PORTSMOUTH, a fast corvette which Hackett had built earlier at the same yard.

By definition, a corvette carried all her cannon on one deck (as contrasted to a frigate, which carried a covered main deck battery and also a secondary battery on her quarterdeck and often her forecastle as well). A *corvette* was also called a *sloop of war*. The two terms were synonymous, although the word *sloop* used in that sense did not indicate a sloop rig. Not only sloops and cutters, but schooners, brigs and brigantines could be rated "sloop" in naval jargon, provided they had a single gun deck. The ship-rigged corvette RANGER was also sometimes referred to as a "ship-sloop".

No plans for RANGER or her sisters[1] have survived. Most historians agree she was probably a scaled-down version of the Hackett-built frigate RALEIGH, plans for which have come down to us. RALEIGH was one of the 13 original frigates authorized by Congress in 1775, and she was launched right next door to RANGER. The

1. After RANGER, Hackett built still a third vessel of the same design, HAMPDEN, not to be confused with the earlier brig of that name under Jones's command, although Hackett's HAMPDEN coincidentally was given a brig rig also, making her one of the largest brigs of her time. Like PORTSMOUTH and RANGER, HAMPDEN proved a fast, handy vessel, and she successfully cruised as a privateer.

British eventually recorded some of RANGER's data, but there has been a good deal of disagreement among historians as to their interpretation. It seems safe to say her length was between 110′ and 116′, with a beam of 28′ or 29′, a draft of about 12′, and a tonnage reckoned somewhere between 308 and 318. Whatever her precise measurements, RANGER was large for a 20-gun ship, and followed the American custom of building warships proportionately heavier and longer for their batteries than their English counterparts. From the late 1600's, the development of ship design in the North American colonies had consisted of generally increasing the size of ships, making them sharper and lowering freeboard, while decreasing topworks.

While in some respects "the Best Cruizer in America" delighted Jones, in others he found RANGER problematic. Her spars were too heavy and too long, which meant she was overcanvased and inevitably was going to be tender. The local Navy Agent, John Langdon, was unco-operative in helping obtain proper spars, or indeed anything else, for that matter. Again, Jones was faced with problems resulting from conflict of interest, and he was in competition with other ships (some of them privateers owned by Langdon) for men, supplies and material. With great difficulty, he eventually overcame these handicaps, more or less, at least enough to get to sea, but he was forced to take on a crew which would frustrate him for the duration of the command.

All of the officers and most of the men whom he was able to sign aboard RANGER were local townsmen, some 150 altogether. They all knew one another, and the officers were beholden for their commissions to Navy Agent Langdon, whose dilatory performance of duty in helping RANGER's outfitting inevitably put him at odds with Jones. *None* of the naval officers forced on Jones had any naval experience whatsoever: no training, no formal discipline. From the outset, there was resentment of Jones, who was not only an outsider, but a Scottish foreigner at that. Jones commanded a local ship with a local clique of officers and a local crew, all of whom were used to deciding their own lot at town meetings, which was more or less how they continued to conduct business throughout RANGER's voyage.

By the end of October, 1777, after many travails, the ship was finally as ready as she was going to be. On November 1, fair stood the wind for France, and so did RANGER, sailing without benefit of a shakedown cruise. She wore a suit of sails made of an inferior jute material called "hessians," which generally was considered more appropriate for gunny sacks than for sails, but it was all Jones had been able to obtain. RANGER's yellow sides were pierced with 20 gunports, behind which were 18 six-pounders. The figurehead of a rifleman ornamented her stem.

Jones carried with him a letter from the Marine Committee of Congress to the American Commissioners in Paris stipulating he was to receive the command of a new 40-gun frigate building in Amsterdam, L'INDIEN. His assignment after carrying various dispatches to France (including news of Burgoyne's surrender at Saratoga) was to relinquish his temporary command of RANGER, and receive new orders from the American plenipotentiaries.

The crossing was blessed by favorable winds. RANGER took two small prizes en route and arrived at Nantes on December 2, 32 days after her departure. Under her oversized rig, the ship's performance had been as cranky as Jones had feared. At Nantes, Jones's first consideration was for his crew. He advanced their pay out of his own pocket, generously reprovisioned, and then set everybody to correcting all of RANGER's problems. Having attended to his crew and his ship, he then departed for Paris.

Captain John Paul Jones reported to First Commissioner Benjamin Franklin, with whom he established an immediate rapport.

Franklin took an immediate liking to the peppery young Jones, and treated him much as an adoptive son, introducing him to Paris society, and to the intricate game of international politics in which France was being nudged closer and closer to war with England. It was a complex game which both sides played. Unfortunately for Jones, the same politics had ensnarled his intended command, L'INDIEN, and neither Jones nor Franklin was able to extricate the new frigate from neutral Holland. In other endeavors, Jones was more successful.

For instance, he formed a relationship in some depth with Madame LeRay de Chaumont, who was the wife of Sieur de Chaumont, through whose hands passed most of France's considerable assistance to the rebellious colonies. When Jones rejoined RANGER at the end of January, 1778, he had obtained the wherewithal to pay for all of the refitting and provisioning of his ship.

RANGER was much admired by the French. Some naval officers who came aboard at Nantes referred to her as *un parfait bijou*, a perfect jewel. By the time of Jones's return, she had undergone extensive work. Her rig had been cut down; her mainmast had been shortened and restepped aft; a whole new set of sails was being completed (her original set was cut up and sold for bread bags); she was completely reballasted, and some 30 tons of lead pigs had been added. All of her ordnance was removed so she could be careened and scraped, and when it was replaced, her log confirmed taking on board 18 six-pounders and five "swiffels." Among Jones's Paris acquisitions was an extraordinary set of orders that instructed him to carry on with RANGER ". . . in the best manner for the cruise you propose, that you proceed with her in the manner you shall judge best for distressing the Enemies of the United States, by sea or otherwise, consistent with the laws of war, and the terms of your commission."

On February 14,[2] her work completed, RANGER sailed for Quiberon Bay, where she arrived the same day. Jones anchored off for an exchange of messages with *Chef d'Escadre* La Motte Piquet, who was there with a French task force of frigates and ships of the line. With the Admiral's assurances of a nine-gun recognition of the Stars and Stripes, RANGER beat past the French fleet in "Very Squaly weather" and received the salute. An earlier version of the American colors had previously received such a salute, but Quiberon Bay marked the first foreign recognition of the national flag of the United States, as adopted by the Congress on June 14, 1777, during the same session in which Jones was posted to RANGER.

Making a series of short hops along the Bay of Biscay's fierce north coast, RANGER experienced back-to-back gales, and her performance convinced Jones that further and even more dramatic alterations were required to the rig. At Brest, her main and mizzen masts were stepped still further aft by some six feet each; the yards were shortened; most of the new sails were recut, and still more ballast was shifted until Jones was at last happy with his ship. As a final touch, RANGER sailed from Brest with a supply of red cloth to stretch over her yellow gunstripe, enabling her to be disguised as a merchantman. It was April 10. Her real cruise had begun at last.

Jones had an ambitious itinerary. Besides the usual business of seeking prizes, he intended to raid the mainland of England and capture some hostages which he hoped would be useful in forcing the exchange of American seamen held in British jails. Also, he wanted to destroy as much shipping as possible in British harbors. After all, the British had burned American seaports; why not fair

2. A week earlier, Franklin and Vergennes had signed the secret treaty of alliance committing France to armed support of the Revolution.

RANGER's flag saluted by La Motte Piquet's squadron at Quiberon Bay

Ashore at St. Mary's Isle

exchange? RANGER's course was made for St. George's Channel.

A brigantine was taken and scuttled, then a 250-ton ship was captured and safely sent into Brest. Heading north into the Irish Sea, "merchantman" RANGER lured a British revenue cutter into cannon range, but the smaller fore-and-aft rigged warship escaped, not unscathed, in a tacking duel which favored the more weatherly little cutter. It was just the sort of tactic Jones had learned so well in PROVIDENCE.

During the next few days, more small vessels were taken and destroyed. From the crew of one capture, Jones learned of the presence in Belfast Lough of a 20-gun ship, DRAKE, which he wanted to attack with armed boats in a cutting-out expedition, but RANGER's crew absolutely demurred. Only grumblingly did they agree to an alternative scheme calling for a surprise attack from RANGER, disguised. The plan was initiated, but it miscarried, due to a stuck anchor, although without blowing RANGER's cover.

A day later, Jones was off Whitehaven, a place he knew well enough to target for a shore raid. Among the crew, however, there was near mutiny. All winter, everyone had been relentlessly worked, first on RANGER's endless modifications, and then in driving the ship through gale after freezing gale. Now Jones wanted them to land on enemy shores and burn shipping — a venture so neatly combining the maximum of danger with the minimum of personal profit that some of the crew did indeed plan a mutiny. A Swedish volunteer tipped the plot to Jones, who stopped it at pistol point at its first manifestation. Eventually managing to scrape together a shore party under his personal command (both lieutenants flatly refused to go), Jones embarked with two boats at midnight, in calm weather. The raiders arrived at the town just before dawn.

Two forts guarded Whitehaven harbor. Jones's boat made for one, and the second boat was sent to surprise the other. Jones's party bloodlessly captured the objective and its defenders. Having spiked the first fort's cannon, Jones proceeded to the other fort, to find out how things were proceeding there. Prudently leaving a loyal midshipman in charge of his boat (which was a good thing, because its crew planned to return to RANGER and leave the Captain ashore), Jones and another midshipman surprised the second fort, decommissioning its guns also. His other boat crew had not been there at all, having decided instead to invade a pub in the town, preferring the unplugging of bottles to the plugging of cannon. While they were engaged on that business, an Englishman of their number slipped away and ran through the streets, banging on doors, giving the alarm that the rebels were coming.

Despite these vicissitudes, Jones managed to direct the torching of some colliers and other craft in the harbor, meanwhile facing an increasing and ominous crowd of townspeople who had begun to gather. Somehow getting his own people regrouped, Jones saw them into the boats and safely off. By 6 a.m. they were back aboard RANGER, and underway for their next frolic ashore, which began four hours later.

By 10 a.m. RANGER had fetched Kirkcudbright Bay, whose waters Jones had sailed as a small boy. The prodigal had returned, but hardly in the biblical sense, to be sure. Here, where he remembered the terrain so well, it was Jones's plan to raid the mansion of the Earl of Selkirk, capture the Earl, and carry him off to France where the Laird could perhaps be traded for some of the captured American sailors whom the British so adamantly had refused to exchange. As a sop for RANGER's reluctant crew, there was hope for booty from the Earl's manor house. RANGER's log entry notes: "At 10 was a Brest of Kilcubra . . . hoisted out the Cutter the Capt. and 12 men went on Shore to Lord Murray and Brought of a small Quantity of Silver Plate."

Soon after landing on the beach at St. Mary's Isle, the armed party (representing itself as a British press gang, thereby scattering the locals) encountered the estate's gardener, from whom they learned that the Earl was away. Disappointed, Jones returned to the beach, reluctantly allowing the rest of his party under Marine Lt. Wallingford to take the family silver from the mansion, which they did, and very politely, too. Jones had made it clear they were permitted to ask for the silver, nothing else, and that they had better stay out of the house and behave themselves. In this, he was obeyed. Notwithstanding the genteel comportment of his crew, Jones was embarrassed enough by the incident that when the spoils of the cruise were divided, he purchased the silver and returned every piece to the Selkirks along with a letter of apology.

Frustrated in his goal at Kirkcudbright Bay, Jones turned back toward Belfast Lough to have another go at H.M.S. DRAKE. The following morning, April 24, DRAKE was encountered standing out in light airs to investigate RANGER. Coyly, Jones lured DRAKE out into open water. RANGER's crew had been sent below, out of the enemy's sight. There, with action imminent, the ship's company held an impromptu town meeting and again made mutinous grumblings, which Jones again squelched. In the meantime he captured an armed boat party sent ahead from DRAKE to learn his identity.

"The action was warm, close and obstinate"

As the day lengthened, the wind finally began to get up. Gradually, DRAKE was allowed to approach. Jones was surprised to find that this enemy ship closely resembled his previous command, ALFRED, right down to the figurehead. Indeed, DRAKE seems to have been a near sister ship to ALFRED.[3] Built at Philadelphia, she had been

3. Jones's account of the action with DRAKE notes she had quarter galleries, so she was not a duplicate of ALFRED in that regard.

Jones departing RANGER for the raid on Whitehaven

purchased into the British Navy and fitted as a corvette, with 20 six-pounders. In his report of the action, Jones described it like this:

> *The Drake hoisted English colours, and at the same instant, the American stars were displayed on board the Ranger. I directed the master to (identify us as) 'the American Continental ship Ranger; that we waited for them, and desired that they would come on; the sun was no little more than an hour from setting, it was therefore time to begin.' The Drake being astern of the Ranger, I ordered the helm up, and gave her the first broadside. The action was warm, close and obstinate. It lasted an hour and four minutes, when the enemy called for quarters; her fore and main-topsail yards being both cut away, and down on the cap; the top-gallant yard and mizen-gaff both hanging up and down along the mast; the second ensign which they had hoisted shot away, and hanging on the quarter gallery in the water; the jib shot away, and hanging in the water; her sails and rigging entirely cut to pieces; her masts and yards all wounded, and her hull also very much galled.*

After the action, DRAKE's decks were described by one witness as "running with blood and rum." (A cask of rum had been gotten up in anticipation of the impending British victory, and it was smashed by a cannon ball.) RANGER's casualties were light. With DRAKE in tow and some 200 prisoners packed like darkling things in the dank depths of the two ships, RANGER at last turned for home, no doubt to everybody's considerable relief. They had all had a busy 48 hours. Both ships safely made port in Brest on May 8, having taken another prize on their way home.

The consequences of RANGER's cruise were far out of proportion to the damage she had done. After all, her prizes were of negligible cost to imperial England. True, the British had lost a few little ships and a tea set and a couple of Whitehaven colliers had been somewhat scorched, but the fires had been extinguished by rain before they could do much damage.

RANGER's great accomplishment was in effectively bringing the war to English soil,[4] creating alarm everywhere. As a result of her activities, insurance rates around the British Isles went up prohibitively, a large squadron of naval vessels was tied down seeking Jones, and militia units throughout England, Ireland, Scotland and Wales were mobilized to protect their shores against his raids. In the tradition of raiders from Drake to von Luckner, Jones in RANGER had made a nuisance of his little vessel far, far out of proportion to her size.

4. England's waters had already been raided by Americans, most notably: Gustavus Conyngham in REVENGE, Samuel Nicholson in DOLPHIN and Lambert Wickes in REPRISAL.

BONHOMME RICHARD

RANGER and her prize arrived at Brest May 8, 1778, and almost immediately Jones, tweaker of lions' tails, found himself bloodied as usual by the jackals of the shore. His crew was sulkier than ever and needed to be victualled and paid. As always, Jones did that out of his own pocket. Then there were the prizes to shepherd through the paperwork so that they could be sold, and there were agents to be dealt with, and prisoners to be tended, and his new command to be actualized, and the bureaucracy to be dealt with at every turn, all in French.

L'INDIEN remained a chimera. In June, she was again promised for sure, and Jones made arrangements whereby RANGER would be turned over to his First Lieutenant, Thomas Simpson. But the frigate remained elusive, and when RANGER sailed under Simpson with a passing American squadron, Jones was again on the beach, where he would chafe as the months passed. Perhaps there would be this or that frigate, perhaps a whole squadron, but the actual offers were disappointing. Jones wrote endless letters to Franklin and Chaumont, and to others; all single-mindedly dedicated to his desire for a worthy command. He became increasingly fractious and irascible. The war was happening without him.

In America, a superior British force ruled the waves wherever it chose to sail, but only as far as the horizon. The British fleet consisted of obsolescent ships, designed for the establishment of an earlier war. They were outclassed by the ships of the French, which were at a zenith of development. The British Navy, while enormous, was stretched thinly to protect its imperial commerce, its colonies and its own shores against American raids or, more dire, French invasion as relations worsened.

The first batch of American frigates had not been overly troublesome to the British. Many had come to grief in various ways, and none had scored any notable triumph. American privateers were a real annoyance to English commerce, and British privateers returned the compliment. In Jones's immediate theater, France was concentrating its naval might, nervously watched by the British channel fleet. Holland, with its own small but highly professional navy, favored the French but clung to an uneasy neutrality. The Dutch would continue to do so until pried loose from that position by Jones himself the following year.

Ship BONHOMME RICHARD, ex DUC DE DURAS

In November, the possibility of a respectable command began to unfold for Jones, a big East Indiaman designed to carry heavy guns. Jones pursued this option, and, in February, 1779, he obtained command of her. She was named DUC DE DURAS. He quickly changed the name to BONHOMME RICHARD in honor of his patron and mentor Benjamin Franklin, whose *Poor Richard's Almanac* had found its way into a popular French translation under the title *Les Maximes du Bonhomme Richard.*

Loaned to Jones by the French, this ship was destined to be the most famous of his commands, and — until quite recently — his most enigmatic.[1] BONHOMME RICHARD, ex DUC DE DURAS, was built in 1765 at *Port de L'Orient*, home port to the *Companie des Indies,* the French East Indies Company, for which she had been designed and built by *Ingenieur-Constructeur* Antoine Groignard as an armed merchant ship of the 900 ton class. Her measurements in English feet (different from the old French feet, or *pieds*) were about 178′ stem to stern, with a beam of 39′ and a maximum draught of 18.6′.

RICHARD was essentially armed as a large frigate, although she had a lower deck that could be pierced to take guns, albeit at the expense of her sailing qualities. As DUC DE DURAS, she had served an active and honorable career transporting passengers and cargo to and from China. She had undergone two major renovations, and had recently been sold out of service. She was purchased by Louis XVI specifically for use as an American warship, with all expenses included.

BONHOMME RICHARD was not totally free of strings. Under the agreement through which she became a warship in the Continental Navy for her final hour, she went to Jones under a kind of lend-lease charter arrangement which made the French regard her much as a privateer, as far as their own bookkeeping was concerned. The major difference was that instead of being issued licenses as a private ship of war, she was put under the national flag and unequivocal command of the United States as a naval vessel.

1. France has twice given BONHOMME RICHARD to the United States: first, in 1779, when the benefactor was Louis XVI, and again in 1987 by historian and 18th century naval architect Jean Boudriot of Paris, who has reconstructed definitively BONHOMME RICHARD in a project which engaged the co-operative efforts of scholars from the United States, France and England. His work is being published simultaneously with this in *John Paul Jones and the* BONHOMME RICHARD (see bibliography).

The ships of the *Companie des Indies* were built and rigged to set specifications, most of which are well documented. The survival of these data, as well as contemporary narratives, have made RICHARD's reconstruction possible, even without her original plans. Generically, she had features typical to French ships of her class. She was long in relationship to her beam, giving her a narrower aspect than her English or European counterparts, and she had a fine run. Viewed in cross-section, RICHARD had a pronounced waterline knuckle, giving her a very conspicuous tumblehome, which increased stability. Below water, the line of the midship frames did not form a smooth, round radius as favored by the English, but made another knuckle, forming two distinct bilge chines.

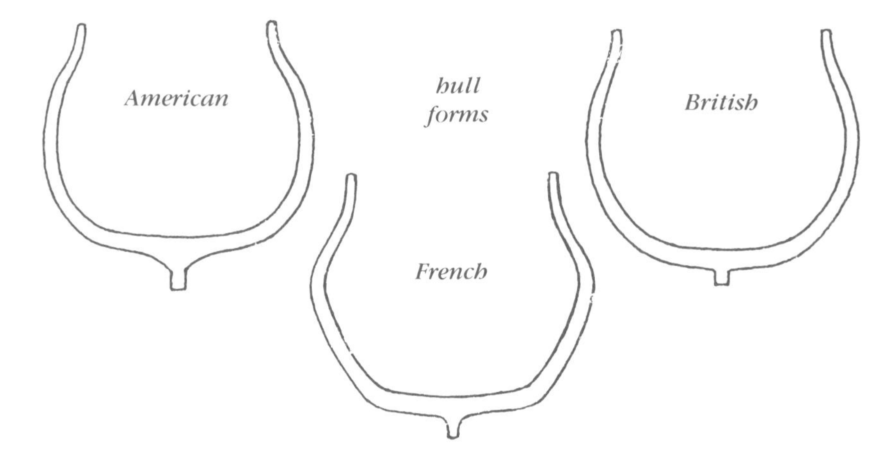

Jones must have enjoyed RICHARD's spacious great cabin, a drawing room and dining room enfolded with windows, bathed in light, 16′ above the water. It opened onto an elegant wrap-around balcony, the gallery, enclosed by gold balustrade and carvings, under arching giltwork. His private sleeping cabin adjoined to starboard. The decks were veneered with parquet; cupboards and furnishings came with the ship, excepting chairs and tables, which the Captain provided. While he might in many ways experience danger and hardship, at least in these quarters the senior officer was blessed with what must surely have been one of the most engaging chambers afloat.

Generations of American historians have exaggerated somewhat BONHOMME RICHARD's difficulties, calling her tired, leaky, rotten, ancient, slow, and so forth. No doubt she had her problems. She was 14 years old when she came to Jones, which for a ship of her time was approximately ripe middle-age. Unquestionably she contained rot, but it must be recalled that like all ships of her time she was tremendously over-built so as not to be too inconvenienced by rot. Rot was a fact of life. If it got too bad somewhere or another, it was fixed, and when it got too bad to fix, the ship was disposed of, if it was a naval vessel. RICHARD's log records that she leaked, especially when she was working hard to weather in a seaway. Again, all ships of that time could be expected to leak when working to weather, without causing undue alarm. That was what pumps were for.

As to her sailing qualities, she certainly was not the racehorse that Jones always wanted. During outfitting at Lorient she got a complete inventory of extra sails and light canvas, including topgallant studding sails and staysails as well as royals that set flying. She carried the seldom used sprit topsail which was a carry-over from the previous century (and was about to vanish), and she had been fitted with a new-style gaff-headed spanker, which was loose-footed.

Unlike ALFRED, BONHOMME RICHARD was a merchantman that had been designed as a warship, and so she was not embarrassed by wind. In fact, she liked good wind and sailed best in lots of it. Captain Landais of ALLIANCE noted on one occasion: "The wind and sea were very high, and the Alliance very crank and dull sailing ship in such a sea; on the other hand the Bonhomme Richard . . . was a better ship in such weather." In favorable circumstances RICHARD was capable of logging 10 knots.

In light airs, she was a clunker. On another occasion Landais reports: ". . . the sea being very smooth; at one o'clock we passed alongside of the Bonhomme Richard as if she had been at anchor; she showed a signal for us to drop astern, we shortened sail and kept only the three topsails on the caps, under which we sailed as fast as the Bonhomme Richard under all her sails."

Her inability to sail in light airs was to be the death of her.

Initially it was planned that RICHARD would be the flagship of a joint French-American amphibious force under the Marquis de LaFayette, who was to lead an expedition to capture an English seaport, possibly Liverpool. The plan never materialized, although as part of RICHARD's fitting-out, her poop deck was extended forward of the mizzen mast so that additional officers could be accommodated in two more cabins.[2] We know details but not the extent of RICHARD's refit. Her lower deck was pierced with additional ports, and two more ports per side were cut into her main deck bulwarks.

Jones needed guns, and he had a hard time finding them. Eventually he was able to obtain for RICHARD a main battery of 28 twelve-pounders. Of these, 12 were called "old", meaning they were cast to the pre-1766 pattern of French naval ordnance, and 16 were called "new," having been cast to the 1766 pattern, which meant they were somewhat shorter and lighter, although both groups were of the same calibre. The heavier "old" guns were placed aft in an effort to lighten ship forward.

In addition, Jones obtained six 18-pounders, which he mounted as far aft as possible, in the gunroom, three per side. The lower deck was pierced with eight ports per side, but the forward five were unarmed, and were at some point caulked closed.

2. A contemporary document refers to this addition as the "round house," a misleading reference. The round house was the area under the cover of the poop deck, which in this case was just extended.

BONHOMME RICHARD, PALLAS and VENGEANCE in a North Sea gale

The upper deck was armed with six French 8-pounders.[3] Initially two of these were placed on the forecastle and later were moved aft, right into the great cabin. The other four were stationed just forward of the break of the poop deck. There is a record of RICHARD's having received 10 four-pounders. Perhaps these were foundry reject tubes taken on with the ballast, or conceivably they could have been swivel howitzers or "coehorns" known to have been used in RICHARD's fighting tops, although a four-pounder would have been a large coehorn. In either case, it is certain that RICHARD had no carriage-mounted four-pounders, and they are not listed as part of her rating of 40 guns.

In addition to her carriage guns, she carried an unknown number of pierriers (small swivel-mounted cannon throwing a shot of one pound or less), and swivel blunderbusses, both of which were employed along the ship's rails like massive shotguns. In her tops were the swivel howitzers. The ship carried a further inventory of small arms for issue to the crew, including muskets, pistols, cutlasses and pikes. Boarding axes were also available as weapons of convenience, although they were primarily intended as emergency tools to fight fires or to cut away lines. Officers were expected to equip themselves with side arms.

3. Jones and others aboard RICHARD referred to these guns as nine-pounders, a discrepancy explained by the fact that the French pound was about 10% larger than the English pound, so a French eight-pounder and the British or American nine-pounder were essentially the same gun.

Cutlasses: American, British, French

BONHOMME RICHARD's complement consisted of 380 officers and men. All of the ship's commissioned naval officers were Americans, some of them just released on exchange from British prisons, such as First Lieutenant Richard Dale and Second Lieutenant Henry Lunt, who was an old PROVIDENCE shipmate of Jones. Two more officers who were to prove invaluable were French: Colonel Paul Chamillard and Lt. Colonel Felix Wuibert, commanding the ship's company of 140 Marines of the *Régiment de Walsh-Sérrant*, mostly Irish volunteers in French service in a "foreign legion" type unit. RICHARD's Marines wore red coats with white linings and waistcoats, blue cuffs and facings — quite a different uniform from the green-coated American Marines of ALLIANCE, in BONHOMME RICHARD's squadron. They were different, too, from the uniforms of the regular French Marines, the *Corps Royal d' Infanterie de la Marine*, with their blue coats and red facings. Jones would one day command 30 of them aboard ARIEL.

A number of contemporary sources remark on another variant uniform style found aboard RICHARD. At Lorient, Jones had purchased for himself and his officers 10 blue uniforms with white facings, waistcoats and breeches, which was the combination of colors worn by the British naval officers. These uniforms were worn to mislead the enemy (which they did), and represented a considerable departure from the regular Continental Navy uniform with its dark blue breeches, blue coat with red facings and red waistcoat. There is evidence both styles were worn aboard RICHARD at various

Jones's Marines: *Régiment de Walsh-Sérrant* *Continental Marine* *Corps Royal d'Infanterie de la Marine*

Phrygian cap

A "coehorn" in the fighting top

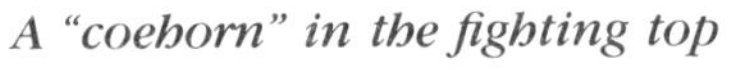

A pierrier

times. All of the surviving portraits of Jones taken from life depict him in the red and blue uniform.[4]

RICHARD's complement of petty officers was American and English, and her crew was made up of Americans, French, English and a sprinkling of other nationalities. In that day, crews wore no uniform, although there were conventions in the dress of common sailors which varied slightly from country to country. The sailor, or *matelot,* sometimes wore a long, heavy tunic, belted like a kilt, or any of a variety of loose-legged pants, or knee breeches. Headgear was eclectic — from cocked or loose-brimmed hats to various styles of knit caps (including the style which survives today and is known as the "watch cap"). Another common style of sailor's cap long pre-dated the Revolution, although it was used as a revolutionary symbol, and that was the Phrygian cap, usually made of leather. In ancient times it denoted its wearer as a free man, not a slave. During both the American and French Revolutions, it was called the "liberty cap."

There were at least 64 French sailors aboard RICHARD, of whom about half were able seamen. The log reports some inevitable trouble between the English and French elements of the crew, although as a whole the indications are that the Americans and Frenchmen got on well enough together. The punishments for minor infractions recorded in the log are no more numerous than on most British warships of that time. Some historians, particularly English ones, have implied RICHARD was some kind of "hell ship," but this would seem

4. James Cheevers has made a study of Paul Jones's uniforms, and it is included in Boudriot's *John Paul Jones and the Bonhomme Richard* (see bibliography). It should perhaps be further noted, there exists a heavy armor breastplate with backplate said to have been worn by Paul Jones, but it is doubtful he ever wore such a thing. None of the surviving 39 eyewitness accounts of the SERAPIS battle mention it, nor does anybody else, although it most certainly would have been noted had Jones ever appeared on the quarterdeck in such an unconventional contraption.

not to have been the case at all. At the beginning of RICHARD's recruiting, some "jail sweepings" were taken aboard, including 12 convicted mutineers, but most of these miscreants were soon sent back ashore. RICHARD took on 38 Portuguese, Danish, and Dutch sailors at the last minute, along with another batch of exchanged American seamen. It was not unusual to find international crews aboard ships. Then as now, sailors were an international lot. This crew was one of Jones's best. Along with difficulties, mostly from the English deserters, there were incidents of humor and laughter recorded.

In a chronicle of war, it is well to remember activities other than fighting and working were encouraged aboard most vessels like RICHARD. Sundays were often times of rest for the crew, with "make and mend," and in good weather music might be allowed on the forecastle. American shipboard musicians played the fife or wooden flute. The favored instruments among French sailors were the cornemuse and hurdy gurdy, or wheel fiddle. Playing in duet, they make a rhythmic, rousing music, quite capable of bringing even Yankees to their feet, dancing. There were ample opportunities for such relaxation during preparations at Lorient.

Music and dancing on BONHOMME RICHARD's forecastle at Lorient

BONHOMME RICHARD, out from Lorient, with LE CERF, PALLAS, and ALLIANCE in the distance.

BONHOMME RICHARD's fitting out, victualling and manning were completed in June. Already, she had become the nucleus of a squadron. Although the earlier plan of a landing at Liverpool had been cancelled, there were active preparations for a grand invasion, with a combined French and Spanish fleet of 64 ships of the line (plus frigates, transports and smaller craft) assembling to carry an army of 40,000 men to Albion's southern shores.

Jones's role in this massive campaign was to attack another area of the coast, by way of diversion. The force under his command consisted of : ALLIANCE, a new American-built 36-gun frigate, mounting 28 twelve-pounders and eight nine-pounders (as a gesture by Congress, she had been given to a French captain, Pierre Landais, who would stand revealed to history as clinically insane); PALLAS, under Denis Cottineau, a French 32-gun frigate with 26 nines and 6 four-pounders; LE CERF, under Sieur Varages, with 16 six-pounders and two long nines, was a very large cutter and carried 157 men — a vessel with common ancestry to PROVIDENCE; last, a brigantine, LA VENGEANCE, carried a crew of 66 and was armed with 12 four-pounder guns.

The squadron sailed June 12 on a preliminary assignment, convoying a group of merchantmen along the French coast. The most noteworthy occurrences included a night collision between RICHARD and ALLIANCE, and a fierce engagement between LE CERF and two British cutters which got the worst of it. As a consequence of the collision, RICHARD'S bowsprit had to be replaced, and ALLIANCE received a new mizzen mast upon return to Lorient.

At last, on August 14, 1779, the squadron set sail from Groix roads. Jones's orders called for a two-month cruise against British commerce. He was to sail around the British Isles and raid the North Sea and Eastern Channel coast, and then to terminate the cruise in Holland, off the Texel. He sailed under a *concordat* stipulating all of his ships as being Continental navy vessels, under American colors

"They slipped their line and pulled for home"

Gunpowder to the British

and navy regulations. All prizes were to be disposed of by the American Commission and the French Minister of Marine, jointly. When the squadron sailed, it was reinforced initially by two French privateers, MONSIEUR and GRANVILLE, both of which would soon go their own way.

There was pleasant late summer weather as the ships moved slowly across the approaches to the western channel, a hunting ground where two good prizes were taken, one of them by MONSIEUR, which then sailed away with it. On August 23, just off the extreme southwest coast of Ireland, another prize was taken by the small boats, as the weather had turned flat calm.

By evening, still becalmed and without steerage way, Jones found his ship being set by a current toward the Skelligs, so he sent seven men in a boat to tow RICHARD's head around on course, putting the vessel in a more favorable position to take advantage of any breeze that might happen by.

It was an oversight. The sailors permitted to volunteer for this duty were all Irish, and with their fair isle in sight through the evening haze, they slipped their line and pulled for home. Another boat with an officer and a dozen men was dispatched in pursuit, but it became lost in fog and ended up ashore the next day. Its crew was captured at once, thereby alerting the enemy to the squadron's presence and its exact force. Jones had sent LE CERF after both boats, and when the weather deteriorated she, too, became detached. That night the remaining privateer, GRANVILLE, departed, as did ALLIANCE, without orders, and PALLAS broke her tiller. On the whole, it was not a particularly auspicious beginning.

With VENGEANCE in company, BONHOMME RICHARD stood west, then north around Ireland, past the Outer Hebrides to Cape Wrath, the squadron's rendezvous point. There, on September 1, they took a valuable ship bound for Canada with uniforms. On the same day, ALLIANCE reappeared with another good prize. That night,

PALLAS came straggling up, and on the following day, VENGEANCE took a brigantine. Off the Shetlands, a couple more small prizes fell prey to ALLIANCE, which thereupon again vanished. RICHARD and consorts PALLAS and VENGEANCE slogged south along Scotland's east coast in heavy weather, taking two colliers on September 14.

It was the Commodore's intention again to raid British soil, this time at Leith, Edinburgh's seaport, which he proposed to take and hold for a ransom of £50,000, under threat of putting the city to the torch. His strategy was to tie up enemy forces and present a distraction from the main thrust at England's south coast by the mighty invasion force under D'Orvilliers. (Jones had no way of knowing that this great effort had already been frustrated by an outbreak of smallpox and typhus in the fleet.)

On the evening of September 16, 1779, BONHOMME RICHARD had been all day slowly beating up the Firth of Forth toward Leith. Armed boat crews stood by while VENGEANCE reconnoitered ahead. PALLAS had become detached pursuing a chase. Word was out that Paul Jones was on the loose again, and Leith, like other seaports, was in a flap and arming to defend itself.

A local Laird, possessing both cannon and ball but no gunpowder, sent a boat out to try to find a Royal Navy frigate patrolling those waters, H.M.S. ROMNEY, for the purpose of obtaining powder and alerting her captain to the danger. The small craft could not locate ROMNEY, but it was able to report to another British warship — BONHOMME RICHARD under British colors, with her English speaking officers in their Royal Navy style uniforms. The boat master was able to impart to the Captain a full report of the defensive preparations underway against the pirate Jones. Jones gave him a cask of powder, received his thanks, and sent him along, to the great merriment and approval of RICHARD's crew.

By the following morning, RICHARD was just beyond cannon range of Leith, and was making preparations to put the Marines ashore, when a sudden, severe westerly gale blew them all the way back out into the North Sea, where Jones abandoned the project. Surprise had been lost.

With PALLAS and VENGEANCE, he sailed south as far as Hull, chasing and taking prizes before turning back to the north on September 22. Just before midnight he encountered prodigal ALLIANCE, which had been missing for the past fortnight. With four vessels again in his squadron, Jones continued on a northerly course toward Flamborough Head and a rendezvous with destiny.

The breeze had died to a whisper

The Commodore on his quarterdeck, cleared for action

There had been showers in the early morning of Thursday, September 23, but the day had become warm and pleasant, with gentle breezes of between four and six knots from the southwest. At 2:30 p.m., sails began blossoming on the horizon to the north, and a half hour later, it was obvious they belonged to a large convoy. Jones fired a gun to signal the general chase, clearing BONHOMME RICHARD for action, setting royals and studding sails.

To the north, Captain Richard Pearson of H.M.S. SERAPIS, a new, 44-gun two decker[5], saw Jones's sails at about the same time. SERAPIS and a 20-gun corvette, COUNTESS OF SCARBOROUGH, were escorting a fleet of 41 merchantment bringing naval supplies from the Baltic to England. Squadron commander Pearson had been alerted by a shore boat that Jones was nearby, and so he trimmed to interpose his men-of-war between the enemy and the convoy, which was ordered north, up the coast, to shelter under the battery at Scarborough. With his two warships, Pearson resolutely tacked south to meet the oncoming, unidentified squadron. Oncoming they were, to be sure, but ever so slowly in the light breeze which backed southerly and decreased as the afternoon wore on.

By 6 p.m., the sun was setting behind the limestone cliffs of Flamborough Head, and Jones ordered his squadron into line of battle. Perhaps having misread Jones's intention, both PALLAS and ALLIANCE did other things entirely, so that SERAPIS met BONHOMME RICHARD alone.

At 7:15, by the light of a rising moon, the two ships approached to within musket shot of one another and Pearson hailed. The breeze had died to a whisper, the worst possible circumstance for the heavy sailing RICHARD, so Jones played to get as close to his enemy as possible. He identified RICHARD as PRINCESS ROYAL, an actual British East Indiaman with some similarity to RICHARD. He stalled until he could stall no longer, at last ordering the British colors down and the American flag up with its red, white and blue stripes and blue canton of 13 white stars. Simultaneously, Pearson ordered his lower-deck ports opened and the 18-pounders run out, revealing for the first time that SERAPIS was a bit more than a frigate. A lone musket popped from RICHARD's tops, and then both ships fired their broadsides into one another.

SERAPIS fired double-shotted guns into RICHARD's hull, and RICHARD's upper batteries fired at SERAPIS' rigging with doublebar shot that snickered through sails and ropes like weed trimmers. Jones's intention was to so damage his opponent's rigging as to neutralize his sailing advantage in the light airs. Sailing on parallel

5. SERAPIS was not a frigate, but a type of small two-decked ship called at that time a "5th rate," an obsolescent ship in an obsolete naval establishment. Nevertheless, she was new, having been commissioned in March, had a copper bottom, sailed well, and was more powerfully armed than RICHARD, with 20 18's, 22 12's, and two sixes on her forecastle. Surviving plans of SERAPIS' sister ship ROEBUCK show us a 165′ ship of 886 tons. SERAPIS was yellow sided with blue quarter and forecastle panels embellished with giltwork.

The two ships gradually closed and fired again

courses toward the northwest, the two ships gradually closed and fired again. On the second broadside, one of the starboard side 18-pounders in RICHARD's gun room burst. It was a devastating explosion that dismounted at least one other gun and killed or wounded 30 men in the confined gun room. When Jones learned what had happened, he ordered the remains of the lower deck battery abandoned. He would fight as a frigate.

Blanketed, SERAPIS lost way, then regained it as RICHARD drew ahead, enabling SERAPIS to wear ship and cross the American's stern. RICHARD was too sluggish to respond, and so took a full, well-aimed raking broadside of round shot and grape through the dissolving glass of her elegant stern galleries; some dozens of iron balls scything at the speed of sound along the whole length of her decks, through her packed rows of men and guns, spraying splinters, some of the 18-pound balls erupting from the bows still travelling. It was a tremendously destructive blow that RICHARD helplessly sustained, and it was followed by two more just like it, as the wind died away entirely for a moment, allowing SERAPIS under backed sails to maintain her lethal position. Then, responding to a puff of breeze more quickly than RICHARD, she got off a fourth, half-raking broadside into RICHARD's quarter.

Having taken almost no fire in exchange, SERAPIS had killed or wounded more than 25% of RICHARD's crew, and disabled many of her remaining guns.

Sailing parallel with RICHARD, SERAPIS at last received some fire, but again it was aimed at her rigging, in a desperate attempt to disable the Englishman aloft. Meanwhile, Pearson's unimpaired and unopposed batteries continued to pour shot into RICHARD's hull. It was 7:45, a half-hour into the battle, and Jones's position was awful at best. He had suffered heavy casualties while inflicting few: his lower deck battery was abandoned; his main deck battery was crippled; his enemy's batteries were undamaged, and his gamble in concentrating fire into SERAPIS' rigging had failed, for the Englishman still outsailed him. Neither ALLIANCE nor PALLAS had engaged, and obviously they could not be counted on. There seemed nothing to prevent SERAPIS from standing off and cutting RICHARD to pieces.

He slowly drove his jibboom into SERAPIS' mainmast

Instead, Pearson tried another maneuver. Drawing ahead, he made a port turn to cross RICHARD's bows, and in so doing, he lost wind and way. Jones was aware his only chance was in grappling and boarding with his still formidable force of Marines. He ordered the helm up and slowly drove his ship's jibboom into SERAPIS' mainmast rigging. It caught there, momentarily making a narrow bridge for a boarding party which was beaten back. Again, RICHARD was taking fire she could not return, so Jones ordered his sails backed, and the two antagonists wrenched free of one another.

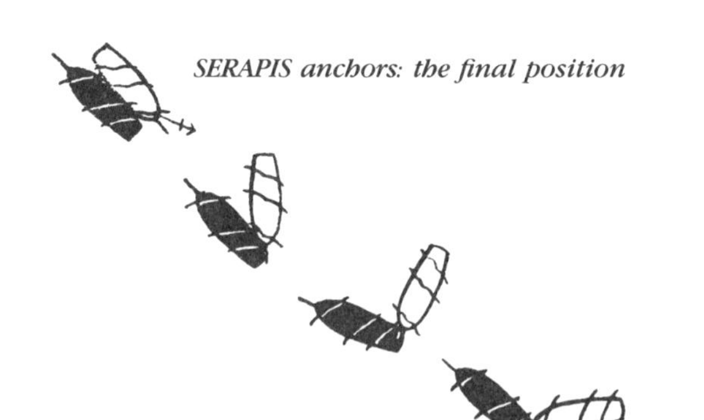

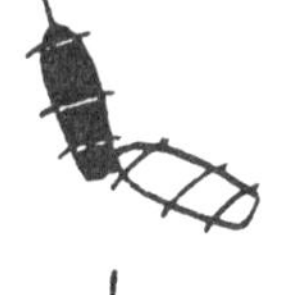

SERAPIS anchors: the final position

8:00 — RICHARD attempts to cross SERAPIS' bow; SERAPIS' bowsprit catches RICHARD's mizzen rigging and pivots; RICHARD grapples

7:50 — RICHARD attempts to board, fails, backs her yards and breaks free

RICHARD raked by SERAPIS

7:20 — Opening guns

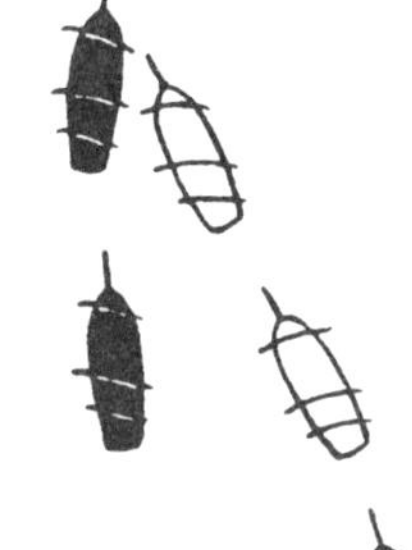

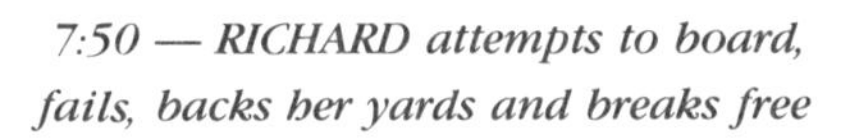

Tenaciously, Jones followed his enemy, whose maneuverability at last was impeded enough to prevent her standing clear. A puff of breeze allowed RICHARD to pull ahead and attempt to cross SERAPIS' bows, but instead of completing the evolution, sluggish RICHARD fell off and caught the enemy's bowsprit aft, where it broke the jackstaff and then hung up in the mizzen rigging. Jones personally seized SERAPIS' severed jibstay, which had fallen across RICHARD's poop, and tied it to his mizzenmast, locking the ships together.

For the moment, SERAPIS was neutralized, with no guns bearing. As the ships drifted in the northwesterly setting current, SERAPIS slowly pivoted, snapping off her jibboom, her stern swinging around, until she faced southeast, side by side with RICHARD, on the opposite heading. She was snared quickly with grappling lines . . . caught.

Seeing his advantage of mobility lost, Pearson tried desperately to cut free, but, according to Midshipman Fanning in command of RICHARD's main top:

> *The enemy's tops being entirely silenced, the men in ours had nothing to do but to direct their whole fire down upon the enemy's decks and forecastle; this we did, and with so much success that in about twenty-five minutes more we had cleared her decks so that not a man on board the SERAPIS was to be seen.*

Below, the cannonade had begun again, with the gun muzzles of the two ships' lower deck batteries nearly touching. SERAPIS now found her previously unused starboard battery engaged, and as the lower deck ports on that side had not been previously opened, those touching RICHARD's sides had to be blown away by their own guns. Above, Pearson ordered SERAPIS' portside anchor dropped, hoping the knot-and-a-half ebb current might wrench free his smothering antagonist. It did not. The two ships remained in their embrace as a bright harvest moon, just shy of full, rose in the northeast to a 45° declination and illuminated the battle and the dark sea with a silvery brilliance, reflecting on the translucent dome of smoke which periodically was lit from within by the red flashes of the guns. The dull thud of the distant cannonade was heard ten miles away, and the scene was viewed by a gathering multitude on the northeast facing cliffs of Flamborough Head.

By 8:30 p.m., the position was thus: RICHARD lay northwest, firmly grappled to SERAPIS, her sails aback in the gentle southerly breeze, her main and lower batteries blasting uninterruptedly into and through RICHARD's hull. Between RICHARD's lower deck and her upper works, nothing lived but the fires.

All of RICHARD's crew not stationed in cockpit or magazines were ordered above, onto RICHARD's high ends, her forecastle and quarterdeck. From there, her only artillery, the two starboard quarterdeck eight-pounders, were scouring SERAPIS' decks. Another eight-pounder was brought over from the port side and captained directly by Jones, firing repeatedly at the scorched yellow trunk of SERAPIS' mainmast, for want of a better target. To this fire was added that of the blunderbusses, pedreros and pierriers along RICHARD's railings, and the plunging spray of canister from the coehorns in the tops. There was also the musketry from aloft and below, and a sporadic hail of combustibles and grenades[6] bouncing down onto SERAPIS' decks. From his main top vantage, Fanning reports:

6. The combustibles were flasks or stinkpots — porcelain or perforated iron spheres packed with compounds that burned fiercely and were difficult to extinguish. Grenades were thick-walled, iron, baseball-sized bombs which were ignited by means of a violently burning wooden fuse of about six seconds, bursting unevenly with an explosion more concussive than that of a modern fragmentation grenade.

. . . By this time, the top-men in our tops had taken possession of the enemy's tops, which was done by reason of the SERAPIS's yards being locked together with ours, that we could with ease go from our main top into the enemy's fore-top, and so on from our fore-top into the SERAPIS's main top. Having knowledge of this, we transported from our own into the enemy's tops — stink pots, flasks, hand grenadoes &c which we threw in among the enemy whenever they made their appearance.

SERAPIS' decks were uninhabitable, and the entire British crew had been ordered below, under the shelter of the ends, from where they could swarm out in force to oppose any attempt at boarding. It is said that brave Pearson never left the deck, keeping it throughout the battle with two Marines, presumably in the shelter of the mizzen mast.

Elsewhere, COUNTESS OF SCARBOROUGH was taking a beating from bigger PALLAS a mile downwind. ALLIANCE, which had been sailing aimlessly around, now approached RICHARD's stern, turned to pass to windward, and at about 9:20 fired a full broadside from her starboard batteries indiscriminately into RICHARD's stern and SERAPIS' bow. Jones ordered the night recognition signals hoisted — a lantern in the shrouds of each mast and two more at the mizzen peak — and received with SERAPIS another broadside from ALLIANCE. These broadsides killed many men aboard both ships. Having thus divested himself, mad Landais sailed off again as mysteriously as he had arrived.

RICHARD's carpenter, inspecting the hold, found there five feet of water, with one of the main pumps shot away. Telling two of the other warrant officers that the ship was settling, he went along with one of them to report the news to the Captain, whom they did not immediately find in the chaos and smoke aft. Thinking all of their officers had been killed, the two warrants went to lower the colors, but found the jackstaff had been shot away. Assuming themselves in command of a sinking, burning ship which was being literally shot in half off a hostile shore, they began calling: "Quarters! Quarters! Our ship is sinking!" Amazed, Jones threw his pistols at them, decking one.

"Have you struck?" called Pearson, "Do you call for quarters?" To which Jones replied:

"No, I'll sink, but I'm damned if I'll strike!" [7] whereupon the firing from both sides was renewed with fury, as was recalled by observers on both sides. A classic stalemate had developed and

7. Or something like that. It has evolved in history to: "I have not yet begun to fight!" The reader is referred to Peter Reaveley's collection of references on the matter, pending publication in Jean Boudriot's *The Bonhomme Richard, 1779,* the separate volume of references accompanying his reconstruction. (See bibliography)

Another eight-pounder was brought over from the port side and captained directly by Jones

Hamilton crawled out along Richard's main yard

solidified. Below, RICHARD was being indeed literally cut in two horizontally by SERAPIS' undamaged and unopposed lower deck battery of 18-pounders; their gun crews were safe enough, sheltered from the rain of death above decks.

To break the deadlock, Pearson ordered a boarding party onto RICHARD. Its rush was met and repelled by a group led by Jones with a boarding pike. And so it went on, at times diminishing as by common consent both sides fought the fires that had begun to take control here and there. At 10 p.m. ALLIANCE again appeared from the south, firing another devastating, inexplicable broadside of grape and canister into both ships, although both were clearly distinguishable in the bright moonlight.

Richard unmistakably was settling lower in the water, for which reason her Master-at-Arms went below and released from the after hold the hundred or so British prisoners confined there, a humanitarian but ill-considered act, as the prisoners undoubtedly could have turned the tide of battle with any leadership. They were eventually rounded up and put to work at the pumps. ALLIANCE vanished again, into the smoke.

The deadly stalemate was about to be resolved from above, by one of Jones's maintopmen, a Scot named Bill Hamilton. Hamilton crawled out along RICHARD's main yard with a sack of grenades. One of these he managed to lob through the main hatch in SERAPIS' waist, so that it bounced down into the lower gun deck area. Detonating there, it touched off a ready box of powder cartridges, creating a flash explosion which ran the length of the crowded deck, killing or wounding some 50 men.

It was the end.

At 10:40, Pearson lowered the red ensign in surrender, and the rest of the guns became silent. An American-French boarding party entered SERAPIS, and there was a fierce skirmish in the waist with a group of British sailors who were not aware their captain had struck.

Pearson boarded BONHOMME RICHARD and presented his sword to Jones, who invited him to his cabin for a glass of wine. Both men must have been dazed, standing with the glasses and wine that somebody somehow found for them, in the smouldering, splintered ruin of what had once been so elegant a cabin. One deck down, below their feet, RICHARD had been so shot away that it was reported a miracle that the entire quarterdeck area of the ship did not collapse into the lower deck. Only a few frames still supported RICHARD's upper works. The stern post and rudder were gone, and, with all pumps going, the water was still slowly rising in the hold. Fires burned unchecked above and below, the most dangerous of them being in the shattered gunroom area, blazing at one point to within three feet of the powder magazine below. The surgeon's cockpit was a charnel house, and the dead and dying men were everywhere. As one survivor remarked: ". . . to see the dead lying in heaps — to hear the wounded and dying — the entrails of the dead scatered promiscuously about, the blood . . . over one's shoes . . . "

Approximately half of BONHOMME RICHARD's crew had been killed or seriously wounded, and it is said that nobody emerged unscathed from that battle. Jones received a bloody head injury.

SERAPIS was also mangled, burning, and making water from being holed at the waterline in the opening salvos. Her rigging had been shredded, and as soon as the ships were separated from one another, her mainmast fell, taking with it the mizzen topmast and fore topgallant mast. SERAPIS also had taken casualties of half her company, with horribly burned men making up a disproportionately large number of her wounded.

The battle had lasted three and a half hours. Now the real work began. The fires were brought gradually under control, wreckage was

The deadly embrace

cleared or sorted and saved, and efforts were made to plug RICHARD's underwater damage. Elsewhere, COUNTESS OF SCARBOROUGH was prize to PALLAS, and VENGEANCE was where she had remained throughout the proceedings, hove to about a mile to windward. ALLIANCE was approaching again, presumably to find out who had won. Soon, Jones had all of their crews working to make his ships functional again.

In the case of BONHOMME RICHARD, it was a futile attempt, as everyone perceived but Jones, who for the rest of the night and the following day directed the exhausted crews in frantic efforts to save his ship. Meanwhile, ALLIANCE and PALLAS took on the prisoners and wounded. As the dawn broke through sea mist, the fires were at last extinguished.

> *We then cleared the ship's decks of the dead and at the rising of the sun, we hove overboard one hundred dead bodies.*

Although RICHARD's hull barely floated, and that by a miracle, her rigging had not taken so much damage as that of SERAPIS, and so sail was got onto her. With ALLIANCE towing SERAPIS, the fleet limped seaward in a faint southerly breeze. In the afternoon Jones transferred his command to SERAPIS. All day, work parties plied through the fleet; all day there was the steady clanking of the pumps, punctuated at irregular intervals by the splash of bodies dropped over the side as the wounded perished.

By evening, the barometer had begun to fall, and a breeze freshened from the south. The last of RICHARD's wounded were transferred. As the sea began at last to get up, the waves licked occasionally into the chasms in her sides, and while the pumps clanked, efforts to save RICHARD became efforts to evacuate her.

The following morning, Saturday, dawned with low clouds making a grey scud overhead and rain squalls were driven by a still rising southerly. By 9 a.m., RICHARD's pumps fell silent, and the last boat pulled away from her. Seeing this, Jones realized that he still had personal possessions in the remains of his cabin, and ordered Midshipman Fanning to make a last trip to RICHARD before she sank, to retrieve his papers and his money.

Cautiously, the boat approached RICHARD, but Fanning saw the water pouring into her shattered sides, and he ordered the boat away. As they cleared, BONHOMME RICHARD listed to port, her head settled, and she sank bows first, with her commission pendant still flying from her mainmast. Paul Jones watched from the deck of SERAPIS, hove-to to windward.

ALLIANCE raking both BONHOMME RICHARD and SERAPIS: 10 p.m.

The last view of BONHOMME RICHARD; SERAPIS hove to, awaiting Fanning's boat

ALLIANCE

Slowly, painfully crippled SERAPIS crossed the North Sea, setting up jury rigged masts and sails. With fair winds, the squadron took eight days to crawl the 200 miles to the Texel, southernmost of the Frisian Islands and Amsterdam's access to the North Sea. There, in the waters of neutral Holland, Jones anchored to lick his wounds and tend to business.

He wanted to get to France as soon as SERAPIS could be repaired. In this he was frustrated, not so much by the sizable British squadron which arrived to blockade him as soon as his whereabouts became known, but by the usual vicissitudes of the shore. There were the wounded; the prisoners; the prizes to be sold; the money to be extracted from somebody or another, and the future disposition of the squadron to be determined. Pearson, so brave in battle, was sulky and aloof as an honored prisoner.[1] Landais was relieved of command and sent to Paris to be dealt with by Franklin. There was bad blood between the crews of BONHOMME RICHARD and ALLIANCE, the latter having shot up the former.

If there was any consolation, it was the reception of his victory. Again, Jones had used his ship to create a stir disproportionate to its strategic significance. In England, he was now quite famous, an early example of press-created celebrity. England has always loved a good pirate, having produced so many of them herself, and Jones was seen by many of the newspapers of the time as a very Robin Hood of pirates. This was not entirely undeserved, for he indeed committed many significant kindnesses to those of his enemy who fell into his hands, and even his personal foes conceded his humanitarianism.

1. Inevitably, Pearson was court-martialled for losing his ship but the court exonerated him and commended him for saving his entire convoy, for in sustaining a tactical loss, he accomplished his strategic objective. He was subsequently knighted for heroism. Hearing of this, Jones made his famous quip: "Next time I'll make a Lord of him."

But, while some in England found his effrontery romantic, the Royal Navy was not amused, having been repeatedly beaten, avoided and/or outmaneuvered by Jones, at his whim. As Morison put it: ". . . he had foiled them and flailed at them and made monkeys of them."

France was delighted. After the fizzle of its grand invasion plan, Jones's victory was succulent news. ". . . scarce anything was talked of at Paris and Versailles but your cool Conduct & persevering Bravery during the terrible Conflict," wrote Franklin to Jones.

In Amsterdam, he was huzzah'd when he made public appearances, and ballads were composed about his exploits. While the crowd acclaimed him as a hero, however, the Netherlands government viewed him as an embarrassment, and it was divided within itself as to whether or not to accede to England's indignant, insistent demands that Jones and his whole fleet be evicted, or at least that the English prizes be repatriated. As political pressures increased, a solution emanated from Paris. It did not please Jones.

Whereas the English regarded him (then as now) as a licensed pirate, the French regarded him (then as now) as an American captain of a French privateer which had flown American colors by French courtesy. From the viewpoint of Paris, it was not an unfair

assessment, and it was the key to appeasing the Netherlands while simultaneously reimbursing France for RICHARD's loss. In short, SERAPIS was declared French property, and with all of the other ships of the squadron except ALLIANCE was put under French command and colors. As French ships, they would be permitted to remain where they were. On November 12, Jones received his re-assignment to ALLIANCE, and on the same day he was requested by the Dutch to leave.

Jones took command of ALLIANCE reluctantly. He had lost his prize two-decker and gotten instead a dirty, demoralized ship, dismal and ill-cared for under her former captain. Her crew was now joined by the Americans from BONHOMME RICHARD, which made an unhappy mixture.

On the other hand, ALLIANCE was a powerful, well conceived and constructed frigate and a good sailer. Probably she was the best ship Jones ever commanded. She had been designed and built at Salisbury, Massachusetts by the Hacketts (RANGER's builders), who had come down from New Hampshire for the project. At her launching in April, 1778, the frigate was christened JOHN HANCOCK, but that was changed to honor the alliance with France.

Although no plans of ALLIANCE have survived, her dimensions are known. She was 178′ overall, which was very long for her 36′ beam and 910 tons. She drew 17′ of water. Jones considered her oversparred, like RANGER, although not so badly. Her foremast no doubt was nearly vertical, or even steeved slightly forward, while her main and mizzen masts raked aft, an American convention. She probably set a small lateen jigger sail on her jackstaff in fair conditions to ease her helm on the wind. She carried a main battery of 28 twelve-pounders and 8 nine-pounders.

For her maiden voyage, ALLIANCE had been given to French captain Pierre Landais as a gesture to France, where she sailed with Lafayette who was returning to plead for more help to the Americans. Thereafter attached to Jones, Landais's crazy maneuvers before and during the SERAPIS battle had earned him the contempt of nearly everybody, and ALLIANCE a doubtful reputation. Now she was his, along with pressure to vacate. To back up the pressure, six Dutch ships of the line sailed into Texel Roads on November 12 and pointedly anchored within cannon range. Just outside, guarding the narrow gateway to the North Sea was the British squadron, thirsting for blood. On November 23, the First Lord of the Admiralty wrote to one of his captains: "For God's sake, get to Sea instantly . . . If you can take Paul Jones, you will be as high in the estimation of the public as if you had beat the Combined fleets."

Jones held his position until he was ready to go, first careening and cleaning ALLIANCE. On December 21 he completed victualling. He remained at anchor over Christmas, awaiting his moment.

It arrived on December 27, 1779, when the wind came hard from the east, a cold wind, pushing His Majesty's hounds offshore, and offering Jones the opportunity to make a dash for Dover Strait. It was the shortest, most direct, and most dangerous route to freedom, for although with a fresh, fair wind it was only a day's sail for fast ALLIANCE, the Strait was a well guarded door. Nor were His Majesty's ships the only hazard, for the waters of the eastern Channel are notoriously treacherous, with fogs, powerful currents, and sand banks.

Jones slipped his cable at 11 a.m. and raced southwest with a beam wind that freshened to storm force, blowing away the foretopsail, biting and tearing at the crew which was time and again sent aloft to shorten frozen canvas. As the long winter night closed in, the laboring ship drove into a narrowing funnel where the ominous banks of the Flemish and English shores converge. At dawn a British fleet was sighted, anchored in the Downs. In the words of Master's Mate John Kilby:

ALLIANCE pursuing an English brig, off Biscay

ALLIANCE

By this time light began to appear and we soon discovered Admiral Hardee's grand fleet of fifty-two ships. Now it was that we had only one chance to escape, and this was by running by one or more of the ships and, of course, receiving their fire. We hauled close on a wind, which was certainly our ship's excellency, for I fully believe that she was the fastest ship on a wind that my foot ever was on board of. In the mean time, we discovered two fifty-gun ships in a press of sail trying to cut us off. The morning was very clear, though the wind blew fresh. We were then under close reefed topsails. Jones ordered out a reef. It was done quickly. He ordered out the second and it was not long before it was done. He ordered out the third, and also that the topsails yards, should be hoisted up taut. It was done. One of the Lieutenants observed to Jones that he was fearful lest we should carry away the mast. Jones answered: 'She shall either carry this sail or drag it.' The ship was then on a taut bow line . . . In four hours we ran every one of the ships hull down, on which we eased sail.

Jones sailed out into the widening waters of the western Channel, chasing ships. He rounded Ushant and headed south toward Cape Finisterre, celebrating his freedom and the new year of 1780 with a little January cruise that yielded an English brig. He reprovisioned at Corunna, and while there began his usual fussing with the rig and trim. He further assessed ALLIANCE's performance on the way back to Lorient, across the Bay of Biscay in contrary gales, and upon arrival in France set the crew to cutting down spars, rearranging ballast, and making the whole panoply of alterations which he deemed necessary. Years later, Captain John Barry commented that Jones had made ALLIANCE the fastest ship in the navy. With the work under way, he went off to Paris to report and begin again the process of dealing with the shore.

This time it wasn't so bad. Jones found himself the toast of the capital. He was invited everywhere, and everywhere honored: decorated by Louis XVI with the Order of Military Merit[2], and presented with a gold sword; as a Mason, he was initiated into the prestigious Lodge of the Nine Sisters; Jean-Antoine Houdon was commissioned to do his portrait bust[3]; and by the ladies of Paris he was lionized. His stay became prolonged.

Meanwhile, trouble began brewing aboard ALLIANCE. Landais had returned to Lorient to find passage to the United States and the court martial he was confident would acquit him. At Lorient he encountered Arthur Lee, another of America's commissioners to France, and the scheming archrival of Franklin. Hating Franklin, Lee made the case to Landais that he was still Captain of ALLIANCE because Franklin was not empowered to relieve him of any command bestowed upon him by Congress. The upshot of it all was that on June 20, as Jones was en route back to his ship from Paris, Lee pulled a coup. With Landais, he boarded ALLIANCE and reinstated her former Captain "by warrant of Congress," and sent ashore many of the loyal old hands from BONHOMME RICHARD. Lee then commandeered hold space in ALLIANCE for his voluminous personal effects (preempting quantities of desperately needed supplies for Washington's army), and they sailed for America.

The French offered to hold ALLIANCE by force, if necessary, but, in truth, Jones seemed relieved to see the last of the fine, fast frigate with its unhappy crew, irrevocably twisted by a dark and insane Captain of which they could not be rid.

2. Equivalent in rank to the English Order of the Knight of the Bath.
3. Of which Jones had at least twenty copies made in plaster to distribute to his friends.

ARIEL

For Jones, there was an immediate new assignment. ARIEL was a Blackwall-built corvette that had been captured from the British in September off the Carolina coast. Her captor was the French frigate L'AMAZONE under the command of the famous explorer Jean Francois Compte de la Pérouse, who had sent his prize back to Lorient where it was purchased by the crown and loaned to Benjamin Franklin. There were quantities of vital military stores awaiting shipment to America, and a shortage of ships to carry them. ALLIANCE had taken some of the matēriel; now Jones in ARIEL was to take more — as much as the ship could carry. She was a warship, not a cargo vessel, and so her capacity was limited by her design. Another limitation was her size.

At 429 tons, ARIEL was a much smaller ship than ALLIANCE. Her overall length of hull was 130′, she had a beam of 30′, and drew 14′ of water. In order to carry a significant quantity of muskets, gunpowder, and uniforms, her battery had to be lightened. Originally, ARIEL had carried 24 nine-pounders. At Lorient, two smaller carriage guns were mounted on her quarterdeck[1], but in order to lighten ship, ten of her nine-pounders were later removed. Her crew was reduced to half-size, which saved room and weight of provisions. Even so, ARIEL could not carry nearly the quantity of cargo waiting on the docks, and so two merchant brigs were chartered to take the remainder.

The usual delays in getting away were used by Jones to modify his new ship according to his own lights and complete his complement, for there were by now only 45 of the old BONHOMME RICHARD gang left, not enough to fill even a half-crew in ARIEL. The difference was made up by a batch of English prisoners who volunteered for the duty to get out of their French jails. Completing the complement were 30 men of the *Corps Royal D'Infanterie de la Marine.*

Jones made continued efforts to obtain a squadron, a fast frigate — anything other than this uninspiring cargo delivery. He was unsuccessful, and ARIEL departed Lorient on September 5, 1780, immediately encountering adverse winds that kept her at anchor in nearby Groix Roads for a month. At last, on October 7, she departed again with a fair breeze. The following day, before she had cleared the Penmarch peninsula and gotten sea room, the wind backed southwest and came on to blow, first to gale and then to storm force and beyond.

Jones shortened canvas to the absolute minimum under which the ship still could sail, clawing desperately to weather. The topgallant masts were struck to decrease windage aloft; her guns were housed, and ARIEL was snugged down for the worst, but the wind

1. By virtue of these quarterdeck guns, probably six-pounders, 24-gun sloop of war ARIEL became a 26-gun frigate, which no doubt explains why some references note her as a corvette, and others as a frigate. Jones called her a sloop of war in one reference, and a frigate in another.

ARIEL in a tempest, clawing off the Penmarch rocks

continued to increase until it became a mighty tempest, like nothing Jones had ever seen. Afterwards, he wrote:

> *. . . till the night of the 7th, I did not fully conceive the awful majesty of tempest and of shipwreck. I can give you no just idea of the tremendous scene that nature then presented, which surpassed the reach even of poetic fancy and the pencil.*

In the worst storm of his life, Jones had under him an overloaded ship which was being so punished that she began opening her seams, and the infamous Penmarch rocks were less than two miles away, directly under her lee. She was rapidly being driven onto them, according to the lead line, for it was night and nothing could be seen. One of her main pumps failed. As the water shoaled, the seas became even more violent, washing completely over the ship. ARIEL first rolled her yardarms into the water, and then went onto her beam ends and would not completely right herself.

Out went the anchors, but even that did not bring the ship head-to-wind, so Jones ordered the foremast cut away. With it gone, the mainmast kicked its butt out of the step, tore loose from its chainplates and also went over the side, taking the mizzen along with it.

Totally dismasted, ARIEL righted and came into the wind where she lay to her anchors for two days while the storm blew itself out and her crew pumped. By October 11, conditions had eased enough to permit a jury mizzenmast to be rigged. The following day her cables were cut, and she limped back to Lorient. Her survival was considered miraculous, for little else had come through the storm, and the coastlines of Europe were littered with the wreckage of hundreds of ships. The Port Captain of Lorient reported on Jones to Paris: "The crew and passengers all credit him with saving the ship."

It took eight weeks to rerig and repair ARIEL, during which time Jones made an inevitable forlorn effort to obtain a more desirable command, but he succeeded only in irritating Franklin. He consoled himself by enjoying to the full a few more golden moments of France's amenities, and sailed again on December 18, heading south until he picked up the northeast trades, then west.

En route, Jones encountered a fast loyalist privateer, the TRIUMPH, mounting four more guns than overladen ARIEL and outsailing her easily. Initially Jones tried to escape, but when that did not work, he hoisted the red ensign, got his French marines out of sight below, and allowed the enemy to approach. Presumably he was wearing again his blue and white uniform, because he was so successful in the role of British Captain that he was able to elicit information during a lengthy conversation, in the course of which he lulled his enemy, edging for a favorable position.

Finding it, he pulled his surprise, running up the Stars and Stripes, crossing TRIUMPH's stern, raking her, then engaging her lee side at close quarters, his trusty RICHARD veterans working their guns with deadly efficiency. Within ten minutes the enemy ship struck her colors in surrender.

As ARIEL was lowering a boat, TRIUMPH pulled a surprise of her own, one worthy of Jones, although it was one Jones wouldn't have contemplated. Suddenly, the defeated ship just departed, leaving in her wake heavy ARIEL, wallowing helplessly while Paul Jones fumed. He had soundly defeated a faster and more powerful enemy warship, and victory had never been so empty.

On February 18, 1781, ARIEL dropped anchor at Philadelphia, where Jones delivered his dispatches, his cargo in time for the Yorktown campaign, and twenty of his English volunteers in irons. They had tried to brew up a little mutiny, not realizing their adversary was a seasoned professional in the mutiny field.

AMERICA

Jones had been gone for three years and three months. He was treated as a returned hero. The faction of his enemies had been mostly squelched before his arrival, Landais having been cashiered. Two days before the signing of the Articles of Confederation on February 29, 1781, Congress moved to allow Jones to receive from its ally the Order of Military Merit granted by Louis XVI, which allowed him to sign his name "Chevalier Paul Jones," and he sometimes did. ARIEL was returned to the French, and Jones began another round of trying to obtain a proper command in which to return to sea.

His honors notwithstanding, there were few commands available. The war at sea was largely up to the French battle fleets of de Grasse and la Touche. The American navy had lost most of its ships, and those that it retained were committed, but there was one uncompleted ship available, and a grand one at that. Jones wanted it, lobbied for it, and got it.

She was the AMERICA, a 74-gun ship of the line of battle, and she was in the final phase of her construction at Portsmouth, New Hampshire. The Chevalier delayed departure only long enough to make a desperate attempt to collect some money from Congress, because he had not been paid since the beginning of the war. Indeed, he personally had spent his prize money on his ships and crews, and was considerably out of pocket. Unreimbursed and unpaid, he departed for New Hampshire, arriving in Portsmouth on August 31, 1781, and took up his command. This entailed supervising his ship's completion, and in many ways it was a replay of his experience with RANGER.

AMERICA had been laid down in May of 1777, the same month in which RANGER was launched, so presumably AMERICA's keel was placed in the same spot just vacated by RANGER in Hackett's yard, on Rising Castle Island in the Piscataqua River between Portsmouth, New Hampshire and what is now Kittery, Maine.

The ship had been on the stocks for four years, largely because of lack of funds and materials. For most of that time she had only 24 carpenters to keep her construction moving along. Jones's old pal John Langdon was still navy agent, no doubt contributing to the delays. Langdon greeted Jones warmly, but they were soon on the outs again, for the same reasons as before.

AMERICA was one of three 74-gun ships authorized by Congress in 1776, and the only one actually built. Her original plans and a set of spar dimensions have survived, as has a half model from Joshua Humphreys, probably her designer. It was to be Jones's third ship from the Hackett yard, and it is thought William Hackett might in some way have had a hand in the ship's design. It was not unusual for a builder to modify a design. When Congress at last became impatient with the delays in her building and appointed Paul Jones her Captain and supervisor, it handed him America's noblest warship of the time, and its only ship of the line.

As a type, the 74-gun ship composed the backbone of all of the great battle fleets of the time. In England's system of rating warships, it was called a third rate. There were bigger warships, the biggest of them being first rates, and awesome vessels they were,[1] mounting a hundred or more cannon. The Spanish first rater SANTISIMA TRINIDAD (1769-1805) at one time mounted 144 guns. A second rate

1. For anyone who is unaware of it, the first rate 104-gun ship HMS VICTORY is fully restored for all to see, a cathedral of oak, at Portsmouth, England. Although she is most famous for her role as Nelson's flagship at Trafalgar, she was launched in 1765 and was Keppel's flagship of the Channel Fleet during the period of Jones's raids around the British Isles. It was his good fortune never to see her. It is our good fortune to be able to see her still.

A French 74-gun ship, two views

mounted 90 guns. Together with the 74's, these were the ships of the line of battle. Ships below third rate were not of the line. In the year of AMERICA's launching, the British Navy List included a total of seventy-three 74-gun ships. The French had more. Historian and 18th century naval architect Jean Boudriot notes: "The art of naval architecture in France reached its pinnacle at the time of the American War of Independence, and that period marked the last time the naval might of France equalled that of England."

After the Revolution, England would gain naval ascendency over France largely because of a fundamental difference in their use of sea power. That is, France employed its navy as a task force, with squadrons or ships activated and deployed for specific chores. The mission of Britannia was to rule the waves, and so its navy kept the sea in order to dominate it. The United States would not have a ship of the line until after the War of 1812.

AMERICA was an impressive 221′ long overall, with a beam of 50.6′ and a draught of 24′. Her tonnage was given as 1,982. She was designed for a battery of 30 long 18-pounders on her lower gun deck, 32 long 12-pounders on her main gun deck, and 14 long nines on her quarterdeck and forecastle.

Rather than accelerating with the arrival of Jones, the construction lagged and lagged until at one point only eight carpenters were at work. There was no paint with which to protect her raw wood from the weather. There were no supplies. Bitterly, Jones complained about the quality of the wood that Langdon was purchasing, and most of it was so freshly cut as to inspire jokes about how the ship would have to be not only painted, but also pruned. Jones noted that Langdon seemed to have no trouble finding supplies and carpenters and good wood for his own personal ship building enterprises. In a Christmas letter he wrote: "I superintended the building, which I found so much more backward than I had expected . . . I expected to have been at sea this winter, but the building does not go on with the vigour I could wish."

Naturally, he made alterations to the design. Apparently he was afraid her masting would suffer as did RANGER's, because he had AMERICA's mainmast step moved aft three full frames. He lengthened her quarterdeck and forecastle, but the only contemporary picture of her does not show that feature. He had two more gunports (one per side) opened in her lower gun deck, and he decreased the size of the quarter galleries.

Captain Jones took particular interest in AMERICA's ornamentation. In our age of grey ships, it is perhaps important at least to attempt to understand the minds of our forebears, who valued many qualities having nothing to do with the pure efficiency we hold so dear. The ornamentation of warships was such a manifestation. Jones designed AMERICA's figurehead himself: the goddess of Liberty, gesticulating upward with her right arm, and with her left holding a shield with 13 silver stars on a blue field. Wooden smoke boiled up around her legs. The ship's stern featured the figures of Tyranny and Oppression chained under a liberty cap, while Neptune and Mars, respectively, adorned the two quarter galleries.

The British were acutely aware of AMERICA. In the spring of 1782, Jones was warned that American intelligence (such as it was) had got wind of a British plan to send an armed expedition in boats into the Piscataqua by night to raid the island and burn the ship. He was aware of considerable enemy activity in the Gulf of Maine, and alarmed enough by the warning to seek a contingent of guards. There were no soldiers available, so Jones formed his own company of recruits from the building crew, arming them at his own expense, and drilling them. He stood the watch every third night, with each of the two master builders taking command on the other nights. It is not recorded which guard commander had the duty the night that the "Large whale boats, with muffled oars, came into the river, mean-

while, full of men . . . and passed and repassed the America in the night; but dared not land on the little island where she was built." Presumably they found themselves facing a battery, for Jones had managed to obtain some guns. In fact, he finally got the battery of 18-pounders which had been cast in France for BONHOMME RICHARD. They had not been completed in time to serve aboard that ship, and so they were transferred to the United States in the hold of ALLIANCE. They eventually found their way at last to Jones, who had also found some 12-pounders, and some swivels.

The guns contributed handsomely to the festivities celebrating the birth of a son to Marie Antoinette of France, a blessed event to America's great ally. It was decreed by Congress as the occasion for a national celebration, and nowhere was it celebrated with more vigor than at Portsmouth on June 20, 1782. AMERICA flew a huge, white Bourbon ensign and other flags, and her guns fired a 21-gun salute to signal the commencement of the day-long party. In Jones's words, ". . . all day, up to midnight, she kept up a rolling fire of musketry and swivel guns . . . and thirteen royal salutes at the toast drunk at a public entertainment, and afterwards çontinued a *feu de joie* until midnight. When it became dark, she was brilliantly illuminated and displayed fire-works."

It was Jones's misfortune that AMERICA was a ship that was out of phase with her moment in history. She had been authorized when it was thought the colonists were going to have to fight their own naval war against England — in the dark days before France had so obligingly taken on that task with huge fleets, at no cost to impoverished Congress. By 1782 everything had changed. The French navy was present in force, war was about to wind down, and AMERICA was even more expensive, relatively, than an entire carrier battle group is today because she was redundant and Congress was bankrupt.

So it came to pass that when a French 74-gun ship struck a rock and sank between Lovell's and Gallop's Islands in Boston harbor, Congress voted to make a present of AMERICA to their ally by way of thanks for all the help. Jones's feelings can be well imagined, but he made no complaint.

He was retained in command to direct the launching, a ticklish operation. AMERICA was the largest ship by far ever to have been constructed at Portsmouth, and there was barely room for her to go into the water between the island and the mainland. In addition, a line of rocks advanced from the ways some 400 feet or so toward the shore. At high tide, the only time of sufficient water, tidal current set directly over the rocks. Normally, pilings would have been planted to bumper the ship and prevent its being swept ashore, but that was impossible in the rocky-bottomed Piscataqua. So Jones invented an ingenious system of anchors and cables which not only held the ship once she was afloat, but were used to check her speed as she slid off the ways into the water. The first effort to launch AMERICA was frustrated when the big ship stuck on the ways. Her cradle was modified, and on November 5, with Jones supervising from a platform, the wedges were knocked out.

Handsomely, the great ship slid down the ways and into the water, her system of checks and cables perfectly securing her. She was warped to her mooring, and there she was ceremoniously presented by Jones to her French captain. The next day, the Chevalier Paul Jones saw her for the last time, the glorious command that might have been. Then his carriage rolled south, taking him to a destiny that held no more commands in the American Navy.

The 74-gun ship AMERICA on the ways at Hackett's yard

EPILOGUE

IRONICALLY, when Paul Jones returned to Philadelphia, he almost got at last that will-o'-the-wisp command: L'INDIEN. She had been obtained in Europe by South Carolina, and named after that state. Currently, she was idle at Philadelphia with a cloud over her ownership. Congressional Marine Agent Robert Morris wanted to secure L'INDIEN/SOUTH CAROLINA for Jones, along with all of the other remnants of the Continental Navy, and send him off to take Bermuda. The plan was frustrated by SOUTH CAROLINA's commander. He sneaked her to sea, out from under the nose of Congress, but he was not as successful at evading the enemy, and was captured before clearing the Delaware Capes.

Again thwarted, Paul Jones requested to be seconded to the French fleet. With it he sailed from Boston to the West Indies, studying fleet tactics and evolutions under the Marquis de Vaudreuil. Word of peace reached him on April Fool's Day, 1783, aboard Vaudreuil's flagship, LE TRIUMPHANT at Puerto Cabello.

Jones returned by packet to Philadelphia, where he lobbied for a ship in which to show the flag of the new nation to Europe, but Congress was divesting itself of naval matters, not taking them on. Jones was more successful at getting an appointment as Naval Agent in Paris, in order to settle up unfinished prize money accounts there. He departed in November, 1783 aboard the packet GENERAL WASHINGTON, an ex-privateer that had been captured by the British, and then recaptured by her present master, Jones's old shipmate Joshua Barney (from the Nassau expedition). The Chevalier disembarked at a south coast fishing village in England to deliver dispatches personally to John Adams, the American Commissioner in London, and he was not arrested as Barney had feared.

Finishing in London, Jones went to Paris, where he spent the better part of three years tenaciously attempting finally to extract from France the money owing for the BONHOMME RICHARD squadron's prizes. During this time, he tried his hand at various mercantile schemes without success, but his socializing met with considerable success, short of getting him a command. Paul Jones bathed in the last twilight of the ancient regime, soon to fall under the shadow of the knife.

In the spring of 1787, he sailed on a packet for New York in order to be on hand for the final adjudication of his Paris accounts. While in New York, he made an unsuccessful try for an admiral's flag, receiving instead an order from Congress for a gold medal honoring his deeds. Commissioned by Jefferson to act as agent again, he returned to Europe by Christmas, this time to settle prize claims in Denmark.

There the Russian Ambassador offered Paul Jones the admiral's flag he craved, but in the service of Catherine II, Empress of Russia, who was looking for a commander to pull together her Black Sea fleet. Currently, it was bottled up in the shallows at the mouth of the Dnieper River, a large, enclosed estuary called the Liman. Catherine wanted the Turkish fleet defeated, she wanted the Liman, and she wanted Constantinople, if possible. And so it came about that the Chevalier Paul Jones, Captain, USN, on extended leave, became Kontradmiral Ivan Pavlovich Jones.

By April, he had concluded his business and obligations, and set out from Denmark for St. Petersburg. In Sweden, his journey was blocked by ice at the Gulf of Bothnia, so he chartered a 30-foot undecked boat and another, smaller craft, with their crews to take him, his servant, and his baggage south. Near Stockholm, he commanded them to steer east, skirting the ice pack, and forced the issue at pistol point. It became a four day ordeal, but eventually Jones was landed at Revel, Esthonia, and he compensated his unwilling boat crew very generously for suffering his act of apparent piracy.

In St. Petersburg, he reported to the Imperial Court and briefly met the cagey, fat, 60 year-old Empress of Russia before proceeding to the Black Sea. He arrived to take up his new command on May 30, 1788.

Prince Potemkin, Commander in Chief of Russia's entire military machine, was directing a land and sea campaign, using three different forces, each with its own command structure: first, a land army; second, a flotilla of some 70 galleys, chaloupes and gunboats,

He forced the issue at pistol point

carrying a second, water-borne army; third, a squadron of larger warships — which was Jones's command. In it was his flagship, VLADIMIR, a 70-gun ship of the line reduced to 24 24-pounders because of the shallows in which it had to work; frigate ALIKSANDR NEVSKII (Jones's spelling), 50 guns, also reduced to half armament; plus seven smaller frigates and a smattering of lesser craft.

This is not the place for a detailed analysis of the Liman campaign. Jones's role was to protect the flank of an army and the large flotilla of small craft. Starting in early June, 1788, his squadron fought a series of battles and skirmishes, using weapons and tactics which were as unfamiliar to the Admiral as the languages spoken around him.

In one action, Jones had to have himself rowed through his squadron so his interpreter could shout the Admiral's orders to his individual ships, there being no signal system. In another, Jones wanted to engage the enemy flagship more closely, but VLADIMIR's captain discerned his Admiral's intention and promptly anchored. The balls provided for the largest cannon in his squadron were too small. The list of problems went on and on, but the worst of them came from his fellow officers. The chain of command was snarled and ambiguous under Potemkin, who apparently told different officers

Flagship VLADIMIR in action at the Liman

different things. The corps of senior officers comprised a competitive cadre of foreign adventurers, soldiers of fortune, and court sycophants who gave Jones their elbows at every opportunity. There were lots of elbows and lots of opportunities.

"I have never suffered so much vexation as in this one Campaign of the Liman, which was nearly the death of me," wrote Jones later.

Ironically, in Russia he had no crew problems, and got on famously with the ordinary sailors of his ship and his squadron. He wrote: "I am delighted with the courage of the Russians, which is the more glorious because it is without show-off." He cracked jokes with his men, ate with them before battle, and once dressed up to look like them, just before having himself rowed by night through the pickets and amongst the enemy's anchored fleet. There, he personally reconnoitered the Turk's strength and disposition, according to the account of Jones by a Cossack sailor named Ivak who was the Admiral's interpreter on his nocturnal exploit.

By the end of October, the Turkish fleet had been pushed from the Liman, and the key Turkish land position was blockaded, meaning Jones's squadron had succeeded in its objective, for which Kontradmiral Jones got little credit and a lot of grief. If he thought he had learned about back stabbing, dirty tricks and internecine politicking in the drawing rooms of Paris and Philadelphia, he was disillusioned in the Ukraine. He took it all with rare patience for nearly half a year, and then inevitably wrote his first snippity letter to Potemkin. It was also his last. Potemkin sacked him, but without telling him so — merely that he would be wanted for the Baltic fleet, to fight the Swedes. He went to St. Petersburg and saw the Empress once more, in June, hoping for his next command. Instead, he got official leave of absence, which, along with a trumped up charge of statutory rape, meant his walking papers.

Jones returned to Europe, making a leisurely tour of several capitals, and settled in an apartment in Paris. There he continued his efforts as usual to obtain a command somewhere. One naval power which he invited to offer him a job was Sweden, but he was not to have the opportunity to fight Russia, so recently his employer. In lingering ill health, he watched as the fury of the French Revolution gathered, consuming the Paris he had loved so much, and at last he died, on the evening of July 18, 1792, in debt and alone, age 45. He was spared by a few weeks the sounds of the tumbrels over the cobblestones, carrying away many of his former friends and benefactors to the guillotine. A few days after his death, there arrived at last his commission as United States Consul to the Barbary States, signed by Washington and Jefferson, among the few who remembered him.

In the absence of any American concern for his funeral, it was arranged by Jones's French friends, who also paid for him to be buried in an alcohol-filled lead coffin so that when his country one day wanted the body returned, it would still be available. That finally

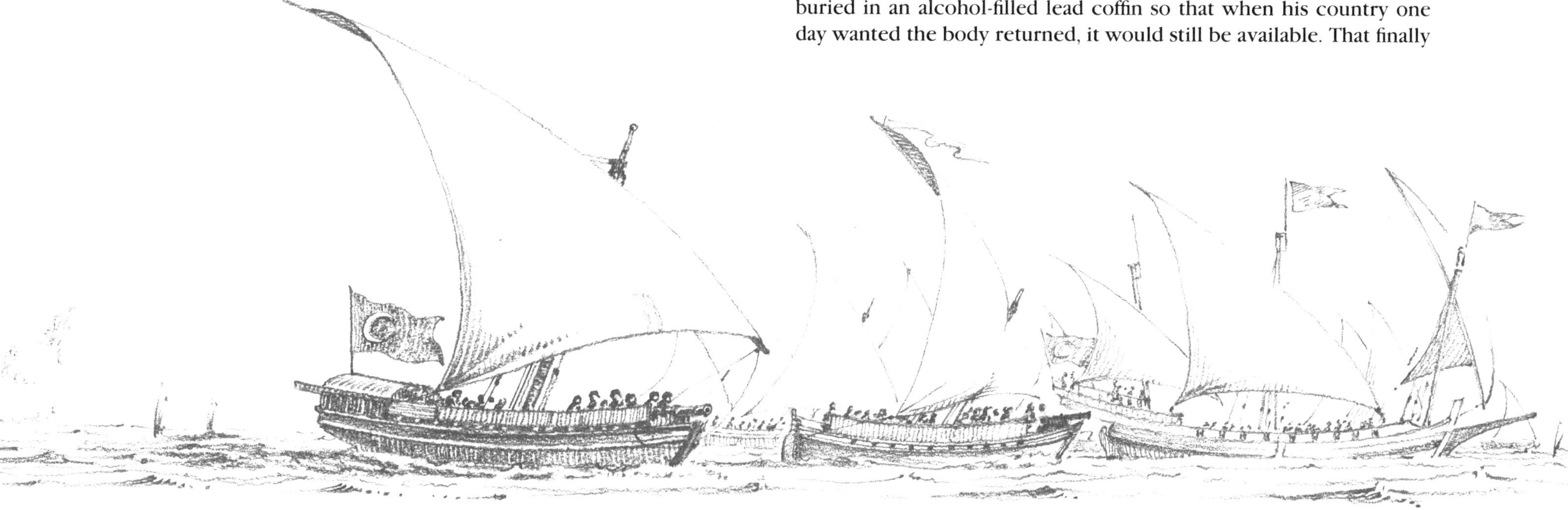

happened in 1905, during a period of intense U.S. naval build-up and politicking.

Jones's mummified remains were located, disinterred, and ceremoniously returned to the U.S., where a mighty fleet that included 11 battleships participated in the fanfare at the body's arrival. Today, the remains of Captain John Paul Jones are interred in their permanent resting place, in a marble sarcophagus not unlike Napoleon's, in the chapel crypt of the United States Naval Academy at Annapolis, Maryland, surrounded by the names of his ships: PROVIDENCE, ALFRED, RANGER, BONHOMME RICHARD, ALLIANCE, ARIEL and AMERICA.

PROVIDENCE

PROVIDENCE continued to be a frisky and particularly troublesome little vessel to the British. While still blockaded in the Providence River, she sortied in June just far enough to give a good pounding with her little cannon balls to the grounded British frigate DIAMOND. In April, she broke through the blockade and sailed east on a cruise, meeting and defeating H.M. Brig LUCY in a bloody engagement during which the brig lost its mainmast.

Returning from Philadelphia with his promotion to Captain, John Peck Rathbun at last got command of the vessel he had taught Paul Jones to sail. Rathbun extensively refitted and repaired her, and made a truly impudent series of cruises. On his first, he engaged simultaneously a 10-gun brig and an 8-gun schooner, badly punishing the brig, and capturing the schooner.

Rathbun's second cruise in PROVIDENCE is surely one of the most extraordinary in the annals of American naval history. In the beginning, off Charleston, he captured a 10-gun Loyalist privateer in a hot fight. Ashore to refit, he learned of a valuable prize laying at Nassau, New Providence, where the defenses were reputedly no stronger than they had been when raided by the Continental squadron in 1776. Under Rathbun, PROVIDENCE now returned alone to Nassau, and with her crew of 50 sailors and Marines recaptured both forts, along with all of their military stores and some American prisoners. For three days Rathbun held the forts, foxing and fending off vastly superior forces on the island while his men worked to secure three prizes taken in the harbor; loading onto PROVIDENCE and the prizes all of the gunpowder and ordnance they could handle, and destroying the rest, including two more prizes that could not be manned. The entire exploit was accomplished without bloodshed.[1] Rathbun's last cruise in PROVIDENCE early in 1779 again netted numerous prizes.

(In 1781, this brilliant officer raided the English Channel in the 20-gun ship WEXFORD, but his luck had run out at last. After a long

1. A good account of this amazing expedition is to be found in Hope Sisson Rider's book *Valour Fore and Aft* (see bibliography), which chronicles the entire Revolutionary War career of sloop PROVIDENCE in one of the best and most interesting accounts of its kind yet published. It is appended with the story of the operational facsimile of PROVIDENCE built at Newport in 1976.

chase and four hours under the guns of a British 32-gun frigate, Rathbun struck his colors. He fell sick in the Old Mill Prison at Plymouth, and died there at age 36.)

In May 1779, under Hoystead Hacker, PROVIDENCE fought her hardest action — a two and a half hour engagement in which she defeated H.M. Brig DILIGENT, with 14 guns and a crew of 95, a greatly superior vessel.

PROVIDENCE's luck ran out in August 1779, when she and nine other craft were trapped in the Penobscot River in Maine by a powerful British force and all of the American vessels were burned by their own crews.

ALFRED

After Jones's departure, ALFRED cruised in company with frigate RALEIGH, capturing some prizes. Returning from a voyage to France, the ships fell in with two British sloops of war which were much more lightly armed than the American ships. Nevertheless, RALEIGH fled from the enemy, leaving slow sailing ALFRED to capture, in one of the American navy's most ignominious moments. ALFRED was taken into British service as an armed transport, in which capacity she served until 1782 when she was sold out of the service, and there the record of her vanishes.

RANGER

After Jones yielded command of RANGER in France, Simpson sailed for American waters with frigates PROVIDENCE (not to be confused with the sloop) and BOSTON. In July, 1779, RANGER was part of a squadron that also included the frigates PROVIDENCE and QUEEN OF FRANCE. Together, they fell in with a convoy off Newfoundland and captured 11 prizes, the sale of which made it the most lucrative capture of the war.

The following year RANGER was at Charleston, South Carolina when that place was taken by the British. Along with frigates BOSTON and PROVIDENCE, she was handed over to the enemy and bought into the Royal Navy. There, she was renamed HALIFAX, and her rig was shortened even further from Paul Jones's final version. In October, 1781, she was sold, and there the record of her vanishes into history's mists.

BONHOMME RICHARD

Too well remembered to be allowed to rest in peace, the bones of BONHOMME RICHARD have magnetized a number of amateur archeologists and others over the years. Indeed, the ship has been sought, both in the water and out of it — at the drawing board and in the library, where Jean Boudriot has created his reconstruction, working with contemporary accounts. Many of these were collected by historian Peter Reaveley, whose words make a fitting epitaph for BONHOMME RICHARD:

> *For more than two hundred years the remains of the valiant ship . . . have rested on the bottom of the North Sea. In the summer of 1980 this amateur historian possibly located her, with the help of some local fishermen who, over the years, occasionally catch their lines and nets on her, sometimes bringing up relics, such as a swivel-mounted blunderbuss. Most years the shifting sands of her resting place cover her. Other years, particularly after strong north-easterly gales, her ancient sides project from the bottom for anyone to examine. Anyone, that is, willing to put down underwater video equipment, or dive to almost 200 feet, in the cold and dark North Sea, with water temperatures around 45° F., in tidal currents of up to 2.5 knots, with sea-bed visibility of three feet. It remains to be seen whether she will ever be examined there by naval archeologists . . . or whether the quarterdeck eight-pounder which Paul Jones himself fired during the action will ever be recovered.*

SERAPIS

At the Texel, command of SERAPIS was given to PALLAS Commandant Cottineau, under whom she returned to Lorient. There, she was sold into merchant service for $48,000, about a sixth of her cost new, and then purchased into the French navy. She was cut down one deck, rearmed, and sent out to the Indian Ocean where she was destroyed by fire in July, 1781, while in port at Madagascar.

ALLIANCE

ALLIANCE departed from France in July of 1780 with Philadelphia her destination under mad Landais. En route, his behavior was so bizarre that he was replaced in command by the First Lieutenant, by common consent of all the officers and passengers. The ship put in at Boston, where Landais was forcibly carried off ALLIANCE by Marines, and he was subsequently dismissed from the service.[3]

3. Landais served briefly in the French Revolutionary navy, but was retired. He settled first in Philadelphia, later New York, and outlived most of the other actors in this chronology, a lonely pensioner at the last, who never ceased explaining to one and all why he had committed his mad acts. He died in 1818.

Her next captain was John Barry, who suppressed a mutiny among her demoralized crew, successfully cruised for prizes, and in 1783 fought the last naval engagement of the Revolutionary War, trouncing the smaller 28-gun SYBIL, bearing a cargo of 100,000 Spanish dollars.

By 1785, ALLIANCE was the last ship in the service of the Continental Navy when she was auctioned off for $26,000 in Philadelphia. She was purchased by Robert Morris for the East Indies trade, and in it she made many fast, profitable voyages, including a record breaking passage to Canton in 1787.

In 1800, ALLIANCE, no doubt the last of the old Revolutionary navy's ships still afloat, went ashore near the mouth of the Delaware, on Petty's Island. There her bones remained for generations, an attraction to sightseers until gradually they were eaten away by the elements and by locals who removed bits of her wood as souvenirs. She seems to have entirely vanished sometime late in the last century, although there may still be buried remnants. LaFayette was presented with a small box made from ALLIANCE's remains, and he was delighted with the relic of the ship that in heroic days had sped him from the rebellious colonies to France. Souvenirs made from her wood are still commonly found around Philadelphia.

ARIEL

ARIEL was attached to Compte d'Estaing's fleet during the attempted capture of Savannah from the British, and then returned to France. She was lost at sea in 1793.

AMERICA

AMERICA floated in the Piscataqua until her new French crew could rig and equip her. She was given the salvaged gear and guns from the sunken LE MAGNIFIQUE, and got to sea the following summer, after the war was over.

Because of her design differences from French ships of her class, she performed differently from them, and so was criticized by her subsequent commanders for difficulty in closely following their fleet evolutions. Worse, in 1786 she was surveyed and found to be *entiērement pourri* — entirely rotten — after only three years of service. The ship was broken up, and the French deferentially named a new French ship AMERICA.

The Pictures

With sources and errata annotated

Note from the artist:

All of the depictions of ships in this book must be viewed as representations. Of all of Jones's ships, the only one whose design has survived is AMERICA, and even with it many ambiguities obtain. Therefore, the process of visually reconstructing these vessels has been one of refining the guesswork, with studies of contemporary sources by this artist and others. The degree of refinement possible in such study is nowhere more evident than in the works of Jean Boudriot, many of which appear in the bibliography. With the exception of his voluminous data used in reconstructing BONHOMME RICHARD, LE CERF, and the French frigates and ships of the line, almost all of the currently known information pertaining to the appearance of Jones's naval commands has been woven into the text, so the reader will perceive the paucity of information available. The most important sources for these data are listed in the bibliography and are annotated below.

In his Introduction to my work, James W. Cheevers has generously referred to it as notably accurate. While it is true that proportions and correct visual relationships (size of men, guns, and equipment relative to each other and to the vessels) and technical accuracy have been an ongoing pursuit throughout the creation of these pictures, they inevitably contain inaccuracies. Unlike a musician's wrong notes, which are blown into the air and forgotten, a graphic artist's errors remain for all to see, pending the disintegration of paper or canvas.

Also, there are inconsistencies, with variations between one picture and the next as to details of the same ship. In many cases, the later pictures correct the earlier ones; in other cases, varying views represent different logical guesses as to how some detail appeared. For instance, in the various views of the BONHOMME RICHARD/SERAPIS battle, RICHARD appears with a different sail combination in each picture; all are logical guesses. Any could be the right one, or none.

I hope the reader views such diversions as I do, as a by-product of education, and will thereby forgive me. Important technical questions (pertaining to rig, hull form or any significant detail) which have been discovered in these studies subsequent to their completion are annotated below, as errata, to avoid promulgation of error. Further questions will no doubt emerge with continued scrutiny, and, as this artist has found flaws in the work of his predecessors, so will this work be found flawed by his successors. It is all part of the process.

WG

THE PLATES (Listed by page number)

PAGE

i. BONHOMME RICHARD AND FLEET, ink, 1975. While the overall aspect of the title ship generally presents a fair visualization of BONHOMME RICHARD, the drawing predates the Boudriot reconstruction and has too many inaccuracies to enumerate. These problems are eliminated in all of the post-1986 renderings of the ships, as listed below.

iv. SLOOP PROVIDENCE PURSUED BY H.M. FRIGATE SOLEBAY, aquarelle, 1982. Preliminary study for the oil, on page 16.

3. BONHOMME RICHARD, pencil, 1973. Inaccurate. See later works.

4. JOHN PAUL AND SCOTTISH SKIFFS, WITH A SOLWAY FIRTH SMACK IN THE BACKGROUND, pencil, 1987.

5. BRIG FRIENDSHIP, pencil, 1987.

6. BRIGANTINE TWO FRIENDS, pencil, 1987.

7. SHIP BETSY, pencil, 1987.

PROVIDENCE—general remarks: Modelist Robert Innes has made a series of widely accepted but nevertheless conjectural models of PROVIDENCE which have found their way into various museums. All have the dolphin striker, although this feature is as of now not known to have existed prior to about 1790. The argument may be fairly made that dolphin strikers were first needed on the early sloops, with their outrageously long jib booms, and through them might well have come into being, but there is no proof of it. Similarly, the use of stern davits at this early date is questionable. Other discrepancies are noted below, as they are reflected in the pictures. Besides the Innes models (which in other regards are acceptable), the primary references for PROVIDENCE are: (Bibliography numbers) 1, 5, 13, 14, 16, 31, 35 (a definitive history of PROVIDENCE), 43, 46, and Chapelle's *History of American Sailing Craft.*

9. SLOOP PROVIDENCE (ex-KATY), ink, 1987.

11. SLOOP PROVIDENCE PREPARING TO ANCHOR, aquarelle, 1981. PROVIDENCE carried 12 four-pounders, not eight as shown. The drawing was made for the Boston Museum of Fine Arts from that institution's Innes model, and all of the model's conjectural details are reflected: the dolphin striker, stern davits, split chain plates, and wooden (as opposed to iron) crutches for the sweeps. The actual sloop would have had two pumps under the break of the

quarterdeck, stanchions for her swivel guns, and probably at least one more gun port per side under the quarterdeck, although the exact disposition of her guns is unknown.

13. SLOOP PROVIDENCE PURSUING A HALIFAX SCHOONER, ink and wash, 1986. The small boat hung from the stern is conjectural.

15. DIAGRAM. PROVIDENCE ELUDING SOLEBAY.

16. SLOOP PROVIDENCE ELUDING H.M. FRIGATE SOLEBAY AND FIRING A SWIVEL CANNON, oil, 1985. Again, the dolphin striker is conjectural. The head of the large square foresail probably would have been laced to a short club, or jackyard, and would have tacked to outhauls. The clews could have been likewise attached to outhauls from a spreader boom. Students are referred to Colan D. Ratliff's essay on the subject in the *Nautical Research Journal,* Vol. 30, #4.

17. JONES, study book sketch, 1986.

ALFRED — The primary references for the reconstructions are: (Bibliography numbers) 13, 14, 23, 28, 30 (by far the most complete study of ALFRED to date), 31, 33, 34 and 42.

18. SHIP ALFRED, pencil, 1987.

21. ALFRED IN COMPANY WITH SLOOP PROVIDENCE, wash, 1987.

RANGER — The general references for the reconstructions are: (Bibliography numbers) 13, 14, 17, 23, 24, 31, 33, 36, 42.

22. RANGER, ink, 1986.

25. RANGER'S FLAG SALUTED BY LA MOTTE PIQUET'S SQUADRON AT QUIBERON BAY, FEBRUARY 14, 1778, ink and wash, 1987. Initial attempts to determine which specific ships were in La Motte Piquet's squadron have to date yielded only the information that it was composed of five ships of the line and three frigates. Hence, the ships of the line are here conjectured as 74's, all of c.1778 design by Jacques-Noel baron Sané, the preeminent French naval architect of that time. The frigate is of the BELLE-POULE class. Pending further research, the name plaque on the stern of the foreground ship has been left blank. As to the flags flown by RANGER, her ensign is shown according to the 1778 description by Franklin, and the commission pendant is the version shown in the French flag sheet of 1781 by Mondhare.

26. RANGER'S SHORE PARTY AT ST. MARY'S ISLE, study book sketch, 1986. The trees were in bud, but not leaf.

27. RANGER vs. DRAKE, aquarelle, 1978. DRAKE's entire stern is highly conjectural here, although she indeed had quarter galleries. Her spanker would have been loose footed, and her red ensign would have been flown from a jackstaff. RANGER's ensign would almost certainly have had red, white and blue stripes, and her fore topsail bowlines would have led to the head of the jib boom, not its doubling.

28. RANGER'S BOATS DEPARTING TO RAID WHITEHAVEN, ink and wash, 1986.

ALFRED (as conjectured in 1975)

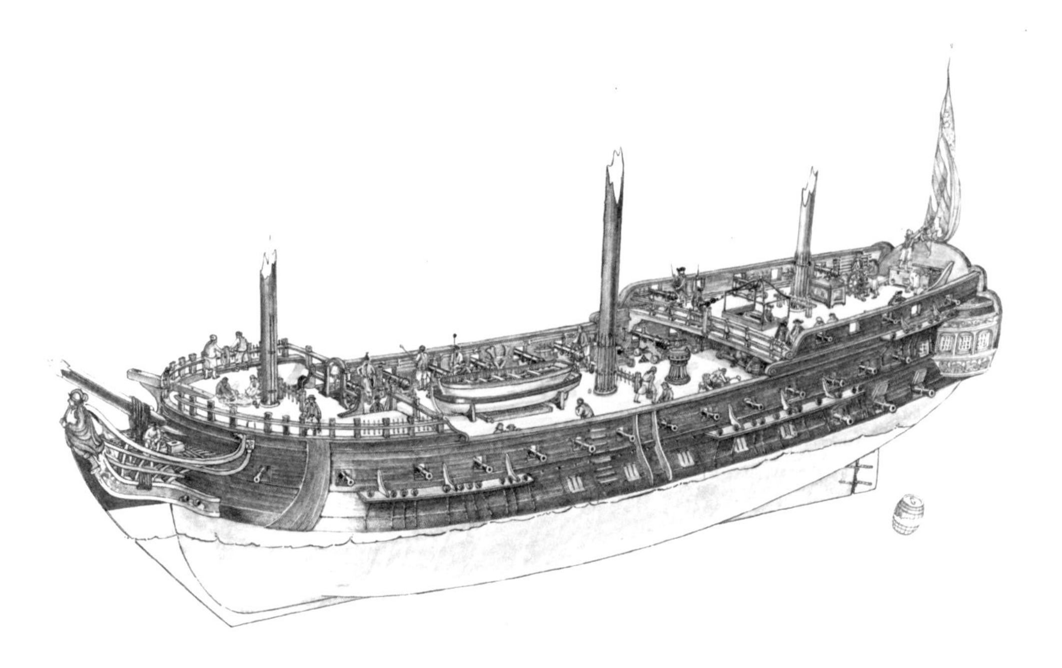

BONHOMME RICHARD — The primary reference for most of the post-1985 depictions of this ship was a detailed model by Richard Boss made from the plans developed by Jean Boudriot, as published in his *John Paul Jones and the Bonhomme Richard* (Bib. #7). The source material for that work is so voluminous that it is being published in a separate volume, to be titled *The Bonhomme Richard, 1779* (Bib. #3) scheduled for publication in 1988. All of the many other bibliographical references pertaining to BONHOMME RICHARD are incorporated into one or the other of those two works.

30. BONHOMME RICHARD, ink, 1986.

31. ENGLISH, FRENCH AND AMERICAN HULL FORMS. Diagram.

32. BONHOMME RICHARD, ink and wash, 1987.

33. PALLAS, VENGEANCE, AND BONHOMME RICHARD IN A NORTH SEA GALE, ink and wash, 1987. The seeming exaggeration of RICHARD's apple cheeks from this bow view was in fact measured from the Boudriot reconstruction, and the tumblehome may be viewed as accurate. The topsails are reefed, and their yards should therefore set some four feet lower on the topmasts. The reconstruction of PALLAS is based upon the plans of LA BELLE-POULE (Bib. #9). VENGEANCE was described as a brigantine, which she probably was, and should not be confused with a cutter, which in French terminology of the time had a *brigantine* main sail.

35. JONES'S MARINES, ink and wash, 1986. A Marine of the *Régiment de Walsh-Sérrant;* a Continental Marine; a Marine of the *Corps Royal d'Infanterie de la Marine.*

36. Three study book sketches (top to bottom): a seaman wearing the Phrygian cap; a sailor firing a swivel cannon; a sailor firing a swivel howitzer, or "coehorn," so called in contemporary naval parlance.

37. MUSIC ON BONHOMME RICHARD'S FORECASTLE, ink and wash, 1987. The scene is conjectural, but not the clothing or the instruments, which are a late 18th-century *cabrette*, and a *vielle à roue.*[1]

38. LE CERF, PALLAS, BONHOMME RICHARD AND ALLIANCE IN THE ENGLISH CHANNEL, IN FINE WEATHER, ink and wash, 1986. Nothing is known of PALLAS, excepting her armament. On the recommendation of Jean Boudriot, the representation used is of the French frigate BELLE-POULE, which was probably similar to PALLAS.

39. BONHOMME RICHARD OFF THE SKELLIGS, DESERTED BY HER IRISH BOAT CREW, ink and wash, 1986. According to Boudriot: no permanent reef pendants would have been fixed in the fore sail reef grommets; the furling pendants should be visible; the water holes in the sprit sail should be slightly further distant from the bolt ropes, and the jib boom cap should be round, not square, as depicted.

40. GUNPOWDER GIVEN TO THE BRITISH BY BONHOMME RICHARD, study book sketch, 1986. Sources differ as to whether it was a bag or a cask containing the 100 pounds of powder Jones gave to the Earl of Fifeshire's boat master.

1. The *cabrette* was the most popular French bagpipe, and it was played throughout the country, although primarily in Berry, Bourgogne, Bourbonaise, and the Auvergne. It is a melodic instrument, with regional variations, but usually found with a long base drone, and a short "tenor" drone (tuned to the 5th note in the octave) parallel to the chanter. There were mouth-blown and bellows-blown varieties of the instrument. Although it is *not* the regional bagpipe of Brittany (from whence BONHOMME RICHARD sailed), it is the instrument most likely to have been found in the hands of a French sailor. Brittany's bagpipe, the *Petit Binou,* would probably not have been tolerated on shipboard because of its remarkably shrill, piercing quality of sound, not unlike a flock of birds in wild distress. The *vielle à roue* (or wheel fiddle) depicted is of late 18th century pattern. These instruments are still played together to this day in some regions of France.

The actual transfer of the powder probably happened just after dark, with BONHOMME RICHARD hove-to in a small but choppy sea.

41. BONHOMME RICHARD APPROACHING SERAPIS AT TWILIGHT, SEPTEMBER 23, 1779, study book sketch, 1986. SERAPIS is shown coming about, and at this point would have been at least another mile distant.

42. JONES ON HIS QUARTERDECK, WITH AN EIGHT-POUNDER MANNED BY FRENCH AND AMERICAN SEAMEN, ink and wash, 1986.

44. OPENING GUNS, BONHOMME RICHARD vs. SERAPIS, aquarelle, 1980. In this early composition, the ships are correctly shown as they were positioned at about the second broadside, around 7:25 p.m., as viewed from the north. The moon is therefore incorrectly placed by approximately 180°. For the correct positioning, subsequently determined, see the diagram on page 46, and the final battle scene on page 53. RICHARD is shown with a studding sail set, plus royals, all of which are conjectural. Such sails were seldom worn into action, although it is recorded in her log that she wore her studding sails and royals during the approach, and there is no mention of their removal. Considering RICHARD's poor sailing qualities in light airs, it is likely Jones carried much more than the usual "fighting" canvas, although just how much more remains anyone's guess. This view of RICHARD was rendered before the Boudriot reconstruction, and while proportionately generally correct, it contains many small technical inaccuracies. Refer to later pictures.

45. BONHOMME RICHARD ABOUT TO COLLIDE WITH SERAPIS, ABOUT 7:45 P.M., aquarelle, 1981. The same remarks annotated to item #44, above, pertain to this picture. RICHARD is notably incorrect in many details, and although SERAPIS is largely correct, she probably would not have been carrying her sprit topsail.

46. DIAGRAM OF THE BATTLE, drawn from the Reaveley reconstruction.

47. BONHOMME RICHARD AND SERAPIS, ANCHORED, ABOUT 8:15 P.M., ink, 1986. An elevated view. Again, the sails worn by RICHARD are necessarily conjectural, and it is perhaps more than likely she would have been flying her fore and main royal sails. Also open to argument is the nest of boats in RICHARD's waist, for it is reasonably believed by some authorities that she was towing them. Again, there is no record. Had she been towing a string of boats, however, RICHARD would have responded even more sluggishly than she did, and whether Jones would have tolerated that is highly questionable.

48. SAILOR LOBBING A GRENADE FROM RICHARD's MAIN YARD, study book sketch, 1986.

49. CAPTAIN JONES SERVING AN EIGHT-POUNDER ON RICHARD's QUARTERDECK, ink and wash, 1984. While this view of a French eight-pounder is approximately correct, the drawing on page 42 is definitive.

50. BONHOMME RICHARD's MAIN TOP IN ACTION, ink and wash, 1986. In order to depict this view of the fighting top, it was necessary to cut away the main topsail's starboard clew tackle so that the sail could blow forward and permit the top to show. This could have happened, but there is no record that it did. The men are correctly proportioned to the top, which was large for a vessel of RICHARD's size. This detail as to the size of the top is supported by historical evidence, and the drawing shows the main doubling and top of a 64-gun ship, with the earlier style cap, which RICHARD is thought to have carried. If the mainsail was in the down position shown, its clew and buntlines would have to have been shot away, for the ship entered the battle with courses clewed up.

51. "THE GUN MUZZLES OF THE TWO SHIPS' LOWER DECK BATTERIES NEARLY TOUCHED" — THE BATTLE AT ABOUT 8:30, ink and wash, 1987. Two additional lights should show at RICHARD's mizzen peak.

53. BONHOMME RICHARD AND SERAPIS, RAKED BY ALLIANCE, 10 P.M., ink and wash, 1987. The view is from the south. SERAPIS is at anchor. RICHARD's damage is depicted from eyewitness descriptions. ALLIANCE is on a northeast heading. The relative position of the moon is correct to the action here, as is the position of the PALLAS/COUNTESS OF SCARBOROUGH action.

54. BONHOMME RICHARD: THE FINAL VIEW, ink and wash, 1986. Several witnesses mentioned the last glimpse of RICHARD as her "flag," still flying. Since her jackstaff had been lost early in the action, the flag last seen must have been the commission pendant at her mainmast truck. The version shown here is from a German flag sheet of about 1783, chosen because it closely agrees with the RICHARD pendant painted by Robert Dodd, as edited by Captain Pearson. SERAPIS is portrayed as she is conjectured to have looked 30 hours after the action.[2]

2. The blown-away lower gun deck ports are being boarded up against the rising southwesterly chop; a jury sail has been rigged to the mizzen yard (which would not ordinarily have carried one); a staysail or jib has been rigged to a new jury mizzen stay, and a new fore sail, fore topsail, and jib have been rigged, replacing those which were cut to rags. The forward bobstay is an obvious jury rig. As RICHARD sinks, SERAPIS is hove-to awaiting the return of Fanning's boat.

SERAPIS —The primary source for the hull reconstruction is the original design of ROEBUCK, prototype for the class of two-deckers which included SERAPIS. The original plans are in the archives at England's National Maritime Museum at Greenwich. Other sources are: (Bibliography numbers) 2, 3, 7, 24 and 37.

ALLIANCE — The design of RALEIGH was instrumental in all of the depictions of ALLIANCE, as they were both designed and constructed by the Hacketts. Additional references: (Bibliography numbers) 11, 13, 23, 31, 33, 34, 37, 42 and the Log of the ALLIANCE.

55. ALLIANCE, ink, 1986.

57. ALLIANCE PURSUING AN ENGLISH BRIG, JANUARY 1780, aquarelle, 1982. ALLIANCE would have been fitted with five topmast shrouds per side, not three as erroneously indicated. The ensign flown by ALLIANCE at the Texel had seven white and six red stripes, with a blue canton of 13 stars arranged in five horizontal rows of three, two, three, two and three stars. What flag she flew later is unknown.

ARIEL — The general references for the reconstructions are: (Bibliography numbers) 11, 16, 23, 24, 33, 42, 46 and 49 (specifically, from #49, AMAZONE defeating ARIEL, *l'Album de L'Admiral Willaumez.*)

59. ARIEL, ink, 1987.

60. ARIEL IN THE GREAT GALE OF OCTOBER, 1780, aquarelle, 1981.

63. A FRENCH 74-GUN SHIP OF THE LINE, ink and wash, 1986. The sole reference for this picture is Boudriot's monumental work, *The Seventy-Four Gun Ship,* four volumes. Bib. #8.

65. CAPTAIN JONES DRILLING THE CARPENTERS, pencil, 1987.

AMERICA — The general references for the reconstructions are: (Bibliography numbers) 13, 23, 31, 33, 34 and 42.

67. AMERICA UNDER CONSTRUCTION ON RISING CASTLE ISLAND, APRIL, 1782, aquarelle, 1985. The drawing incorporates the various changes probably made by Jones to the original design, but does not illustrate raised midship bulwarks, a debatable feature. Also open to argument is whether or not she would have worn a protective roof, or if other vessels would have been under construction within this view. Both questions are a research project for some future historian.

69. JONES EN ROUTE FROM SWEDEN TO RUSSIA, THROUGH BROKEN ICE IN THE BALTIC, pencil, 1987. The vessels depicted are two Swedish lapstraked, double ended, sprit rigged craft. These were the most common Swedish small craft of the day. (Ref. Bib. #16 and #19.)

70. FLAGSHIP VLADIMIR AND SQUADRON IN ACTION IN THE LIMAN, pencil, 1987. (Ref. *Svenska Flottans Historia,* Vol. II, Stockholm, 1943.) A definitive view of these ships awaits the attention of some future scholar with access to the Russian naval archives.

73. SLOOP PROVIDENCE BURNING, pencil, 1987.

75. THE BONES OF ALLIANCE, pencil, 1987.

77. ALFRED, aquarelle, 1975. An early effort with numerous technical errors. Refer to previous references.

78. BONHOMME RICHARD, aquarelle, 1979. This picture must be regarded as a fiction, having been commissioned and made to the specifications of a researcher whose work has been subsequently revealed by Boudriot and others as founded on incomplete information. The reader is referred to the earlier references on BONHOMME RICHARD.

84. AN AMERICAN AND A FRENCH SAILOR CARRYING A BUDGE BUCKET, pen and wash, 1986. The budge bucket held grenades, combustibles and sometimes powder charges for the swivels. It had a leather purse top with drawstrings to keep out rain, spray or sparks.

BIBLIOGRAPHY

The following bibliography lists the primary references for the enclosed work, and among them they contain nearly all of the technical information drawn upon for text and graphics. Some historic artists are listed in this bibliography because of the great value of their visual information pertaining to early sailing craft. Alphabetically, these are: Baugean, Charles Brooking, Robert Cleveley, E.W. Cooke, John Sell Cotman, Robert Dodd, Antoine-Léon Morel-Fatio, Francis Holman, William John Huggins, Philip James de Loutherbourg, the Ozanne Family, Nicholas Pocock, the Roux Family, Robert Salmon, Dominic and John Thomas Serres, Clarkson Stanfield, Willem Van de Velde (Elder and Younger) and George C. Wales.

EYEWITNESS NARRATIVES:
Aboard ALLIANCE: John Buckley, Master; James Bragg, Master Carpenter; Thomas Berry, Seaman; James Degge, First Lieutenant; Pierre Landais, Captain; John Larchar, Master's Mate; Matthew Parke, Captain, American Marines; Arthur Robinson, Midshipman; John Spencer, Lieutenant Colonel.
Aboard BONHOMME RICHARD: Robert Coram, Midshipman; Paul Chamillard, Colonel, Volunteer; Thomas Chase, Midshipman; Richard Dale, First Lieutenant; Nathaniel Fanning, Midshipman; Beaumont Groube, Midshipman; John Paul Jones, Captain; John Kilby, Gunner's Mate; John Linthwaite, Midshipman; Henry Lunt, Second Lieutenant; Thomas Lundy, Midshipman; John Mayrant, Midshipman; Mathew Mease, Purser; Eugene McCarthy, Second Lieutenant, French Royal Marines, Regiment Irelandais de Comte de Walsh-Serrant (Joseph Phillippe Walsh, Comte de Serrant); Thomas Potter, Midshipman; Edward Stack, First Lieutenant, French Royal Marines; Samuel Stacey, Master; Benjamin Stubbs, Midshipman; Antoine Wuibert, Lieutenant Colonel, Volunteer.
Ashore: Mitchell Graham, Captain, British Royal Navy, and William Maling, Customs Officer, Bridlington.
Aboard COUNTESS OF SCARBOROUGH: Thomas Piercy, Commander.
Aboard PALLAS: Denis Cotineau, Captain.
Aboard SERAPIS: Ship's Log; Robert Ozard, Sailmaker; Richard Pearson, Captain; Michael Stanhope, Second Lieutenant; and John Wright, First Lieutenant.
Aboard VENGEANCE: Philippe Ricot, Captain.

1. Baugean, Jean, *Collection de Toutes Les Espèces de Bâtimens de Guerre et de Bâtimens Marchands,* Paris: 1826.
2. BONHOMME RICHARD, *The Log of the Bonhomme Richard,* Mystic, CT: Marine Historical Association, Inc., 1936.
3. Boudriot, Jean, *The Bonhomme Richard 1779,* (from pre-publication notes; publication scheduled 1988).
4. __________, *Compagnie des Indies 1720-1770,* Paris: Boudriot, 1983.
5. __________, Conversations with, 1984-1986.
6. __________, *Cutter Le Cerf 1779-1780,* translated by H. Bartlett Wells, Paris: A.N.C.R.E.
7. __________, *John Paul Jones and the Bonhomme Richard,* translated by David H. Roberts, Annapolis, MD: Naval Institute Press, 1987.
8. __________, *The Seventy-Four Gun Ship,* translated by David H. Roberts, Annapolis, MD: Naval Institute Press, 1986.
9. __________, and Berti, Hubert, *La Belle-Poule, 1765,* Paris: A.N.C.R.E.
10. Bouquet, Michael, *Westcountry Sail: Merchant Shipping 1840-1960,* Devon: David & Charles, Ltd., 1971.
11. Bradford, James C., Notes: pre-microfilm excerpts from the papers of John Paul Jones, subsequently published by Texas A&M University, 1987.
12. Carr, Frank G.G., *The Medley of Mast and Sail,* United Kingdom: Teredo Books Ltd., 1976.
13. Chapelle, Howard I., *The History of the American Sailing Navy,* New York: W.W. Norton & Co., 1949.
14. __________, *The History of American Sailing Ships,* New York: W.W. Norton & Co., 1935.
15. __________, *The Search For Speed Under Sail 1700-1855,* New York: W.W. Norton & Co., 1967.
16. Chapman, Fredrik Henrik af, *Architectura Navalis Mercatoria,* Stockholm: 1768.
17. Cooper, James Fenimore, *The History of the Navy of the United States of America,* Philadelphia: Lea & Blanchard, 1839.
18. Falconer, William, *An Universal Dictionary of the Marine,* London: T. Cadell, 1780.

19. Fjellsson, Sigvard, *Kungsviken,* Föreningen Allmoge Båtar, Träbiten #38, Sweden, 1982.

20. Gilkerson, William, *Fighting Small Arms of the Revolutionary War Navy,* MS Catalog of Exhibition at U.S. Naval Academy Museum, Annapolis, Maryland: 1987.

21. __________ , *Maritime Arts by William Gilkerson,* Salem, Mass.: Peabody Museum of Salem, 1981.

22. __________ , "The Ships of John Paul Jones", *Sea History,* No. 12, Fall, 1978.

23. Jones, John Paul, *John Paul Jones' Memoir of the American Revolution, Presented to King Louis XVI of France,* translated and edited by Gerard W. Gawalt, Washington, D.C.: Library of Congress, 1979.

24. Lees, James, *The Masting and Rigging of English Ships of War 1625-1860,* Annapolis, MD: Naval Institute Press, 1979.

25. Lescallier, C., *Vocabulaire des Termes de Marine,* Paris: L'An VIII.

26. Lloyd's Registers, 1764-1785, Lloyd's Underwriters, England: (reprint) Gregg Press Inc., 1965.

27. MacGregor, David R., *Fast Sailing Ships 1775-1875, Their Design and Construction,* Switzerland: Edita Lausanne, 1973.

28. __________ , *Merchant Sailing Ships 1775-1815,* United Kingdom: Argus Books Ltd., 1980.

29. Martin, Tyrone, Commander, USN (Ret.), with Wm. Gilkerson, "Top Guns in the U.S. Sailing Navy", *Man at Arms,* July/Aug. 1987.

30. McCusker, John J., Jr., "The Continental Ship ALFRED", *Nautical Research Journal,* V.13, Nos.2,3, Washington, D.C.: Nautical Research Guild, Autumn, 1965.

31. Millar, John Fitzhugh, *Early American Ships,* Williamsburg, VA: Thirteen Colonies Press, 1986. (Artist's Note: This informative book attempts to index and expose the historically noted ships of the Revolutionary War period, among others, a monumental task inevitably containing a great many holes and flaws. Nevertheless, it represents the only known collection of contemporary depictions identifying specific vessels, in a series of vignettes redrawn from prints, paintings, drawings, and engravings of the period. The value of this kind of visual reference is left to the student's judgment, but it cannot be disregarded.)

32. Mollo, John, and McGregor, Malcolm, *Uniforms of the American Revolution,* New York: MacMillan Publishing Co., 1975.

33. Morison, Samuel Eliot, *John Paul Jones, A Sailor's Biography,* Boston: Little, Brown & Co., 1959.

34. Naval History Division, U.S.N., *Naval Documents of the American Revolution,* Volumes I-IX, Washington, D.C.: Department of the Navy, 1964-1986.

35. Ozanne, *Marine Militaire,* Paris: 1762.

36. RANGER, Log of the, MS. copy.

37. Reaveley, Peter, MS. Account of the BONHOMME RICHARD/SERAPIS Battle, (subsequently excerpted into Boudriot #7), 1987.

38. Rider, Hope S., *Valour Fore and Aft,* Annapolis, MD: Naval Institute Press, 1978.

39. Roberts, David, Correspondence with, 1986-1987.

40. __________ , "Vocabulaire de Marine, A French-English Vocabulary of Maritime Words and Phrases," MS., 1987. (Pending publication.)

41. Rubin, Norman, BONHOMME RICHARD research MS.

42. Sands, Robert C., *Life and Correspondence of John Paul Jones,* New York: A. Chandler, 1830.

43. Smith, Captain John, *The Sea-Mans Grammar and Dictionary . . . and the Practical Navigator and Gunner,* London: 1692.

44. Smith, Edgar Newbold, *American Naval Broadsides,* New York: Philadelphia Maritime Museum and Clarkson N. Potter, Inc., 1974.

45. Smyth, H. Warington, *Mast and Sail in Europe and Asia,* Edinburgh and London: Wm. Blackwood & Sons Ltd., 1929.

46. Steel, David, *The Elements and Practice of Naval Architecture,* London: 1805.

47. de Tousard, Louis, *The Naval Gunner,* Philadelphia: C. & A. Conrad & Co., 1809.

48. University of East Anglia, School of Environmental Sciences, Historical Weather Department, Norwich, Norfolk, England, MS: "General Synoptic Weather Analysis and Sea-State Analysis for Flamborough Head, Yorkshire, for the period Thursday, September 23, 1779, to Saturday, September 25, 1779."

49. Vichot, Jacques, *L'Oeuvre des Ozanne,* Paris: Editions Neptunia des Amis des Musées de la Marine, 1967-1971.

Acknowledgements

The vision, expertise, and generosity of Jean Boudriot have provided impetus and inspiration to this work beyond measure, and to him it is dedicated.

This book owes its existence to the sponsorship of the United States Naval Academy Museum and its staff, and most particularly to the Beverley R. Robinson Collection's curator, Sigrid Trumpy, who uncorked this illustrated history by initiating a "catalogue" of the pictures to be hung in the Museum's exhibition entitled *The Ships of John Paul Jones,* which in turn was initiated by Senior Curator James W. Cheevers, whose Introduction begins these pages. The entire project has been under the aegis of Director William W. Jeffries, and its conception may be said to have been sparked many years ago by Curator of Ship Models Robert Sumrall, whose early collaboration in the effort to penetrate the BONHOMME RICHARD enigma ultimately led to its solution by others.

Those others are: Peter Reaveley, whose formidable research provided Jean Boudriot in Paris with the English language data which was essential in finding the reincarnation of DUC DE DURAS into BONHOMME RICHARD; England's David Roberts, whose energy, knowledge and linguistic skills bridged the barriers between 18th century French and modern English to technical perfection; Tom Gillmer of Annapolis, Dean of America's designers of historic ships, who found the time to draft the initial orthographic drawings of BONHOMME RICHARD. The eventual reconstruction of this elusive ship, which had successfully defied so many previous efforts over two centuries, depended upon each of these men, for each one played a crucial role. The results were translated from paper into form by modelist Richard Boss, of Anacortes, Washington, whose work created the studio model which enabled its portrayal herein.

Others who have been instrumental specifically to the BONHOMME RICHARD project include: Philip K. Lundeberg, Curator Emeritus, Division of Naval History, National Museum of History and Technology, Smithsonian Institution, Washington, D.C.; Norman Rubin, whose initial systematic sleuthing debunked the pictures and models previously held as authoritative; Dr. James C. Bradford of the Department of History at Texas A & M University, for pertinent notes from his subsequently published *John Paul Jones Papers;* the staff of the Naval Historical Center, Department of the Navy, whose great helpfulness to colleague Peter Reaveley has ricocheted repeatedly into this work, and whose series *Naval Documents of the American Revolution* has been crucial in a variety of ways; to translator Monique Bradley for her ongoing efforts with the correspondence between Massachusetts and France; to Britt Noree for her talents as interpreter in Paris, and to Sarinda Parsons for the translation of Jean Boudriot's Foreword to this book.

This book as a whole owes thanks in specific or general ways to: Philip Annis, Assistant Director Emeritus of England's National Maritime Museum at Greenwich; Jan Piet Puype, librarian of the Scheepvaartmuseum, Amsterdam; John Hamilton, Curator of the Museum of Our National Heritage, Lexington, Mass.; those in the Engineering Division of the Sippican Co., Marion, Mass., who have furthered the work in ways technical and otherwise; Peter Stanford, N.M.H.S., for encouragement over the years; historians John F. Millar and Merritt A. Edson, who have responded willingly to agitated phone calls, sometimes at odd hours, in pursuit of sources to sundry arcane data.

For assistance in sorting out questions pertaining to the fighting small arms of the Revolutionary War navy, thanks go to Norman Flayderman, William Bass, James Wertenberger, Howard L. Blackmore, George C. Neumann, Robert M. Reilly, and again Jean Boudriot.

Last, very special thanks to those friends who have assisted with their special skills in the production of this work: Hope Sisson Rider, Commander Tyrone G. Martin, U.S.N. (Retired) and again Peter Reaveley, all of whom are talented writers and historians, and who have at the author's request edited the text of this book to its great benefit; Walter Tower for design assistance and ongoing education; Mark Sexton of the Peabody Museum of Salem for most of the photography from which the pictures were reproduced; Christine Parker for processing mountains of manuscript with wise council,

There are no words to express my gratitude to my beloved wife, Kerstin, air spirit, for the fresh breeze she has brought daily through the hermit's cave into which this work has increasingly forced me for nearly two years. Last, for sundry greater and lesser insights, a tip of the tricorne to Jackson and to pirate Anna.

This book has been co-published by the U.S. Naval
Academy Museum's Beverley R. Robinson Collection, and
the Naval Institute Press. The type is Garamond.
The book was designed by the author
with assistance from Walter Tower.
Four thousand copies have been printed at the Nimrod Press, Boston.
Twenty-eight are numbered
and specially bound for the author's use.